Chapter 3

Recording Media Sales On the basis of data collected between 1982 and 1997, manufacturer's shipments of vinyl record albums (LPs and EPs) declined according to the function $A(t) = 482.794e^{-0.351t}$ million albums, where t is the number of years after 1980.

a. Graph the function A for $0 \le t \le 20$.

b. If album sales continue according to the pattern indicated in the graph, what sales level are vinyl record albums approaching?

c. According to the function, when did vinyl record album sales fall to 5 million?

d. Why are the sales decreasing?

(Source: *Statistical Abstract of the U.S., 1998*)

Chapter 4

Immigration The percent of the U.S. population that is foreign born is given by the graph below. The data can be modeled by the function $P = 0.00006465x^3 - 0.0074x^2 + 0.0734x + 14.053$, where x is the number of years from 1900.

a. Graph this function on the window [0, 100] by [0, 20].

b. What does the model give as the percent in 1960? How does this compare with the data?

c. Use graphical or numerical methods to find the years after 1900 during which the percent of the U.S. population that is foreign born is 7%.

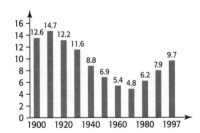

Chapter 5

Concerta The graphs that compare the weekly market share of the two drugs Concerta and Ritalin are shown in Figure 5.7. If the linear equation $y = 0.690x + 2.4$ models the percent of market share of Concerta and $y = -0.072x + 7.7$ models the percent of market share of Ritalin, where x is the number of weeks since Concerta was made available, find the number of weeks past the release date that the weekly market share of Concerta reached that of Ritalin.

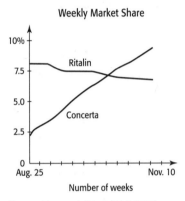

Weekly Market Share

(Source: *Newsweek*, December 4, 2000)

Chapter 6

Salaries Suppose that you are offered two identical jobs, one paying a starting salary of $40,000 with yearly raises of $2000 and a second one paying a starting salary of $36,000 with yearly raises of $2400.

a. Which job will be paying more in 5 years? By how much?

b. Which job will be paying more in 10 years? By how much?

c. What factor is important in deciding which job to take?

College Algebra in Context

with Applications for the Managerial, Life, and Social Sciences

Ronald J. Harshbarger
University of South Carolina — Beaufort

Lisa S. Yocco
Georgia Southern University

PEARSON
Addison Wesley

Boston San Francisco New York
London Toronto Sydney Tokyo Singapore Madrid
Mexico City Munich Paris Cape Town Hong Kong Montreal

Publisher: Greg Tobin
Senior Acquisitions Editor: Anne Kelly
Editorial Assistant: Cecilia Stashwick
Associate Project Editor: Joanne Ha
Senior Production Supervisor: Karen Wernholm
Production Coordination: Nesbitt Graphics, Inc.
Senior Marketing Manager: Becky Anderson
Marketing Coordinator: Carolyn Buddeke
Associate Producer: Sharon Smith
Software Development: David Malone and Marty Wright
Senior Manufacturing Buyer: Evelyn Beaton
Rights and Permissions Advisor: Dana Weightman
Text Design, Composition, and Illustrations: Nesbitt Graphics, Inc.
Cover Design: Studio Montage
Cover Photo: © PhotoDisc volume 1, Business and Industry

Library of Congress Cataloging-in-Publication Data
Harshbarger, Ronald J., 1938-
    College algebra in context with applications for the managerial, life, and social sciences /
Ronald J. Harshbarger, Lisa S. Yocco.
        p. cm.
    Includes bibliographical references and index.
    ISBN 0-201-72954-7
    1. Algebra. 2. Social sciences—Mathematics. I. Yocco, Lisa S. II. Title.

QA152.2.H376 2003
512—dc21   2003049782

3 4 5 6 7 8 9 10 — VH — 07060504

Contents

Preface

College Algebra in Context is designed for a course in algebra that is based on real life applications from the management, life, and social sciences, and on data analysis and modeling. The text is designed to show students how to analyze, solve, and interpret problems in this course, future courses, and in future careers. The text features a constructive chapter-opening Algebra Toolbox, which reviews previously-learned algebra concepts by presenting the prerequisite skills needed for successful completion of the chapter. In addition, the authors emphasize how students can use their new knowledge in a variety of calculus courses by devoting a section to calculus preparation at the end of the text.

The text is application driven and uses numerous real data problems to motivate interest in the skills and concepts of algebra. Modeling is introduced early, in the discussion of linear functions and in the discussion of quadratic and power functions. Additional models are introduced as exponential, logarithmic, logistic, cubic, and quartic functions are discussed. Mathematical concepts are introduced informally with the emphasis on applications. Each chapter contains real data problems and extended application projects that can be solved by students working collaboratively.

Students can take advantage of available technology (such as graphing calculators or spreadsheet software) to solve applied problems that are drawn from real life situations. There are examples of how technology can be used throughout the text, and step-by-step solutions using the technology are given in the accompanying technology manual. Technology is used to check answers of solved problems, to study function types, and to solve equations graphically and numerically.

Features of the text include the following:

- Early on the text discusses most types of functions, so that students can look at real application data with an understanding of what type of function describes a given data set.

  Basic functions and quadratic functions are discussed after the study of linear functions. Exponential, logarithmic, and logistic functions follow immediately.

- The development of algebra is motivated by, and used to find, the solution of real data based applications.

  Real-life problems are used to demonstrate the need for specific algebraic concepts and techniques. Each section begins with a motivational problem that presents a real life setting for that topic. The problem is then solved after the necessary skills are learned in that section. The aim is to prepare students to solve problems of all types by first introducing them to various functions and encouraging them to take advantage of available technology. Special business and finance sections are included in the text. These sections can be used to demonstrate the applications of functions to the business world, or they can be omitted without loss of continuity. At the heart of this text is its emphasis on problem solving in meaningful contexts.

- The use of technology has been integrated into the text.

  The text discusses the use of graphing calculators and/or computers, but there are no specific technology requirements. As a new calculator or spreadsheet skill becomes useful in a section, the keystrokes or commands required are given in the **GRAPHING CALCULATOR AND EXCEL MANUAL** that accompanies the text. The text indicates where calculators and spreadsheets can be used to solve problems. Technology is used to enhance and support learning when appropriate, not to supplant learning.

- The first five chapters begin with an Algebra Toolbox section that provides the prerequisite skills needed for the successful completion of the chapter.

  Topics discussed in the Toolbox are topics that are prerequisite to a college algebra course (usually found in a chapter R or appendix of a college algebra text). The topics are introduced "just in time" to be used in the chapter under consideration.

- The text ends with a Preparing for Calculus section that shows how algebra skills from the first four chapters are used in a calculus context.

  Many students have difficulty in calculus because they have trouble applying algebra skills to calculus. This section reviews earlier topics and shows how they apply to the development and application of calculus.

- Many problems posed in the text are multilevel.

  Many problems require thoughtful, real world answers under varying conditions rather than just a numerical answer. Questions like "when will this model no longer be valid," "what additional limitations must be placed on your answer," and "interpret your answer" are commonplace in the text.

- The text encourages collaborative learning.

  One or more Extended Applications at the end of each chapter provide opportunities for students to work together to solve real problems that involve the use of technology and frequently require modeling. The multilevel nature of the problems makes it desirable for students to collaborate in their solutions.

- The text encourages students to improve communication skills and research skills.

  The Extended Applications/Group Projects require written reports and frequently require use of the Internet or library. Some Extended Applications require students to use the literature or the Internet to find a graph or table of discrete data describing an issue. They are then required to make a scatter plot of the data, determine the function type that is the best fit for the data, create the model, discuss how well the model fits the data, and discuss how it can be used to analyze the issue.

- Each chapter has a Chapter Summary and a Chapter Review.

  The Chapter Summary lists the key terms and formulas discussed in the chapter, with section references, and a Chapter Review gives additional review problems. Each chapter ends with one or more Extended Applications that require students to solve multilevel problems involving real data or situations.

Supplements

For the Student

STUDENT'S SOLUTIONS MANUAL, by Brent Griffin (Floyd College), provides selected solutions to the Algebra Toolbox sets, Skills Check problems, and exercises. (*ISBN 0-321-17016-4*)

GRAPHING CALCULATOR AND EXCEL MANUAL, by Florence Chambers (Southern Maine Community College), follows the sequence of topics presented in the text and enhances the study of these topics with different sections giving the keystrokes for specific calculator types and with EXCEL commands. (*ISBN 0-321-17012-1*)

ADDITIONAL DRILL AND EXERCISE MANUAL, by Jay Abramson (Arizona State University), provides additional skill and testing preparation for students. *(ISBN 0-321-22027-7)*

A REVIEW OF ALGEBRA, by Heidi Howard (Florida Community College — Jacksonville), offers additional support for those students in need of further algebra review. *(ISBN 0-201-77347-3)*

ADDISON-WESLEY MATH TUTOR CENTER The Addison-Wesley Math Tutor Center is staffed by qualified math and statistics instructors who provide students with tutoring on examples and odd-numbered exercises from the textbook. Tutoring is available via toll-free telephone, fax, email, or the Internet. White Board technology allows tutors and students to actually see the problems worked while they "talk" in real time over the Internet during tutoring sessions.

An access card is required. For more information, go to www.aw.com/tutorcenter.

For the Instructor

INSTRUCTOR'S SOLUTIONS MANUAL, by Brent Griffin (Floyd College), contains detailed solutions to all of the Algebra Toolbox sets, Skills Check problems, exercises, and Extended Applications. (*ISBN 0-321-17423-2*)

INSTRUCTOR'S TESTING MANUAL, by Demetria Neal (Gwinnett Technical College), provides tests for each chapter of the text. Answer keys are included. *(ISBN 0-321-17250-7)*

TESTGEN-EQ WITH QUIZMASTER-EQ TestGen enables instructors to build, edit, print, and administer tests using a computerized bank of questions developed to cover all the objectives of the text. Instructors can modify test bank questions or add new questions by using the built-in question editor, which allows users to create graphs, import graphics, insert math notation, and insert variable numbers or text. Tests can be printed or administered online via the Web or other network. TestGen comes packaged with QuizMaster, which allows students to take tests on a local area network. The software is available on a dual-platform Windows/Macintosh CD-ROM. *(ISBN 0-321-17252-3)*

Acknowledgments

Many individuals contributed to the development of this textbook. We would like to thank the following reviewers, whose comments and suggestions were invaluable in preparing this text:

Jay Abramson, *Arizona State University*
Khadija Ahmed, *Monroe County Community College*
Jamie Ashby, *Texarkana College*
Sohrab Bakhtyari, *St. Petersburg College*
Jean Bevis, *Georgia State University*
Thomas Bird, *Austin Community College*

Len Brin, *Southern Connecticut State University*
Marc Campbell, *Daytona Beach Community College*
Florence Chambers, *Southern Maine Community College*
Floyd Downs, *Arizona State University*
Aniekan Ebiefung, *University of Tennessee at Chattanooga*
Toni W. Fountain, *Chattanooga State Technical Community College*
John Gosselin, *University of Georgia*
Linda Green, *Santa Fe Community College*
Richard Brent Griffin, *Floyd College*
Deborah Hanus, *Brookhaven College*
Steve Heath, *Southern Utah University*
Todd A. Hendricks, *Georgia Perimeter College*
Suzanne Hill, *New Mexico State University*
Sue Hitchcock, *Palm Beach Community College*
Arlene Kleinstein, *State University of New York — Farmingdale*
Danny Lau, *Kennesaw State University*
Ann H. Lawrence, *Wake Technical Community College*
Kitt Lumley, *Columbus State University*
Antonio Magliaro, *Southern Connecticut State Universit*
Beverly K. Michael, *University of Pittsburgh*
Demetria Neal, *Gwinnett Technical College*
Malissa Peery, *University of Tennessee*
Ingrid Peterson, *University of Kansas*
Beverly Reed, *Kent State University*
Michael Rosenthal, *Florida International University*
Jacqueline Underwood, *Chandler-Gilbert Community College*
Denise Widup, *University of Wisconsin — Parkside*
Sandi Wilbur, *University of Tennessee — Knoxville*

Many thanks to Janis Cimperman, Danny Lau, Patricia Nelson, Deana Richmond, and Lauri Semarne for checking the accuracy of this text, and to Bridgett Dougherty of Nesbitt Graphics for assistance with production. A special thanks goes to the Addison-Wesley team for their assistance, encouragement, and direction throughout this project: Greg Tobin, Anne Kelly, Becky Anderson, Joanne Ha, Karen Wernholm, Joe Vetere, Barbara Atkinson, and Sarah Santoro.

Ronald J. Harshbarger
Lisa S. Yocco

CHAPTER

1

Functions, Graphs, and Models; Linear Functions

Federal government analysts predicted in 1995 that there would be a federal budget surplus in 2000 (and there was) and that the Social Security Trust Fund (from which social security is paid to retirees) would be empty in 2038. BMW, the most profitable car company in the world, introduced the new Mini Cooper car in 2002 and predicted that it would break even on its development costs in seven years. How do the government and companies make predictions of this type? When real-world information is collected as data, mathematical functions can be created that model the data, and the models can then be used to solve problems. Mathematical models are also created from known facts and formulas. In this text we will solve real data problems in business, economics, biology, sociology, and the physical sciences.

The table below presents the main concepts and skills discussed in this chapter, along with some of their applications.

		TOPICS	APPLICATIONS
1.1	Functions and Graphs	Determining graphs, tables and equations that represent functions; finding domains and ranges, and evaluating functions	Body temperature, personal computers, stock market, men in the workforce
1.2	Graphs of Functions; Mathematical Models	Graphing and evaluating functions with technology; graphing and evaluating mathematical models	General Electric revenues, cost-benefit, crime rates, bankruptcies, U.S. executions
1.3	Linear Functions	Identifying and graphing linear functions; finding and interpreting intercepts and slopes; finding constant rates of change	Ritalin use, loan balances, temperature measurements, Sheriff's department patrol cars
1.4	Equations of Lines; Constant Rates of Change	Writing equations of lines; finding constant and average rates of change	Depreciation, inmate population, service call charges, Toyota sales, population growth
1.5	Algebraic and Graphical Solutions of Linear Equations	Relating solutions, zeros, and x-intercepts; solving linear equations; finding functional forms of equations in two variables	Prison sentence length, credit card debt, simple interest
1.6	Fitting Lines to Data Points; Modeling Linear Functions	Linear regression; modeling with linear functions; applying linear models	SAT scores, health service employment, carbon monoxide emissions, U.S. population
1.7	Models in Business and Economics	Modeling revenue, cost, and profit; finding break-even; finding marginal revenue, cost and profit, supply and demand; finding market equilibrium	Break-even in production and sales, marginal revenue and cost, market equilibrium
1.8	Solutions of Linear Inequalities	Solving linear inequalities with analytical and graphical methods; solving double inequalities	Profit, body temperature, apparent temperature, course grades, expected prison sentences

ALGEBRA *Toolbox*

The algebra toolbox is designed to review prerequisite skills needed for success in each chapter. This toolbox discusses the real numbers, the coordinate system, algebraic expressions, equations, inequalities, absolute values, and subscripts.

Sets

A **set** is a well-defined collection of objects, such as a set of books or the set of odd numbers. There are two ways to define a set. One way is by listing the **elements** (or **members**) of the set (usually between braces). For example, we may say that a set A contains 2, 3, 5, and 7 by writing $A = \{2, 3, 5, 7\}$. To say that 5 is an element of the set A, we write $5 \in A$. To indicate that 6 is not an element of the set, we write $6 \notin A$.

If all the elements of the set can be listed, the set is said to be a **finite set**. If all elements of a set cannot be listed, the set is called an **infinite set**. To indicate that the set continues infinitely, we use three dots. For example, $N = \{1, 2, 3, 4, \dots\}$ is an infinite set called the **natural numbers**.

Another way to define a set is to give its description. For example, we may write $\{x \mid x \text{ is a math book}\}$ to define the set of math books. This is read as "the set of all x such that x is a math book." $N = \{x \mid x \text{ is a natural number}\}$ defines the natural numbers, which was also defined by $N = \{1, 2, 3, 4, \dots\}$ above. A set that contains no elements is called an **empty set**, and is denoted by $\varnothing$.

EXAMPLE 1

Write the following sets in two ways.

a. The set A containing the natural numbers less than 7.

b. The set B of natural numbers that are at least 7.

Solution

a. $A = \{1, 2, 3, 4, 5, 6\}, \quad A = \{x \mid x \in N, x < 7\}$

b. $B = \{7, 8, 9, 10, \dots\}, \quad B = \{x \mid x \in N, x \geq 7\}$ ■

The relations that can exist between two sets follow.

RELATIONS BETWEEN SETS

1. Sets X and Y are **equal** if they contain exactly the same elements.

2. Set A is called a **subset** of set B if each element of A is an element of B. This is denoted $A \subseteq B$.

3. If sets C and D have no elements in common, they are called **disjoint**.

EXAMPLE 2

For the sets $A = \{x \mid x \leq 9, x \text{ is a natural number}\}$, $B = \{2, 4, 6\}$, $C = \{3, 5, 8\}$:

a. Which of the sets A, B, and C are subsets of A?

b. Which pairs of sets are disjoint?

c. Are any of these three sets equal?

Solution

a. Every element of B is contained in A and every element of C is contained in A. Thus both sets B and C are subsets of A. Because every element of A is contained in A, A is a subset of A.

b. Sets B and C have no elements in common, so they are disjoint.

c. No pairs of sets have exactly the same elements, so none of these three sets are equal. ∎

The Real Numbers

Because most of the mathematical applications you will encounter in an applied nontechnical major use real numbers, the emphasis in this text is the **real number system**.* Real numbers can be rational or irrational. **Rational numbers** include integers, fractions containing integers, and decimals that either terminate or repeat. Some examples of rational numbers are

$$-9, \quad \frac{1}{2}, \quad 0, \quad 12, \quad -\frac{4}{7}, \quad 6.58, \quad -7.333$$

Irrational numbers are real numbers that are not rational. Some examples of irrational numbers are π (a number familiar to us from the study of circles), $\sqrt{2}$, $\sqrt[3]{5}$, and $\sqrt[3]{-10}$. These numbers are described in Table 1.1.

TABLE 1.1

Types of Real Numbers	Descriptions
Natural numbers	1, 2, 3, 4, . . .
Integers	Natural numbers, zero, and the negatives of the natural numbers: . . . , -3, -2, -1, 0, 1, 2, 3, . . .
Rational numbers	All numbers that can be written in the form $\frac{p}{q}$, where p and q are both integers with $q \neq 0$. Rational numbers can be written as terminating or repeating decimals.
Irrational numbers	All real numbers that are not rational numbers. Irrational numbers cannot be written as terminating or repeating decimals.

We can represent real numbers on a **real number line**. Exactly one real number is associated with each point on the line, and we say there is a one-to-one correspondence between the real numbers and the points on the line. That is, the real number line is a graph of the real numbers (see Figure 1.1).

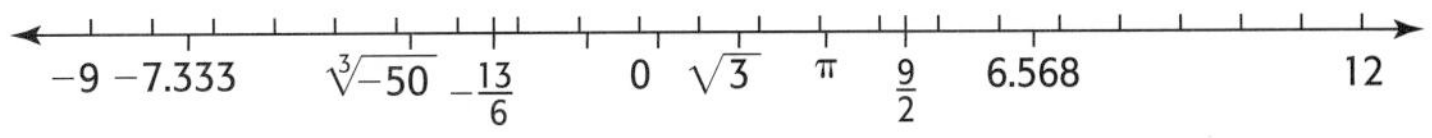

FIGURE 1.1

---

*The complex number system will be discussed in Section 4.5.

Notice the number π on the real number line in Figure 1.1. This special number, which can be approximated by 3.14, results when the circumference of (distance around) any circle is divided by the diameter of the circle. Another special real number is e, which is approximately 2.71828; it is denoted by

$$e \approx 2.71828.$$

We will discuss this number, which is important in financial and biological applications, later in the text.

Inequalities and Intervals on the Number Line

An **inequality** is a statement that one quantity is greater (or less) than another quantity. We say that a is less than b (written $a < b$) if the point representing a is to the left of the point representing b on the real number line. We may indicate that the number a is greater than or equal to b by writing $a \geq b$. The subset of real numbers x that lie between a and b (excluding a and b) can be denoted by the **double inequality** $a < x < b$ or by the **open interval** (a, b). This is called an open interval because neither of the endpoints is included in the interval. The **closed interval** $[a, b]$ represents the set of all real numbers satisfying $a \leq x \leq b$. Intervals containing one endpoint, such as $[a, b)$ or $(a, b]$, are called **half-open intervals**. We can represent the inequality $x \geq a$ by the interval $[a, \infty)$, and we can represent the inequality $x < a$ by the interval $(-\infty, a)$. Note that ∞ and $-\infty$ are not numbers, but ∞ is used in $[a, \infty)$ to represent the fact that x increases without bound and $-\infty$ is used in $(-\infty, a)$ to indicate that x decreases without bound. Table 1.2 shows the graphs of different types of intervals.

TABLE 1.2

Interval Notation	Inequality Notation	Verbal Description	Number Line Graph
(a, ∞)	$x > a$	x is greater than a	
$[a, \infty)$	$x \geq a$	x is greater than or equal to a	
$(-\infty, b)$	$x < b$	x is less than b	
$(-\infty, b]$	$x \leq b$	x is less than or equal to b	
(a, b)	$a < x < b$	x is between a and b, not including either a or b	
$[a, b)$	$a \leq x < b$	x is between a and b, including a but not including b	
$(a, b]$	$a < x \leq b$	x is between a and b, not including a but including b	
$[a, b]$	$a \leq x \leq b$	x is between a and b, including both a and b	

Note that open circles may be used instead of parentheses and closed circles may be used instead of brackets in the number line graphs.

EXAMPLE 3 Intervals

Describe the interval corresponding to each of the inequalities in parts (a) – (e), then graph the inequality.

a. $-1 \le x \le 2$ **b.** $2 < x < 4$ **c.** $-2 < x \le 3$ **d.** $x \ge 3$ **e.** $x < 5$

Solution

a. $[-1, 2]$

b. $(2, 4)$

c. $(-2, 3]$

d. $[3, \infty)$

e. $(-\infty, 5)$

Absolute Value

The distance the number a is from 0 on a number line is the **absolute value** of a, denoted by $|a|$. The absolute value of any nonzero number is positive, and the absolute value of 0 is 0. For example, $|5| = 5$ and $|-8| = 8$. Note that if a is a nonnegative number, then $|a| = a$, but if a is negative, then $|a|$ is the positive number $-a$. Formally, we say

$$|a| = \begin{cases} a \text{ if } a \ge 0 \\ -a \text{ if } a < 0 \end{cases}$$

For example, $|5| = 5$ and $|-5| = -(-5) = 5$.

Algebraic Expressions

In algebra we deal with a combination of real numbers and letters. Generally, the letters are symbols used to represent unknown quantities or fixed but unspecified constants. Letters representing unknown quantities are usually called **variables** and letters representing fixed but unspecified numbers are called **literal constants**. An expression created by performing additions, subtractions, or other arithmetic operations with one or more real numbers and variables is called an **algebraic expression**. Unless otherwise specified, the variables represent real numbers for which the algebraic expression is a real number. Examples of algebraic expressions include

$$5x - 2y, \frac{3x - 5}{12 + 5y}, 7z + 2.$$

A term of an algebraic expression is the product of one or more variables and a real number; the real number is called a **numerical coefficient**, or simply a **coefficient**. A constant is also considered a term of an algebraic expression and is called a **constant term**. For instance, the term $5yz$ is the product of the factors 5, y, and z; this term has coefficient 5. An expression containing a finite number of additions, subtractions, and multiplications of constants and nonnegative integer powers of variables is called a **polynomial**. A polynomial cannot contain negative powers of variables, variables in a denominator, or variables inside a radical. The expressions $5x - 2y$ and $7z^3 + 2y$ are polynomials, but $\dfrac{3x - 5}{12 + 5y}$ and $3x^2 - 6\sqrt{x}$ are not polynomials.

Terms that contain exactly the same variables with exactly the same exponents are called **like terms**. For example, $3x^2y$ and $7x^2y$ are like terms, but $3x^2y$ and $3xy$ are not. We can *simplify* an expression by adding or subtracting the coefficients of the like terms. For example, the simplified form of

$$3x + 4y - 8x + 2y \ \text{ is } \ -5x + 6y.$$

and the simplified form of

$$3x^2y + 7xy^2 + 6x^2y - 4xy^2 - 5xy \ \text{ is } \ 9x^2y + 3xy^2 - 5xy.$$

Removing Parentheses

We often need to remove parentheses when simplifying algebraic expressions and when solving equations. Removing parentheses frequently requires use of the **distributive property**, which says that for real numbers a, b, and c, $a(b + c) = ab + ac$. Care must be taken to avoid mistakes with signs when using the distributive property. Multiplying a sum in parentheses by a negative number changes the sign of each term in the parentheses. For example,

$$-3(x - 2y) = -3(x) + (-3)(-2y) = -3x + 6y \ \text{ and } \ -(3xy - 5x^3) = -3xy + 5x^3.$$

We add or subtract (**combine**) algebraic expressions by combining the like terms. For example, the sum of the expressions $5sx - 2y + 7z^3$ and $2y + 5sx - 4z^3$ is

$$(5sx - 2y + 7z^3) + (2y + 5sx - 4z^3) = 5sx - 2y + 7z^3 + 2y + 5sx - 4z^3$$
$$= 10sx + 3z^3.$$

and the difference of these two expressions is

$$(5sx - 2y + 7z^3) - (2y + 5sx - 4z^3) = 5sx - 2y + 7z^3 - 2y - 5sx + 4z^3$$
$$= -4y + 11z^3.$$

Equivalent Equations

An **equation** is a statement that two quantities or expressions are equal. To **solve** an equation means to find the value(s) of the variable(s) that makes the equation a true statement. For example, the equation $3x = 6$ is true when $x = 2$, so 2 is a solution of this equation. Equations of this type are sometimes called **conditional equations**, because they are true only for certain values of the variable. Equations that are true for all values of the variables for which both sides are defined are called **identities**. For example, the equation $7x - 4x = 5x - 2x$ is an identity.

We frequently can solve an equation for a given variable by rewriting the equation in an equivalent form whose solution is easy to find. Two equations are **equivalent** if and only if they have the same solutions. The following operations give equivalent equations:

PROPERTIES OF EQUATIONS

1. **Addition Property**　Adding the same number to both sides of an equation gives an equivalent equation. For example, $x - 5 = 2$ is equivalent to $x - 5 + 5 = 2 + 5$ or to $x = 7$.

2. **Subtraction Property**　Subtracting the same number from both sides of an equation gives an equivalent equation. For example, $z + 12 = -9$ is equivalent to $z + 12 - 12 = -9 - 12$ or to $z = -21$.

3. **Multiplication Property**　Multiplying both sides of an equation by the same nonzero number gives an equivalent equation. For example, $\frac{y}{6} = 5$ is equivalent to $6\left(\frac{y}{6}\right) = 6(5)$ or to $y = 30$.

4. **Division Property**　Dividing both sides of an equation by the same nonzero number gives an equivalent equation. For example, $17x = -34$ is equivalent to $\frac{17x}{17} = \frac{-34}{17}$ or to $x = -2$.

EXAMPLE 4　　Solving Equations

Solve the following equations.

a. $3x = 6$　　**b.** $4x - 5 = 7 + 2x$　　**c.** $8x + 12 = 25 - 5x$

Solution

a. Dividing both sides of the equation by 3 gives an equivalent equation whose solution is evident.

$$\frac{3x}{3} = \frac{6}{3}$$

$$x = 2$$

Thus $x = 2$ is the solution to the original equation.

b. Adding 5 to both sides of the equation and subtracting $2x$ from both sides of the equation, then dividing by 2 gives the solution.

$$4x - 5 = 7 + 2x$$

$$4x - 5 + 5 = 7 + 2x + 5$$

$$4x = 12 + 2x$$

$$4x - 2x = 12 + 2x - 2x$$

$$2x = 12$$

$$\frac{2x}{2} = \frac{12}{2}$$

$$x = 6$$

Thus $x = 6$ is the solution to the original equation.

c. Adding $5x$ to both sides of the equation and subtracting 12 from both sides gives an equivalent equation.

$$8x + 12 = 25 - 5x$$
$$8x + 12 + 5x = 25 - 5x + 5x$$
$$13x + 12 = 25$$
$$13x + 12 - 12 = 25 - 12$$
$$13x = 13$$

Dividing both sides by 13 gives the solution to the original equation.

$$\frac{13x}{x} = \frac{13}{13}$$
$$x = 1$$ ∎

Subscripts

We sometimes need to distinguish between two y-values and/or x-values in the same problem or on the same graph, or to designate literal constants. It is often convenient to do this by using **subscripts**. For example, if we have two fixed but unidentified points on a graph, we can represent one point as (x_1, y_1) and the other as (x_2, y_2). Subscripts also can be used to designate different equations entered in graphing utilities; for example, $y = 2x - 5$ may appear as $y_1 = 2x - 5$ when entered in the equation editor of a graphing calculator.

The Coordinate System

Much of our work in algebra involves graphing. In order to graph in two dimensions, we use a rectangular coordinate system, or **Cartesian coordinate system**. Such a system allows us to assign a unique point in a plane to each ordered pair of real numbers. We construct the coordinate system by drawing a horizontal number line and a vertical number line so that they intersect at their origins (see Figure 1.2). The point of intersection is called the **origin** of the system, the number lines are called the coordinate **axes**, and the plane is divided into four parts called **quadrants**. We call the horizontal axis

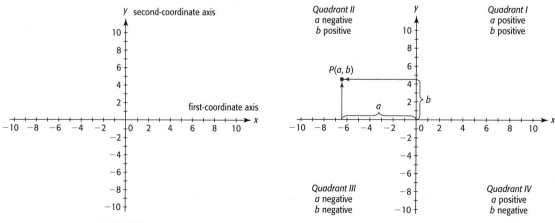

FIGURE 1.2   **FIGURE 1.3**

the **x-axis** (or the first-coordinate axis) and the vertical axis the **y-axis** (or the second-coordinate axis).

The ordered pair (a, b) represents the point P that is a units from the y-axis (right if a is positive, left if a is negative) and b units from the x-axis (up or down, depending on whether b is positive or negative). The values of a and b are called the **rectangular coordinates** of the point. Figure 1.3 shows point P with coordinates (a, b). The point is in the second quadrant, where $a < 0$ and $b > 0$.

Toolbox EXERCISES

1. Write "the set of all natural numbers less than 9" in two different ways.

2. Is it true that $\frac{1}{2} \in N$, if N is the set of natural numbers?

3. Is the set of integers a subset of the set of rational numbers?

4. Are sets of rational numbers and irrational numbers disjoint sets?

Identify the sets of numbers in Exercises 5–7 as one or more sets of integers, rational numbers, and/or irrational numbers.

5. $\{5, 2, 5, 8, -6\}$

6. $\left\{\dfrac{1}{2}, -4, \dfrac{5}{3}, 1\dfrac{2}{3}\right\}$

7. $\left\{\sqrt{3}, \pi, \dfrac{\sqrt[3]{2}}{4}, \sqrt{5}\right\}$

In Exercises 8–10, express each interval or graph as an inequality.

8. ![number line with open circle at −3]
   -3

9. $[-3, 3]$

10. $(-\infty, 3]$

In Exercises 11–13, express each inequality or graph in interval notation.

11. $x \leq 7$

12. $3 < x \leq 7$

13. ![number line with open circle at 4]
    4

In Exercises 14 and 15, graph the inequality or interval on a real number line.

14. $(-2, \infty)$

15. $5 > x \geq 2$

In Exercises 16 and 17, plot the points on a coordinate plane.

16. $(-1, 3)$

17. $(4, -2)$

Find the absolute values in Exercises 18 and 19.

18. $|-6|$

19. $|7 - 11|$

For each algebraic expression in Exercises 20 and 21, give the coefficient of each term and give the constant term.

20. $-3x^2 - 4x + 8$

21. $5x^4 + 7x^3 - 3$

22. Find the sum of $z^4 - 15z^2 + 20z - 6$ and $2z^4 + 4z^3 - 12z^2 - 5$.

23. Simplify the expression
    $3x + 2y^4 - 2x^3y^4 - 119 - 5x - 3y^2 + 5y^4 + 110$.

Remove the parentheses and simplify in Exercises 24–27.

24. $4(p + d)$

25. $-2(3x - 7y)$

26. $-a(b + 8c)$

27. $4(x - y) - (3x + 2y)$

Solve the equations in Exercises 28–32.

28. $3x = 6$

29. $\dfrac{x}{3} = 6$

30. $x + 3 = 6$

31. $4x - 3 = 6 + x$

32. $3x - 2 = 4 - 7x$

1.1   Functions and Graphs

One indication of illness in children is body temperature. There are two common measures of temperature, Fahrenheit (° F) and Celsius (° C), with temperature measured in Fahrenheit degrees in the United States and in Celsius degrees in many other countries of the world. Suppose you know that a child's normal body temperature is 98° F and his or her temperature is now 37° C. Does this indicate that the child is ill? To help decide this, we could find the Fahrenheit temperature that corresponds to 37° C. We could do this easily if we knew the relationship between Fahrenheit and Celsius temperature scales. In this section, we will see that the relationship between these measurements can be defined by a **function** and that functions can be defined numerically, graphically, or algebraically. We also explore how this and other functions can be applied to help solve problems that occur in real-world situations.

Function Definitions

There are several techniques to show how Fahrenheit degree measurements are related to Celsius degree measurements.

One way to show the relationship between Celsius and Fahrenheit degree measurements is by listing some Celsius measurements and the corresponding Fahrenheit measurements. These measurements, and any other real-world information collected in numerical form are called **data.** These temperature measurements can be shown in a table (see Table 1.3).

TABLE 1.3

Celsius degrees (°C)	−20	−10	−5	0	25	50	100
Fahrenheit degrees (°F)	−4	14	23	32	77	122	212

This relationship is also defined by the set of ordered pairs

$$\{(-20, -4), (-10, 14), (-5, 23), (0, 32), (25, 77), (50, 122), (100, 212)\}.$$

We can picture the relationship between the measurements with a graph. Figure 1.4(a) shows a **scatter plot** of the data, that is, a graph of the ordered pairs as points.

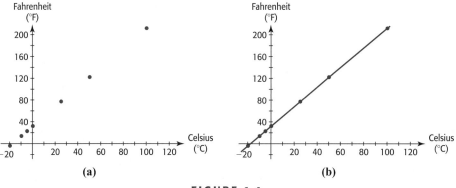

FIGURE 1.4

The points on the graph in Figure 1.4(a) can be connected by the line whose equation is

$$F = \frac{9}{5}C + 32,$$

where F is the temperature measured in degrees Fahrenheit and C is the temperature measured in degrees Celsius. This line, shown in Figure 1.4(b), contains the points on the scatter plot and other points that satisfy the relationship between the measures. This relationship defines Fahrenheit (F) as a **function** of Celsius (C), with **input** C and **output** F. It is called a function because each input C results in exactly one output F.

FUNCTION

A function is a rule or correspondence that assigns to each element of one set (called the **domain**) exactly one element of a second set (called the **range**).

The function may be defined by a set of ordered pairs, a table, a graph, or an equation.

As the definition indicates, functions can be represented by tables and graphs as well as by equations. Arrow diagrams that show how each individual input results in exactly one output can also represent functions. Each arrow in Figure 1.5(a) and in Figure 1.5(c) goes from an input to exactly one output, so each of these diagrams defines a function. On the other hand, the arrow diagram in Figure 1.5(b) does not define a function because one input, 8, results in two different outputs, 6 and 9.

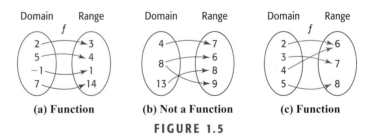

(a) Function   (b) Not a Function   (c) Function

FIGURE 1.5

DOMAIN

The domain of a function is the set of all inputs.

RANGE

The range of a function is the set of outputs.

How a function is defined determines its domain and range. For instance, the domain of the function defined by Table 1.3 or by the scatter plot in Figure 1.4(a) is the finite set $\{-20, -10, -5, 0, 25, 50, 100\}$ with all values measured in degrees Celsius, and the range is the set $\{-4, 14, 23, 32, 77, 122, 212\}$ with all values measured in degrees Fahrenheit. This function has a finite number of inputs in its domain.

The function $F = 9/5C + 32$ does not have specific inputs and outputs stated, so we may use any real number that makes sense for the domain if the resulting output is a real number. Thus the domain for the function $F = 9/5C + 32$ is the set of all real numbers. Some functions defined by equations are restricted by the context in which

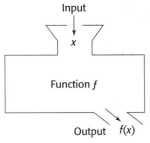

FIGURE 1.6

they are used. For example, in measuring the temperature of water, the domain of this function is limited to the real numbers from 0 to 100 and the range is limited to the real numbers from 32 to 212, because water changes state for other temperatures.

If x represents any element in the domain, then x is called the **independent variable** and if y represents an output of the function from an input x, then y is called the **dependent variable**. Figure 1.6 shows a general "function machine" in which the input is called x, the rule is denoted by f, and the output is symbolized by $f(x)$. The symbol $f(x)$ is read "f of x."

Tests for Functions

Functions play an important role in the solution of mathematical problems. Also, in order to graph a relationship using graphing calculators and computer software, it is usually necessary to express the association between the variables in the form of a function. It is therefore essential for you to recognize when a relationship is a function. Recall that a function is a rule or correspondence that determines exactly one output for each input.

EXAMPLE 1   Recognizing Functions

For each of the following, determine whether or not the indicated relationship represents a function. Explain your reasoning. For each function that is defined, give the domain and range.

a. The number N of personal computers in use worldwide determined by the year x, as defined in Table 1.4.

TABLE 1.4

x Year	N Worldwide Personal Computers (millions)
1991	129.4
1992	150.8
1993	177.4
1994	208.0
1995	245.0
2000 (projected)	535.6

(Source: *The Time Almanac, 1999.*)

b. The daily profit P (in dollars) from the sale of x tons of a particular product is shown in Figure 1.7. The points graphed in the figure are listed next to the scatter plot.

c. The number of tons x of a particular product sold determined by the profit P that is made from the sale of the product, as shown in the table in Figure 1.7.

d. W is a person's weight in pounds during the nth week of a diet for $n = 1$ and $n = 2$.

e. $y^2 = 3x - 3$ (x is the input).

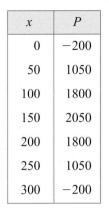

x	P
0	-200
50	1050
100	1800
150	2050
200	1800
250	1050
300	-200

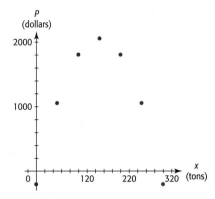

FIGURE 1.7

Solution

a. For each year (input) listed in Table 1.4, only one value is given for the number of computers used worldwide (output), so Table 1.4 represents N as a function of x. The set $\{1991, 1992, 1993, 1994, 1995, 2000\}$ is the domain. The range is $\{129.4, 150.8, 177.4, 208, 245, 535.6\}$ million.

b. Each input x corresponds to only one daily profit P, so this scatter plot represents P as a function of x. The domain is $\{0, 50, 100, 150, 200, 250, 300\}$ tons, and the range is $\{-200, 1050, 1800, 2050\}$ dollars. For this function, x is the independent variable and P is the dependent variable.

c. The number of tons x of a particular product sold is not a function of the profit P that is made, because some values of P result in two values of x. For example, a profit of \$1800 corresponds to both 100 tons of product and 200 tons of product.

d. A person's weight varies during any week; for example, a woman may weigh 122 lb on Tuesday and 121 lb on Friday. Thus there is more than one output (weight) for each input (week) and this relationship is not a function.

e. This indicated relationship between x and y is not a function because there can be more than one output for each input. For instance, the rule $y^2 = 3x - 3$ determines both $y = 3$ and $y = -3$ for the input $x = 4$. Note that if we solve this equation for y, we get $y = \pm\sqrt{3x - 3}$, so two values of y will result for any value of $x > 1$. ∎

Another way to determine whether an equation defines a function is to inspect its graph. If y is a function of x, no two distinct points (x, y) on the graph of $y = f(x)$ can have the same first coordinate. There are two points, $(4, 3)$ and $(4, -3)$, on the graph of

$$y^2 = 3x - 3,$$

shown in Figure 1.8(a) on page 14, so the equation does not represent y as a function of x (as we concluded in Example 1). In general, no two points of the graph can lie on the same vertical line if the relationship is a function.

VERTICAL LINE TEST

A set of points in a coordinate plane is the graph of a function if and only if no vertical line intersects the graph in more than one point.

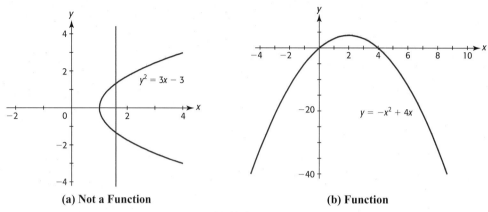

(a) Not a Function   **(b) Function**

FIGURE 1.8

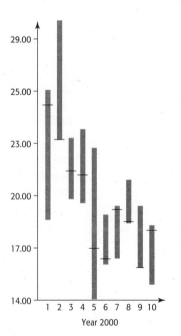

FIGURE 1.9 Coca-Cola Stock

(Source: 2000 Annual Report, Coca-Cola Enterprises)

When we look at the graph of

$$y = -x^2 + 4x$$

in Figure 1.8(b), we can see that any vertical line will intersect the graph in at most one point for the portion of the graph that is visible. If we know that the graph will extend indefinitely in the same pattern, we can conclude that no vertical line will intersect the graph in two points and so the equation $y = -x^2 + 4x$ represents a function.

The graph in Figure 1.9 gives the stock prices for the first 10 months of 2000 for Coca-Cola Enterprises. The graph shows that the price of a share of Coca-Cola Enterprises stock during each of the first 10 months of 2000 is not a function. The vertical bar above each month shows that the stock has many prices between its monthly high and low. Because of the fluctuations in price during each month, the *price* of Coca-Cola Enterprises stock during these months is not a function of the month.

Functional Notation

We can use the functional notation $y = f(x)$, read "y equals f of x," to indicate that the variable y is a function of the variable x. For specific values of x, $f(x)$ represents the resulting outputs, or y-values. In particular, the point $(a, f(a))$ lies on the graph of $y = f(x)$ for any number a in the domain of the function.

Thus if

$$f(x) = 4x^2 - 2x + 3,$$

then

$$f(3) = 4(3)^2 - 2(3) + 3 = 33$$
$$f(-1) = 4(-1)^2 - 2(-1) + 3 = 9$$

We can also use the functional notation $F(C) = \dfrac{9}{5}C + 32$ to write the Fahrenheit temperature measurement as a function of the Celsius temperature measurement. From the equation, values from Table 1.5, or points on Figure 1.10, we see that

$$F(50) = 122, F(0) = 32, \text{ and } F(25) = 77.$$

TABLE 1.5

Celsius degrees (°C)	−20	−10	−5	0	25	50	100
Fahrenheit degrees (°F)		−4	14	23	32	77	122	212

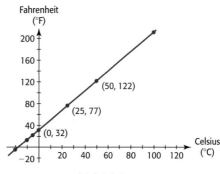

FIGURE 1.10

EXAMPLE 2   Functional Notation

Figure 1.11 shows the graph of

$$f(x) = 2x^3 + 5x^2 - 28x - 15.$$

a. Use the points shown on the graph to find $f(-2)$ and $f(4)$.

b. Use the equation to find $f(-2)$ and $f(4)$.

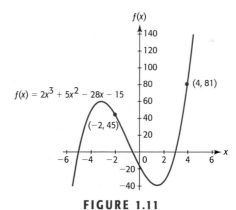

FIGURE 1.11

Solution

a. By observing Figure 1.11, we see that the point $(-2, 45)$ is on the graph of $f(x) = 2x^3 + 5x^2 - 28x - 15$, so $f(-2) = 45$. We also see that the point $(4, 81)$ is on the graph, so $f(4) = 81$.

b. $f(-2) = 2(-2)^3 + 5(-2)^2 - 28(-2) - 15 = 45.$

$f(4) = 2(4)^3 + 5(4)^2 - 28(4) - 15 = 81.$

Note that the values found by substitution agree with the y-coordinates of the points on the graph.   ∎

EXAMPLE 3   Men in the Workforce

The points on Figure 1.12 give the number of men in the workforce (in millions) as a function g of the year for selected years t from 1890 to 1990.

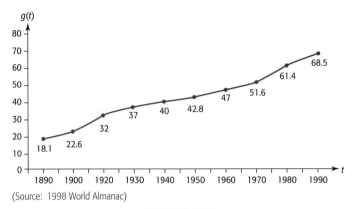

(Source: 1998 World Almanac)

FIGURE 1.12

a. Find and interpret $g(1940)$.

b. What is the input t if the output is $g(t) = 51.6$ million men?

c. What can be said about the number of men in the workforce during this period of time?

d. What is the maximum number of men in the workforce during the period shown in the graph?

Solution

a. The point above 1940 has coordinates $(1940, 40)$, so $g(1940) = 40$. This means that there were 40 million men in the workforce in 1940.

b. The point $(1970, 51.6)$ occurs on the graph, so $g(1970) = 51.6$ and $t = 1970$ is the input when the output is $g(t) = 51.6$.

c. The number increases during the period 1890 to 1990.

d. The maximum number of men in the workforce during the period is 68.5 million, in 1990. ∎

Domains and Ranges

Recall that if the domain and range of a function are not specified, it is assumed that the domain consists of all real number inputs that result in real number outputs in the range, and that the range is a subset of the real numbers. For the functions that we will be studying, we have the following.

DOMAIN

If the domain of a function is not specified, we assume that it includes all real numbers except those that give nonreal or undefined outputs. Examples of values that are not in the domain of a function are

- Values that result in a denominator of 0, and

- Values that result in an even root of a negative number.

We can use the graph of a function to find or to verify its domain and range. We can usually see the interval(s) on which the graph exists and therefore agree on the subset of the real numbers for which the function is defined. This set is the domain of the function. We can also usually determine if the outputs of the graph form the set of all real numbers or some subset of real numbers. This set is the range of the function.

EXAMPLE 4 Domains and Ranges

For each of the following functions determine the domain. Determine the range of the function in parts (a) and (b).

a. $y = 4x^2$ **b.** $y = \sqrt{4 - x}$ **c.** $y = 1 + \dfrac{1}{x - 2}$

Solution

a. Because any real number input for x, when squared and multiplied by 4, results in a real number output for y, we conclude that the domain is the set of all real numbers. Because

$$y = 4x^2$$

cannot be negative for any value of x that is input, the range is the set of all nonnegative real numbers ($y \geq 0$). This can be confirmed by looking at the graph of this function, shown in Figure 1.13.

b. Because

$$y = \sqrt{4 - x}$$

cannot be a real number if $4 - x$ is negative, the only values of x that give real outputs to the function are values that are less than or equal to 4, so the domain is $x \leq 4$. Because $\sqrt{4 - x}$ (the principal square root) can never be negative, the range is $y \geq 0$. A table of selected input values and graph of this function, as shown in Figure 1.14(a), confirm this.

c. Because the denominator of the fractional part of this function will be 0 when $x = 2$, and the outputs for every other value of x are real, the domain of this function contains all real numbers except 2. The graph in Figure 1.14(b) confirms this conclusion. ∎

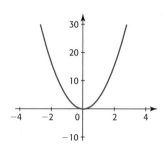

FIGURE 1.13

x	y
2	$\sqrt{2}$
3	1
4	0
4.001	Undefined (not real)

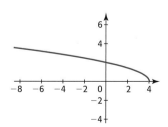

FIGURE 1.14(a)

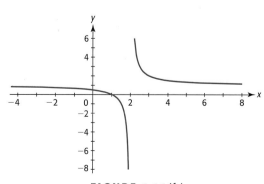

FIGURE 1.14(b)

1.1 SKILLS CHECK

Use the values in the table below in Exercises 1–5.

x	-9	-5	-7	6	12	17	22	16	18
y	5	6	7	4	9	9	10	19	15

1. Explain why the relationship shown by the table describes y as a function of x.

2. State the domain and range of the function.

3. If the function defined by the table is denoted by f, so that $y = f(x)$, is $f(-5)$ an input or an output of this function?

4. If $y = f(x)$ is defined by the table, find $f(-9)$ and $f(17)$.

5. Does the table describe x as a function of y? That is, would each input y result in a unique output x?

6. Determine if the graph in the figure represents y as a function of x. Explain your reasoning.

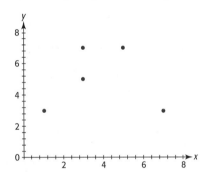

7. For each of the functions $y = f(x)$ described below, find $f(2)$.

   a.

x	$f(x)$
-1	5
0	7
1	2
2	-1
3	-8

   b. $y = 10 - 3x^2$

c.

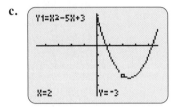

Determine if each graph in Problems 8 and 9 indicates that y is a function of x.

8.

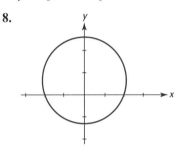

9.

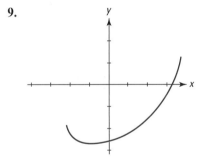

10. If $R(x) = 5x + 8$, find (a) $R(-3)$ (b) $R(-1)$ (c) $R(2)$

11. If $C(s) = 16 - 2s^2$, find (a) $C(3)$ (b) $C(-2)$ (c) $C(-1)$

12. Does the table below define y as a function of x? If so, give the domain and range of f.

x	y
-1	5
0	7
1	2
2	-1
3	-8

13. Which of the following arrow diagrams defines a function?

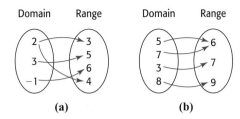

| (a) | (b) |

14. Which of the following sets of ordered pairs defines a function?
  a. $\{(1, 6), (4, 12), (4, 8), (3, 3)\}$
  b. $\{(2, 4), (3, -2), (1, -2), (7, 7)\}$

15. Does $x^2 + y^2 = 4$ describe y as a function of x?

16. Does $x^2 + 2y = 9$ describe y as a function of x?

17. Write an equation to represent the function described by the statement: "The perimeter p of a circle is found by multiplying 2π times the radius r."

Define, in your own words, the following terms:

18. function

19. domain of a function

20. range of a function

21. Write a verbal statement to represent the function $D = 3E^2 - 5$.

1.1 EXERCISES

In Exercises 1–5, determine whether the given relationship defines a function.

1. *Stock Prices*
  a. The price p at which IBM stock can be bought on a given day x.
  b. The closing price p of IBM stock on a given day x.

2. *Weight*
  a. The weight of a man each time he steps on a scale.
  b. The weight of a man in 2002.

3. *Groceries* The average price of iceberg lettuce in U.S. cities, in cents per pound, for selected months during 1999, defined by the table.

Month	Average Price (cents per pound)
February	66
March	77
April	75
May	69
June	65

(Source: Bureau of Labor Statistics)

4. *Life Insurance*
  a. The monthly rate for a $100,000 life insurance policy determined by age for males aged 26–32 (as shown in the table.)

  b. Ages that get a $100,000 life insurance policy for a monthly rate of $11.81 (from the table).

Age (years)	26	27	28	29	30	31	32
Premium (dollars per month)	11.72	11.81	11.81	11.81	11.81	11.81	11.81

5. *Income* The average income of male workers with given years of education, as defined by the graph in the figure.

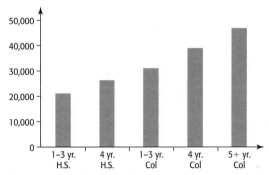

(Source: South Carolina Statistical Abstract, 1998)

6. *Temperature* The graph in the figure is a graph of $T = 0.43m + 76.8$, which gives the temperature T (in degrees Fahrenheit) inside a concert hall m minutes after a 40-minute power outage during a summer rock concert. Is the temperature a function of the time?

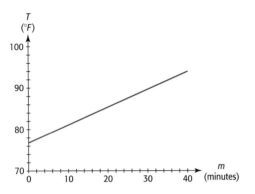

7. *Seawater Pressure* In seawater, the pressure p is related to the depth d according to the model

$$p = \frac{18d + 496}{33},$$

where d is the depth in feet and p is in pounds per square inch. Is p a function of d?

8. *Depreciation* A business property valued at $300,000 is depreciated over 30 years by the straight-line method, so that its value x years after the depreciation began is

$$V = 300,000 - 10,000x$$

Is the value of the property a function of the number of years?

9. *Weight* During the first two weeks in May, a man weighs himself daily (in pounds) and records the data in the table below.
   a. Does the table define weight as a function of the day in May?
   b. What is the domain of this function?
   c. What is the range?
   d. During what day(s) did he weigh most?
   e. During what day(s) did he weigh least?
   f. He claimed to be on a diet. What is the longest period of time during which his weight decreased?

May	1	2	3	4	5	6	7
Weight (lbs)	178	177	178	177	176	176	175

May	8	9	10	11	12	13	14
Weight (lbs)	176	175	174	173	173	172	171

10. *Test Scores*
    a. Is the average score on the final exam in a course a function of the average score on a placement test for the course, if the table below defines the relationship?
    b. Is the average score on a placement test a function of the average score on the final exam in a course, if the table below defines the relationship?

Average score on math placement test (percent)	81	75	60	90	75
Average score on final exam in algebra course (percent)	86	70	58	95	81

11. *Car Financing* If a couple finances an automobile costing $35,000 by making equal monthly payments over 5 years at an interest rate of 10%, the balance owed at the end of t years after purchase is given in the table below.

t (years)	$B(t)$ Balance (dollars)
1	29,321
2	23,047
3	16,115
4	8458.60
5	0

(Source: Sky Financial Mortgage Tables)

a. What is the balance owed by the couple at the end of 3 years?
b. What is $B(2)$ if we represent the balance at the end of t years as $B(t)$? Write a sentence that explains its meaning.
c. At the end of what year does the couple owe $23,047?
d. What is t if $B(t) = 8458.60$?

12. *Mortgage* A couple can afford $800 per month to purchase a home. As indicated in the table below, if they can get an interest rate of 7.5%, the number of years t that it will take to pay off the mortgage is a function of the dollar amount A of the mortgage for the home they purchase.

Amount A (dollars)	t (years)
40,000	5
69,000	10
89,000	15
103,000	20
120,000	30

(Source: Comprehensive Mortgage Tables (Publication No. 492). Financial Publishing Co.)

a. If the couple wishes to finance $103,000, for how long must they make payments? Write this correspondence in functional form if $t = f(A)$.

b. What is $f(89,000)$? Write a sentence that explains its meaning.

c. What value of A makes $f(A) = 30$ true?

d. Does $f(3 \cdot 40,000) = 3 \cdot f(40,000)$? Explain your reasoning.

13. *Working Age* The projected ratio of the working-age population (25- to 64-year-olds) to the elderly shown in the figure below defines the ratio as a function of the year shown. If this function is defined as $y = f(t)$ where t is the year, use the graph to answer the following:

a. What is the projected ratio of the working-age population to the elderly population in 1995?

b. Estimate $f(2005)$ and write a sentence that explains its meaning.

c. What is the domain of this function?

d. Is the projected ratio of the working-age population to the elderly increasing or decreasing over the domain shown in the figure?

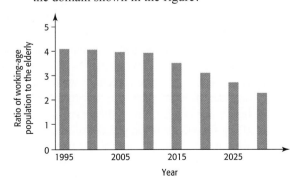

(Source: *Newsweek,* December 1999)

14. *Women in the Workforce* The points on the figure below give the number (in millions) of women in the workforce as a function f of the year for selected years from 1890 to 1990.

a. How many women were in the labor force in 1960?

b. Estimate $f(1890)$ and write a sentence that explains its meaning.

c. What is the domain of this function if we consider only the indicated points?

d. Is the number of women in the workforce increasing or decreasing over the domain shown in the graph?

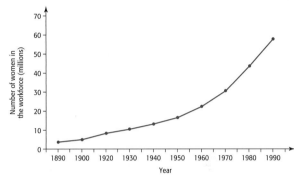

(Source: 1998 *World Almanac*)

15. *Gun Crime* The table below gives the total number of firearm crimes as a function f of the year t.

a. What is $f(1990)$?

b. What is the domain of this function?

c. What is the maximum number of firearm crimes during this period of time? In what year did it occur?

Year	Total Firearm Crimes	Year	Total Firearm Crimes
1985	340,942	1992	565,575
1986	376,064	1993	581,697
1987	365,709	1994	542,529
1988	385,934	1995	504,421
1989	410,039	1996	458,458
1990	492,671	1997	414,530
1991	548,667	1998	364,776

(Source: U.S. Department of Justice)

16. *Movie Attendance* The movie admissions (in billions) in the U.S. for selected years is given in the table on page 22 as a function of the number of years from 1980.

a. What is the domain of this function?

b. What is the range of this function?

c. What is the output when the input is 10, and what does this mean?

d. What can be said about the number of movie admissions over this period of time?

Years from 1980	Movie Admissions (billions of people)
0	1.02
5	1.06
10	1.10
15	1.26
18	1.48

(Source: Motion Picture Association of America)

17. *Farm Workers* The points on the figure below show the percent *p* of U.S. workers in farm occupations for selected years *t*.
   a. Is the percent of U.S. workers in farm occupations a function of the year?
   b. What is $f(1840)$ if $p = f(t)$?
   c. What is *t* if $f(t) = 27$?
   d. Write a sentence explaining the meaning of $f(1960) = 6.1$.
   e. Describe the change in the percent of U.S. workers in farm occupations.

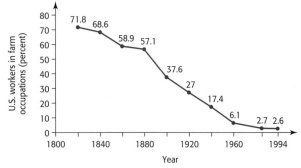

(Source: U.S. Department of Agriculture)

18. *Internet Use* The figure below gives the number of millions of U.S. homes using the Internet for the years 1995 to 1998. If the number of millions of U.S. homes is the function $f(x)$, where *x* is years
   a. How many homes used the Internet during 1995?
   b. Find $f(1997)$ and explain its meaning.
   c. In what year did 31.7 million homes use the Internet?
   d. Is this function increasing or decreasing? What do you think has happened to home use of the Internet since 1998?

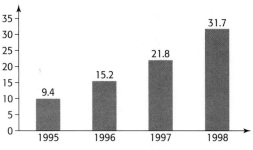

(Source: *Cyber Atlas*)

19. *Teen Pregnancy* The points on the figure below give the number of pregnancies per 1000 U.S. girls (ages 15–19) for the years 1980 through 1996.
   a. Find the output when the input is 1995 and explain its meaning.
   b. For what years was the rate 113?
   c. During what years did the pregnancy rate increase?
   d. For what year is the pregnancy rate at its maximum?

Teen Pregnancy Rate per 1000 Girls

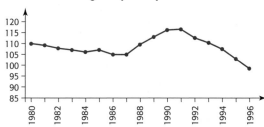

(Source: The National Campaign to Prevent Teen Pregnancy)

20. *Educational Attainment* The table below shows the percentage of 25- to 29-year-old U. S. white, black, and Hispanic males who are identified in selected years between 1971 and 1999 as having completed at least some college.
   a. If the function *f* is the percent of white males who are identified in year *x* as having completed at least some college, find $f(1995)$ and $f(1999)$.
   b. If the function *g* is the percent of Hispanic males who are identified in year *x* as having completed at least some college, find $g(1995)$ and explain what it means.
   c. If the function *h* is the percent of black males who are identified in year *x* as having completed at least some college, find $h(1983)$ and $h(1999)$.
   d. For what value of *x* is $f(x) = 51.5$? Write the result in a sentence.

Year	White male	Black male	Hispanic male
1971	50.2	29.0	38.3
1973	53.0	33.5	39.4
1975	57.3	41.0	50.4
1977	59.9	44.2	42.6
1979	59.4	40.7	50.7
1981	54.1	43.0	41.7
1983	53.4	42.0	41.1
1985	52.5	42.4	45.9
1987	51.5	38.4	46.3
1989	53.4	42.2	44.8
1991	54.7	38.3	40.9
1993	60.3	43.6	46.1
1995	62.6	51.2	48.0
1997	66.9	50.2	51.9
1999	66.1	52.1	47.7

(Source: *March Current Population Surveys*, U.S. Department of Commerce, Bureau of the Census)

21. *Body-Heat Loss* The description of body-heat loss due to convection involves a coefficient of convection K_c, which depends on wind speed s according to the equation $K_c = 4\sqrt{4s + 1}$.
a. Is K_c a function of s?
b. What is the domain of the function defined by this equation?
c. What restrictions do the physical nature of the model put on the domain?

22. *Cost-Benefit* Suppose that the cost C (in dollars) of removing p percent of the particulate pollution from the smokestack of a power plant is given by

$$C(p) = \frac{237{,}000p}{100 - p}$$

a. Use the fact that percent is measured between 0 and 100, and find the domain of this function.
b. Evaluate $f(60)$ and $f(90)$.

23. *Postal Restrictions* Some postal restrictions say that the size of the largest box that can be shipped has its length (longest side) plus its girth (distance around the box in the other two dimensions) equal to 108 inches.

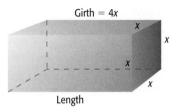

If a box has a square cross section that is x inches on each side, then the volume of the box is given by $V = x^2(108 - 4x)$ cubic inches.
a. Find $V(12)$ and $V(18)$.
b. What restrictions must be placed on x to satisfy the conditions of this model?
c. Create a table of functional values to investigate the value of x that maximizes the volume. What are the dimensions of the box that has the maximum volume?

24. *Height of a Bullet* The height of a bullet shot into the air is given by $S(t) = -4.9t^2 + 98t + 2$, where t is the number of seconds after it is shot and $S(t)$ is in meters.
a. Find $S(0)$ and interpret it.
b. Find $S(9)$, $S(10)$, and $S(11)$.
c. What appears to be happening to the bullet at 10 seconds? Evaluate the function at some additional times near 10 seconds to confirm your conclusion.

1.2 Graphs of Functions; Mathematical Models

The process of translating real-world information into a mathematical form so that it can be applied and then interpreted in the real-world setting is called **modeling**. As we most often use it in this text, a **mathematical model** is a functional relationship (usually in the form of an equation) that includes not only the function rule but also descriptions of all involved variables and their units of measure. For example, the function $F = 9/5\,C + 32$ was used to describe the relationship between temperature scales in Section 1.1. The **model** that describes how to convert from one measuring scale to another must include the equation ($F = 9/5\,C + 32$) and a description of the variables (F is the temperature measure in degrees Fahrenheit and C is the temperature measure in degrees Celsius).

Suppose the General Electric Company had published only the equation

$$R(x) = 10.027x - 19{,}935.193$$

in its 1998 annual report to inform its stockholders regarding its annual revenues. The stockholders would have had no idea what the equation meant or how it could be used; that is, the equation is meaningless without further explanation. A statement such as: "For the years between 1994 and 1998, General Electric's annual revenue is given by

$$R = 10.027x - 19{,}935.193 \text{ billion dollars in year } x"$$

allows the reader to understand what the equation means and how it can be used. This statement is a **model** of General Electric's revenue for this period. (Source: Hoover's Online Capsules.)

In this section, we will graph functions with technology and use technology to graph and evaluate mathematical models.

Graphs of Functions

If an equation defines y as a function of x, we can sketch the graph of the function by plotting enough points to determine the shape of the graph and then drawing a line or curve through the points. This is called the **point-plotting method** of sketching a graph.

EXAMPLE 1 Graphing an Equation by Hand

a. Graph the equation $y = x^2$ by drawing a smooth curve through points determined by integer values of x between 0 and 3.

b. Graph the equation $y = x^2$ by drawing a smooth curve through points determined by integer values of x between -3 and 3.

Solution

a. We use the values in Table 1.6 to determine the points. The graph of the function drawn through these points is shown in Figure 1.15(a).

b. We use Table 1.7 to find the additional points. The graph of the function drawn through these points is shown in Figure 1.15(b). ∎

TABLE 1.6

x	0	1	2	3
$y = x^2$	0	1	4	9
Points	$(0, 0)$	$(1, 1)$	$(2, 4)$	$(3, 9)$

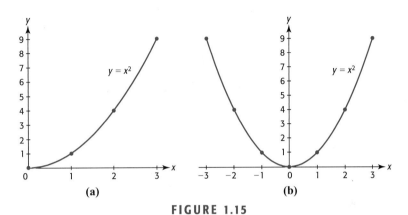

(a)   (b)

FIGURE 1.15

TABLE 1.7

x	-3	-2	-1	0	1	2	3
$y = x^2$	9	4	1	0	1	4	9
Points	$(-3, 9)$	$(-2, 4)$	$(-1, 1)$	$(0, 0)$	$(1, 1)$	$(2, 4)$	$(3, 9)$

Although neither graph in Figure 1.15 shows all of the points satisfying the equation $y = x^2$, the graph in Figure 1.15(b) is a much better representation of the function (as we will learn later). When enough points are connected to determine the shape of the graph, the important parts of the graph, and to suggest what the unseen parts of the graph look like, the graph is called **complete**.

COMPLETE GRAPH

A graph is a complete graph if it shows the basic shape of the graph and important points on the graph (including points where the graph crosses the axes and points where the graph turns), and suggests what the unseen portions of the graph will be.

Different **viewing windows** give different views of a graph. There are usually many different viewing windows that give complete graphs for a particular function, but some viewing windows do not show all the important parts of a graph. Being able to draw complete graphs, with or without the aid of technology, improves with experience and requires knowledge of the shape of the basic function(s) involved in the graph. As we learn more about types of functions, we will be better able to determine when a graph is complete. In the meantime, viewing windows will be determined by the limitations of the application being considered or will be specified in the problem.

Graphing with Technology

Obtaining a sufficient number of points to sketch a graph by hand can be time consuming. Computers and graphing calculators have graphing utilities that can be used to plot many points quickly, thereby showing the graph with minimal effort. However, graphing with technology involves much more than pushing a few buttons. Unless you know what to enter into the calculator or computer (i.e., the domain of the function and sometimes, the range), you cannot see or use the graph that is drawn. We suggest the following steps for graphing most functions with a graphing utility.

USING A GRAPHING CALCULATOR TO DRAW A GRAPH

1. Write the function with x representing the independent variable and y representing the dependent variable. Solve for y, if necessary.

2. Enter the function in the equation editor of the graphing utility. Use parentheses* as needed to ensure the mathematical correctness of the expression.

3. Activate the graph by pressing the ZOOM or GRAPH key. Most graphing utilities have several preset **viewing windows** under ZOOM , including the **standard viewing window** that gives the graph of a function on a coordinate system in which the x-values range from -10 to 10 and the y-values range from -10 to 10.

4. To see parts of the graph of a function other than those shown in a standard window, set the x- and y-boundaries of the viewing window before pressing GRAPH . Viewing window boundaries are discussed below and in the technology supplement. As you gain more knowledge of graphs of functions, determining viewing windows that give complete graphs will be less complicated.

---

*Parentheses should be placed around numerators and/or denominators of fractions, fractions that are multiplied by a variable, exponents consisting of more than one symbol, and in other situations where the order of operations needs to be indicated.

Although the standard viewing window is a convenient window to use, it may not show the desired graph. To see parts of the graph of a function other than those that might be shown in a standard window, we change the x- and y-boundaries of the viewing window. The values that define the viewing window can be set manually or by using the ZOOM keys. The boundaries of a viewing window are:

x_{min}: the smallest value on the x-axis (the left boundary of the window)

y_{min}: the smallest value on the y-axis (the bottom boundary of the window)

x_{max}: the largest value on the x-axis (the right boundary of the window)

y_{max}: the largest value on the y-axis (the top boundary of the window)

x_{scl}: the distance between ticks on the x-axis (helps visually find x-intercepts and other points.)

y_{scl}: the distance between ticks on the y-axis (helps visually find y-intercept and other points.)

When showing viewing window boundaries on calculator graphs in this text, we write them in the form:

$$[x_{min}, x_{max}] \text{ by } [y_{min}, y_{max}].$$

To illustrate these ideas, consider the graph of $y = x^3 - 3x^2 - 13$. A graph of this function in the standard viewing window appears to be a line. See Figure 1.16(a). By manually setting the viewing window as follows: $x_{min} = -10$, $x_{max} = 10$, $y_{min} = -25$, $y_{max} = 10$ (denoted $[-10, 10]$ by $[-25, 10]$), we obtain the graph in Figure 1.16(b). This window gives a better picture of the actual curvature of the graph of this equation. As we learn more about functions, we will see that the graph in Figure 1.16(b) is a complete graph of this function.

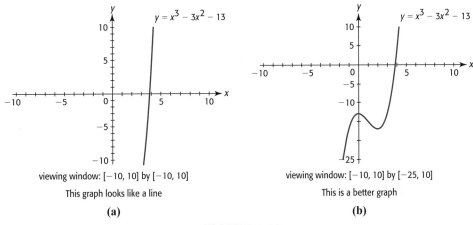

viewing window: $[-10, 10]$ by $[-10, 10]$

This graph looks like a line

(a)

viewing window: $[-10, 10]$ by $[-25, 10]$

This is a better graph

(b)

FIGURE 1.16

EXAMPLE 2   Cost-Benefit

Suppose that the cost C of removing p percent of the pollution from drinking water is given by the model

$$C = \frac{5350p}{100 - p} \text{dollars.}$$

a. Use the restriction on p to determine the limitations on the horizontal-axis values (which are the x-values on a calculator).

b. Graph the function on the viewing window $[0, 100]$ by $[0, 50,000]$. Why is it reasonable to graph this model on a viewing window with the limitation $C \geq 0$?

c. Find the point on the graph that corresponds to $p = 90$. Explain the coordinates of this point.

Solution

a. Because p represents the percent of pollution removed, it is limited to values from 0% to 100%. However, 100% of the pollution cannot be removed, so p is restricted to $0 \leq p < 100$ for this model.

b. The graph of the function is shown in Figure 1.17(a), with x representing p and y_1 representing C. The interval $[0, 100]$ contains all the possible values of p. The value of C is bounded below by 0 because it represents the cost of removing the pollution, which cannot be negative.

c. We can find (or estimate) the output of a function $y = f(x)$ at specific inputs with a graphing utility. We do this with TRACE , CALC , VALUE , or TABLE . Using

TRACE or CALC, VALUE with the x-value 90 gives the point (90, 48,150) (see Figure 1.17(b)). Because the values of p are represented by x-values and the values of C are represented by y, the coordinates of the point tell us that the cost of removing 90% of the pollution from the drinking water is $48,150. Figure 1.17(c) shows the value of y for $x = 90$ and other values in a calculator table.

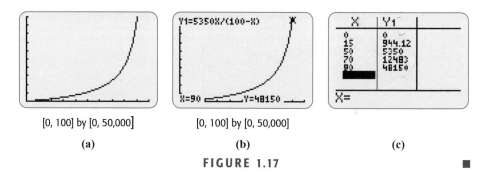

[0, 100] by [0, 50,000]	[0, 100] by [0, 50,000]	
(a)	**(b)**	**(c)**

FIGURE 1.17   ■

Spreadsheet Solution

We have briefly described how to use a graphing calculator to graph. Computer software of several types can be used to create more accurate and better-looking graphs. Software such as Scientific Notebook, Maple, and Mathematica can be used to create graphs, and **spreadsheets** like Excel can also be used. Table 1.8 shows a spreadsheet with the output of $C = \dfrac{5350p}{100 - p}$ at $x = 90$ and at other values of x, and Figure 1.18 shows the graph of this function on Excel.

TABLE 1.8

	A	B
1	p	C
2	0	0
3	10	594.4444
4	20	1337.5
5	30	2292.857
6	40	3566.667
7	50	5350
8	60	8025
9	70	12483.33
10	80	21400
11	90	48150

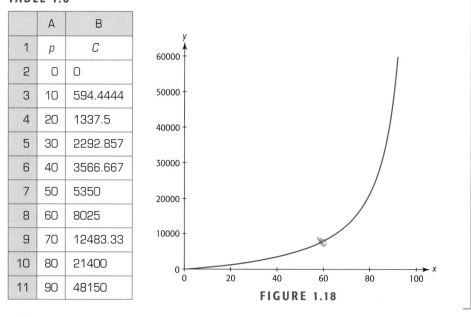

FIGURE 1.18

Mathematical Models

A mathematical model can sometimes provide an exact description of a real situation (such as the Celsius/Fahrenheit model discussed at the beginning of this section), but a model frequently provides only an approximate description of a real-world situation. In using a model that approximates real data, it is important to balance two often-competing goals: accuracy and simplicity. We now use (and later create) functions that accurately model the data and yet are simple enough to perform mathematical operations on them with relative ease.

EXAMPLE 3   General Electric Revenue

- -

Use the model $R(x) = 10.03x - 19,935.19$, where R is General Electric's revenue in billions of dollars for year x between 1994 and 1998, to find General Electric's revenue corresponding to the years 1995, 1996, and 1998. (Source: Hoover's Online Capsules.)

Solution

The revenue for each of these years can be found by evaluating the function at $x = 1995$, 1996, and 1997 manually, with a calculator, or with an electronic spreadsheet. (A spreadsheet, such as *Excel*, would be especially useful in evaluating a function at a large number of values of x.) Table 1.9 gives the year inputs and the revenue outputs that result from substituting each of the years for x in the given equation for R. Note that the annual revenue value for year x can be denoted by $R(x)$, so that $R(1995) = 74.66$.

TABLE 1.9

Year	1995	1996	1998
G.E. revenue (billions of dollars)	74.66	84.69	104.75

Thus the revenue is *estimated* to be \$74.66 billion in 1995, \$84.69 billion in 1996, and \$104.75 billion in 1998. Unless we know that the function is an exact fit to the data it is modeling, the outputs are only estimates of the actual data.   ■

TECHNOLOGY NOTE

Table 1.10 shows a **spreadsheet** with these inputs and outputs for the function shown in Example 3. When evaluating a function with a spreadsheet, we use the cell location of the data to represent the variable. Thus if 1995 is in cell A2, then $R(1995)$ can be found in cell B2 by typing 10.03*A2 − 19,935.19 in cell B2. By using the fill-down capacity of the spreadsheet, we can obtain all the functional values shown in column B.

TABLE 1.10   SPREADSHEET

	A	B	
1	x	$R(x) = 10.03x - 19,935.19$	
2	1995	74.66	
3	1996	84.69	
4	1998	104.75	
5			

Aligning Data

When finding a model to fit a set of data, it is often easier to use aligned inputs rather than the actual data values. **Aligned inputs** are simply input values that have been converted to smaller numbers by subtracting the same number from each input. For instance, instead of using x as the actual year in the following example, it is more convenient to use an aligned input that is the number of years after 1950.

EXAMPLE 4   Voting

Between 1950 and 1996, the percent of the voting population who voted in presidential elections is given by

$$f(x) = 65.40 - 0.36x$$

where x is the number of years after 1950. (Source: Federal Election Commission.)

a. What are the values of x that correspond to the years 1960 and 1996?

b. Find $f(10)$ and explain its meaning.

c. If this model is accurate for 1984, find the percent of the voting population who voted in the 1984 presidential election.

Solution

a. Because 1960 is 10 years after 1950, $x = 10$ corresponds to the year 1960 in the aligned data. Using similar reasoning, 1996 corresponds to $x = 46$.

b. An input value of 10 represents 10 years after 1950, which is the year 1960. To find $f(10)$, we substitute 10 for x to obtain $f(10) = 65.40 - 0.36(10) = 61.8$. One possible explanation of this answer is this: Approximately 62 percent of the voting population voted in the 1960 presidential election.

c. To find the percent in 1984, we first calculate the aligned input for the function. Because 1984 is 34 years after 1950, we use $x = 34$. Thus, we find

$$f(34) = 65.40 - 0.36(34) = 53.16.$$

So, if this model is accurate for 1984, approximately 53 percent of the voting population voted in the 1984 presidential election.   ■

Finding the functional values (y-values) for selected inputs (x-values) can be useful when setting viewing windows for graphing utilities.

TECHNOLOGY NOTE

Once the input values for a viewing window have been selected, TRACE , VALUE , or TABLE can be used to find enough output values to determine a y-view that gives a complete graph.*

---

*It is occasionally necessary to make more than one attempt to find a window that gives a complete graph.

EXAMPLE 5   Bankruptcies

The number of bankruptcies per year of publicly traded companies during the years 1990 to 1999 can be modeled by

$$y = 2.8447x^2 - 24.4023x + 128.9364$$

where x is the number of years after 1990.

a. Choose an appropriate viewing window for the 10-year period and graph the function on a graphing calculator.

b. Use the model to estimate the number of bankruptcies of publicly traded companies in 1996.

Solution

a. The viewing window should include values of x that represent the years 1990 to 1999, so the x-view should include the aligned values $x = 0$ to $x = 9$. The y-view should not include negative numbers and should include the outputs obtained from the inputs between 0 and 9. We can use TRACE , CALC , VALUE , or TABLE with some or all of the integers 0 through 9 to find corresponding y-values (regardless of how the y-view is set when we are evaluating).* These evaluations indicate that the y-view should include values from about 76 through about 140. The graph of the function with the boundaries $x_{min} = 0$, $x_{max} = 9$, $y_{min} = 66$ and $y_{max} = 150$ is shown in Figure 1.19(a).

b. The year 1996 is represented by $x = 6$. Evaluating the function with TRACE or CALC , VALUE on the graph (see Figure 1.19(b)) gives an output of approximately 85 when $x = 6$. Figure 1.19(c) shows the values of the function at $x = 0, 9$, and 6 using TABLE . Thus we estimate that the number of bankruptcies of publicly traded companies in 1996 was 85.

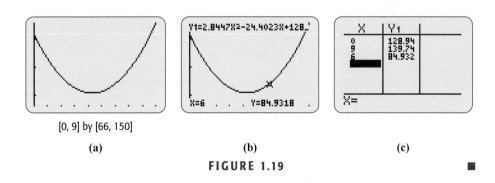

[0, 9] by [66, 150]

(a)   (b)   (c)

FIGURE 1.19   ■

Excel can also be used to graph and evaluate the function $y = 2.8447x^2 - 24.4023x + 128.9364$, discussed in Example 5. The Excel graph is shown in Figure 1.20, and an Excel spreadsheet with values of the function at selected values of x is shown in Table 1.11.

---

* TRACE may only give an approximate value of a function. On some calculators, a function can be evaluated exactly by pressing TRACE , the x-value, and ENTER or by pressing CALC , VALUE , the x-value, and ENTER .

TABLE 1.11

	A	B
	x	f(x)
1	0	128.9364
2	1	107.3788
3	2	91.5106
4	3	81.3318
5	4	76.8424
6	5	78.0424
7	6	84.9318
8	7	97.5106
9	8	115.7788
10	9	139.7364

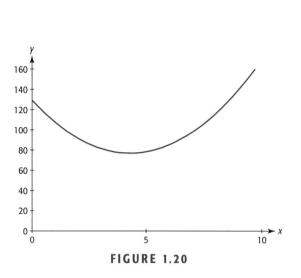

FIGURE 1.20

Graphing Data Points

Graphing utilities can also be used to create **lists** of numbers and to create graphs of the data stored in the lists. To see how lists and scatter plots are created, consider the following example.

EXAMPLE 6   U.S. Executions

Table 1.12 gives the number of executions in the United States for the decades beginning in 1930. Column 1 contains the decades. The data is aligned in column 2 to give the number of years from 1900 to the *end* of the given decade. Column 3 gives the number of executions in the given decade.

TABLE 1.12

Decade	Years from 1900 to End of Decade	Number of Executions
1930–1939	40	1690
1940–1949	50	1284
1950–1959	60	717
1960–1969	70	191
1970–1979	80	3
1980–1989	90	117
1990–1999	100	518

(Source: "The Death Penalty in the U.S.," www.clarkprosecutor.org)

a. Enter the number of years from 1900 to the end of the decade in list L1 and the number of executions in that decade in list L2.

b. Use a graphing command* to create the graph of these data points (called a scatter plot).

Solution

a. Figure 1.21(a) shows the data from column 2 of Table 1.12 in L1 and the data from column 3 of the table in L2.

b. The scatter plot is shown in Figure 1.21(b), with each entry in L1 represented by an x-coordinate on the graph and the corresponding element in L2 represented by the corresponding y-coordinate on the graph. The window can be set automatically or manually to include x-values from 40 to 100 and y-values from 3 to 1690.

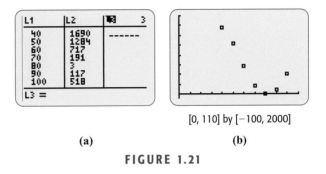

[0, 110] by [−100, 2000]

(a)   (b)

FIGURE 1.21   ■

Figure 1.22 shows the values from Table 1.12 in an Excel spreadsheet and the scatter plot (graph) of the data.

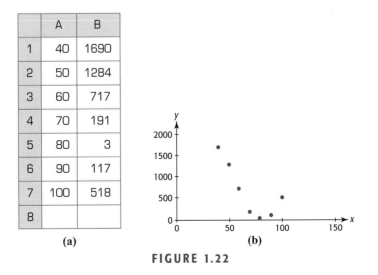

	A	B
1	40	1690
2	50	1284
3	60	717
4	70	191
5	80	3
6	90	117
7	100	518
8		

(a)   (b)

FIGURE 1.22

---

*Many calculators have a ⎍STAT PLOT⎍ command, and Excel has an ⎍XY⎍ (Scatter) command.

1.2 SKILLS CHECK

1. Complete the table of values for the function $y = x^3$ and then graph the equation by hand by drawing a smooth curve through the points.

x	-3	-2	-1	0	1	2	3
$y = x^3$							
(x, y)							

2. Graph the function $y = x^3$ with a graphing utility using a standard viewing window and compare your graph to the graph you obtained in Exercise 1.

In Exercises 3–6, graph the functions with a graphing utility using a standard viewing window.

3. $y = x^2 - 5$

4. $y = 4 - x^2$

5. $y = x^3 - 3x^2$

6. $y = \dfrac{9}{x^2 + 1}$

For Exercises 7–10, graph the given function with a graphing calculator using (a) the standard viewing window and (b) the specified window. Which window gives a better view of the graph of the function?

7. $y = x + 20$ with $x_{min} = -10$, $x_{max} = 10$, $y_{min} = -10$, $y_{max} = 30$

8. $y = x^3 - 3x + 13$ with $x_{min} = -5$, $x_{max} = 5$, $y_{min} = -10$, $y_{max} = 30$

9. $y = \dfrac{0.4(x - 0.1)}{x^2 + 300}$ on $[-20, 20]$ by $[-0.02, 0.02]$

10. $y = -x^2 + 20x - 20$ on $[-10, 20]$ by $[-20, 90]$

For Exercises 11 and 12, find an appropriate viewing window for the function, using the given x-values. Then graph the function.

11. $y = x^2 + 50$, for x-values between -3 and 3

12. $y = (x - 28)^3$, for x-values between 25 and 31

13. Find a complete graph of $y = 10x^2 - 90x + 300$. (A complete graph of this function shows a turning point.)

14. Use a calculator or a spreadsheet to find $S(t)$ for the values of t given in the following table.

t	$S(t) = 5.2t - 10.5$
12	
16	
28	
43	

15. Enter the data into lists and graph the scatter plot of the data, using the window $[0, 110]$ by $[0, 600]$.

x	20	30	40	50	60	70	80	90	100
y	500	320	276	80	350	270	120	225	250

Use the table below in Exercises 16–18.

x	1	3	6	8	10
y	-10	0	15	25	35

16. Use a graphing utility to graph the points (x, y) from the table, using the viewing window $[0, 12]$ by $[-12, 40]$.

17. Use a graphing utility to graph the equation $y = 5x - 15$ on the same set of axes as the data in the table, using the viewing window $[0, 12]$ by $[-12, 40]$.

18. Do the data points fit on the graph of the equation? What is the connection between the values of x and y in the table?

For Problems 19 to 21, use the function $L = 35t^2 + 740t + 1207$, which is related to state lotteries with t equal to the number of years after 1980.

19. What are the values of t that represent the years 1982, 1988, and 2000?

20. $L = f(4)$ gives the value of L for what year? What is $f(4)$?

21. **a.** Rewrite the function $L = 35t^2 + 740t + 1207$ with x as the independent variable and y as the dependent variable.

    b. What x-min and x-max should be used to set a viewing window so that x represents 1980–1997?

1.2 EXERCISES

----

1. *Earnings and Minorities* According to the U.S. Equal Employment Opportunity Commission, the relation between the median annual salaries of minorities and whites can be modeled by the function $M = 0.96W - 1.23$, where M and W represent the median annual salary (in thousands of dollars) for minorities and whites, respectively.
   a. Find the value of M when $W = 80$. Explain what this means.
   b. What minority salary corresponds to a median white salary of $100,000, according to this model?
   (Source: *Statistical Abstract of the U.S., 1993*)

2. *Seawater Pressure* In seawater, the pressure p is related to the depth d according to the model
   $$p = \frac{18d + 496}{33},$$
   where d is the depth in feet and p is in pounds per square inch.
   a. Find the value of p when $d = 22$. Explain what this means.
   b. What is the pressure at a depth of 66 feet?

3. *Welfare Cases* The average number of welfare cases during 1993–1998 in Niagara, Canada, is given by the model $y = -112x^2 - 107x + 15,056$, where x is the number of years after 1990.
   a. What are the values of x that correspond to the years 1994 and 1998?
   b. Find the value of y when $x = 8$. Explain what this means.
   c. How many welfare cases were there in 1995, according to this model?
   (Source: Regional Niagara Social Services Dept, April 25, 2000)

4. *Women in the Workplace* The number y (in thousands) of women in the workforce is given by the function
   $$y = 7x^2 - 19.88x + 409.29,$$
   where x is the number of years after 1900.
   a. Find the value of y when $x = 44$. Explain what this means.
   b. Use the model to find the number of women in the workforce in 1980.
   (Source: Bureau of the Census, U.S. Dept of Commerce)

5. *Height of a Ball* If a ball is thrown into the air at 64 feet per second from the top of a 100-foot-tall build-ing, its height can be modeled by the function $S = 100 + 64t - 16t^2$, where S is in feet and t is in seconds.
   a. Graph this function on a viewing window $[0, 6]$ by $[0, 200]$.
   b. Find the height of the ball 1 second after it is thrown and 3 seconds after it is thrown. How can these values be equal?
   c. Find the maximum height the ball will reach.

6. *Depreciation* A business property valued at $300,000 is depreciated over 30 years by the straight-line method, so that its value x years after the depreciation began is
   $$V = 300,000 - 10,000x.$$
   a. Graph this function on a viewing window $[0, 30]$ by $[0, 300,000]$.
   b. What is the value 10 years after the depreciation is started?

7. *Earnings and Gender* A model that relates the median annual salary (in thousands of dollars) of females, F and males, M in the United States is given by $F = 0.84M - 1.36$.
   a. Use a graphing utility to graph this function on the viewing window $[0, 85]$ by $[0, 65]$.
   b. Use the graphing utility to find the median female salary that corresponds to a male salary of $63,000.
   (Source: U.S. Equal Opportunity Commission, 1993)

8. *Education Spending* Federal spending (in billions of dollars) for public education during the years 1996–2001 can be modeled by the function $S = 3.32x + 23.16$, where x is the number of years after 1990.
   a. Use a graphing utility to graph this function on the viewing window $[0, 11]$ by $[0, 60]$.
   b. Find the federal spending for public education in 2001.
   (Source: U.S. Dept of Education)

9. *Medical School* Based on data collected in the spring of each year between 1988 and 1997, the number of students (in thousands) of osteopathic medicine in the United States can be described by
   $$S = 0.027t^2 - 4.85t + 218.93$$
   where t is the number of years after 1980.
   a. Graph this function with the viewing window $[0, 17]$ by $[0, 300]$.

b. Use technology to find S when t is 15.

c. Use the model to estimate the number of osteopathic students in 1995.
(Source: *Statistical Abstract of the U.S.*, 1998)

10. *State Lotteries* The cost (in millions of dollars) of prizes and expenses for state lotteries between 1980 and 1997 can be described by $L = 35.3t^2 + 740.2t + 1207.2$ where t is the number of years after 1980.

a. Graph this function with the viewing window [0, 17] by [1200, 12,000].

b. Use technology to find L when t is 16.

c. What was the cost of prizes and expenses for state lotteries in 1996?
(Source: *Statistical Abstract of the U.S.*, 1998)

11. *Tax Burden* Using data from the Internal Revenue Service, the per capita tax burden B (in hundreds of dollars) can be described by $B(t) = 20.37 + 1.83t$ where t is the number of years after 1980.

a. Graph this function with technology using a viewing window with $t \geq 0$ and $B(t) \geq 0$.

b. Did the tax burden increase or decrease?
(Source: Internal Revenue Service)

12. *Crime* The rate (number per 100,000 people) of juvenile arrests for crimes is given by

$$f(x) = -0.027x^2 + 5.69x + 51.15$$

where x is the number of years after 1950.
(Source: Federal Bureau of Investigation)

a. Use technology to graph this model with the viewing window [0, 50] by [0, 300].

b. This viewing window shows the graph for what time period?

c. Did the rate of juvenile arrests increase or decrease over this period?

d. Use this model to find the rate of juvenile arrests in 1960 and in 1998.
(Source: Federal Bureau of Investigation)

13. *Unemployment Rate* The U.S. unemployment rate for the years 1982 to 1996 can be modeled by

$$u(x) = -0.0012x^4 + 0.033x^3 - 0.23x^2$$
$$- 0.35x + 11.32 \text{ percent}$$

where x is the number of years after 1980.

a. What inputs correspond to the years 1982 through 1996?

b. What outputs for $u(x)$ could be used to estimate the U. S. unemployment rate, which is a percent?

c. Based on your answers to parts (*a*) and (*b*), choose an appropriate window and graph the equation on a graphing utility.

d. Graph the function again with a new window that gives a graph nearer the center of the screen.

e. Use the function to estimate the unemployment rate in 1993.
(Source: Bureau of Labor Statistics and U.S. Department of Labor)

14. *Cocaine Use* The percent of high school seniors during the years 1975 through 1996 who have ever used cocaine can be described by

$$y = 0.0094x^3 - 0.36x^2 + 3.35x + 8.53$$

where x is the number of years after 1975.

a. What inputs correspond to the years 1975 through 1996?

b. What outputs for y could be used to estimate the percent of seniors who have ever used cocaine?

c. Based on your answers to parts (*a*) and (*b*), choose an appropriate window and graph the equation on a graphing utility.

d. Graph the function again with a new window that gives a graph nearer the center of the screen.

e. Use this function to estimate the percent in 1990.
(Source: 1998 *World Almanac*)

15. *Runway Near-Hits* The number of runway near-hits from 1990–2000 is shown in the table below.

a. Create a new table with x representing the number of years after 1990 and y representing the number of near-hits.

b. Use a graphing utility to graph the data from the new table as a scatter plot.

Years	Runway Near-Hits	Years	Runway Near-Hits
1990	281	1996	275
1991	242	1997	292
1992	219	1998	325
1993	186	1999	321
1994	200	2000	421
1995	240		

(Source: Federal Aviation Administration)

16. *U.S. Population* The total population of the U.S. for selected years from 1950–1995 is shown in the table below, with the population given in thousands.

a. According to this table, what was the U.S. population in 1970?

b. Create a new table with x representing the number of years after 1950 and y representing the number of millions, to three decimal places.

c. Use a graphing utility to graph the data from the new table as a scatter plot.

Year	Population (thousands)	Year	Population (thousands)
1950	152,271	1975	215,973
1955	165,931	1980	227,726
1960	180,671	1985	238,466
1965	194,303	1990	249,948
1970	205,052	1995	263,044

(Source: U.S. Census Bureau)

17. *Grades* The course grade for a student is found by adding the percent grade for each of three tests plus twice the percent grade on the final test, divided by 5.

  a. Create a model that gives the course grade y (as a percent) as a function of the final test grade x (as a percent) for a student whose three test grades are 78%, 83%, and 91%.

  b. Enter the function that models the course grade (percent) of this student into your graphing utility to find the course grade if the final test grade is 94%.

  c. If the highest possible grade on any test is 100%, determine if it is possible for this student to earn a course grade of 92%.

18. *Interest* The simple interest I on an investment is equal to the principal P times the annual interest rate r times the time t the money is invested, in years.

  a. Write the equation that models the interest as a function of the number of years invested if $2000 is invested at 10% per year.

  b. Enter the function that models the interest into your graphing utility to find the interest if $2000 is invested at 10% for 5 years, and for 9 months.

19. *Crickets* The number of times per minute n that a cricket chirps can be modeled as a function of the Fahrenheit temperature T. The data can be approximated by the function

$$n = \frac{12T}{7} - \frac{52}{7}$$

The number of chirps cannot be negative, so this function cannot model the chirps for all values of T. Test integer values of T to find the values of T for which this function fails to model the data because the model gives a negative number of chirps.

1.3 Linear Functions

There is concern among health care professionals that too many children are now taking Ritalin and that some young people are using Ritalin as a recreational drug. The increase in the use of Ritalin is evident by looking at the graph in Figure 1.23, which shows Ritalin consumption in grams per 100,000 persons as a function of the number of years since 1990. Because the graph of the function that models the consumption is a line, the function is called a **linear function**. The function that models the consumption of Ritalin over the years 1990 to 1996 is

$$y = 225.304x + 493.432$$

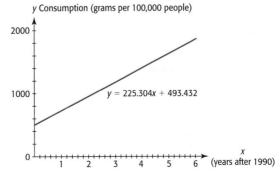

(Source: Index of Leading Cultural Indicators)

FIGURE 1.23

where x is the number of years after 1990 and y is measured in grams per 100,000 people. The average rate at which the use of Ritalin was increasing over the years 1990 to 1996 is constant and is the same as the slope of the line in Figure 1.23.

In this section, we investigate linear functions; discuss slope, intercepts, and average rate of change; and learn how to write equations of lines.

Linear Functions

A function whose graph is a line is a **linear function**. The line depicting Ritalin use is increasing; that is, the line rises between 1990 and 1996. In general, as the input of a linear function gets larger, the output can increase (the line rises), the output can decrease (the line falls), or the output can stay the same (the line is horizontal).

LINEAR FUNCTION

A linear function is a function that can be written in the form $f(x) = ax + b$ where a and b are constants.

If x and y are in separate terms in an equation and are each to the first power (and not in a denominator), we can rewrite the equation relating them in the form

$$y = ax + b$$

for some constants a and b, and the original equation is a linear equation that represents a linear function. For example,

$$2x + 5y = 6$$

is a linear equation in two variables and represents a linear function. However,

$$3x + 2xy = 2$$

is not a linear equation. If no restrictions are stated or implied by the nature of the problem situation, and if its graph is not a horizontal line, both the domain and range of a linear function consist of the set of all real numbers. Note that an equation of the form $x = d$ is not a function.

EXAMPLE 1   **LINEAR FUNCTIONS**

Determine whether each equation represents a linear function. If so, give the domain and range.

a. $0 = 2x - y + 1$ **b.** $y = 5$ **c.** $xy = 2$

Solution

a. The equation

$$0 = 2x - y + 1$$

does represent a linear function, because each of the variables x and y appear to the first power and each is in a separate term. We can solve this equation for y, getting

$$y = 2x + 1,$$

showing y as a function of x. Because any real number can be multiplied by 2 and increased by 1, and the result is a real number, both the domain and range consist of the set of all real numbers.

b. The equation $y = 5$ is in the form $y = ax + b$, where $a = 0$ and $b = 5$, so it represents a linear function. The domain is the set of all real numbers (because $y = 5$ regardless of what x we choose), and the range is the set containing 5.

c. The equation

$$xy = 2$$

does not represent a linear function because x and y are not in separate terms, so the equation cannot be written in the form $y = ax + b$. ∎

Intercepts

The points where a graph crosses or touches the x-axis and y-axis are called the **x-intercepts** and **y-intercepts**, respectively, of the graph. For example, Figure 1.24 shows that the graph of the linear function $2x - 3y = 12$ crosses the x-axis at $(6, 0)$, so the x-intercept is $(6, 0)$. The graph crosses the y-axis at $(0, -4)$, so the y-intercept is $(0, -4)$.

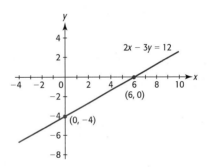

FIGURE 1.24

In this text, we will use the widely accepted convention that the x-coordinate of the x-intercept may also be called the x-intercept, and that the y-coordinate of the y-intercept may also be called the y-intercept.* The procedure for finding intercepts is a direct result of these definitions.

FINDING INTERCEPTS ALGEBRAICALLY

To find the y-intercept of a graph of $y = f(x)$, set $x = 0$ in the equation and solve for y.

To find the x-intercept(s) of the graph of $y = f(x)$, set $y = 0$ in the equation and solve for x.

---

*We usually call the horizontal axis intercept(s) the x-intercept(s) and the vertical axis intercept(s) the y-intercept(s), but realize that symbols other than x and y can be used to represent the input and output. For example, if $p = f(q)$, the vertical intercept is called the p-intercept and the horizontal intercept is called the q-intercept.

For example, the y-intercept of the graph of $2x - 3y = 12$ in Figure 1.24 can be found by substituting 0 for x in the equation and solving for y.

$$2(0) - 3y = 12$$
$$-3y = 12$$
$$y = -4$$

Similarly, the x-intercept can be found by substituting 0 for y in the equation and solving for x.

$$2x - 3(0) = 12$$
$$2x = 12$$
$$x = 6$$

The graph of a linear function has one y-intercept and has one x-intercept unless the graph is a horizontal line. The intercepts of the graph of the linear function are often easy to calculate, and plotting these two points and connecting them with a line gives the graph.

TECHNOLOGY NOTE

When graphing with a graphing utility, finding or estimating the intercepts of a line can help set the viewing window for the graph of a linear equation.

Evaluating Linear Models

We sometimes use linear functions to model applications even though some points on the line do not fit the application. In the following example, a linear function is used to model an application.

EXAMPLE 2   Loan Balance

A business property is purchased with a promise to pay off a $60,000 loan plus the $16,500 interest on this loan by making 60 monthly payments of $1275. The amount of money, y, remaining to be paid on $76,500 (the loan plus interest) is reduced by $1275 each month. Although the amount of money remaining to be paid changes every month, it can be modeled by the linear equation

$$y = 76,500 - 1275x$$

where x is the number of monthly payments made. This equation is a linear function that models the application. We recognize that only integer values of x from 0 to 60 apply to this application.

a. Find the x-intercept and the y-intercept of the graph of this linear equation.

b. Interpret the intercepts in the context of this problem situation.

c. How should x and y be limited in this model so that they make sense in the application?

d. Use the intercepts and the results of part (c) to sketch the graph of the given equation.

Solution

a. To find the *x*-intercept, set $y = 0$ and solve for *x*.

$0 = 76{,}500 - 1275x$

$1275x = 76{,}500$

$x = \dfrac{75{,}500}{1275} = 60$

Thus, 60 is the *x*-intercept.

To find the *y*-intercept, set $x = 0$ and solve for *y*.

$y = 76{,}500 - 1275(0)$

$y = 76{,}500$

Thus, 76,500 is the *y*-intercept.

b. The *x*-intercept corresponds to the number of months that must pass before the amount owed is $0. Therefore, a possible interpretation of the *x*-intercept is "The loan is paid off in 60 months." The *y*-intercept corresponds to the total (loan plus interest) that must be repaid 0 months after purchase; that is, when the purchase is made. Thus, the *y*-intercept tells us "A total of $76,500 must be repaid."

c. We know that the total time to pay the mortgage is 60 months. A value of *x* larger than 60 will result in a negative value of *y*, which makes no sense in the application, so *x* varies from 0 to 60. The output, *y*, is the total amount owed at any time during the loan. The amount owed cannot be less than 0, and the value of the loan plus interest will be at its maximum, 76,500, when time is 0. Thus, the values of *y* vary from 0 to 76,500.

d. The graph intersects the horizontal axis at (60, 0) and intersects the *y*-axis at (0, 76,500), as indicated in Figure 1.25. Because we know that the graph of this equation is a line, we can simply connect the two points to obtain this first-quadrant graph.

Figure 1.25 shows the graph of the function on a viewing window determined by the context of the application.

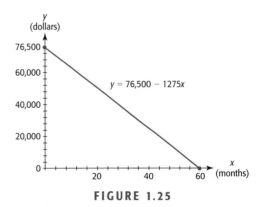

FIGURE 1.25

Zero of a Function

Because the horizontal-axis intercepts of a function are found by setting the output variable equal to zero, these intercept(s) are often called **zeros** of the function. Some functions can have more than one *x*-intercept (zero), but a nonconstant linear function has one *x*-intercept and thus one zero.

ZERO OF A FUNCTION

Any number *a* for which $f(a) = 0$ is called a **zero** of the function *f(x)*. In this case, *a* is an *x*-intercept of the graph of the function.

Some graphs do not have two intercepts. For example, vertical lines (except for the line lying on the *y*-axis) have only one *x*-intercept and no *y*-intercept. (See Figure 1.26(a).) Horizontal lines (except for the line lying on the *x*-axis) have only one *y*-intercept and no *x*-intercept. (See Figure 1.26(b).) Lines that pass through the origin have the origin as both the *x*-intercept and the *y*-intercept. (See Figure 1.26(c).) In such cases, additional points are useful in graphing a linear function.

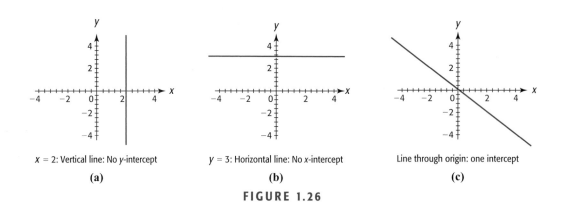

x = 2: Vertical line: No *y*-intercept

(a)

y = 3: Horizontal line: No *x*-intercept

(b)

Line through origin: one intercept

(c)

FIGURE 1.26

Slope of a Line

Consider two stairways: one goes from the park entrance to the shogun shrine at Nikko-Japan and rises vertically 1 foot for every 1 foot of horizontal increase, while the second goes from the Tokyo subway to the street and rises vertically 20 cm for every 25 cm of horizontal increase. To find which set of stairs would be easier to climb, we can find the steepness, or **slope** of each stairway. If a board is placed along the steps of each stairway, its slope is a measure of the incline of the stairway.

$$\text{Shrine stairway steepness} = \frac{\text{vertical increase}}{\text{horizontal increase}} = \frac{1 \text{ foot}}{1 \text{ foot}} = 1$$

$$\text{Subway stairway steepness} = \frac{\text{vertical increase}}{\text{horizontal increase}} = \frac{20 \text{ cm}}{25 \text{ cm}} = 0.8$$

The shrine stairway has a slope that is larger than that of the subway stairway, so it is steeper than the subway stairway. Thus, the subway steps would be easier to climb. In general, we define the slope of a line as follows.

<hr />

SLOPE OF A LINE

The slope of a line is defined as

$$\text{slope} = \frac{\text{vertical change}}{\text{horizontal change}} = \frac{\text{rise}}{\text{run}}$$

The slope can be found by using any two points on the line (see Figure 1.27). If a nonvertical line passes through the two points, P_1 with coordinates (x_1, y_1) and P_2 with coordinates (x_2, y_2), its slope, denoted by m, is found by using

$$m = \frac{y_2 - y_1}{x_2 - x_1}$$

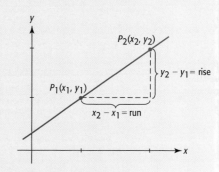

FIGURE 1.27

The slope of a vertical line is undefined, because $x_2 - x_1 = 0$ and division by 0 is undefined.

<hr />

The slope of any given nonvertical line is a constant. Thus, the same slope will result regardless of which two points on the line are used in its calculation.

EXAMPLE 3   Calculating the Slope of a Line

- -

a. Find the slope of the line passing through the points $(-3, 2)$ and $(5, -4)$. What does the slope mean?

b. Find the slope of the line joining the x-intercept point and y-intercept point in the mortgage situation of Example 2.

Solution

a. We choose one point as P_1 and the other point as P_2. While it does not matter which point is chosen to be P_1, it is important to keep the correct order of the terms in the numerator and denominator of the slope formula. Letting $P_1 = (-3, 2)$ and $P_2 = (5, -4)$, and substituting in the slope formula gives

$$m = \frac{-4 - 2}{5 - (-3)} = \frac{-6}{8} = -\frac{3}{4}.$$

Note that letting $P_1 = (5, -4)$ and $P_2 = (-3, 2)$ gives the same slope:

$$m = \frac{2 - (-4)}{-3 - 5} = \frac{6}{-8} = -\frac{3}{4}.$$

A slope of $-\dfrac{3}{4}$ means that, from a given point on the line, by moving 3 units down and 4 units to the right or by moving 3 units up and 4 units to the left, we arrive at another point on the line.

b. Because $x = 60$ is the x-intercept in part (b) of Example 2, $(60, 0)$ is a point on the graph. The y-intercept is $y = 76{,}500$, so $(0, 76{,}500)$ is a point on the graph. Recall that x is measured in months and that y has units of dollars in the real-world setting (that is, the *context*) of Example 2. Substituting in the slope formula, we obtain

$$m = \frac{76{,}500 - 0}{0 - 60} = \frac{76{,}500}{-60} = -1275$$

This slope means that the amount owed decreases by \$1275 each month.   ■

     As Figure 1.28(a) to (d) indicates, the slope describes the direction of a line as well as the steepness.

THE RELATION BETWEEN ORIENTATION OF A LINE AND ITS SLOPE

1. The slope is *positive* if the line *rises* upward toward the right.

2. The slope is *negative* if the line *falls* downward toward the right.

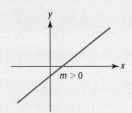

FIGURE 1.28(a)

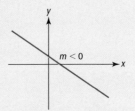

FIGURE 1.28(b)

3. The slope of a *horizontal line* is 0 because a horizontal line has a vertical change (rise) of 0 between any two points on the line.

4. The slope of a *vertical line* does not exist because a vertical line has a horizontal change (run) of 0 between any two points on the line.

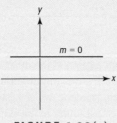

FIGURE 1.28(c)

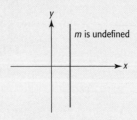

FIGURE 1.28(d)

     Remembering how the orientation of a line is related to its slope can help you check that you have the correct order of the points in the slope formula. For instance, if you find that the slope of a line is positive and you see that the line falls downward from left to right when you observe its graph, you know that there is a mistake either in the slope calculation or in the graph that you are viewing.

Slope and *y*-Intercept of a Line

There is an important connection between the slope of the graph of a linear equation and its equation when it is written in the form $y = f(x)$. To investigate this connection, we can graph the equation

$$y = 3x + 2$$

with a graphing utility (see Figure 1.29(a)). By using the TABLE feature of the graphing utility for values of x equal to 0, 1, 2, 3, and 4, we can see that each time that x increases by 1, y increases by 3 (see Figure 1.29(b)). Thus the slope is

$$\frac{\text{change in } y}{\text{change in } x} = \frac{3}{1} = 3.$$

From this table, we see that the y-intercept is 2, the same value as the constant term of the equation. Observe also that the slope of the graph of $y = 3x + 2$ is the same as the coefficient of x.

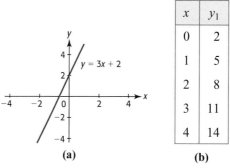

x	y_1
0	2
1	5
2	8
3	11
4	14

(a)   **(b)**

FIGURE 1.29

In general, the graph of the linear function

$$y = ax + b$$

has y-intercept b because $y = b$ when $x = 0$. In addition, each increase of x by 1 (from x to $x + 1$) gives an increase in y from $ax + b$ to $y = a(x + 1) + b$. Thus the slope of this line is equal to

$$\frac{\text{Change in } y}{\text{Change in } x} = \frac{a(x + 1) + b - (ax + b)}{1} = ax + a + b - ax - b = a$$

so a is the slope of the graph of $y = ax + b$. Because we denote the slope of a line by m, we formalize our result as follows.

SLOPE AND *Y*-INTERCEPT OF A LINE

The slope of the graph of the equation $y = mx + b$ is m, and the y-intercept of the graph is b.

Thus when the equation of a linear function is written in the form $y = mx + b$ or $f(x) = mx + b$, we can "read" the values of the slope and y-intercept of its graph.

EXAMPLE 4 Loan Balance

As we saw in Example 2, the amount of money y remaining to be paid on the loan of $60,000 with $16,500 interest is

$$y = 76,500 - 1275x$$

where x is the number of months the mortgage has been paid.

a. What is the slope and y-intercept of the graph of this function?

b. How does the amount owed on the loan change as the number of months increases?

Solution

a. Writing this equation in the form $y = mx + b$ gives $y = -1275x + 76,5000$. The coefficient of x is -1275, so the slope is $m = -1275$; the constant term is 76,500, so the y-intercept is $b = 76,500$.

b. The slope of the line indicates that the amount owed decreases by 1275 dollars each month. This can be verified from the graph of the function and the table of sample inputs and outputs in Figure 1.30.

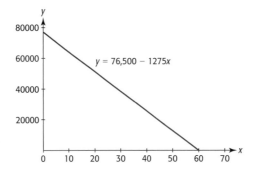

Months (x)	Amount Owed (y)
1	75,225
2	73,950
3	72,675
4	71,400
5	70,125
6	68,850
7	67,575
8	66,300
9	65,025
10	63,750

FIGURE 1.30 ■

Constant Rates of Change

The function $y = 76,500 - 1275x$ whose graph is shown in Figure 1.30, gives the amount owed as a function of the months remaining on the mortgage. The coefficient of x, -1275, indicates that for each additional month, the value of y changes by -1275 (see the table in Figure 1.30). That is, the amount owed decreases at the **constant rate** -1275 each month. Note that the **constant rate of change** of this linear function is equal to the **slope** of its graph. This is true for all linear functions.

CONSTANT RATE OF CHANGE

The rate of change of the linear function $y = mx + b$ is the constant m, the slope of the graph of the function.

This means that if a function is linear, the rate of change of the outputs with respect to the inputs is equal to the slope of the line that is the graph of the function. Note: The rate of change in an applied context should include appropriate units of measure.

EXAMPLE 5 Ritalin

The graph in Figure 1.31 shows Ritalin consumption in grams per 100,000 persons as a function of the number of years after 1990. The function that models the consumption of Ritalin over the years 1990 to 1996 is

$$y = 225.304x + 493.432$$

where x is the number of years after 1990 and y is measured in grams per 100,000 people.

a. What is the slope of the graph of this model?

b. What is the rate at which Ritalin consumption grew during this period? (Source: Index of Leading Cultural Indicators.)

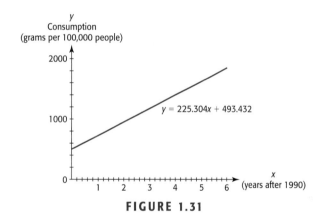

FIGURE 1.31

Solution

a. The coefficient of x in the linear function is 225.304, so the slope of the line is 225.304.

b. The slope of the line in Figure 1.31 equals the rate of change of the function, so the use of Ritalin increased at a rate of 225.304 grams per 100,000 people each year. ∎

Special Linear Functions

A special linear function that has the form $y = c$, where c is a real number, is called a **constant function**. The graph of the constant function $y = 3$ is shown in Figure 1.32(a). The temperature inside a sealed case containing an Egyptian mummy in a museum is a constant function of time because the temperature inside the case never changes. Notice that even though the range of a constant function consists of a single value, the input of a constant function is any real number or any real number that makes sense in the context of an applied problem.

Another special linear function is the **identity function**

$$y = x,$$

which is a linear function of the form $y = mx + b$ with slope $m = 1$ and y-intercept $b = 0$. For the general identity function $f(x) = x$, the domain and range are each the set of all real numbers. A graph of the identity function f is shown in Figure 1.32(b).

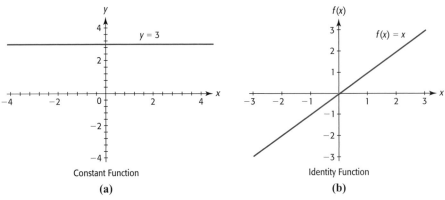

Constant Function
(a)

Identity Function
(b)

FIGURE 1.32

EXAMPLE 6 Patrol Cars

The Beaufort County Sheriff's office assigns a patrol car to each of its deputy sheriffs, who keeps the car 24 hours a day. As the population of the county grew, the number of deputies and the number of patrol cars also grew. The data in Table 1.13 could describe how many patrol cars were necessary as the number of deputies increased. Write the function that models the relationship between the number of deputies and patrol cars.

TABLE 1.13

Number of deputies	10	50	75	125	180	200	250
Number of patrol cars in use	10	50	75	125	180	200	250

Solution

For each input in Table 1.13, the output equals the input, so the function that models the number of patrol cars in use is

$$p(x) = x$$

where x is the number of deputies. This function is called the identity function because the output is the same as the input. ∎

1.3 SKILLS CHECK

1. Which of the following functions are linear?

   a. $y = 3x^2 + 2$ **b.** $3x + 2y = 12$

   c. $y = \dfrac{1}{x} + 2$

2. Find the slope of the line through $(4, 6)$ and $(28, -6)$.

3. Find the slope of the line through $(8, -10)$ and $(8, 4)$.

4. Find the slope of the line through $(-6, 5)$ and $(-2, 5)$.

5. If a line is horizontal, then its slope is _____ . If a line is vertical, then its slope is _____ .

6. Based on the graph of the line, determine whether the slope of the graph of the line is positive, negative, 0, or undefined.

a.

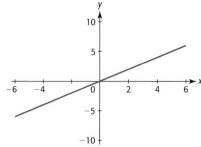

b.

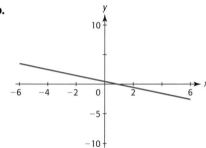

c.

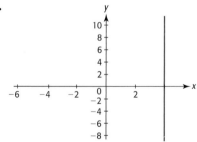

d.

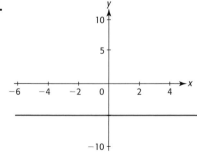

In Exercises 7–10, (a) find the x- and y-intercepts of the graph of the given equation, if they exist, and (b) graph the equation.

7. $5x - 3y = 15$ **8.** $x + 5y = 17$

9. $3y = 9 - 6x$ **10.** $y = 9x$

For Exercises 11–14, give the slope of the line (if it exists) and the y-intercept (if it exists).

11. $y = 4x + 8$ **12.** $3x + 2y = 7$

13. $5y = 2$ **14.** $x = 6$

For each of the functions in Exercises 15–17, do the following:

a. Find the slope and y-intercept (if possible) of the graph of the function.

b. Determine if the graph is rising or falling.

c. Graph each function on a window with the given x-range and a y-range that shows the graph.

15. $y = 4x + 5; [-5, 5]$

16. $y = 0.001x - 0.03; [-100, 100]$

17. $y = 50,000 - 100x; [0, 500]$

18. Rank the functions in Exercises 15–17 in order of increasing steepness.

For each of the functions in Exercises 19 and 20, find the rate of change.

19. $y = 4x - 3$ **20.** $y = \dfrac{1}{3}x + 2$

21. a. What is the slope of the identity function?
    b. What is the rate of change of the identity function?

22. a. What is the slope of the constant function $y = k$?
    b. What is the rate of change of a constant function?

In Exercises 23 and 24, find the zero of the function, if possible.

23. $f(x) = 3x - 12$ **24.** $f(x) = 7 - 2x$

25. Which intercept of the graph of a function gives an input that makes the function 0?

1.3 EXERCISES

In Exercises 1–4, determine whether or not the function is linear. If it is, tell whether the line that is the graph of the function rises or falls as the input increases.

1. *Reading tests* The average reading score of 17-year-olds on the National Assessment of Progress tests is given by $y = 0.155x + 244.37$ points, where x is the number of years past 1970.
   (Source: U.S. Dept. of Education)

2. *Women in the workforce* The number y (in thousands) of women in the workforce is given by the function $y = 7x^2 - 19.88x + 409.29$, where x is the number of years from 1900.
   (Source: Bureau of the Census, U.S. Dept of Commerce)

3. *Marriage Rate* When x is the number of years past 1950, the percent of unmarried women who get married each year is given by $M(x) = -0.762x + 85.284$.
   (Source: Index of Leading Cultural Indicators)

4. *Voters* The percent of the population voting in presidential elections is given by $V(t) = -0.356t + 65.404$, where t is the number of years after 1950.
   (Source: Federal Election Commission)

5. *Marijuana Use* The percent p of high school seniors using marijuana daily can be related to x, the number of years after 1990, by the equation $30p - 19x = 30$.
   a. Find the x-intercept of the graph of this function.
   b. Find and interpret the p-intercept of the graph of this function.
   c. Graph the function, using the intercepts. What values of x on the graph represent years 1990 and after?

6. *Depreciation* An $828,000 building is depreciated for tax purposes by its owner, using the straight-line depreciation method. The value of the building after x months of use is given by $y = 828,000 - 2300x$ dollars.
   a. Find and interpret the y-intercept of the graph of this function.
   b. Find and interpret the x-intercept of the graph of this function.
   c. Use the intercepts to graph the function for non-negative x- and y-values.

7. *Life Insurance* The monthly rates for a $100,000 life insurance policy for males aged 27–32 are shown in the table.

Age (years), x	27	28	29
Premium (dollars per month), y	11.81	11.81	11.81

Age (years), x	30	31	32
Premium (dollars per month), y	11.81	11.81	11.81

   a. Could the data in this table be modeled by a constant function or an identity function?
   b. Write an equation whose graph contains the data points in the table.
   c. What is the slope of the graph of the function found in part (b)?
   d. What is the rate of change of the data in the table?

8. *Eating Asparagus* The per capita consumption of asparagus between 1990 and 1996 is shown in the table.

Year	1990	1991	1992	1993
Asparagus consumption (pounds per person)	0.6	0.6	0.6	0.6

Year	1994	1995	1996
Asparagus consumption (pounds per person)	0.6	0.6	0.6

(Source: *Statistical Abstract of the U.S.*, 1998)

   a. Sketch the data as a scatter plot, with y equal to the consumption and x equal to the number of years.
   b. Could the data in the table be modeled by a constant function or an identity function?
   c. Write the equation of a function that fits the data points.
   d. Sketch a graph of the function you found in part (c) on the same axes as the scatter plot.

9. *Internet Recruiting* The percentage of Fortune Global 500 firms that actively recruited workers on the Internet from 1998 through 2000 can be modeled by $P(x) = 26.5x - 194.5$ percent, where x is the number of years after 1990.
   a. What is the slope of the graph of this function?
   b. Interpret the slope as a rate of change.
   (Source: *Time*, August 16, 1999)

10. *Cigarette Use* For the years 1975–1991, the percent p of high school seniors who have tried cigarettes can be modeled by $p = 75.4509 - 0.7069t$, where t is the number of years after 1975.

   a. Is the rate of change of the percent positive or negative?

   b. What does the answer to part (a) tell us about the percent of seniors who tried cigarettes during this period?

11. *Crickets* The number of times per minute n that a cricket chirps can be modeled as a function of the Fahrenheit temperature T. The data can be approximated by the function

$$n = \frac{12T}{7} - \frac{52}{7}$$

   a. Is the rate of change of the number of chirps positive or negative?

   b. What does this tell us about the relationship between temperature and the number of chirps?

12. *Tax Burden* The per capita tax burden T (in hundreds of dollars) can be described by $T(t) = 20.37 + 1.834t$, where t is the number of years after 1980:

   a. What is the slope of the graph of this function?

   b. What is the rate of growth of the per capita tax burden per year?

   c. How does the answer to part (b) compare to the slope of the graph of the function?

   (Source: Internal Revenue Service)

13. *Earnings and Minorities* According to the U.S. Equal Employment Opportunity Commission, the relation between the median annual salaries of minorities and whites can be modeled by the function $M = 0.959W - 1.226$, where M and W represent the median annual salary (in thousands of dollars) for minorities and whites, respectively.

   a. Is this function a linear function?

   b. What is the slope of the graph of this function?

   c. Interpret the slope as a rate of change.

   (Source: *Statistical Abstract of the U.S.,* 1993)

14. *Voting* Between 1950 and 1996, the percent of the voting population who voted in presidential elections is given by $p = 65.4042 - 0.3552x$, where x is the number of years after 1950.

   a. What is the slope of the graph of this function? Write a sentence interpreting this value.

   b. What is the annual rate of change in the percent of the eligible population who voted in presidential elections between 1950 and 1996?

   c. How does the rate of change compare to the slope of the graph of this function?

   (Source: Federal Election Commission)

15. *Marijuana Use* The percent p of high school seniors using marijuana daily can be modeled by $30p - 19x = 30$, where x is the number of years after 1990.

   a. Use this model to determine the slope of the graph of this function if x is the independent variable.

   b. What is the rate of change of the percent of high school seniors using marijuana per year?

   (Source: Index of Leading Cultural Indicators)

16. *Seawater Pressure* In seawater, the pressure p is related to the depth d according to the model $33p - 18d = 496$, where d is the depth in feet and p is in pounds per square inch.

   a. What is the slope of the graph of this function?

   b. Interpret the slope as a rate of change.

17. *Advertising Impact* An advertising agency has found that when it promotes a new product in a city of 575,000 people, the weekly rate of change R of the number of people who are aware of it x weeks after it is introduced is given by $R = 3500 - 70x$. Find the x- and R-intercepts, then graph the function on a viewing window that is meaningful in the application.

18. *ATM Transactions* The dollar volume of transactions at automatic teller machines (ATMs) is modeled by $D(x) = 0.137x - 5.09$ billion dollars, where x is the number of terminals (in thousands).

   a. What was the value of the transactions when 50,000 ATMs were available, according to this model?

   b. This model approximates the dollar value of transactions and does not fit the data exactly. The equation cannot model the data for values of x that give negative dollar values. Test integer values of x to find the values for which this function fails to model the data because it gives negative dollar values.

   (Source: Electronic Funds Transfer Association, 1993).

1.4 Equations of Lines; Rates of Change

If a \$750,000 building were depreciated by the straight-line method over a period of 25 years, then the value of the property would be reduced by 1/25 of the original value, or by the **constant rate** of 30,000, each year. We can use this information to write the linear equation that models the value y of the property as a function of the number of years x that it is depreciated, because the rate of change is constant. In this section, we learn how to write the equation of a linear function from information about the line, such as the slope and a point on the line or two points on the line. We also discuss **average rates of change** for data that can be modeled by nonlinear functions, slopes of secant lines, and how to create linear models that approximate data that is "nearly" linear.

Writing Equations of Lines

Creating a linear model from data involves writing a linear equation that describes the mathematical situation. If we know two points on a line or the slope of the line and one point, we can write the equation of the line.

Recall that if a linear equation has the form $y = mx + b$, then the coefficient of x is the slope of the line that is the graph of the equation, and the constant b is the y-intercept of the line. Thus if we know the slope and the y-intercept of a line, we can write the equation of this line.

SLOPE-INTERCEPT FORM OF THE EQUATION OF A LINE

The slope-intercept form of the equation of a line with slope m and y-intercept b is

$$y = mx + b$$

EXAMPLE 1   Appliance Repair

An appliance repairman charges \$60 for a service call plus \$25 per hour for each hour spent on the repair. If a linear function models his service call charges, write the equation of the function.

Solution

Let x represent the number of hours spent on the appliance repair, and let y dollars be the service call charge. The slope of the line is the amount that the charge increases for every hour of work done, so the slope is 25. Because \$60 is the basic charge before any time is spent on the repair, 60 is the y-intercept of the line. Substituting these values in the slope-intercept form of the equation gives the equation of the linear function modeling this situation. When the repairman works x hours on the service call, the charge is

$$y = 25x + 60 \text{ dollars.} \qquad \blacksquare$$

We next consider a form of an equation of a line that can be written if we know the slope and a point. If the slope of a line is m, then the slope between a fixed point (x_1, y_1) and *any other point* (x, y) on the line is also m. That is,

$$m = \frac{y - y_1}{x - x_1}$$

Solving for $y - y_1$ (that is, multiplying both sides of this equation by $x - x_1$) gives the **point-slope form** of the equation of a line.

POINT-SLOPE FORM OF THE EQUATION OF A LINE

The equation of the line with slope m that passes through a known point (x_1, y_1) is

$$y - y_1 = m(x - x_1)$$

EXAMPLE 2 Using a Point and Slope to Write an Equation of a Line

Write equations for the lines that pass through the point $(-1, 5)$ and have

a. slope $\dfrac{3}{4}$ **b.** slope 0 **c.** undefined slope

Solution

a. We are given $m = \dfrac{3}{4}$, $x_1 = -1$, and $y_1 = 5$. Substituting in the point-slope form, we obtain

$$y - 5 = \frac{3}{4}(x - (-1))$$

$$y - 5 = \frac{3}{4}(x + 1)$$

$$y - 5 = \frac{3}{4}x + \frac{3}{4}$$

$$y = \frac{3}{4}x + \frac{23}{4}$$

Figure 1.33 shows the graph of this line.

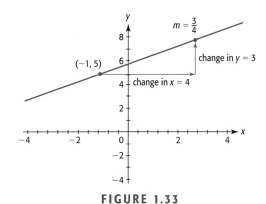

FIGURE 1.33

b. If $m = 0$, the point-slope form gives us the equation

$$y - 5 = 0(x - (-1))$$

$$y = 5$$

Because the output is always the same value, the graph of this linear function is a horizontal line. See Figure 1.34(a).

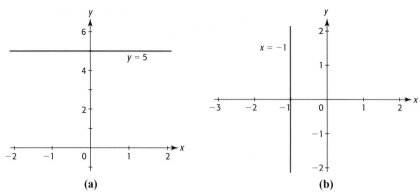

FIGURE 1.34

c. Because m is undefined, we cannot use the point-slope form to write the equation of this line. Lines with undefined slope are vertical lines. Every point on the vertical line through $(-1, 5)$ has an x-coordinate of -1. Thus the equation of the line is $x = -1$. Note that this equation does not represent a function. See Figure 1.34(b). ∎

In part (a) of Example 2, we found that the equation of the line was

$$y = \frac{3}{4}x + \frac{23}{4}$$

By multiplying both sides of this equation by 4 and writing the new equation with x and y on the same side of the equation, we have a new form of the equation:

$$4y = 3x + 23, \text{ or } 3x - 4y = -23.$$

This equation is called the **general form** of the equation of the line in Figure 1.33.

GENERAL FORM OF THE EQUATION OF A LINE

$ax + by = c$ where a, b, and c are real numbers, with a and b not both equal to 0.

If we know two points on a line, we can find the slope of the line and use either of the points and this slope to write the equation of the line.

EXAMPLE 3   Using Two Points to Write an Equation of a Line

Write the equation of the line that passes through the points $(-1, 5)$ and $(2, 4)$.

Solution

Because we know two points on the line, we can find the slope of the line.

$$m = \frac{4 - 5}{2 - (-1)} = \frac{-1}{3} = -\frac{1}{3}$$

We can now substitute one of the points and the slope in the point-slope form to write the equation. Using the point $(-1, 5)$ gives

$$y - 5 = -\frac{1}{3}(x - (-1)), \text{ or } y - 5 = -\frac{1}{3}x - \frac{1}{3}, \text{ so } y = -\frac{1}{3}x + \frac{14}{3}$$

Using the point $(2, 4)$ gives

$$y - 4 = -\frac{1}{3}(x - 2)$$

$$y - 4 = -\frac{1}{3}x + \frac{2}{3}$$

$$y = -\frac{1}{3}x + \frac{14}{3}$$

Notice that both equations are the same regardless of which of the two given points is used in the point-slope form. ∎

EXAMPLE 4   Inmate Population

The number of people (in millions) in prisons or jails in the United States grew at a constant rate from 1990 to 2000, with 1.15 million people incarcerated in 1990 and 1.91 million incarcerated in 2000.

a. Write the equation that models the number N of prisoners as a function of the year x.

b. The Bureau of Justice Statistics projects that 2.29 million people will be incarcerated in 2005. Does your model agree with this projection? (Source: Bureau of Justice Statistics)

Solution

a. The rate of growth is constant, so the points fit on a line, and a linear function can be used to find the model for the number of incarcerated people as a function of the years. The rate of change (and slope of the line) is given by

$$m = \frac{1.91 - 1.15}{2000 - 1990} = \frac{.76}{10} = 0.076$$

Substituting in the point-slope form of the equation of a line (with either point) gives the equation of the line that contains the two points, and thus models the application.

$$N - 1.91 = 0.076(x - 2000), \text{ or}$$

$$N = 0.076x - 150.09$$

b. Substituting 2005 for x in the function gives

$$N = f(2005) = 0.076(2005) - 150.09 = 2.29.$$

Thus the projection that 2.29 million people will be incarcerated in 2005 fits this model. ∎

In summary, these are the forms we have discussed for the equation of a line.

FORMS OF LINEAR EQUATIONS

General form:	$ax + by = c$	where a, b, and c are real numbers, with a and b not both equal to 0.
Point-slope form:	$y - y_1 = m(x - x_1)$	where m is the slope of the line and (x_1, y_1) is a point on the line.
Slope-intercept form:	$y = mx + b$	where m is the slope of the line and b is the y-intercept.
Vertical line:	$x = a$	where a is a constant, and a is the x-coordinate of any point on the line. The slope is undefined.
Horizontal line:	$y = b$	where b is a constant, and b is the y-coordinate of any point on the line. The slope is 0.

Constant Rates of Change

As we saw in Section 1.3, if a function is linear, the rate of change of the outputs with respect to the inputs is a constant equal to the slope of the graph of the function. Thus, we can write the equation of the linear function that fits a set of real data if the rate of change remains constant for the data.

EXAMPLE 5 Depreciation

If a \$750,000 building were depreciated by the straight-line method over a period of 25 years, then the value of the property would be reduced by $\dfrac{1}{25}$ of the original value each year. Thus the value would be reduced by

$$\frac{1}{25}(750{,}000) = 30{,}000$$

each year. Its value after the first year is $750{,}000 - 30{,}000 = 720{,}000$, and its value at the end of each 5-year period is given in Table 1.14. Write the equation of the function that gives the value of the building after x years, for x from 0 to 25.

TABLE 1.14

Years after depreciation began	5	10	15	20	25
Value of building (dollars)	600,000	450,000	300,000	150,000	0

Solution

Because the building depreciates by the same amount each year, the rate of change of the value of the property is constant, so the value of the property can be modeled as a linear function of the number of years after the depreciation started. Because the value of the building is $750,000 at the beginning of the depreciation period (when time equals 0), the *y*-intercept of the graph is 750,000. The value of the building decreases by $30,000 each year, so the rate of change (and slope of the line) is

$$m = -30,000.$$

Thus the equation of the value of the building can be modeled by the function

$$y = -30,000x + 750,000,$$

where *y* is in dollars and *x* is in years. ∎

First Differences

The following example shows that if the changes in outputs (called *first differences*) in a set of data are constant when the changes in inputs are constant, then we can find a linear equation that models the data.

EXAMPLE 6   Future Value of an Investment

If $1000 is invested at 6% simple interest, the future value *S* in *t* years is given in Table 1.15.

a. Is the rate of change of the future value constant?

b. Can the future value be modeled by a linear function?

c. Use the rate of change and the value of *S* when *t* = 0 to write the equation that gives the future value as a function of the time in years.

TABLE 1.15

Year (*t*)	0	1	2	3	4	5
Future value (*S*)	$1000	$1060	$1120	$1180	$1240	$1300

Solution

a. Each input value is one more (year) than the previous value, and the future value increases by the same amount for each year, so the rate of change of the future value is constant. The future value increases by $60 per year. Consider the following table showing the results we get when we subtract the outputs from one year to the next (called the first differences of outputs).

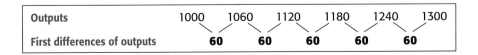

b. Because the rate of change is constant (equal to 60), the data can be modeled by a linear function.

c. The constant rate of change, 60, is equal to the slope of the line that is the graph of the function. Because the value of the investment is $1000 when *t* = 0, the *y*-intercept of the graph of the function is 1000. By using the slope-intercept form of the equation of a line, the equation of this function is

$$S = 60t + 1000,$$

where S is the future value in t years.   ∎

Average Rate of Change

For a function whose graph is not a line, the slope may vary from point to point on the curve. The calculation and interpretation of the slope of a curve are topics you will study if you take a calculus course. However, there is a quantity called the **average rate of change** that can be applied to any function relating two variables.

In general, we can find the average rate of change of a function between two input values if we know how much the function outputs change between the two input values.

AVERAGE RATE OF CHANGE

The average rate of change of $f(x)$ with respect to x over the interval from $x = a$ to $x = b$ (where $a < b$) is calculated as

$$\text{average rate of change} = \frac{\text{change in } f(x) \text{ values}}{\text{corresponding change in } x \text{ values}} = \frac{f(b) - f(a)}{b - a}$$

How is average rate of change over some interval of points related to the slope of a line connecting the points? For any function, the average rate of change between two points is the slope of the line joining the two points. Such a line is called a **secant line**.

EXAMPLE 7   Toyota Sales
- -

The total Toyota hybrid vehicle sales for the years between 1997 and 2001 can be approximated by the model

$$S(x) = 2821x^3 - 75{,}653x^2 + 674{,}025x - 1{,}978{,}335,$$

where x is the number of years after 1990. A graph of $S(x)$ is shown in Figure 1.35(a). Because $S(7) = 446$, there were 446 units sold in 1997; $S(9) = 16{,}506$, so the number sold in 1999 was 16,506. (Source: www.toyota.com)

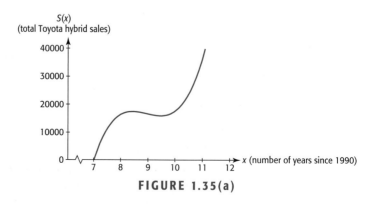

FIGURE 1.35(a)

a. Find the average rate of change of Toyota hybrid sales between 1997 and 1999.

b. Interpret your answer to part (a).

c. What is the relationship between the slope of the secant line joining the points (7, 446) and (9, 16,506) and the answer to part (a)?

Solution

a. We find the average rate of change between these two points to be

$$\frac{16{,}506 - 446}{9 - 7} = \frac{16{,}060 \text{ units}}{2 \text{ years}} = 8030 \text{ units per year}$$

b. A possible interpretation is this: On average between 1997 and 1999, 8030 Toyota hybrid units were sold per year.

c. The slope of this line, shown in Figure 1.35(b), is the same as the average rate of change of the sales between 1997 and 1999 found in part (a).

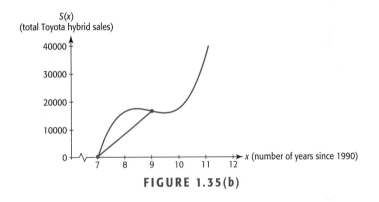

FIGURE 1.35(b)   ∎

Approximately Linear Data

Real-life data are rarely perfectly linear, but some sets of real data points lie sufficiently close to a line that two points can be used to create a linear function that models the data. Consider the following example.

EXAMPLE 8   Population Growth

Many southern states have had large increases of retirement-aged people, but the population of people under age 65 has also increased significantly. Table 1.16 and Figure 1.36(a) show the number of residents under age 65 in South Carolina. The growth of this population is approximately linear, growing from 76,353 in 1990 to 94,721 in 1999.

a. Create the scatter plot for this data.

b. Find the average rate of change of the under-age-65 population between 1990 and 1999.

c. Write the equation of the line determined by this rate of change and one of the two points.

TABLE 1.16

Year (N)	Population (p)
1990	76,353
1991	77,900
1992	77,800
1993	82,100
1994	83,900
1995	86,000
1996	89,700
1997	90,500
1998	92,800
1999	94,721

(Source: *Island Packet*, April 22, 2000)

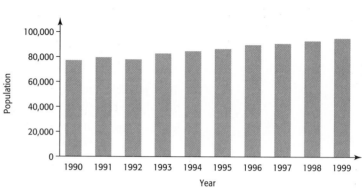

FIGURE 1.36(a)

Solution

a. The scatter plot is shown in Figure 1.36(b). The graph shows that the data do not fit a linear function, but that a linear function could be used as an approximate model for the data.

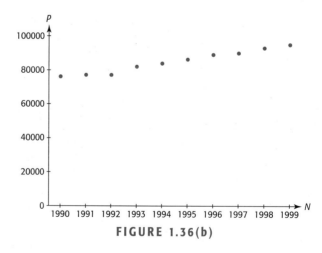

FIGURE 1.36(b)

b. Using N to represent the years and p to represent the population, the average rate of change of the population between 1990 and 1999 is

$$\frac{\text{change in } p}{\text{change in } N} = \frac{94{,}721 - 76{,}353}{1999 - 1990} = \frac{18{,}368}{9} = 2040.9 \text{ people per year}$$

This value means that the population has grown an average of approximately 2041 people per year between 1990 and 1999.

c. Because the population growth is *approximately* linear, the average rate of change of the under-age-65 population between 1990 and 1999 can be used as the slope of a linear function describing this growth. A point on the line is (1990, 76,353), so we use this point, the slope from part (b), and the point-slope form to write the equation of the line.

$$p - 76{,}353 = 2041(N - 1990)$$
$$p - 76{,}353 = 2041N - 4{,}061{,}590$$
$$p = 2041N - 3{,}985{,}237$$

Thus the equation $p = 2041N - 3{,}985{,}237$ models the population of residents under 65 as a function of the year.   ∎

The equation developed in Example 8 approximates a model for the data, but it is not the best possible linear model for the data. In Section 1.6 we will see how technology can be used to find the linear function that is the best fit for a set of data of this type.

1.4 SKILLS CHECK

1. Write the equation of the line that has slope 4 and y-intercept $\frac{1}{2}$.

2. Write the equation of the line that has slope $\frac{1}{3}$ and y-intercept 3.

3. Write the equation of the line through the point $(-1, 4)$ having slope $m = 5$.

4. Write the equation of the line through the point $(-1, 4)$ having slope $m = 0$.

5. Write the equation of a line that has slope $\frac{-3}{4}$ and passes through $(4, -6)$.

6. Write the equation of the line having x-intercept -5 and y-intercept 4.

7. Write the equation of the line that passes through $(-1, 3)$ and $(2, 6)$.

8. Write the equation of the line through the point $(-1, 4)$ having slope m undefined.

9. Write the equation of the vertical line through the point $(9, -10)$.

10. Write the equation of the horizontal line through the point $(9, -10)$.

For Exercises 11 and 12, write the equation of the line whose graph is shown.

11.

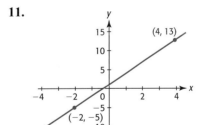

12.

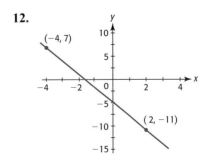

13. Find the rate of change of the function whose graph is shown in Exercise 11.

14. Find the rate of change of the function whose graph is shown in Exercise 12.

15. What is the rate of change of the function $y = 300 - 15x$?

16. What is the rate of change of the function $y = 300x - 15$?

17. For the function $y = x^2$, compute the average rate of change between $x = -1$ and $x = 2$.

18. For the function $y = x^3$, compute the average rate of change between $x = -1$ and $x = 2$.

19. **a.** Compute the first differences for the data in the table.
    b. Would you say that a linear function could be used to model the data? Explain.
    c. Write the equation of the linear function that fits the data.

x	10	20	30	40	50
y	585	615	645	675	705

In Exercises 20 and 21, use the table below that gives a set of input values, x, and the corresponding outputs, y, that satisfy a function.

x	1	7	13	19
y	−.5	8.5	17.5	26.5

20. Verify that the values satisfy a linear function.

21. Write the equation of the linear function that fits this data.

1.4 EXERCISES

1. *Utility Charges* Palmetto Electric determines its monthly bills for residential customers by charging a base price of $8.95 plus an energy charge of 9.35 cents for each kilowatt hour (KWh) used. Write an equation for the monthly charge y (in dollars) as a function of x, the number of KWh used.

2. *Phone Bills* For interstate calls, AT&T charges 7 cents per minute plus a base charge of $4.95 each month. Write an equation for the monthly charge y as a function of the number of minutes of use.

3. *Depreciation* A business uses a straight-line depreciation to determine the value y of a piece of machinery over a 10-year period. Suppose the original value (when $t = 0$) is equal to $36,000 and its value is reduced by $3600 each year. Write the linear equation that models the value y of this machinery at the end of year t.

4. *Sleep* Each day, a young person should sleep 8 hours plus $\frac{1}{4}$ hour for each year the person is under 18 years of age. Write the equation relating hours of sleep y to age x, for $6 \leq x \leq 18$.

5. *Population Growth* Assume the growth of the population of Del Webb's Sun City Hilton Head community was linear from 1996 to 2000, with a population of 198 in 1996 and a rate of growth of 705 per year.
   a. Write an equation for the population P of this community as a function of the number of years x after 1996.

   b. Use the function to predict the population in 2002.
      (Source: Island Packet, March 2000)

6. *SAT Scores* The composite SAT score for the Beaufort County School District was 952 points in 1994, and the average rate of increase was 0.51 point per year. Write a linear function that models the SAT scores y as a function of x, the number of years after 1994.
   (Source: Island Packet, October 1999)

7. *Depreciation* A business uses a straight-line depreciation to determine the value y of an automobile over a 5-year period. Suppose the original value (when $t = 0$) is equal to $26,000 and the salvage value (when $t = 5$) is equal to $1000.
   a. By how much has the automobile depreciated over the 5 years?
   b. By how much is the value of the automobile reduced at the end of each of the 5 years?
   c. Write the linear equation that models the value s of this automobile at the end of year t.

8. *Retirement* For Pennsylvania state employees for whom the average of the three best yearly salaries is $75,000, the retirement plan gives an annual pension of 2.5% times 75,000, multiplied by the number of years of service. Write the linear function that models the pension P in terms of the number of years of service, y.
   (Source: Pennsylvania State Retirement Fund)

9. *Blood Alcohol Percent* The table below gives the number of drinks and the resulting blood alcohol percent for a 90-lb woman. ("One drink" is equal to 1.25

oz of 80 proof liquor, 12 oz of regular beer, or 5 oz of table wine, and many states have set .08% as the legal limit for driving under the influence.)

a. What is the rate of change in blood alcohol percent per drink for a 90-lb woman?

b. Write the equation of the function that models the blood alcohol percent as a function of the number of drinks.

Number of Drinks	0	1	2	3	4
Blood Alcohol Percent	0	.05	.10	.15	.20

Number of Drinks	5	6	7	8	9
Blood Alcohol Percent	.25	.30	.35	.40	.45

(Source: Pennsylvania Liquor Control Board)

10. *Drinking and Driving* The following table gives the number of drinks and the resulting blood alcohol percent for a 180-lb man legally considered driving under the influence (DUI).

a. Is the average rate of change of the blood alcohol percent with respect to the number of drinks a constant? What is it?

b. Use the rate of change and one point determined by a number of drinks and the resulting blood alcohol percent to write the equation of a linear model for this data.

Number of Drinks	5	6	7	8	9	10
Blood Alcohol Percent	.11	.13	.15	.17	.19	.21

(Source: Pennsylvania Liquor Control Board)

11. *Men in the Workforce* The number of men in the workforce (in millions) for selected decades from 1890 to 1990 is shown in the figure below. The decade is defined by the year at the beginning of the decade and $g(t)$ is defined by the average number of men (in millions) in the workforce during the decade (indicated by the point on the graph within the decade). This data can be approximated by the linear model determined by the line connecting (1890, 18.1) and (1990, 68.5).

a. Write the equation of the line connecting these two points to find a linear model for this data.

b. Does this line appear to be a reasonable fit to the data points?

c. How does the slope of this line compare with the average rate of change in the function during this period?

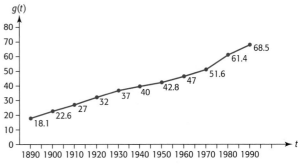

(Source: 1998 *World Almanac*)

12. *Farm Workers* The figure below shows the percent p of U.S. workers in farm occupations for selected years t.

a. Write the equation of a line connecting the points (1820, 71.8) and (1994, 2.6), with values rounded to two decimal places.

b. Does this line appear to be a reasonable fit to the data points?

c. Interpret the slope of this line in terms of rate of change of the percent of farm workers.

d. Can the percent p of U.S. workers in farm occupations continue to fall at this rate for 20 more years? Why or why not?

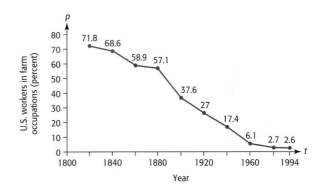

(Source: U.S. Department of Agriculture)

13. *Education Spending* The figure below gives the amount (in billions of dollars) spent on education during the years 1996–2001.

a. What is the average annual rate of change in spending between 1996 and 2001?

b. If a line were drawn connecting the data points (1996, 23) and (2001, 40.1), what is the slope of the line?

c. Is the rate of change in spending for education the same each year?

d. Can this data be modeled exactly by a linear equation? Explain.

Education spending grows
Amount spent (in billions):

1996 $23
1997 $26.6
1998 $29.9
1999 $33.5
2000 $35.6
2001 $40.1

(Source: U.S. Department of Education)

14. *Enrollment Projection* The figure below shows enrollment projections for the three schools. (Outputs are measured in students.)
   a. What is the slope of the line joining the two points in the figure that show the Beaufort enrollment projections?
   b. Find the average rate of change in projected enrollment for students in Beaufort schools between 2000 and 2005.
   c. How should this information be used when planning future school building in Beaufort?

Enrollment projections by area

Projections show southern Beaufort County schools growing fastest.*

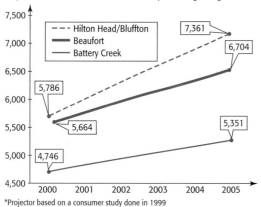

*Projector based on a consumer study done in 1999

(Source: Beaufort County School District)

15. *Teenage Mothers* The percent of births to teenage mothers that are out-of-wedlock can be approximated by a linear function of the number of years after 1950. The percent was 15 in 1960 and 76 in 1996.
   a. What is the slope of the line joining the points (10, 15) and (46, 76)? Round the answer to two decimal places.
   b. What is the average rate of change in the percent of teenage out-of-wedlock births over this period?
   c. What does this average rate of change tell about the out-of-wedlock births to teenage mothers?

   d. Write the equation of the line joining the two given points. Round the values to two decimal places.
   (Source: Father Facts)

16. *Voting* The percent of eligible people voting in presidential elections can be expressed as a linear function of the years 1960–1996. The percent was 63.1 in 1960 and 55.1 in 1992.
   a. What is the slope of the line joining the given points?
   b. What is the average rate of change in the percent voting in these elections? Interpret this value.
   c. Write the equation of the line joining these two given points.
   (Source: Federal Election Commission)

17. *Prison population* The graph showing the total number of prisoners in state and federal prisons for the years 1960 through 1997 is shown in the figure. There were 212,953 prisoners in 1960 and 1,197,590 in 1997.
   (Source: U.S. Bureau of Justice Statistics)
   a. What is the average rate of growth of the prison population from 1960 to 1997?
   b. What is the slope of the line connecting the points satisfying the conditions above?
   c. Write the equation of the secant line joining these two points on the curve.
   d. Can this secant line be used to make a good estimate of total number of prisoners in 1998?
   e. Points associated with what two years could be used to write a linear equation that will give a better estimate of total number of prisoners in 1998?

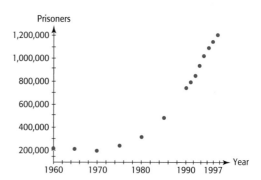

18. *Investment* The graph of the future value of an investment of $1000 for x years earning interest at a rate of 8% compounded continuously is shown in the figure. The $1000 investment is worth approximately $1083 after one year and about $1492 after 5 years.
   a. What is the average rate of change of the future value over the 4-year period?
   b. Interpret this rate of change.
   c. What is the slope of the line connecting the points satisfying the conditions above?

d. Write the equation of the secant line joining the two given points on the curve.

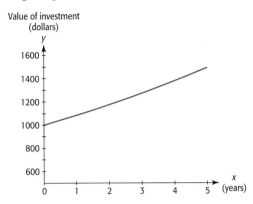

19. *Working Age* The scatter plot below projects the ratio of the working-age population to the elderly.
  a. Does the data appear to fit a linear function?
  b. The data points shown in the scatter plot from 2010 to 2030 are projections made from a mathematical model. Do those projections appear to be made with a linear model? Explain.
  c. If the ratio is projected to be 3.9 in 2010 and 2.2 in 2030, what is the average annual rate of change of the data over this period of time?
  d. Write the equation of the line joining these two points.
  e. Use the graph to estimate when the linear model is no longer valid.

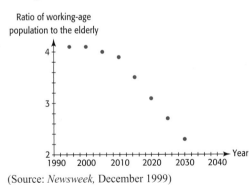

(Source: *Newsweek*, December 1999)

20. *Women in the Workforce* The number of women in the workforce (in millions) for selected years from 1890 to 1990 is shown in the figure.
  a. Would the data in the scatter plot be modeled well by a linear function? Why or why not?
  b. The number of women in the workforce was 3.704 million in 1890 and 16.443 million in 1950. What is the average rate of change in the number of women in the workforce during this period? Round the answer to four decimal places.

c. If the number of women in the workforce was 16.443 million in 1950 and 59.531 million in 1990, what is the average rate of change in the number of women in the workforce during this period?
  d. Is it reasonable that these two average rates of change are different? How can you tell this from the graph?
  (Source: *Newsweek*, December 1999)

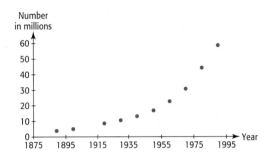

21. *U.S. Population* The total population of the U.S. for selected years from 1950 to 1995 is shown in the table below, with the population given in thousands. Using *x* to represent the number of years from 1950 and *y* to represent the number of millions of people:
  a. Find the average annual rate of change in population during these years, with the appropriate units.
  b. Write the equation of the line connecting the points associated with 1950 and 1995.
  c. Use the equation to estimate the population in 1975. Does it agree with the population shown in the table?
  d. Why might the values be different?

Year	Population (thousands)	Year	Population (thousands)
1950	152,271	1975	215,973
1955	165,931	1980	227,726
1960	180,671	1985	238,466
1965	194,303	1990	249,948
1970	205,052	1995	263,044

(Source: U.S. Census Bureau)

22. *Smoking Cessation* The table below gives the percent of people over age 20 who once smoked and quit smoking between 1965 and 1990.
  a. Create a new table with the data aligned so that the years are represented by the number of years past 1965.

b. An equation that models the data in the new table is $y = 0.723x + 29.246$. Show the graph of the data and the equation on the same set of axes.

c. Do the graphs indicate that a linear model fits the data reasonably well?

Year	People who have quit smoking (percent)	Year	People who have quit smoking (percent)
1965	29.8	1979	39
1966	29.3	1980	38.9
1970	35.2	1983	41.8
1975	36	1985	45
1976	37	1987	44.9
1977	36.6	1990	48.4
1978	38.3		

(Source: Indiana Tobacco Control Center, U.S. Centers for Disease Control)

23. *Taxes* The table below shows some sample incomes and the income tax due for each income, and the figure shows a graph of the data points.

a. Is the rate of change of tax with respect to income bracket constant for the incomes shown in the table?

b. What is the rate of change in tax per $1 in income?

c. Write a linear equation to represent income tax due as a function of taxable income for the incomes in the table.

d. Verify that the model fits the data by evaluating the function at $x = 30,100$ and at $x = 30,300$ and comparing the resulting y-values with the income tax due for these taxable incomes.

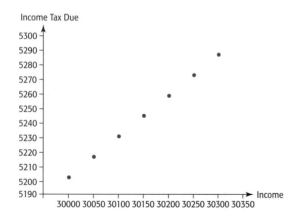

Taxable Income	Income Tax Due
$30,000	$5203
30,050	5217
30,100	5231
30,150	5245
30,200	5259
30,250	5273
30,300	5287

(Source: U.S. Federal tax table for 1997)

24. *Social Agency* A social agency provides emergency food and shelter to two groups of clients. The first group has x clients who need an average of $300 for emergencies, and the second group has y clients who need an average of $200 for emergencies. The agency has $100,000 to spend for these two groups.

a. Write the equation that gives the number of clients who can be served in each group.

b. Find the y-intercept and the slope of the graph of this equation. Interpret each value.

c. If 10 clients are added from the first group, what happens to the number served in the second group?

25. *Depreciation* For tax purposes, business real estate can be depreciated over 30 years by the straight-line method. This means that the value of a $900,000 building will drop by 1/30th of its value each year. Create a model for the value of the building s as a function of the year t.

26. *Measurement Conversion* The number of inches in a measurement of length is directly proportional to the number of yards in the measurement, and the number of inches in a yard is 36.

a. Write a function that can be used to convert yards to inches.

b. Use this function to find the number of inches in the length of a football field.

27. *Utility Costs* An electric utility company determines the monthly cost for a residential customer by adding an energy charge of 10.86 cents per kilowatt-hour to its base charge of $9.86 per month.

a. What would be the monthly cost if no electricity were used?

b. By what rate does the cost for electricity change per kilowatt-hour?

c. What is the slope of the line that represents the cost for this utility?

d. Write a linear equation that models the monthly charge as a function of the number of kilowatt-hours used, representing the monthly cost in dollars by C and the number of kilowatt-hours by K.

28. *Depreciation* Suppose a business property that has a basis (original value) of $480,000 is completely depreciated (worth $0) in 20 years using the straight-line method.

a. What is the rate of change in the value y of the property per year?

b. What are the slope and the y-intercept of the graph of the equation that gives the value of the property as a function of the number of years that it has been depreciated?

c. Write the equation that gives the value y of the property t years after the depreciation begins.

d. Use the function to find the value of the property 2 years after the depreciation begins.

1.5 Algebraic and Graphical Solutions of Linear Equations

Key Concepts

- Solution of linear equations algebraically

- Graphical solution of linear equations
  The x-intercept method
  The intersect method

- Solution of equations for a specified linear variable
  Solution of equations for the variable y

The average sentence length and the average time served in state prisons for various crimes in 1996 is shown in Table 1.17. This data can be used to create a linear function which is an approximate model for the data:

$$y = 0.55x - 2.886.$$

This function describes the 1996 mean time y served in prison for a crime as a function of the mean sentence length x, where x and y are each measured in months. Assuming that this model remains valid after 1996, it predicts that the mean time served on a five-year (60 months) sentence is $0.55(60) - 2.886 \approx 30$ months; that is, approximately $2\frac{1}{2}$ years.

TABLE 1.17

Average Sentence (months)	Average Time Served
62	30
85	45
180	95
116	66
92	46
61	33
56	26

(Source: U.S. Department of Justice, "Truth in Sentencing in State Prisons," January 1999)

To find the sentence that would give an expected time served of 10 years (120 months), we use algebraic or graphical methods to solve the equation

$$120 = 0.55x - 2.866$$

for x. In this section, we use additional algebraic methods and graphical methods to solve linear equations in one variable.

Algebraic Solution of Linear Equations

We can use algebraic, graphical, or a combination of algebraic and graphical methods to solve linear equations. Sometimes it is more convenient to use algebraic solution methods rather than graphical solution methods, especially if an exact solution is desired. The steps used to solve a linear equation in one variable algebraically follow.

STEPS FOR SOLVING A LINEAR EQUATION IN ONE VARIABLE

1. If a linear equation contains fractions, multiply both sides of the equation by a number that will remove all denominators from the equation. If there are two or more fractions, use the least common denominator (LCD) of the fractions.

2. Remove any parentheses or other symbols of grouping.

3. Perform any additions or subtractions to get all terms containing the variable on one side and all other terms on the other side of the equation. Combine like terms.

4. Divide both sides of the equation by the coefficient of the variable.

5. Check the solution by substitution in the original equation. If a real-world solution is desired, check the algebraic solution for reasonableness in the real-world situation.

EXAMPLE 1   Algebraic Solutions

a. Solve for x: $\dfrac{2x - 3}{4} = \dfrac{x}{3} + 1$ **b.** Solve for y: $y - \dfrac{1}{2}\left(\dfrac{y}{2}\right) = -6$

Solution

a.
$$\frac{2x - 3}{4} = \frac{x}{3} + 1$$

$$12\left(\frac{2x - 3}{4}\right) = 12\left(\frac{x}{3} + 1\right) \qquad \text{Multiply both sides by the LCD, 12.}$$

$$3(2x - 3) = 12\left(\frac{x}{3} + 1\right) \qquad \text{Simplify the fraction on the left.}$$

$$6x - 9 = 4x + 12 \qquad \text{Remove parentheses using the distributive property.}$$

$$2x = 21 \qquad \text{Subtract 4x from both sides and add 9 to both sides.}$$

$$x = \frac{21}{2} \qquad \text{Divide both sides by 2.}$$

Check the result: $$\frac{2\left(\frac{21}{2}\right) - 3}{4} \overset{?}{=} \frac{\frac{21}{2}}{3} + 1$$

$$\frac{18}{4} \overset{?}{=} \frac{21}{6} + 1$$

$$\frac{9}{2} = \frac{9}{2}$$

b. $y - \frac{1}{2}\left(\frac{y}{2}\right) = -6$

$\quad\quad y - \frac{y}{4} = -6$ Remove parentheses first to find the LCD.

$\quad 4\left(y - \frac{y}{4}\right) = 4(-6)$ Multiply both sides by the LCD, 4.

$\quad\quad 4y - y = -24$ Remove parentheses using the Distributive Property.

$\quad\quad\quad 3y = -24$ Combine like terms.

$\quad\quad\quad\quad y = -8$ Divide both sides by 3.

Check the result: $$-8 - \frac{1}{2}\left(\frac{-8}{2}\right) \overset{?}{=} -6$$

$$-8 - \left(\frac{-8}{4}\right) \overset{?}{=} -6$$

$$-6 = -6 \quad\blacksquare$$

As the next example illustrates, we solve an application problem that is set in a real-world context by using the same solution methods. However, you must remember to include units of measure with your answer and check that your answer makes sense in the problem situation.

EXAMPLE 2 **Credit-Card Debt**

It is much harder for people to pay off credit-card debts in a reasonable period of time because of high interest rates. The interest paid on a $10,000 debt over 3 years is approximated by

$$y = 175.393x - 116.287 \text{ dollars}$$

when the interest rate is $x\%$. What is the interest rate if the interest is $1637.60? (Source: Consumer Federation of America)

Solution

To answer this question, we solve the linear equation

$$1637.60 = 175.393x - 116.287$$

$$1753.887 = 175.393x$$

$$x = 9.9998$$

Thus the interest is $1637.60 if the interest rate is approximately 10%.

(Note that if you check the approximate answer, you are checking only for the reasonableness of the estimate.) ■

Solutions, Zeros, and *x*-Intercepts

Most, but not all, of the equations we work with represent functions. When we are given an input, we *evaluate* the function (that is, find the output) by substituting the input value into the equation. When given an output, we can *solve* the equation to find the input. We know that the *x*-intercepts of the graph of $y = f(x)$ are the zeros of the function, so we have the following.

The following three concepts are numerically the same:

The *x*-intercepts of the graph of $y = f(x)$
The zeros of the function f
The real solutions to the equation $f(x) = 0$

The following example illustrates the relationships that exist among *x*-intercepts of the graph of a function, zeros of the function, and solutions to associated equations.

EXAMPLE 3 Relationships Among *x*-Intercepts, Zeros, and Solutions

For the function $f(x) = 13x - 39$, find

a. $f(3)$ **b.** The zero of $f(x) = 13x - 39$

c. The *x*-intercept of the graph of $y = 13x - 39$

d. The solution to the equation $13x - 39 = 0$

Solution

a. Evaluate the output $f(3)$ by substituting the input 3 for *x* in the function:

$$f(3) = 13(3) - 39 = 0$$

b. Because $f(x) = 0$ when $x = 3$, we say that 3 is a zero of the function.

c. The *x*-intercept of a graph occurs at the value of *x* where $y = 0$, so the *x*-intercept is $x = 3$.

The only *x*-intercept is $x = 3$ because the graph of $f(x) = 13x - 39$ is a line that crosses the *x*-axis at only one point. (See Figure 1.37(a)).

d. The solution to the equation $13x - 39 = 0$ is $x = 3$ because $13x - 39 = 0$ gives $13x = 39$ or $x = 3$. ■

We can use [TRACE], [CALC], [VALUE], or [TABLE] on a graphing utility to check the reasonableness of a solution. A graph of $y = 13x - 39$ with *x* between −5 and 5 and *y* between −10 and 10 is given in Figure 1.37(a). We can use this graph to confirm that $y = 0$ when $x = 3$. A spreadsheet for $y_1 = 13x - 39$, shown in Figure 1.37(b), also confirms that $f(3) = 0$.

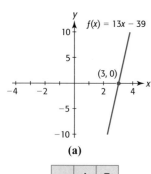

	A	B
1	x	y
2	3	0

(b)

FIGURE 1.37

Graphical Solution of Linear Equations

We can also view the graph or use TRACE on a graphing utility to obtain a quick estimate of an answer. TRACE may or may not provide the exact solution to an equation, but
it will provide an estimate of the solution. If your calculator or computer has a graphical
or numerical solver, the solver can be used to obtain the solution to the equation
$f(x) = 0$. Recall that an x-intercept of the graph of $y = f(x)$, a zero of $f(x)$, and the solution to the equation $f(x) = 0$ are all different names for the same input value. If the graph
of the linear function $y = f(x)$ is not a horizontal line, we can find the one solution to the
nonconstant linear equation $f(x) = 0$ as described below. We call this solution method
the *x-intercept method*.

SOLVING A LINEAR EQUATION USING THE *X*-INTERCEPT METHOD WITH GRAPHING UTILITIES

1. Rewrite the equation to be solved with 0 (and nothing else) on one side of the
   equation.

2. Enter the nonzero side of the equation found in the previous step in the equation editor of your graphing utility and graph the line in an appropriate viewing window. Be certain that you can see the line cross the horizontal axis on
   your graph.

3. Find the x-intercept by inspection with TRACE or by using the graphical
   solver that is often called ZERO or ROOT . The x-intercept is the value of x
   that makes the equation equal to zero, so it is the solution to the equation. The
   value of x found by this method is sometimes a decimal approximation of the
   exact solution rather than the exact solution.

EXAMPLE 4   Graphical Solution

Solve $\dfrac{2x - 3}{4} = \dfrac{x}{3} + 1$ for x using the x-intercept method.

Solution

To solve the equation using the x-intercept method, we first rewrite the equation with 0
on one side.

$$0 = \frac{x}{3} + 1 - \frac{2x - 3}{4}$$

Next, enter the right-hand side of the equation,

$$\frac{x}{3} + 1 - \frac{2x - 3}{4},$$

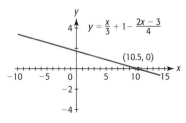

in the equation editor of your graphing utility as

$$y_1 = x/3 + 1 - (2x - 3)/4$$

and graph this function. Use parentheses as needed to preserve the order of operations.
You should obtain a graph similar to the one seen in Figure 1.38, which was obtained

FIGURE 1.38

using the viewing window $[-10, 15]$ by $[-5, 5]$. However, any graph in which you can see the x-intercept will do. Using $\boxed{\text{ZERO}}$ or $\boxed{\text{ROOT}}$ gives the x-intercept of the graph, 10.5 (See Figure 1.38.) This is the value of x that makes $y = 0$, and thus the original equation true, so it is the solution to this linear equation.

You can determine if $x = 10.5 = \dfrac{21}{2}$, found graphically, is the exact answer by substituting this value into the original equation. Both sides of the original equation are equal when $x = 10.5$, so it is the exact solution. ∎

We can also use the intersection method described below to solve linear equations:

> ## SOLVING A LINEAR EQUATION USING THE INTERSECTION METHOD
>
> 1. Enter the left side of the equation as y_1 and the right side as y_2. Graph both of these equations.
> 2. Find the point of intersection of the two graphs. This is the point where $y_1 = y_2$. The x-value of this point is the value of x that makes the two sides of the equation equal, so it is the solution to the original equation.

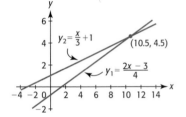

FIGURE 1.39

The equation in Example 4,

$$\frac{2x - 3}{4} = \frac{x}{3} + 1,$$

can be solved using the intersection method. Enter the left side as

$$y_1 = (2x - 3)/4 \text{ and the right side as } y_2 = (x/3) + 1.$$

Graphing these two functions gives the graphs in Figure 1.39. The point of intersection is found to be $(10.5, 4.5)$. So, the solution is $x = 10.5$ (as we found in Example 4).

EXAMPLE 5   Criminal Sentences

Solving $120 = 0.55x - 2.886$, where x is measured in months, gives the sentence for a crime that would give an expected time served of 10 years. Solve this equation by using the

a. x-intercept method.

b. intersection method.

Solution

a. We solve the linear equation $120 = 0.55x - 2.886$ by using the x-intercept method as follows:

Rewrite the equation in a form with 0 on one side:

$$0 = 0.55x - 2.886 - 120$$

Combine the constant terms to obtain

$$0 = 0.55x - 122.886$$

Enter $y_1 = 0.55x - 122.886$ and graph this equation. (See Figure 1.40.) Using ZERO gives an x-intercept of approximately 223. (See Figure 1.40.) Thus if a prisoner receives a sentence of 223 months, we would expect him or her to serve 120 months, or 10 years.

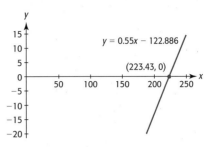

FIGURE 1.40

b. To use the intersection method, enter $y_1 = 0.55x - 2.886$ and $y_2 = 120$, graph these equations, and find the intersection of the lines. Figure 1.41 shows the point of intersection.

Note that the zero found in Figure 1.40 and the x-coordinate of the point of intersection in Figure 1.41 agree and are both approximations of the exact solution, which could be found with algebraic methods. In this application the approximate solution, $x \approx 223$ months, is a more useful way of expressing this answer.

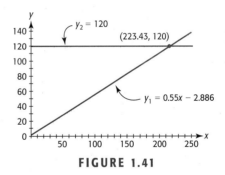

FIGURE 1.41   ■

Spreadsheet Solution

We can use the **Goal Seek** feature of Excel to find the x-intercept of the graph of a function; that is, the zeros of the function. To solve

$$120 = 0.55x - 2.886,$$

we enter the nonzero side of $0 = 0.55x - 122.886$ as shown in Table 1.18. Using **Goal Seek** with cell B2 set to value 0 gives the value of x that solves the equation as shown in Table 1.19.

TABLE 1.18

	A	B
1	x	y
2	1	= 0.55*A2 − 122.886

TABLE 1.19

	A	B
1	x	y
2	223.43	0

Formulas; Solving an Equation for a Specified Linear Variable

Suppose we want to solve an equation for one variable if two or more variables are in the equation. If that variable is to the first power in the equation, we can solve for that variable by treating the other variables as constants and using the same steps that we used to solve a linear equation in one variable. This is sometimes useful, because it is necessary to get equations in the proper form to enter them in a graphing utility.

STEPS FOR SOLVING AN EQUATION FOR A SPECIFIED LINEAR VARIABLE

1. If the equation contains fractions, multiply both sides of the equation by the least common denominator to remove all denominators in the equation.

2. Remove any parentheses or other symbols of grouping.

3. Perform any additions or subtractions to get all terms containing the variable to be solved for on one side and all other terms on the other side of the equation. Combine like terms.

4. If more than one term contains the variable to be solved for, factor out the variable.

5. Divide both sides of the equation by the coefficient of the variable (which may be an expression).

EXAMPLE 6   Simple Interest

The formula for the future value of an investment of $\$P$ at simple interest rate r for t years is $A = P(1 + rt)$. Solve the formula for r, the interest rate.

Solution

$$A = P(1 + rt)$$

$$A = P + Prt \qquad \text{Remove parentheses.}$$

$$A - P = Prt \qquad \text{Get the term containing } r \text{ by itself on one side of the equation.}$$

$$r = \frac{A - P}{Pt} \qquad \text{Divide both sides by } Pt. \qquad ■$$

EXAMPLE 7   Writing an Equation in Functional Form and Graphing Equations

Solve each of the following equations for y, so that y is expressed as a function of x. Then graph the equation on a graphing utility with a standard viewing window.

a. $2x - 3y = 12$ **b.** $x^2 + 4y = 4$

Solution

a. Because y is linear (to the first power) in the equation, we solve the equation for y using linear equation solution methods:

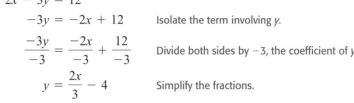

$$2x - 3y = 12$$

$-3y = -2x + 12$ Isolate the term involving y.

$\dfrac{-3y}{-3} = \dfrac{-2x}{-3} + \dfrac{12}{-3}$ Divide both sides by -3, the coefficient of y.

$y = \dfrac{2x}{3} - 4$ Simplify the fractions.

To graph the equation on a graphing utility, enter the function

$$y = \dfrac{2x}{3} - 4$$

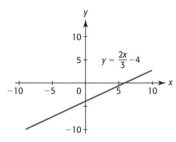

FIGURE 1.42

and use the standard viewing window to obtain the graph shown in Figure 1.42.

b. The variable y is to the first power in the equation

$$x^2 + 4y = 4,$$

so we solve for y by using linear solution methods. (Note that to solve this equation for x would be more difficult; this method will be discussed in Chapter 2.)

$4y = -x^2 + 4$ Isolate the term containing y.

$\dfrac{4y}{4} = \dfrac{-x^2}{4} + \dfrac{4}{4}$ Divide both sides by 4, the coefficient of y.

$y = \dfrac{-x^2}{4} + 1$ Simplify the fractions.

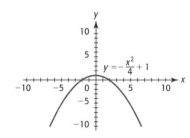

FIGURE 1.43

To graph the equation on a graphing utility, enter the function $y = \dfrac{-x^2}{4} + 1$ and use the standard viewing window to obtain the graph shown in Figure 1.43. ■

1.5 SKILLS CHECK

In Exercises 1–8, solve the equations for x (a) algebraically and (b) with a graphing utility.

1. $5x - 14 = 23 + 7x$

2. $3x - 2 = 7x - 24$

3. $3(x - 7) = 19 - x$

4. $x - \dfrac{5}{6} = 3x + \dfrac{1}{4}$

5. $3x - \dfrac{1}{3} = 5x + \dfrac{3}{4}$

6. $\dfrac{5(x - 3)}{6} - x = 1 - \dfrac{x}{9}$

7. $\dfrac{1}{3}(x - 4) = 2(5 + x)$

8. $5.92t = 1.78t - 48.1$

For Exercises 9–12, find

a. the solution to the equation $f(x) = 0$
b. the x-intercept of the graph of $y = f(x)$
c. the zero of $f(x)$

9. $f(x) = 32 + 1.6x$ **10.** $f(x) = 15x - 60$

11. $f(x) = \dfrac{3}{2}x - 6$ **12.** $f(x) = \dfrac{x - 5}{4}$

In Exercises 13 and 14, you are given a table showing input and output values for a given function $y_1 = f(x)$. Using this table, find (if possible)

a. the x-intercept of the graph of $y = f(x)$
b. the y-intercept of the graph of $y = f(x)$
c. the solution to the equation $f(x) = 0$.

13.

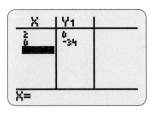

14.

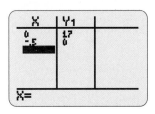

In Exercises 15 and 16, you are given the graph of a certain function $y = f(x)$ and the zero of that function. Using this graph, find

a. the x-intercept of the graph of $y = f(x)$
b. the solution to the equation $f(x) = 0$.

15.

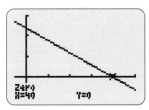

16.

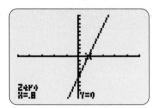

17. Solve $A = P(1 + rt)$
(a) for t
(b) for P.

18. Solve $V = \dfrac{1}{3}\pi r^2 h$ for h.

19. Solve $4(a - 2x) = 5x + \dfrac{c}{3}$ for x.

20. What are the slope and y-intercept of the graph of $x = -4y + 1$?

In Exercises 21–24, solve the equations for y, and graph them with a standard window on a graphing utility.

21. $5x - 3y = 5$ **22.** $3x + 2y = 6$

23. $x^2 + 2y = 6$ **24.** $4x^2 + 2y = 8$

1.5 EXERCISES

----

1. *Depreciation* An $828,000 building is depreciated for tax purposes by its owner using the straight-line depreciation method. The value of the building, y, after x months of use is given by $y = 828,000 - 2300x$ dollars. After how many years will the value of the building be $690,000?

2. *Temperature Conversion* The equation $5F - 9C = 160$ gives the relationship between Fahrenheit and Celsius temperature measurements. What Fahrenheit measure is equivalent to a Celsius measurement of $20°$?

3. *Game Show Question* The following question was worth $32,000 on the game show *Who Wants to Be a Millionaire?*: "At what temperature are the Fahrenheit and Celsius temperature scales the same?" Answer this question.

4. *Sales Tax* The total cost of a new automobile, including a 6% sales tax on the price of the automobile, is $29,998. How much of the total cost of this new automobile is sales tax?

5. *Investments* The future value of a simple interest investment is given by $S = P(1 + rt)$. What principal P must be invested for $t = 5$ years at the simple interest rate $r = 10\%$ so that the future value grows to $9000?

6. *Grades* To earn an A in a course, a student must get an average score of at least 90 on 5 tests. If her first 4 test scores are 92, 86, 79, and 96, what score does she need on the last test to obtain a 90 average?

7. *Grades* To earn an A in a course, a student must get an average score of at least 90 on 3 tests and a final exam. His final exam score counts double and his lowest score is removed if the final score is higher than the lowest score. If his first 3 test scores are 86, 79, and 96, what is the lowest score he needs on the final exam to obtain a 90 average?

8. *Temperature-Humidity Index* The temperature-humidity index I is given by $I = t - 0.55(1 - h)(t - 58)$ where t is the air temperature in degrees Fahrenheit and h is the relative humidity expressed as a decimal. If the air temperature is 80 degrees, find the humidity h (as a percent) that gives an index value of 79.

9. *U.S. Civilian Population* The population of the U.S., excluding Armed Forces, can be modeled for the years 1970 to 1998 by the function $p = 2351x + 201,817$, where p is in thousands of people and x is in years after 1970. During what year does the model estimate the population to be 258,241,000?
(Source: www.census.gov/statab)

10. *Earnings and Minorities* The median annual salary (in thousands of dollars) of minorities M is related to the median salaries of whites W by $M = 0.959W - 1.226$. What is the median annual salary for whites when the median annual salary for minorities is $50,560?
(Source: *Statistical Abstract of the U.S.*, 1993)

11. *Reading Score* The average reading score on the National Assessment of Progress tests is given by $y = 0.155x + 255.37$, where x is the number of years after 1970. In what year would the average reading score be 259.4, if this model is accurate?

12. *Marriage Rate* The percent of unmarried women who get married in any particular year is given by $y = -0.0762x + 8.5284$, where x is the number of years after 1950. During what year does this model predict the percent will be 6.09?
(Source: Index of Leading Cultural Indicators)

13. *Tobacco Judgment* As part of the largest-to-date damage award in a jury trial in history, the July 2000 penalty handed down against the tobacco industry included a $74 billion judgment against Philip Morris. If this amount was 94% of this company's 1999 revenue, how much was Philip Morris's 1999 revenue?
(Source: *Newsweek*, July 24, 2000)

14. *Tobacco Judgment* As part of the largest-to-date damage award in a jury trial in history, the July 2000 penalty handed down against the tobacco industry included a $36 billion judgment against R. J. Reynolds. If this amount was 479% of this company's 1999 revenue, how much was R. J. Reynolds 1999 revenue?
(Source: *Newsweek*, July 24, 2000)

15. *Presidential Elections* The percent of the population voting in presidential elections has been estimated to be $p = 65.4042 - 0.3552x$, where x is the number of years after 1950.
(Source: Federal Election Commission)
    a. During what year does this model predict that the percent voting in a presidential election will be 49%?
    b. Of those persons eligible to vote, 53% cast a ballot in the 2000 presidential election. Did this model accurately describe the percent voting in the 2000 election? What does this answer tell you about using models for predictions?

16. *Tax Burden* The per capita tax burden B can be described by $B(t) = 20.37 + 1.834t$ hundred dollars, where t is the number of years after 1980. In what year does this model indicate that the per capita tax burden was $4788?
(Source: Internal Revenue Service)

17. *Internet Brokerage Accounts* The number of brokerage accounts on the Internet in year t can be modeled by $B(t) = 3.303t - 6591.560$ million accounts. According to the model in what year were there 14.44 million accounts?
(Source: Gomez Advisors)

18. *Marijuana Use* The percent p of high school seniors using marijuana daily can be related to x, the number of years after 1990, by the equation $30p - 19x = 1$, where x is the number of years after 1990. According to the model, during what year was the percent using marijuana daily equal to 3.2%?
(Source: Index of Leading Cultural Indicators)

19. *Internet Workers* The percent of Fortune Global 500 firms that actively recruited workers on the Internet between 1998 and 2000 can be modeled by

$P(x) = 26.5x - 62$ percent, where x is the number of years after 1995. In what year does this model indicate that 44% of the firms recruited on the Internet?
(Source: iLogos.com)

20. *Cigarette Advertising* The total U.S. cigarette advertising and promotional expenditures from 1975 to 1996 can be modeled by the equation $y = 277.318x - 1424.766$, where y is measured in millions of dollars and x is the number of years after 1970. If this model remains accurate, in what year did the spending exceed 6 billion dollars?
(Source: Federal Trade Commission)

21. *Inmates* The total number of inmates in custody between 1990 and 1998 in state and federal prisons is approximately given by $y = 68.476x + 728.654$ thousand prisoners, where x is the number of years after 1990. In what year will the number of inmates be 797,130?
(Source: Bureau of Justice Statistics)

22. *Out-of-Wedlock Births* The percent of all teen mothers who are unmarried can be given by the equation $y = 1.78x - 3.998$, where x is the number of years after 1950. If this model remains accurate, in what year will the percent have reached 80?
(Source: Father Facts)

23. *Personal income* U.S. personal income increased between 1993 and 1997 according to the model $I(x) = 341.28x + 4459.78$ billion dollars, where x is the number of years since 1990. In what year does this model predict that U.S. personal income would reach $7190 billion if it is accurate beyond 1997?
(Source: ECONOMAGIC)

24. *Marriage Rates* The marriage rate (per 1000 women) can be given by the equation $y = -0.762x + 85.284$, where x is the number of years after 1950. If this model remains valid, what is the first year in which the marriage rate fell below 45 per 1000?
(Source: Index of Leading Cultural Indicators)

25. *Mobile Phone Subscribers* The number of mobile-phone subscribers (in millions) between 1995 and 1999 can be modeled by $S(x) = 11.75x + 32.95$, where x is the number of years after 1995. In what year does this model indicate that there were 68,200,000 subscribers?
(Source: *Newsweek*, March 20, 2000)

26. *Mobile Phone Bills* The average mobile phone local monthly bill between 1995 and 1998 can be modeled by $B(x) = -3.963x + 51.172$ dollars, where x is the number of years after 1995. If the model remains accurate beyond 1998, in what year does it indicate that the monthly bill is $35.32?
(Source: *Newsweek*, March 20, 2000)

27. *Investment* A retired couple has $240,000 to invest. They chose one relatively safe investment fund that has an annual yield of 8% and another riskier investment that has a 12% annual yield. How much should they invest in each fund to earn $23,200 per year?

28. *Salaries* A man earning $4000 per month has his salary reduced by 5% because of a reduction in his company's market. A year later, he receives a 15% raise from the reduced salary.
   a. What is his new salary?
   b. What percent is this raise over the $4000 salary?

29. *Wildlife Management* In wildlife management, the capture-mark-recapture technique is used to estimate the population size of certain types of fish or animals. To estimate the population, we enter information in the equation

$$\frac{\text{total in population}}{\text{total number marked}} = \frac{\text{total number in 2nd capture}}{\text{number found marked in 2nd capture}}$$

Suppose that 50 sharks are caught along a certain shoreline and marked, and then released. If a second capture of 50 sharks gives 20 sharks that have been marked, what is the resulting population estimate?

30. *Social Agency* A social agency provides emergency food and shelter to two groups of clients. The first group has x clients who need an average of $300 for emergencies and the second group has y clients who need an average of $200 for emergencies. The agency has $100,000 to spend for these two groups.
   a. Write an equation that describes the number of clients who can be served with the $100,000.
   b. Solve this equation for y to write the equation in slope-intercept form.
   c. If 10 clients are added to the first group, what happens to the number served in the second group?

31. *Medication* Suppose that combining x cubic centimeters of a 20% concentration of a medication and y cc of a 5% concentration of the medication gives $(x + y)$ cc of a 15.5% concentration. If 7 cc of the 20% concentration are added, by how much must the amount of 5% concentration be increased to keep the same concentration?

1.6 Fitting Lines to Data Points: Modeling Linear Functions

Key Concepts

- Fitting lines to data points
- Linear regression (least squares method)
  Modeling data
  The best-fit line
  Extrapolation
- Applying models
  Extrapolation
  Interpolation
- Goodness of fit
  Constant first differences

The College Board began reporting SAT scores with a different scaling system in 1996, leading to what appeared to be significantly higher scores. The College Board created conversion charts to compare the scores under the old scaling system with the scores under the new scaling system. Table 1.20 gives the average total SAT scores in the old scaling system and in the new scaling system for the years 1970 through 1998. We can determine the relationship between the two scales by creating an equation that gives the new scale as a function of the old scale and use this equation to estimate a "new scale" score from a given "old scale" score.

TABLE 1.20

Year	Average Total SAT Score (Old Scale)	Average Total SAT Score (New Scale)
1970	948	1049
1975	906	1000
1980	890	994
1985	906	1009
1990	900	1001
1991	896	999
1992	899	1001
1993	902	1003
1994	902	1003
1995	910	1010
1996	911	1014
1997	915	1016
1998	916	1017

(Source: College Board, Index of Leading Cultural Indicators)

Plotting points where the horizontal-axis coordinate represents the old scale scores and the vertical-axis coordinate represents the new scale scores gives the scatter plot shown in Figure 1.44(a). These points appear to come close to following a linear pattern, and indeed the data points fit rather well on the line with equation

$$y = 0.97x + 128.3829$$

where x is the old scale score and y is the new scale score. Figure 1.44(a) on page 80 shows the data points and Figure 1.44(b) shows the graph of the linear model drawn on the scatter plot of the data. Note that the points do not all lie on the line, but they do lie close to the line. But how do we find the linear equation that models these data? In this section, we will learn how to create linear models from data points, how to use graphing utilities to create linear models, and how to use the model to answer questions about the data.

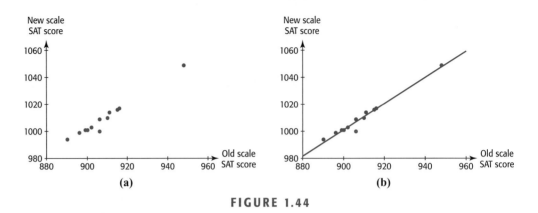

FIGURE 1.44

Fitting Lines to Data Points

When real-world information is collected as numerical information called *data*, technology can be used to determine the pattern exhibited by the data (provided that a recognizable pattern exists). These patterns can often be described by mathematical functions. Before discussing methods of creating linear models for real-world data, we compare the graphs of several functions and a data set to see which of the graphs visually gives the best fit for the data. For now, we informally define the "best fit" line as the one that appears to come the closest to all the data points.

EXAMPLE 1   Health Service Employment
- -

Table 1.21 gives the number of full- and part-time employees in offices and clinics of dentists for selected years between 1970 and 1998. (Source: U.S. Bureau of the Census)

TABLE 1.21

Year	1970	1975	1980	1985	1990	1995	1998
Employees (in thousands)	222	331	415	480	580	644	666

a. Draw a scatter plot of the data with the *x*-value of each point representing the number of years after 1970 and the *y*-value representing the number of dental employees (in thousands) corresponding to that year.

b. Individually graph each of the following three equations on the same graph as the scatter plot, and determine which of the lines appears to give the best fit to the data.

$$y_1 = -12x + 660$$
$$y_2 = 13x + 220$$
$$y_3 = 16x + 242$$

Solution

a. By using x as the number of years after 1970, we have aligned the data with $x = 0$ representing 1970, $x = 1$ representing 1971, and so forth. Enter into the lists of a graphing utility (or an Excel spreadsheet) the aligned input data representing the years in Table 1.21 and the output data representing the number of employees shown in the second row of Table 1.21. Figure 1.45(a) shows the lists containing the data. The graph of these data points is shown in Figure 1.45(b). The window for this scatter plot can be set manually or with a command on the graphing utility such as $\boxed{\text{ZOOMSTAT}}$, which is used to automatically set the window and display the graph.

L1	L2
0	222
5	331
10	415
15	480
20	580
25	644
28	666

(a)

Dental employees (in thousands)

(b)

FIGURE 1.45

b. The graphs of the data and

$$y_1 = -12x + 660,$$

shown in Figure 1.46(a), show that the line is falling (its slope is negative), but the data points follow a rising pattern. Thus, an equation with a positive slope would be a better fit. Figure 1.46(b) shows that the graph of

$$y_2 = 13x + 220$$

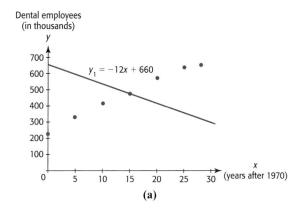

(a)

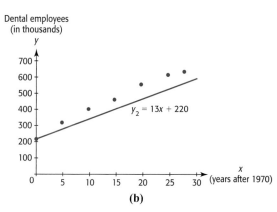

(b)

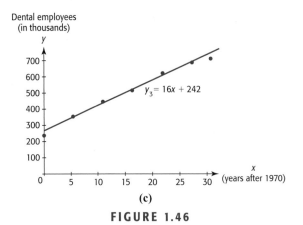

(c)

FIGURE 1.46

is a better fit to the data than the graph in Figure 1.46(a), but a line with a steeper (larger) slope would be better. The graph of

$$y_3 = 16x + 242$$

is the line that appears to fit better than either of the other two lines. (See Figure 1.46(c).) Thus, of the three linear functions given to model the number of full and part-time employees in dentists' offices and clinics, the best is $y_3 = 16x + 242$ thousand employees, where x is the number of years past 1970. ■

In Example 1 it was not difficult to choose which of the three given lines fits the data the best. However, is the line in Figure 1.46(c) the best-fitting line of all possible lines that could be drawn on the scatter plot? How do we determine the equation of the best-fit line? We now discuss answers to these questions.

Linear Regression

The points determined by the data in Table 1.21 do not all lie on a line, but we can determine the equation of the line that is the "best fit" for these points by using a procedure called **linear regression** (or the **least squares method**). This procedure defines the "best fit line" as the line for which the sum of the squares of the vertical distances from the data points to the line is a minimum. For this reason, the linear regression procedure is also called the least squares method.

The vertical distance between a data point and the corresponding point on a line is simply how much the line misses going through the point, that is, the difference in the outputs of the data point and the point on the line. If we call this difference in outputs d_i (where i takes on the values from 1 to n for n data points), the least squares method* requires that

$$d_1{}^2 + d_2{}^2 + d_3{}^2 + \ldots + d_n{}^2$$

be as small as possible. The line for which this sum of squared differences is as small as possible is called the *linear regression line* or *least squares line* and is the one that we consider to be the **best-fit line** for the data.

---

*The sum of the squared differences is often called SSE, the *sum of squared errors*. The development of the equations that lead to the minimum SSE is a calculus topic.

To illustrate the linear regression process, consider again the two lines in Figures 1.46(b) and 1.46(c) and the data points in Table 1.21. Figures 1.47 and 1.48 indicate the vertical distances for each of these lines. Below is the calculation of the sum of squared differences for each line.

Figure 1.47:

$$d_1^2 + d_2^2 + d_3^2 + \ldots + d_7^2 = (222 - 220)^2 + (331 - 285)^2 + (415 - 350)^2$$
$$+ (480 - 415)^2 + (580 - 480)^2$$
$$+ (644 - 545)^2 + (666 - 584)^2$$
$$= 4 + 2116 + 4225 + 4225 + 10{,}000$$
$$+ 9801 + 6724$$
$$= 37{,}095$$

Figure 1.48:

$$d_1^2 + d_2^2 + d_3^2 + \ldots + d_7^2 = (222 - 242)^2 + (331 - 322)^2 + (415 - 402)^2$$
$$+ (480 - 482)^2 + (580 - 562)^2 + (644 - 642)^2$$
$$+ (666 - 690)^2 = 400 + 81 + 169 + 4 + 324$$
$$+ 4 + 576 = 1558$$

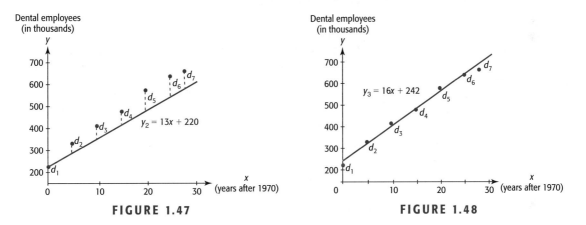

FIGURE 1.47 **FIGURE 1.48**

It is not difficult to see that the vertical distances are smaller for the graph of

$$y_3 = 16x + 242.$$

We see numerically from the sum of squared differences calculations, as well as visually, from Figures 1.47 and 1.48, that the line

$$y_3 = 16x + 242$$

is the line that seems to better model the data when given a choice of the lines y_2 and y_3. However, is this line the best-fit line for these data? Remember that the best-fit line must have the smallest sum of squared differences of *all* lines that can be drawn through the data! We will see that the line which is the best fit for these data points, rounded off to three decimal places, is

$$y = 15.910x + 242.758.$$

The sum of squared differences for the regression line is approximately 1550.46, which is slightly smaller than 1558.

Development of the formulas that give the best-fit line for a set of data is beyond the scope of this text, but graphing calculators, computer programs, and spreadsheets

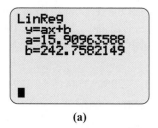

(a)

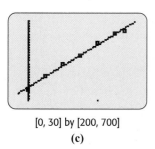

(b)

[0, 30] by [200, 700]

(c)

FIGURE 1.49

have built-in formulas or programs that give the equation of the best-fit line. That is, we can use technology to find the best linear model for the data. The calculation of the sum of squared differences for the dental employees data was given only to illustrate what the best-fit line means. You will not be asked to calculate, nor do we use from this point on, the value of the sum of squared differences in this text.

We illustrate the use of technology to find the regression line by returning to the data of Example 1. To find the equation that gives the number of employees in dental offices as a function of the years past 1970, we use the following steps:

MODELING DATA

Step 1: Enter the data into lists of a graphing utility.

Step 2: Create a scatter plot of the data to see if a linear model is reasonable. The data should appear to follow a linear pattern with no distinct curvature.

Step 3: Use the graphing utility to obtain the linear equation that is the best fit for the data. (Figure 1.49(a) shows the equation for Example 1, which can be approximated by $y = 15.910x + 242.758$.)

Step 4: Graph the linear function (unrounded) and the data points on the same graph to see how well the function describes the data. (The equation entered into an equation editor and the graph of the data and the best-fit line for Example 1 are shown in Figures 1.49(b) and 1.49(c), respectively.)

Step 5: Report the equation and/or numerical results in a way that makes sense in the context of the problem, with the appropriate units, and with the variables identified.

Figure 1.49(c) shows that

$$y = 15.910x + 242.758$$

models the number of full and part-time employees in dentists' offices and clinics, where y is in thousands of employees and x is the number of years past 1970.

The screens for the graphing utility that you use may vary slightly from those given in this text. Also, the regression line you obtain is dependent on your particular technology and may have some decimal places slightly different than those shown here.

When modeling a set of data, it is important to be careful when rounding coefficients in equations and when rounding during calculations. We will use the following guidelines in this text.

TECHNOLOGY NOTE

After a model for a data set has been found, it can be rounded for reporting purposes. However, do not use a rounded model in calculations, and do not round answers during the calculation process unless otherwise specified. When the model is used to find numerical answers, the answers should be rounded in a way that agrees with the context of the problem, and with no more accuracy than the original output data.

EXAMPLE 2   Carbon Monoxide Emissions

The amount of carbon monoxide emissions in the United States for 1986 through 1995 is shown in Table 1.22.

TABLE 1.22

Years	Carbon Monoxide Emission (tons)	Years	Carbon Monoxide Emission (tons)
1986	109,199	1991	93,376
1987	108,012	1992	94,043
1988	115,849	1993	94,133
1989	103,144	1994	98,779
1990	100,650	1995	92,099

(Source: Environmental Protection Agency)

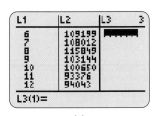

(a)

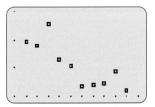

[5, 16] by [90,000, 120,000]

(b)

FIGURE 1.50

Align the data given in the table so that x represents the number of years after 1980, and

a. Create a scatter plot of the data, where x is the number of years after 1980.

b. Create a linear equation that is the best fit for the data, where y is in tons and x is the number of years after 1980. Write the linear model.

c. Graph the equation of the linear model on the same graph with the scatter plot and discuss how well the model fits the data.

d. Compare the changes in the yearly emissions and the slope of the best-fit line.

Solution

a. Align the data with $x = 0$ representing 1980, so that x is the number of years after 1980. Using y as the corresponding carbon monoxide emission in tons, from Table 1.22, enter the data in the lists of a graphing utility. Figure 1.50(a) shows a partial list of the data (the first seven points). The scatter plot of all the data points from Table 1.22 is shown in Figure 1.50(b).

b. The points in the scatter plot in Figure 1.50(b) do not exhibit a nice linear pattern, but the downward trend and no apparent curvature indicate that a line could possibly be used to model the data. The equation of the line that is the best fit for the data points is

$$y = -2192.0485x + 123{,}944.9091$$

(rounding off the decimals to four places). Remember that a *model* gives not only the equation but also a description of the variables and their units of measure. A linear *model* for the data in Table 1.22 is

$$y = -2192.0485x + 123{,}944.9091$$

tons of carbon monoxide emission in the United States between 1986 and 1995, inclusive, where x is the number of years after 1980.

c. Enter the unrounded equation into the equation editor and graph the model along with the data points to observe that the line follows the general downward trend indicated by the data. (See Figures 1.51(a) and 1.51(b).) Note that not all the points lie

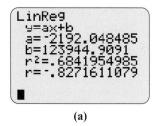

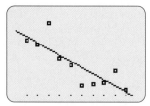

(a)

[5, 16] by [90,000, 120,000]

(b)

FIGURE 1.51

on the graph of the equation, even though this is the line that is the best fit for the data. If carbon monoxide emissions had decreased by exactly the same amount each year, then every data point would have been on the regression line.

d. Overall, the carbon monoxide emissions are decreasing, which corresponds to the fact that the slope of the regression line is negative. The slope indicates that between 1986 and 1995, the average rate of decrease in emissions is about 2192 tons per year, which seems reasonable for the pattern indicated by the data. However, because not all the data points lie on the line, the change from one point to another does not agree exactly with the slope. For example, from 1986 to 1987 the decrease was 1187 tons. ■

Modeling with Linear Functions

We sometimes use linear equations to model applications even though some points on the line do not fit the application. For example, suppose a yogurt shop charges $2 for a plain cone and $0.75 for each additional topping, with up to 5 toppings. The total price of the cone, y, is determined by the number of toppings, x. The relationship between these variables can be expressed by a table with x-values from 0 to 5 or the graph (scatter plot) of this **discrete function** (with a finite number of inputs) shown in Figure 1.52(a). For ease of discussion, we can also model this application with the linear function $y = 2 + 0.75x$, with no restrictions on the domain, because the points from the discrete function fit on the graph of this **continuous function**. We use the term **continuous function** to describe a function whose inputs can be any real number or any real number between two specified values (informally, a continuous function is a function whose graph can be drawn without lifting pen from paper). Figure 1.52(b) shows the graph of $y = 2 + 0.75x$, which is the continuous model of this discrete application. In the context of this application, x and y only make sense when they are both nonnegative, so we draw the graph in the first quadrant.

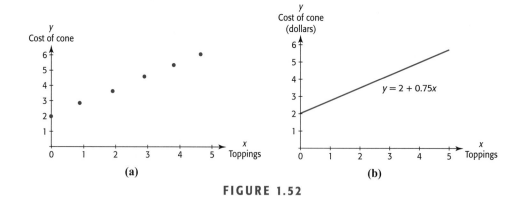

(a)

(b)

FIGURE 1.52

Applying Models

Once a model is developed, it can be used to answer questions about the data. For instance, if we assume that the model developed in Example 2 is valid for years beyond

1995, the carbon monoxide emissions level for 1997 can be predicted by substituting $x = 17$ into the model. Evaluating the unrounded model* gives a predicted emission level of approximately 86,680 tons. This value is an **extrapolation** from the model, because it comes from an input outside the given data points. Note that extrapolated outputs may not be realistic and that predictions are less reliable as inputs get farther from the given data points. Caution must always accompany an extrapolation from a model.

Using the model to estimate a value between two values given in the data is called **interpolation**. If a continuous function is used to model data in an application that only has a discrete interpretation, then we cannot use the model to estimate outputs for every real number input. For example, the number of shirts sold each day at a large department store as a function of the day on which the shirts are sold might be modeled by an equation whose graph is continuous, but a noninteger number input does not make sense in this situation. Only whole number inputs make sense for the days on which the shirts are sold. Nevertheless, in situations like this, we can use continuous models to solve problems, but then give a discrete interpretation to the solution.

Another question arises whenever data that involves time is used. If we label the x-coordinate of a point on the input axis as 1999, what time during 1999 do we mean? Does 1999 refer to the beginning of the year, the middle of the year, its end, or some other point in time? Because most data from which the functions in applications are derived represent end-of-year totals, we make the following convention when modeling, unless otherwise specified.

> **A point on the input axis indicating a time refers to the end of the time period.**

For instance, the point representing the year 1999 means "at the end of 1999." Notice that this instant in time also represents the beginning of the year 2000. If, for example, data is aligned with x equal to the number of years past 1990, then any x-value greater than 9 and less than or equal to 10 represents some time in the year 2000. If a point represents anything other than the end of the period, this information will be clearly indicated.

EXAMPLE 3   U.S. Population

The total population of the United States for selected years from 1950 to 1995 is shown in Table 1.23, with the population given in thousands.

a. Align the data to represent the number of years after 1950, and draw a scatter plot of the data.

b. Write the best-fit linear model for the data.

c. Use the model found in part (b) to estimate the population in 1982 and in 2000.

d. Graph the linear function and the data points on the same graph, and discuss how well the function models the data.

---

*Even though we write rounded models found from data given in this text, all calculations are done using the unrounded model found by the graphing utility, unless otherwise specified.

TABLE 1.23

Years	(Population in thousands)
1950	152,271
1955	165,931
1960	180,671
1965	194,303
1970	205,052
1975	215,973
1980	227,726
1985	238,466
1990	249,948
1995	263,044

(Source: U.S. Census Bureau)

Solution

a. The aligned data has $x = 0$ representing 1950, $x = 5$ representing 1955, and so forth, so that $x = 45$ represents 1995. Figure 1.53(a) shows the first seven entries. The scatter plot of the data is shown in Figure 1.53(b). The coordinates of the first two points in the scatter plot are (0, 152,271) and (5, 165,931).

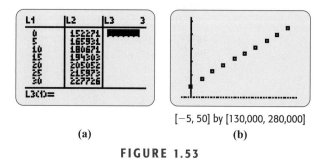

(a)   [−5, 50] by [130,000, 280,000]
                       (b)

FIGURE 1.53

b. The equation of the best-fit line is found by using linear regression in the graphing utility. Rounding off the decimals to three places, the linear model for the U.S. population is

$$y = 2406.353x + 155,195.564 \text{ thousands}$$

where x is the number of years after 1950. The model is valid between (the end of the year) 1950 and (the end of the year) 1995.

c. Entering 32 to represent 1982 gives the value 232,199, so the model estimates that the population was 232,199 thousand, or 232,199,000 in 1982. The model predicts that the population in 2000, with $x = 50$, to be 275,513 thousand, or 275,513,000. The actual U.S. population was 281,421,996 in the 2000 census, so the prediction from the model is within 2% of the actual data.

d. Enter the complete (unrounded) equation in the graphing utility (Figure 1.54(a)) and graph it along with the scatter plot. As seen in Figure 1.54(b), the graph of the best-fit line is very close to the data points. However, the points do not all fall on the line because the U.S. population did not increase by exactly the same amount each year.

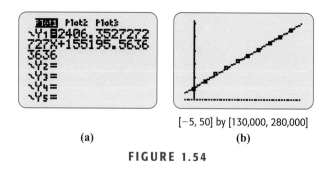

(a)   [−5, 50] by [130,000, 280,000]
                       (b)

FIGURE 1.54   ■

Spreadsheet Solution

We can also use spreadsheets to fit the linear function that is the best fit for the data. Table 1.24 shows the Excel spreadsheet for the data of Example 3. Selecting the cells containing the data, using **Chart Wizard** to get the scatter plot of the data, and selecting **Add Trendline** gives the equation of the linear function that is the best fit for the data, along with the scatter plot and the graph of the best-fitting line. The equation and graph of the line are shown in Figure 1.55.

TABLE 1.24

	A	B
1	Year x	Population (thousands) y
2	0	152,271
3	5	165,931
4	10	180,671
5	15	194,303
6	20	205,052
7	25	215,973
8	30	227,726
9	35	238,466
10	40	249,948
11	45	263,044

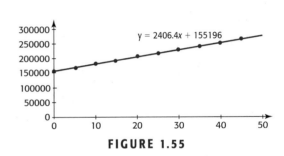

$y = 2406.4x + 155196$

FIGURE 1.55

Goodness of Fit

Because the rates of change are constant for linear functions, it is appropriate to model a set of data with a linear function if, when successive inputs differ by the same amount, the change in successive outputs from one data point to the next is constant or nearly constant. Let's consider the data in Table 1.25, which gives the conversion from selected Celsius temperatures to Fahrenheit temperatures. Because the Celsius temperatures (inputs) differ by a constant amount (10), we can compare the (first) differences of the outputs to see if they are constant. If we subtract successive Fahrenheit temperatures (see Table 1.26), we see that all the differences are the same constant (18), which indicates that the data can be modeled exactly by a linear function. (We know that this linear model exists; its equation is $F = 9/5 \, C + 32$.)

TABLE 1.25

Celsius degrees (°C)	−20	−10	0	10	20	30	40
Fahrenheit degrees (°F)	−4	14	32	50	68	86	104

TABLE 1.26

Outputs	
First differences of outputs	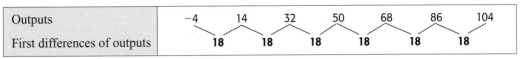

Consider again Example 3 where we modeled the U.S. population for selected years. Looking at Table 1.27, notice that for uniform inputs the first differences of the outputs appear to be relatively close to the same constant, especially compared to the size of the population. (If these differences were closer to a constant, the fit would be better.)

TABLE 1.27

Uniform Inputs in Years	Outputs (Population in thousands)	First Differences in Output (Difference in population)
1950	152,271	
1955	165,931	**13,660**
1960	180,671	**14,740**
1965	194,303	**13,632**
1970	205,052	**10,749**
1975	215,973	**10,921**
1980	227,726	**11,753**
1985	238,466	**10,740**
1990	249,948	**11,482**
1995	263,044	**13,096**

So how "good" is the fit of the linear model $y = 2406.353x + 155{,}195.564$ to the data in Example 3? Based on observation of the graph of the line and the data points on the same set of axes (see Figure 1.54(b) and Figure 1.55), it is reasonable to say that the regression line provides a very good fit, but not an exact fit, to the data.

CONSTANT FIRST DIFFERENCES

If the first differences of data outputs are constant (for uniform inputs), a linear model can be found that fits the data exactly. If the first differences are "nearly constant," a linear model can be found that is an approximate fit for the data.

The "goodness of fit" of a line to a set of data points can be observed from the graph of the line and the data points on the same set of axes, and it can be measured if your graphing utility computes the **correlation coefficient**. The correlation coefficient is a number r, $-1 \leq r \leq 1$, that measures the strength of the linear relationship that exists between the two variables. The closer that $|r|$ is to 1, the more closely the data points fit on the linear regression line. (There is no linear relationship between the two variables if $r = 0$.) Positive values of r indicate that the output variable increases as the input variable increases, and negative values of r indicate that the output variable decreases as the input variable increases. But the *strength* of the relationship is indicated by how close $|r|$ is to 1. For the data of Example 3, computing the correlation coefficient gives $r = .9988$, which means that the linear relationship is strong and that the linear model is an excellent fit for the data. When a calculator feature or computer program is used to fit a linear model to data, the resulting equation can be considered a best fit for the data. As we will see later in this text, other mathematical models may be better fits to some sets of data, especially if the first differences of the outputs are not close to being constant.

1.6 SKILLS CHECK

Report models to three decimal places unless otherwise specified. Use unrounded models to graph and calculate unless otherwise specified.

Discuss whether the data shown in the scatter plots in the figures for Exercises 1 and 2 should be modeled by a linear function.

1.

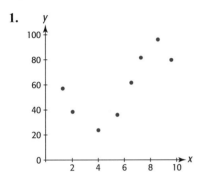

2.

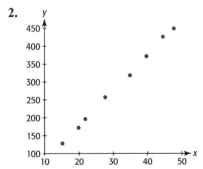

Create a scatter plot for each of the sets of data in Exercises 3 and 4.

3.

x	y
1	4
3	7
5	10
7	13
9	16

4.

x	y
1	1
2	3
3	6
5	1
7	9
9	2
12	6

5. Can the scatter plot in Exercise 3 be fit exactly or only approximately by a linear function? How do you know?

6. Can the scatter plot in Exercise 4 be fit exactly or only approximately by a linear function? How do you know?

7. Find the linear function that is the best fit for the data in Exercise 3.

8. Find the linear function that is the best fit for the data in Exercise 4.

Use the data in the following table for Exercises 9–12.

x	5	8	11	14	17	20
y	7	14	20	28	36	43

9. Construct a scatter plot of the data in the table.

10. Determine if the points plotted in Exercise 9 appear to lie near some line.

11. Create the linear equation that models the data in the table.

12. Use the function $y = f(x)$ created in Exercise 11 to evaluate $f(3)$ and $f(5)$.

Use the data in the table for Exercises 13–16.

x	2	5	8	9	10	12	16
y	5	10	14	16	18	21	27

13. Construct a scatter plot of the data in the table.

14. Determine if the points plotted in Exercise 13 appear to lie near some line.

15. Write the linear equation that best fits the data in the table.

16. Use the rounded function $y = f(x)$ that was reported in Exercise 15 to evaluate $f(3)$ and $f(5)$.

17. Determine which of the equations, $y = -2x + 8$ or $y = -1.5x + 8$, is the better fit for the data points $(0, 8)$, $(1, 6)$, $(2, 5)$, $(3, 3)$.

18. Without graphing, determine which of the following data sets are exactly linear, approximately linear, or nonlinear.

a.

x	1	2	3	4	5
y	5	8	11	14	17

b.

x	1	2	3	4	5
y	2	5	10	17	26

c.

x	1	2	3	4	5
y	6	7	12	14	18

19. Why can't first differences be used to tell if the following data is linear? $(1, 3)$, $(4, 5)$, $(5, 7)$, $(7, 9)$

1.6 EXERCISES

Report models to three decimal places unless otherwise specified. Use unrounded models to graph and calculate unless otherwise specified.

1. *Working Age* The scatter plot in the figure below projects the ratio of the working-age population to the elderly.
   a. Does the scatter plot in the figure represent discrete or continuous data?
   b. Would a linear function be a good model for the data points shown in the figure? Explain.
   c. Could the data shown for the years beyond 2010 be modeled by a linear function?

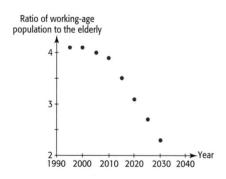

(Source: *Newsweek*, December 1999)

2. *Women in the Workforce* The number of women in the workforce for selected years from 1890 to 1990 is shown in the figures below.

   a. Does the scatter plot in the figure define the number of working women as a discrete or continuous function of the year?
   b. Does the graph of $y = W(x)$ shown in the figure define the number of working women as a discrete or continuous function of the year?
   c. Would the data in the scatter plot be better modeled by a linear function rather than by the nonlinear function $y = W(x)$? Why or why not?
   (Source: *Newsweek*, December 1999)

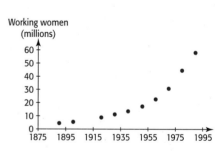

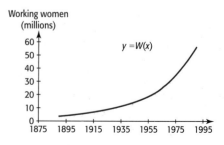

3. *Life Insurance* The monthly rates for a $100,000 life insurance policy for males aged 27–32 are shown in the table below.
   a. Is the data in this table discrete or continuous?
   b. Write the best-fit linear model for the data in the table.

Age (years)	27	28	29
Premium (dollars per month)	11.81	11.81	11.81

Age (years)	30	31	32
Premium (dollars per month)	11.81	11.81	11.81

4. *Lettuce Prices* The average price $p(x)$ of iceberg lettuce, in cents per pound, in U.S. cities between February and June during 1999 is a function of the month x in which the average price was determined. Does the point $(5.75, p(5.75))$ have a valid meaning in this situation? If so, interpret the coordinates of the point. If not, explain why not.

5. *Employees* The function $y = 15.910x + 242.758$ models the number of full and part-time employees in dentists offices and clinics, where y is in thousands of employees and x is the number of years after 1970. This model was found using data for the years between 1970 and 1998.
   (Source: U. S. Bureau of the Census)
   a. What does this model estimate as the number of employees in 1993? Is this interpolation or extrapolation?
   b. What does this model estimate as the number of employees in 2000? Is this interpolation or extrapolation?

6. *Internet Recruiting* The percentage of Fortune Global 500 firms that actively recruited workers on the Internet between 1998 and 2000 can be modeled by $P(x) = 26.5x - 194.5$ percent where x is the number of years after 1990.
   (Source: *Time*, August 16, 1999)
   a. Is this model discrete or continuous?
   b. Does the input $x = 9$ have a valid meaning in this situation? If so, interpret $P(9)$. If not, explain why not.
   c. Does the input $x = 9.4$ have a valid meaning in this application? If so, interpret $P(9.4)$. If not, explain why not.

7. *Smoking Cessation* The data in the table below gives the percent of adults over age 20 who once smoked and quit smoking between 1965 and 1990.
   a. Using an input equal to the number of years after 1960, write a linear model for the data.
   b. In what year does the model indicate that the output is 39 percent?
   c. Discuss the reliability of the answer to part (b).

Number of years after 1960	Adults who have quit smoking (percent)
5	29.8
6	29.3
10	35.2
15	36
16	37
17	36.6
18	38.3
19	39
20	38.9
23	41.8
25	45
27	44.9
30	48.4

(Source: Indiana Tobacco Control Center, U.S. Centers for Disease Control)

8. *Smoking Cessation* The table on page 94 gives the percent of people over age 20 who once smoked and quit smoking between 1965 and 1990.
   a. Align the data to the number of years after 1965, and fit a linear model to these data.
   b. Show the graph of the data and the equation on the same set of axes.
   c. Compare the model in part (a) with the one developed in Exercise 7. What parts of the equations are the same and what parts are different?

Year	Adults who have quit smoking (percent)
1965	29.8
1966	29.3
1970	35.2
1975	36
1976	37
1977	36.6
1978	38.3
1979	39
1980	38.9
1983	41.8
1985	45
1987	44.9
1990	48.4

(Source: Indiana Tobacco Control Center,
U.S. Centers for Disease Control)

9. *Postal Rates* The table below gives the local postal
rates as a function of the weight of the printed matter
that is mailed.
 a. Write the linear equation that models the data.
 b. Compare the outputs from the equation with the
data outputs for several different weights. How
well does the model fit the data?

Weight (lb)	Postal rate (dollars)	Weight (lb)	Postal rate (dollars)
1.5	1.14	9	1.43
2	1.16	10	1.47
3	1.20	11	1.51
4	1.24	12	1.55
5	1.28	13	1.59
6	1.31	14	1.63
7	1.36	15	1.67
8	1.39		

(Source: 1999 *World Almanac*)

10. *SAT Scores* The average math SAT scores in Beaufort
County, South Carolina, are given in the table below.
Write the equation of the line that is the best fit for
these data, aligning the data so that $x = 0$ in 1990.
Compare the outputs from the equation with the original data for each of the years 1994 to 1999 and determine the year in which the model output is closest to
the data value.

Year	Average Math SAT Score
1994	472
1995	464
1996	470
1997	471
1998	473
1999	470

(Source: Beaufort County School Report
1998–1999, *The Island Packet*)

11. *Postal Rates* Consider the data and the rounded function that models the data in Exercise 9.
 a. What does the model predict the rate to be for 16
lbs. of bound printed matter?
 b. If the postal rate is $1.55, what is the estimated
weight?
 c. What does the model indicate that the increase in
postal rate will be for each increase of 1 lb in
weight?
 d. What does the data give as the increase in postal
rate for each pound increase in weight for weights
from 9 lb to 14 lb?

12. *Jail Population* The figure gives the average daily
number of inmates in the Beaufort County, South
Carolina, Detention Center.
 a. Align the data to the number of years after 1990,
and create a linear equation that models the data.
 b. Assuming that the model is still appropriate, use
the linear function to predict the population in the
Beaufort County jail in 2005.
 c. In what year does the model indicate that the average daily number of inmates is 115?

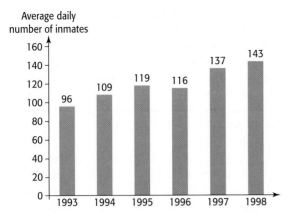

Average daily number of inmates

(Source: Beaufort County Detention Center, *The Island Packet*, October 1999)

13. *Marriage Rate* The marriage rate (per 1000 unmarried women) is given by the table below for the years 1960 to 1997.
    a. Write the linear equation that models the marriage rate as a function of years after 1950.
    b. What does the model indicate the marriage rate to be in 1985?
    c. For what year does the model indicate that the rate fell below 50 per 1000?
    d. Discuss the reliability of your answer to part (c).

Year	Marriage Rate (marriages per 1000 unmarried women)
1960	73.5
1970	76.5
1980	61.4
1990	54.5
1991	54.2
1992	53.3
1993	52.3
1994	51.5
1995	50.8
1996	49.7
1997	49.4

(Source: Index of Leading Cultural Indicators)

14. *Volunteers in Medicine* Retired physicians and nurses donate services to treat patients in the Volunteers in Medicine Clinic on Hilton Head Island. The numbers treated free of charge during each of the years 1995 to 1999 are given in the table below.
    a. Write the equation of the line that is the best fit for the data, aligning the input to be the number of years after 1990.
    b. What does the model indicate as the average rate of increase of patients per year from 1995 to 1999?

Year	Patients Treated
1995	3106
1996	3417
1997	4133
1998	4246
1999	4400

(Source: "Volunteers in Medicine," *The Island Packet*, November 21, 1999)

15. *Education* The numbers of children up to 21 years of age that are served by educational programs for the disabled for the years between 1988 and 1996 are given in the table below.
    a. Write a linear equation that models these data, aligning the input to be the number of years after 1980 and using as output the number of thousands of children served.
    b. What does the model indicate is the average rate of increase in the number of children served per year from 1988 to 1996?

Year	Disabled Children Served (in thousands)
1988	4446
1989	4544
1990	4641
1991	4771
1992	4949
1993	5125
1994	5309
1995	5378
1996	5573

(Source: Office of Special Education and Rehabilitative Services of the U.S. Department of Education)

16. *Giving at IBM* The amount of individual donations to nonprofit organizations and educational institutions by employees at IBM, for the years 1994 to 1998, is shown in the table below.

    a. Write the linear equation that is the best fit for these data, with x equal to the number of years after 1990.

    b. If this model is valid after 1998, for what year does the model predict that giving will reach $50 million?

Year	Donations (million dollars)
1994	35.6
1995	35.9
1996	36.7
1997	39.2
1998	43.9

(Source: 1998 IBM Annual Report)

17. *Salaries* The table below gives the median annual income of female workers with 4 years of college for the years 1979 to 1989.

Year	Salary (dollars)	Year	Salary (dollars)
1979	13,441	1985	21,589
1980	15,143	1986	22,412
1981	16,322	1987	23,399
1982	17,405	1988	25,187
1983	18,452	1989	26,709
1984	20,257		

(Source: U.S. Bureau of Census)

    a. Align the data to represent the number of years after 1970, and draw a scatter plot of the data.

    b. Write the best-fit linear model for the data.

    c. Graph the linear function and the data points on the same graph, and discuss how well the function models the data.

    d. Classify each of the following as discrete or continuous.

      i) The data in the table.

      ii) The scatter plot in part (a).

      iii) The model in part (b).

18. *Gross National Product* The data in the table below and the graph in the figure below describe the gross national product y (the value of all goods and services) for the United States for the years 1980 through 1997.

    a. Write the linear equation that models the gross national product as a function of the number of years after 1980. Round to two decimal places.

    b. Is the model with equation in part (a) discrete or continuous?

    c. According to the rounded model, what is the annual increase in the gross national product?

    d. What does the model estimate the gross national product to be in 1994? Is this extrapolation or interpolation?

Year	Gross National Product (billions of dollars)
1980	2742.1
1985	4053.6
1988	4908.2
1990	5666.4
1992	6255.5
1995	7270.6
1997	8060.1

(Source: *Statistical Abstract of the U.S., 1995, 1998*)

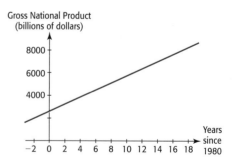

19. *World Cigarette Production* The table on page 97 gives the per capita world production of cigarettes for selected years from 1950 to 1992.

    a. Find the equation of the line that is the best fit for this data. Align the data so that $x = 0$ in 1950.

    b. Is the model with equation in part (a) discrete or continuous?

    c. Does the model indicate that world cigarette production was increasing or decreasing between 1950 and 1992?

d. Do you think the model with equation from part (a) would apply for the years 1993 to 2003? Why or why not?

Year	World Production (cigarettes per person)
1950	660
1955	691
1960	708
1965	767
1970	840
1975	916
1980	985
1985	1000
1990	1017
1992	984

(Source: The National Clearinghouse on Tobacco and Health)

20. *Consumer Spending* The consumer spending per person per year for television subscription video services is given in the table below.

  a. Find the equation of the line that gives the per capita yearly consumer spending as a function of the number of years after 1990.

  b. Use the equation to determine the year in which the per capita yearly consumer spending for these television services first exceeds $250.

  c. What assumption must be made for the answer to part *b* to be considered valid?

Year	Consumer Spending (dollars per person per year)
1992	101.39
1993	109.02
1994	110.89
1995	125.54
1996	140.37
1997	155.70
1998	168.75
1999	178.23
2000	187.60
2001	196.62

(Source: *Statistical Abstract of the U.S.*, 1998)

21. *Internet Brokerage Accounts* The number of Internet brokerage accounts is given in the table below.

  a. Write the linear equation that models the number of Internet brokerage accounts as a function of the number of years after 1990.

  b. According to the model, what is the annual increase in the number of accounts?

  c. Use the model to estimate when the number of accounts will be 20 million.

  d. What happened in 2001 that may make this model invalid for years after 2001?

Year	Internet Brokerage Accounts (millions of accounts)
1996	1.5
1997	4.1
1998	7.1
1999	10.5
2000	14.0
2001	18.0

(Source: *Time*, May 31, 1999)

22. *Prison Sentences* The National Center for Policy Analysis calculates expected prison time by multiplying the probabilities of being arrested, of being prosecuted, and of being convicted by the length of sentence if convicted. The expected time served (in days) for each serious crime committed is shown in the table below.

  a. Use the data to create a linear equation to model expected prison time in days for a serious crime as a function of the number of years after 1960.

  b. Use the model to find the expected prison time in 1975.

Year	Expected Prison Time (days)
1970	10.1
1980	10.6
1985	13.2
1990	18.0
1992	18.5
1993	17.2
1994	19.5
1995	20.2
1996	21.7

(Source: National Center for Policy Analysis)

1.7 Models in Business and Economics

Key Concepts

- Total revenue
- Total cost
- Profit
- Marginal profit
- Marginal revenue
- Marginal cost
- Break-even

Ace Electric produces an electric component that sells for $9.95. Producing and selling this product involves a monthly fixed cost of $1999.50, and the cost of producing each component is $3.50. To find the number of units that Ace Electric needs to produce each month to break even (that is, have its revenue equal its cost), we can use the given information to model the revenue, cost, and profit functions for this product, and then determine the number of units that gives revenue equal to the cost, or equivalently, a profit of $0.

Some real-life situations can be modeled by finding functions to fit data points, and some can be modeled by using factual information about the situation and known formulas. Because a number of business applications occur frequently, functions have been developed to model them. Examples include models for total revenue, total cost, and profit. Other models are used so frequently that they are written as formulas. Examples of such models include the formulas for the future value of an investment at simple interest and at compound interest, future value of annuities, and payments for amortization of a loan.

Revenue, Cost, and Profit

If a company sells x units of a product for p dollars per unit, then the **total revenue** for this product can be modeled by the linear function

$$R(x) = px.$$

For example, if the Telstar computer is sold for $899, then the revenue for x computers can be modeled by the function

$$R(x) = 899x \text{ dollars.}$$

Note that this is a linear function; its graph is shown in Figure 1.56.

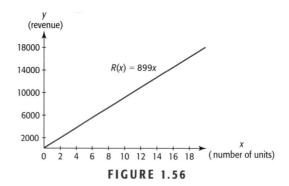

FIGURE 1.56

The **total cost** of producing and selling a product involves two parts, the **fixed costs** and the **variable costs**. **Fixed costs** include such things as rent, utilities, and equipment, and remain constant regardless of the number of units produced. **Variable costs** are those directly related to the number of units produced. Thus the cost (frequently called the total cost) is found by using the formula

$$\text{Cost} = \text{variable costs} + \text{fixed costs.}$$

For example, if the monthly cost of producing the Telstar computer is $249 per unit, with a fixed cost of $13,000, then the cost of producing x computers is

$$C(x) = 249x + 13,000 \text{ dollars.}$$

This also is a linear function, and its graph is shown in Figure 1.57.

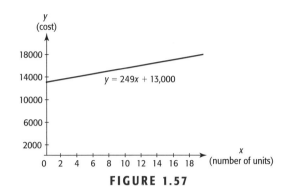

FIGURE 1.57

> The **profit** that a company makes on its product is found by subtracting the total cost of production from the total revenue for the product:
>
> $$P(x) = R(x) - C(x)$$

Thus the profit for x Telstar computers is given by

$$P(x) = 899x - (249x + 13,000) \quad \text{or} \quad P(x) = 650x - 13,000 \text{ dollars.}$$

Note that each of these three quantities, total revenue, total cost, and profit, is a function of the number of units that are produced and sold.

EXAMPLE 1   Revenue, Cost, and Profit
- -

Ace Electric produces an electric component that sells for $9.95. Producing and selling this product involves a monthly fixed cost of $1999.50, and the cost of producing each component is $3.50.

a. Write the equations that model total revenue and total cost as functions of the units produced for this electric component during a month.

b. Write the equation that models the profit as a function of the units produced for this component during a month.

c. Find the total revenue, total cost, and profit for Ace Electric if 3000 components are produced and sold in a month.

Solution

a. When x components are produced and sold, the total revenue from the sale of x units is 9.95 times x, so it can be modeled by $R(x) = 9.95x$ dollars. The monthly total cost can be modeled by adding the variable cost, $3.50x$, to the fixed cost, 1999.50:

$$C(x) = 1999.50 + 3.50x \text{ dollars.}$$

b. The profit for the production and sale of this electric component can be modeled by

$$P(x) = R(x) - C(x) = 9.95x - (1999.50 + 3.50x) \text{ dollars, or by}$$

$$P(x) = 6.45x - 1999.50 \text{ dollars}$$

c. The total revenue, total cost, and profit from the sale of 3000 components are, respectively,

$$R(3000) = 9.95(3000) = 29,850 \text{ dollars}$$

$$C(3000) = 1999.50 + 3.50(3000) = 12,499.50 \text{ dollars}$$

$$P(3000) = 6.45(3000) - 1999.50 = 17,350.50 \text{ dollars}$$

The graphs of the revenue, cost, and profit functions are shown in Figure 1.58.

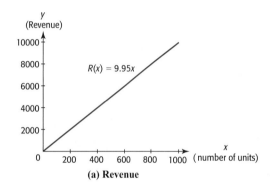

(a) Revenue

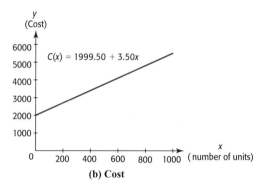

(b) Cost

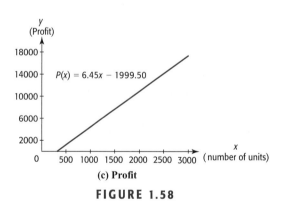

(c) Profit

FIGURE 1.58   ■

Break-Even Analysis

A company is said to **break even** from the production and sale of a product if the total revenue equals the total cost; that is, if $R(x) = C(x)$. Because profit $P(x) = R(x) - C(x)$, we can also say that the company breaks even if the profit for the product is zero.

EXAMPLE 2   Break-Even
- -

Determine the number of units that Ace Electric (for the situation described in Example 1) must produce and sell to break even by solving

a. $R(x) = C(x)$

b. $P(x) = 0$

Solution

a. For the production and sale of the components of Example 1, $R(x) = 9.95x$ and $C(x) = 1999.50 + 3.50x$, so we set these equal to obtain

$$9.95x = 1999.50 + 3.50x$$
$$6.45x = 1999.50$$
$$x = 310$$

This means that the company will break even if 310 components are produced and sold.

b. The profit function is $P(x) = 6.45x - 1999.50$. Solving $P(x) = 0$ gives

$$6.45x - 1999.50 = 0$$
$$6.45x = 1999.50$$
$$x = 310$$

Thus, we reach the same conclusion as we did in part (a): The company will have to produce 310 units in order to break even. (Evaluating $P(310)$ shows that producing 310 units would result in a profit of $0.) ∎

Marginal Cost, Revenue, and Profit

Whenever the total revenue, total cost, and profit are modeled by linear functions, the slopes of their graphs give the constant rate of change of the functions. In the case of these business functions, the constant rates of change are called **marginals**. The slope of the graph of the profit function represents the rate of change of the profit with respect to the number of units sold. This rate of change is called the **marginal profit** (denoted $\overline{MP}$), and it gives the profit from the sale of one additional unit. For the situation given in Example 1, with profit function $P(x) = 6.45x - 1999.50$, the marginal profit is 6.45 dollars per component.

The constant rate of change of the cost function with respect to the number of units sold (which equals the slope of the graph of the cost function) is called the **marginal cost** ($\overline{MC}$). The total cost for the product in Example 1 is $C(x) = 1999.50 + 3.50x$, and the marginal cost is 3.50 dollars per component. Similarly, the rate of change (and the slope of the graph) of the revenue function with respect to the number of units sold is the **marginal revenue** ($\overline{MR}$). The total revenue for the product in Example 1 is $R(x) = 9.95x$, and the marginal revenue is 9.95 dollars per component.

EXAMPLE 3 **Marginal Cost**

- -

Suppose the function that models the weekly total cost (in dollars) for the production of 19-inch televisions is $C(x) = 58.20x + 4850$, where x is the number of sets produced.

a. What is the marginal cost for this product?

b. What does it mean?

Solution

a. The cost function has the form

$$C(x) = mx + b, \text{ with slope } m = 58.20,$$

so the marginal cost is $\overline{MC} = 58.20$.

b. Because the marginal cost is the constant rate of change of this linear function, production of each additional television set will cost $58.20, regardless of the level of production.   ■

Note that when total cost functions are linear, the marginal cost is the same as the variable cost. This is not the case if the functions are not linear, as we will see later.

EXAMPLE 4   Modeling Profit

Suppose that the weekly profit for a product is a linear function of the number of units produced and sold, and marginal profit is $80. If the profit is $1310 when 32 units are produced and sold, write the function that models the profit.

Solution

The marginal profit is the rate of change of a linear function, so it is the slope of its graph, that is, $m = 80$. If we let x represent the number of units produced and sold and let P represent the profit, then the point (32, 1310) lies on the graph of the weekly profit function. Using the point-slope form of a linear equation, we can write the equation:

$$P - 1310 = 80(x - 32).$$

Solving the equation for P gives

$$P = 80x - 1250.$$

Thus the weekly profit is $P = 80x - 1250$ dollars, where x is the number of units produced and sold.   ■

EXAMPLE 5   Modeling Revenue

Suppose that the total revenue for a product is a linear function of the number of units produced and sold.

a. If the revenue is $437 when 23 units are produced and sold, and the revenue is $627 when 33 units are produced and sold, write the equation that models the total revenue.

b. Interpret the function in terms of the price.

Solution

a. The information that revenue $R = 437$ when $x = 23$ and $R = 627$ when $x = 33$ tells us that the points (23, 437) and (33, 627) lie on the graph of the revenue function. Then the slope of the graph is

$$m = \frac{627 - 437}{33 - 23} = \frac{190}{10} = 19.$$

Using this slope and one of the points, (23, 437), gives the equation of the revenue function:

$$R - 437 = 19(x - 23)$$
$$R - 437 = 19x - 437$$
$$R = 19x$$

Note that using the other point would result in the same equation.

b. Because revenue can be found by multiplying the price times the number of units sold, and because x represents the number of units sold, the price is 19 dollars per unit. ∎

Supply, Demand, and Market Equilibrium

The willingness of consumers to buy a product is called the demand for the product, and demand is almost always related to the price of the product. The **law of demand** states that the quantity of a product that is demanded will increase as the price decreases. Figure 1.59(a) shows the graph of a typical linear demand function. Note that although the quantity demanded is thought of as a function of price, economists have traditionally graphed demand functions with the quantity on the horizontal axis and the price on the vertical axis, and we will follow this tradition. Linear equations relating price p and quantity demanded q can be solved for either p or q, and we will have occasion to use the equations in both forms.

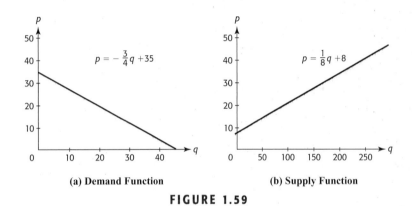

(a) Demand Function   (b) Supply Function

FIGURE 1.59

Just as a consumer's willingness to buy is related to price, a manufacturer's willingness to sell is related to price. The **law of supply** states that the quantity supplied for sale will increase as the price of a product increases. Figure 1.59(b) shows the graph of a typical linear supply function, with price placed on the vertical axis. Note that negative quantities and negative prices have no meaning, so graphs of supply and demand functions are restricted to the first quadrant.

EXAMPLE 6   Demand

An electronics store indicates that it will buy 100 stereo systems if the price is $140 per system and 60 if the price is $180 per system. Assuming that the demand for these systems is linear, write the equation that models the demand.

Solution

The information that is given tells us that the points (100, 140) and (60, 180) lie on the graph of the demand function, where the first coordinate is the number of units q that are demanded at price p, and the second coordinate is the price p. Then the slope of the graph is

$$m = \frac{180 - 140}{60 - 100} = \frac{40}{-40} = -1$$

The linear demand equation can be written using the point-slope form. Note that we use the point (100, 140) in the point-slope form, but that the point (60, 180) would give the same result:

$$p - 140 = -1(q - 100)$$
$$p = -q + 240$$

Thus the price is given by $p = -q + 240$, where q is the number of units demanded.

■

Market equilibrium is said to occur when the quantity of a commodity demanded is equal to the quantity supplied. If we graph the supply function and the demand function on the same axes with the same units, market equilibrium occurs at the point where the graphs intersect. The price at this point is called the **equilibrium price**, and the quantity at this point is called the **equilibrium quantity**. In Figure 1.60, the demand function is $p = -5q + 450$ and the supply function is $p = 2q + 170$. The equilibrium point is (40, 250).

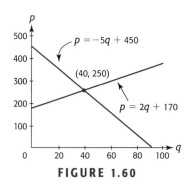

FIGURE 1.60

EXAMPLE 7   Market Equilibrium

Suppose the manufacturer of the stereo systems of Example 6 will deliver 100 systems if the price is $220 per system and 120 systems if the price is $280 per system.

a. Assuming that the supply for these systems is linear, write the equation that models the supply.

b. At what price will the quantity of stereo systems demanded equal the quantity supplied?

c. Check your answer by graphing both functions on the same axes.

Solution

a. The information that is given tells us that the points (100, 220) and (120, 280) lie on the graph of the supply function, where the first coordinate is the number of units q that will be supplied at price p, and the second coordinate is the price p. Then the slope of the graph is

$$m = \frac{280 - 220}{120 - 100} = \frac{60}{20} = 3$$

Thus the linear supply equation can be written using the point-slope form, with slope 3 and the point (100, 220):

$$p - 220 = 3(q - 100)$$
$$p = 3q - 80$$

Thus the price is given by $p = 3q - 80$, where q is the number of units that will be supplied.

b. Recall that the demand for this product, found in Example 6, is $p = -q + 240$. The equations that model the supply and demand functions are both solved for p:

Supply: $p = 3q - 80$
Demand: $p = -q + 240$

and the quantity supplied equals the quantity demanded where the prices are equal. Thus we can set the functions equal to each other and solve for q:

$$3q - 80 = -q + 240$$
$$4q = 320$$
$$q = 80$$

The quantity supplied equals the quantity demanded at $q = 80$, when the price is

$$3(80) - 80 = 160 \text{ dollars.}$$

c. The graphs of the two functions are shown in Figure 1.61. The point at which the two graphs intersect is (80, 160), which confirms that supply equals demand when the price is $160.

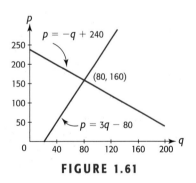

FIGURE 1.61

1.7 SKILLS CHECK

1. If $f(x) = 45x$ and $g(x) = 4x + 5600$, what is $f(x) - g(x)$?

2. Is the function $C(x) = 62x + 1800$ a linear function?

3. What is the rate of change of the function $C(x) = 56x + 89$?

4. Show that $R(100) - C(100) = 0$ if $R(x) = 79x$ and $C(x) = 9x + 7000$.

5. If $p_1 = 29x + 27$ and $p_2 = 5457 - x$, find the value of x for which $p_1 = p_2$.

1.7 EXERCISES

1. *Total Revenue* A brand of computer sells for $1297 per unit.
   a. Write the function that models the total revenue *R* if *x* units are sold.
   b. What will be the revenue if 300 computers are sold?
   c. What is the slope of the graph of the total revenue function?

2. *Total Revenue* A satellite system sells for $295 per unit.
   a. Write the function that models the total revenue R if *x* units are sold.
   b. What will be the revenue if 600 systems are sold?
   c. What is the *x*-intercept of the graph of the total revenue function?

3. *Total Cost* A manufacturer of computers has monthly fixed costs of $32,000 and variable costs of $432 per computer.
   a. Find the function that models the total monthly cost *C* of this system if *x* units are produced.
   b. What is the total cost of production of 300 computers in one month?
   c. What is the total cost if no units are produced in a month?

4. *Total cost* A manufacturer of a satellite systems has monthly fixed costs of $87,500 and variable costs of $87 per system.
   a. Find the function that models the total monthly cost *C* of this system, if *x* units are produced.
   b. What is the total cost of producing 600 satellite systems in one month?
   c. Where is the fixed cost located on the graph of the total cost function?

5. *Cost and Revenue* A couple decides to open a new business. They rent a building, buy equipment to manufacture beach balls, hire one person to sell the beach balls, and begin operations.
   a. List three items that contribute to the fixed cost.
   b. If the total cost results in a cost of $2.10 per beach ball and each ball is sold for $5.95, how much profit is made from the manufacture and sale of 350 balls?

6. *Cost and Revenue* The band booster's club for a high school decides to sell hot dogs and sodas at the homecoming game in order to help raise money for new uniforms.
   a. List three items that are included in the variable costs.
   b. If hot dogs sell for $1.50 each and sodas sell for $1.00 each, what is the club's revenue from the sale of 250 hot dogs and 300 sodas?

7. *Profit* AT&T Corporation reported these figures for the quarter ending in September 1999: revenue equal to $16,270.0 million and cost of goods sold (including fixed cost) equal to $943.0 million. What was AT&T's profit for the quarter ending September 1999?
   (Source: Hoover's Online Capsules)

8. *Profit* Office Depot Inc. reported a revenue of $2578.5 million and a profit of $650.0 million for the quarter ending in September 1999. What were Office Depot's total costs for the quarter ending September 1999?
   (Source: Hoover's Online Capsules)

9. *Profit* Suppose that the total weekly cost for the production and sale of a bicycle is $C(x) = 23x + 3420$ dollars and that the total revenue is given by $R(x) = 89x$ dollars, where *x* is the number of bicycles.
   a. Write the equation of the function that models the weekly profit from the production and sale of *x* bicycles.
   b. Graph this function on a viewing window with *x* between 0 and 200.
   c. What is the profit on the production and sale of 150 bicycles?
   d. What is the slope of the graph of the profit function?

10. *Profit* Suppose that the total monthly cost for the production and sale of television sets is $C(x) = 189x + 5460$ and that the total revenue is given by $R(x) = 988x$, where *x* is the number of televisions and $C(x)$ and $R(x)$ are in dollars.
   a. Write the equation of the function that models the weekly profit from the production and sale of *x* television sets.
   b. Graph this function on a viewing window with *x* between 0 and 200.
   c. What is the profit on the production and sale of 80 television sets?
   d. What is the *y*-intercept of the graph of the profit function, and what does it mean?

11. *Profit* A manufacturer of satellite systems has monthly fixed costs of $32,000 and variable costs of $432 per system, and it sells the systems for $592 per unit.
   a. Write the function that models the profit *P* from the production and sale of *x* units of the system.
   b. What is the profit if 600 satellite systems are produced and sold in one month?

c. At what rate does the profit grow as the number of units increases?

12. *Profit* A manufacturer of computers has monthly fixed costs of $87,500 and variable costs of $87 per computer, and it sells the computer for $295 per unit.
  a. Write the function that models the profit P from the production and sale of x computers.
  b. What is the profit if 700 computers are produced and sold in one month?
  c. What is the y-intercept of the graph of the profit function? What does it mean?

13. *Marginal Cost* Suppose that the monthly total cost for the manufacture of golf balls is $C(x) = 3450 + 0.56x$, where x is the number of balls produced each month.
  a. What is the slope of the graph of the total cost function?
  b. What is the marginal cost for the product?
  c. Interpret the marginal cost for this product.

14. *Marginal Cost* Suppose that the monthly total cost for the manufacture of 19-inch television sets is $C(x) = 2546 + 98x$, where x is the number of TVs produced each month.
  a. What is the slope of the graph of the total cost function?
  b. What is the marginal cost for the product?
  c. Interpret the marginal cost for this product.

15. *Marginal Revenue* Suppose that the monthly total revenue for the manufacture of golf balls is $R(x) = 1.60x$, where x is the number of balls sold each month.
  a. What is the slope of the graph of the total revenue function?
  b. What is the marginal revenue for the product?
  c. Interpret the marginal revenue for this product.

16. *Marginal Revenue* Suppose that the monthly total revenue for the manufacture of television sets is $R(x) = 198x$, where x is the number of TVs sold each month.
  a. What is the slope of the graph of the total revenue function?
  b. What is the marginal revenue for the product?
  c. Interpret the marginal revenue for this product.

17. *Profit* The total revenue for a product is given by $R(x) = 56x$ and the total cost is given by $C(x) = 5060 + 37x$, where x is the number of units produced and sold.
  a. Write the equation that describes the profit for this product.
  b. Find the marginal revenue, marginal cost, and marginal profit for the product.

18. *Profit* The total revenue for a product is given by $R(x) = 1832x$ and the total cost is given by $C(x) = 12,207 + 893x$, where x is the number of units produced and sold.
  a. Write the equation that describes the profit for this product.
  b. Find the marginal revenue, marginal cost, and marginal profit for the product.

19. *Break-Even* The total revenue for a product is given by $R(x) = 56x$ and the total cost is given by $C(x) = 5054 + 37x$, where x is the number of units produced and sold. Find the number of units that gives break-even.

20. *Break-Even* The total revenue for a product is given by $R(x) = 1832x$ and the total cost is given by $C(x) = 12,207 + 893x$, where x is the number of units produced and sold. Find the number of units that gives break-even.

21. *Supply and Demand* The demand by consumers for a certain pair of shoes is given by $p = 100 - \frac{5}{2}q$ and the willingness of manufacturers to supply these shoes is given by $p = 10 + 2q$.
  a. Compare the quantity demanded and supplied when the price is $60. Will there be a surplus or a shortfall?
  b. What price will give market equilibrium?

22. *Supply and Demand* The demand by consumers for a certain reclining chair is given by $p = -2q + 312$ and the willingness of manufacturers to supply this product is given by $p = 2 + 8q$.
  a. Compare the quantity demanded and supplied when the price is $202. Will there be a surplus or a shortfall?
  b. What price will give market equilibrium?

23. *Market equilibrium* A retail chain will buy 900 cordless phones if the price is $10 each and will buy 400 if the price is $60. A wholesaler will supply 700 phones at $30 each and 1400 at $50 each. Assuming that the supply and demand functions are linear, find the market equilibrium point and explain what it means.

24. *Market equilibrium* A retail chain will buy 800 televisions if the price is $350 each and will buy 1200 if the price is $300. A wholesaler will supply 700 of these televisions at $280 each and 1400 at $385 each. Assuming that the supply and demand functions are linear, find the market equilibrium point and explain what it means.

1.8 Solutions of Linear Inequalities

For a certain product, the weekly revenue and the weekly cost are given by

$$R(x) = 40x \text{ and } C(x) = 20x + 1600, \text{ respectively,}$$

where x is the number of units produced and sold. For what levels of production will a profit result?

Profit will occur when revenue is greater than cost. So we find the level of production and sale x that gives a profit by solving the **linear inequality**

$$R(x) > C(x), \text{ or } 40x > 20x + 1600.$$

In this section, we will solve linear inequalities of this type analytically and graphically.

Algebraic Solution of Linear Inequalities

An **inequality** is a statement that one quantity or expression is greater than, less than, greater than or equal to, or less than or equal to another.

> ### LINEAR INEQUALITY
>
> A linear inequality (or first-degree inequality) in the variable x is an inequality that can be written in the form $ax + b > 0$ where $a \neq 0$.
> (The inequality symbol can be $>$, $\geq$, $<$, or $\leq$.)

The inequality $4x + 3 < 7x - 6$ is a linear inequality (or first-degree inequality) because the highest power of the variable (x) is one. The values of x that satisfy the inequality form the solution set for the inequality. For example, 5 is in the solution set of this inequality because substituting 5 into the inequality gives

$$4 \cdot 5 + 3 < 7 \cdot 5 - 6, \text{ or } 23 < 29$$

which is a true statement. On the other hand, 2 is not in the solution set because

$$4 \cdot 2 + 3 \not< 7 \cdot 2 - 6.$$

Solving an inequality means finding its solution set. The solution to an inequality can be written as an inequality or in interval notation. The solution can also be represented by a graph on a real number line.

Two inequalities are *equivalent* if they have the same solution set. As with equations, we can find solutions to inequalities by finding equivalent inequalities from which the solutions can be easily seen. We use the following properties to reduce an inequality to a simple equivalent inequality.

Properties of Inequalities	Examples	
Substitution Property The inequality formed by substituting one expression for an equal expression is equivalent to the original inequality.	$7x - 6x \geq 8$ $x \geq 8$	$3(x - 1) < 8$ $3x - 3 < 8$
Addition and Subtraction Properties The inequality formed by adding the same quantity to (or subtracting the same quantity from) both sides of an inequality is equivalent to the original inequality.	$x - 6 < 12$ $x - 6 + 6 < 12 + 6$ $x < 18$	$3x > 13 + 2x$ $3x - 2x > 13 + 2x - 2x$ $x > 13$
Multiplication Property I The inequality formed by multiplying (or dividing) both sides of an inequality by the same *positive* quantity is equivalent to the original inequality.	$\frac{1}{3}x \leq 6$ $3\left(\frac{1}{3}x\right) \leq 3(6)$ $x \leq 18$	$3x > 6$ $\frac{3x}{3} > \frac{6}{3}$ $x > 2$
Multiplication Property II The inequality formed by multiplying (or dividing) both sides of an inequality by the same *negative* number and reversing the inequality symbol is equivalent to the original inequality.	$-x > 7$ $-1(-x) < -1(7)$ $x < -7$	$-4x \geq 12$ $\frac{-4x}{-4} \leq \frac{12}{-4}$ $x \leq -3$

Of course, these properties can be used in combination to solve an inequality. This means that the steps used to solve a linear inequality are the same as those used to solve linear equations, except that the inequality symbol is reversed if both sides are multiplied (or divided by) a negative number.

STEPS FOR SOLVING A LINEAR INEQUALITY ALGEBRAICALLY

1. If a linear inequality contains fractions with constant denominators, multiply both sides of the inequality by a number that will remove all denominators in the inequality. If there are two or more fractions, use the least common denominator (LCD) of the fractions.
2. Remove any parentheses by multiplication.
3. Perform any additions or subtractions to get all terms containing the variable on one side and all other terms on the other side of the inequality. Combine like terms.
4. Divide both sides of the inequality by the coefficient of the variable. *Reverse the inequality symbol if this number is negative.*
5. Check the solution by substitution or with a graphing utility. If a real-world solution is desired, check the algebraic solution for reasonableness in the real-world situation.

EXAMPLE 1   Solution of a Linear Inequality

Solve the inequality $3x - \dfrac{1}{3} \le -4 + x$.

Solution

$$3x - \frac{1}{3} \le -4 + x$$

Multiplying both sides by 3 gives $3\left(3x - \dfrac{1}{3}\right) \le 3(-4 + x)$

Removing parentheses gives $9x - 1 \le -12 + 3x$

Performing additions and subtractions to both sides to get the variables on one side and the constants on the other side gives

$$6x \le -11$$

Dividing both sides by the coefficient of the variable gives

$$x \le -\frac{11}{6}$$

The solution set contains all real numbers less than or equal to $-\dfrac{11}{6}$. The graph of the solution set $\left(-\infty, -\dfrac{11}{6}\right]$ is shown in Figure 1.62.

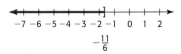

$$-\tfrac{11}{6}$$

FIGURE 1.62   ∎

EXAMPLE 2   Profit

For a certain product, the weekly revenue and the weekly cost are given by

$$R(x) = 40x \text{ and } C(x) = 20x + 1600, \text{ respectively,}$$

where x is the number of units produced and sold. For what levels of production will a profit result?

Solution

Profit will occur when revenue is greater than cost. So we find the level of production and sale that gives a profit by solving the linear inequality $R(x) > C(x)$, or $40x > 20x + 1600$.

$40x > 20x + 1600$	
$20x > 1600$	Subtracting 20x from both sides
$x > 80$	Dividing both sides by 20

Thus a profit occurs if more than 80 units are produced and sold.   ∎

EXAMPLE 3   Body Temperature
- -

A child's health is at risk when his or her body temperature is 103°F or higher. What Celsius temperature reading would indicate that a child's health is at risk?

Solution

A child's health is at risk if $F \geq 103$, and $F = \frac{9}{5}C + 32$, where F is the temperature in degrees Fahrenheit and C is the temperature in degrees Celsius. Substituting $\frac{9}{5}C + 32$ for F, we have

$$\frac{9}{5}C + 32 \geq 103$$

Now we solve the inequality for C.

$$\frac{9}{5}C + 32 \geq 103$$

$9C + 160 \geq 515$	Multiplying both sides by 5 to clear fractions.
$9C \geq 355$	Subtracting 160 from both sides
$C \geq 39.\overline{4}$	Dividing both sides by 9

Thus, a child's health is at risk if his or her Celsius temperature is approximately 39.4° or higher.   ■

Graphical Solution of Linear Inequalities

In Section 1.5 we used graphical methods to solve linear equations. In a similar manner, graphical methods can be used to solve linear inequalities. We will illustrate both the intersection of graphs method and the x-intercept method below.

Intersection Method

To solve an inequality by the intersection method, we use the following steps.

STEPS FOR SOLVING A LINEAR INEQUALITY WITH THE INTERSECTION METHOD

1. Set the left side of the inequality equal to y_1 and set the right side equal to y_2, and graph the equations using your graphing utility.

2. Choose a viewing window that contains the point of intersection, and find the point of intersection, with x-coordinate a. This is the value of x where $y_1 = y_2$.

3. The values of x that satisfy the inequality represented by $y_1 < y_2$ are those values for which the graph of y_1 is below the graph of y_2. The values of x that satisfy the inequality represented by $y_1 > y_2$ are those values for which the graph of y_1 is above the graph of y_2.

To solve the inequality

$$5x + 2 < 2x + 6$$

by using the intersection method, let

$$y_1 = 5x + 2 \text{ and } y_2 = 2x + 6,$$

enter y_1 and y_2, and graph the equations using a graphing utility (see Figure 1.63). Figure 1.63 shows that the point of intersection occurs at $x = \frac{4}{3}$. Some graphing utilities show this answer in the form $x = 1.3333333$. The exact x-value $\left(x = \frac{4}{3}\right)$ can also be found by solving the equation $5x + 2 = 2x + 6$ algebraically. Figure 1.63 shows that the graph of y_1 is below the graph of y_2 when x is less than $\frac{4}{3}$. Thus, the solution to the inequality is $x < \frac{4}{3}$, which can be written in interval notation as $\left(-\infty, \frac{4}{3}\right)$.

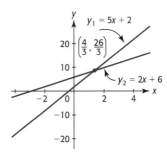

FIGURE 1.63

EXAMPLE 4   Intersection Method of Solution

Solve $\dfrac{5x + 2}{5} \geq \dfrac{4x - 7}{8}$ using the intersection of graphs method.

Solution

Enter the left side of the inequality as $y_1 = (5x + 2)/5$, enter the right side of the inequality as $y_2 = (4x - 7)/8$, graph these lines, and find their point of intersection. As seen in Figure 1.64, the two lines intersect at the point where $x = -2.55$ and $y = -2.15$.

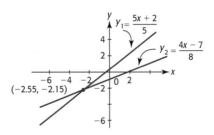

FIGURE 1.64

The solution to the inequality is the x-interval for which the graph of y_1 is above the graph of y_2, or the x-value for which the graph of y_1 intersects the graph of y_2. Figure 1.64 indicates that this is the interval to the right of and including the input value of the point of intersection of the two lines. Thus the solution is $x \geq -2.55$ or $[-2.55, \infty)$. ∎

x-Intercept Method

To use the x-intercept method to solve the inequality, we use the following steps.

> **SOLVING LINEAR INEQUALITIES WITH THE *x*-INTERCEPT METHOD**
>
> 1. Rewrite the inequality with all nonzero terms on one side of the inequality and combine like terms, getting $f(x) > 0$, $f(x) < 0$, $f(x) \leq 0$, or $f(x) \geq 0$.
>
> 2. Graph the nonzero side of this inequality. (Any window in which the x-intercept can be clearly seen is appropriate.)
>
> 3. Find the x-intercept of the graph to find the solution to the equation $f(x) = 0$. (The exact solution can be found algebraically.)
>
> 4. Use the graph to determine where the inequality is satisfied.

To use the x-intercept method to solve the inequality $5x + 2 < 2x + 6$, we rewrite the inequality with all nonzero terms on one side of the inequality and combine like terms.

$$5x + 2 < 2x + 6$$

$3x - 4 < 0$ Subtracting $2x$ and 6 from both sides of the inequality

Graphing the nonzero side of this inequality as the linear function $f(x) = 3x - 4$ gives the graph in Figure 1.65. Finding the x-intercept of the graph (see Figure 1.65) gives the solution to the equation $3x - 4 = 0$. The x-intercept (and zero of the function) is $x = 4/3$.

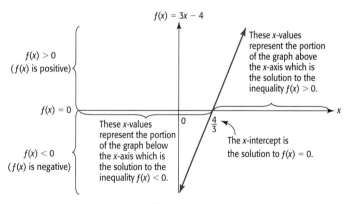

FIGURE 1.65

We now want to find where $f(x) = 3x - 4$ is less than zero. Notice that the portion of the graph *below* the x-axis gives $3x - 4 < 0$. Thus the solution to $3x - 4 < 0$, and thus to $5x + 2 < 2x + 6$, is $x < 4/3$ or $(-\infty, 4/3)$.

EXAMPLE 5   Apparent Temperature

During the summer of 1998, Dallas, Texas endured 29 consecutive days where the temperature was at least 110°F. On many of these days, the combination of heat and humidity made it feel even hotter than it was. When the temperature is 100°F, the apparent temperature A (or heat index) depends on the humidity h (expressed as a decimal) according to

$$A = 90.2 + 41.3h.$$

For what humidity levels is the apparent temperature at least 110°F? (Source: W. Bosch and C. Cobb, "Temperature-Humidity Indices," *UMAP Journal,* Fall 1989)

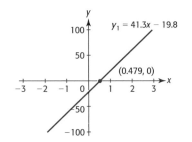

FIGURE 1.66

Solution

If the apparent temperature is at least 110°F, the inequality to be solved is

$$A \geq 110 \quad \text{or}$$
$$90.2 + 41.3h \geq 110.$$

Rewriting this inequality with 0 on the right side gives $41.3x - 19.8 \geq 0$. Entering $y_1 = 41.3x - 19.8$ and graphing gives the graph in Figure 1.66. The x-intercept of the graph is (approximately) 0.479.

   The x-interval where the graph is on or above the x-axis is the solution we seek, so the solution to the inequality is $[0.479, \infty)$. However, humidity is limited to 100%, so the solution is $0.479 \leq h \leq 1.00$, and we say that the humidity is between 47.9% and 100%, inclusive.   ■

Double Inequalities

The inequality $0.479 \leq h \leq 1.00$ in Example 5 is a **double inequality**. A double inequality represents two inequalities connected by the word "and" or "or." The inequality $0.479 \leq h \leq 1.00$ is a compact way of saying $0.479 \leq h$ and $h \leq 1.00$. Double inequalities can be solved analytically or graphically, as illustrated in the following example. Note that any arithmetic operation is performed to *all three* parts of a double inequality.

EXAMPLE 6   Course Grades

A student has taken four exams and has earned grades of 90%, 88%, 93%, and 85%. What grade must the student earn on the final test so that his course average is a B (that is, so his average is at least 80% and less than 90%)?

Analytical Solution

To receive a B, the final test score, represented by x, must satisfy

$$80 \leq \frac{90 + 88 + 93 + 85 + x}{5} < 90.$$

Solving this inequality gives

$$80 \leq \frac{356 + x}{5} < 90$$

$$400 \leq 356 + x < 450 \qquad \text{Multiplying all three parts by 5}$$

$$44 \leq x < 94 \qquad \text{Subtracting 356 from all three parts}$$

Thus he will receive a grade of B if his final test score is at least 44 but less than 94.

Graphical Solution

To solve this inequality graphically, we assign the left side of the inequality to y_1, the middle to y_2, and the right side to y_3, and graph these equations to obtain the graph in Figure 1.67.

$$y_1 = 80$$
$$y_2 = (356 + x)/5$$
$$y_3 = 90$$

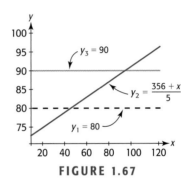

FIGURE 1.67

We seek the values of x where the graph of y_2 is above or on the graph of y_1 and below the graph of y_3. The left endpoint of this x-interval occurs at the point of intersection of y_2 and y_1, and the right endpoint of the interval occurs at the intersection of y_2 and y_3. These two points can be found using the intersection method. Figure 1.68 shows the points of intersection of these graphs.

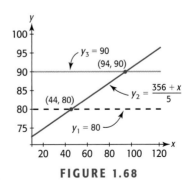

FIGURE 1.68

The x-values of the points of intersection are 44 and 94, so the solution to $80 \leq \dfrac{356 + x}{5} < 90$ is $44 \leq x < 94$, which agrees with our analytical solution.   ∎

EXAMPLE 7   Expected Prison Sentences

For the years 1990 through 1996, the mean (expected) time y served in prison for a serious crime can be approximated by a function of the mean sentence length x, with $y = 0.55x - 2.886$, where x and y are measured in months. If this model still applies, to how many months should a judge sentence a convicted criminal so that the criminal will be expected to serve between 37 and 78 months? (Source: National Center for Policy Analysis)

Solution

We seek values of x that give y-values between 37 and 78, so we solve $37 \le 0.55x - 2.886 \le 78$ for x.

$$37 \le 0.55x - 2.886 \le 78$$
$$37 + 2.886 \le 0.55x \le 78 + 2.886$$
$$39.886 \le 0.55x \le 80.886$$
$$73 \le x \le 147 \quad \text{to the nearest month}$$

Thus the judge could impose a sentence of 73 to 147 months if she wants the criminal to actually serve between 37 and 78 months.   ■

1.8 SKILLS CHECK

In Exercises 1–9, solve the inequalities both algebraically and graphically. Draw a number line graph of each solution.

1. $3x - 7 \le 5 - x$ **2.** $2x + 6 < 4x + 5$

3. $4 - 3x \ge 4x + 5$ **4.** $5(2x - 3) > 4x + 6$

5. $4x + 1 < -\dfrac{3}{5}x + 5$ **6.** $4x - \dfrac{1}{2} \le -2 + \dfrac{x}{3}$

7. $\dfrac{x - 5}{2} < \dfrac{18}{5}$ **8.** $\dfrac{3(x - 6)}{2} \le \dfrac{2x}{5} - 12$

9. $2.2x - 2.6 \ge 6 - 0.8x$

10. Solve the inequality $7x + 3 < 2x - 7$ by the intersection method.

11. Use the x-intercept method to solve the inequality $5x + 4 \ge 8x - 28$.

12. Solve for x: $17 \le 3x - 5 < 31$.

13. Solve for x: $37.002 \le 0.554x - 2.886 \le 77.998$.

14. Solve for x: $70 \le \dfrac{60 + 88 + 73 + 65 + x}{5} < 80$.

1.8 EXERCISES

1. *Blood Alcohol Percent* The blood alcohol percent p of a 220-lb male is a function of the number of 12.5-oz drinks, and the percent at which a person is legally intoxicated (and guilty of DUI if driving) is 0.1 percent or higher (see the following figure).

a. Use an inequality to indicate the percent of alcohol in the blood when a person is considered legally intoxicated.

b. If x is the number of drinks imbibed by a 220-lb male, write an inequality that gives the number of

drinks that will cause him to be legally intoxicated.
(Source: Pennsylvania Liquor Control Board)

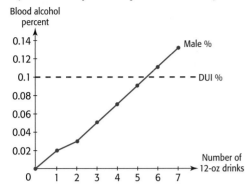

2. *Depreciation* Suppose a business purchases equipment for $12,000 and depreciates it over five years with the straight-line method until it reaches its salvage value of $2000 (see the figure below). Assuming that the depreciation can be for any part of a year:
   a. Write an inequality that indicates that the depreciated value V of the equipment is less than $8000.
   b. Write an inequality that describes the time t during which the depreciated value is at least half of the original value.

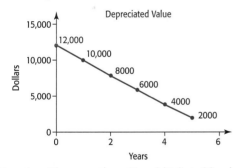

3. *Freezing* The equation $F = 9/5\,C + 32$ gives the relationship between temperatures measured in degrees Celsius and degrees Fahrenheit. We know that a temperature at or below 32°F is "freezing." Use an inequality to represent the corresponding "freezing" Celsius temperature.

4. *Stock Market* Susan Mason purchased 1000 shares of stock for $22 per share, and 3 months later it had dropped by 20%. What is the minimum percent increase required for her to make a profit?

5. *Job Selection* A job candidate is given the choice of two positions, one paying $3100 per month and the other paying $2000 per month plus a 5% commission on all sales made during the month. What amount must the employee sell in a month for the second position to be more profitable?

6. *Grades* If Jill Ball has a course average score between 80 and 89, she will earn a grade of B in her algebra course. Suppose that she has four exam scores of 78, 69, 92, and 81 and that her teacher said the final exam score has twice the weight of the other four exams. What range of scores on the final exam will result in Jill earning a grade of B?

7. *Marijuana Use* The percent p of high school seniors who use marijuana daily is given by $30p - 19x = 1$, where x is the number of years after 1990. If this model is accurate, what is the range of daily use of marijuana for high school seniors between 1996 and 2000?
(Source: Index of Leading Cultural Indicators)

8. *Cigarettes* Data from 1950 to 1992 indicates that world cigarette production can be modeled by $y = 9.3451x + 649.3385$ cigarettes per person per year, where x is the number of years after 1950. According to this function, what is the range of the per capita world cigarette production between 1950 and 1992 to two decimal places?
(Source: The National Clearinghouse on Tobacco and Health)

9. *SAT Scores* The College Board began reporting SAT scores with a new scale in 1996, with the new scale score y defined as a function of the old scale score x by the equation $y = 0.97x + 128.3829$. Suppose a college requires a new scale score greater than or equal to 1000 to admit a student. What old score values would be equivalent to the new scores that would result in admission to this college?

10. *Cigarettes* Data from 1950 to 1992 indicates that world cigarette production can be modeled by $y = 9.3451x + 649.3385$ cigarettes per person per year, where x is the number of years after 1950. If the model is accurate, for what years was the per capita world production of cigarettes below 1000 cigarettes? Note that your answer should be discretely interpreted.
(Source: The National Clearinghouse on Tobacco and Health)

11. *HID headlights* The new high intensity discharge (HID) lights that contain xenon gas and are installed in some new cars have an expected life of 1500 hours. Because a complete system costs $1000, it is hoped that these lights will last for the life of the car. Suppose that the actual life of the lights could be 10% longer or shorter than the advertised expected life. Write an inequality that gives the range of life of these new lights.
(Source: *Automobile*, July 2000)

12. *Prison Sentences* The mean time in prison for a crime, y, can be found as a function of the mean sentence length, x, using the equation $y = 0.554x - 2.886$, where x and y are measured in months. How many months should a judge sentence a convicted criminal if she wants the criminal to actually serve between 4 and 6 years?
(Source: Index of Leading Cultural Indicators)

13. *Marriage Rate* According to data from the Index of Leading Cultural Indicators, the marriage rate (marriages per 1000 unmarried women) can be described by $y = -0.763x + 85.284$, where x is the number of years after 1950. For what years does this model indicate that the marriage rate
    a. was above 50 marriages per 1000 unmarried women?
    b. will be below 45 marriages per 1000 unmarried women?

14. *Earnings and Minorities* According to the U.S. Equal Employment Opportunity Commission, the relation between the median annual salaries of minorities and whites can be modeled by the function $M = 0.959W - 1.226$, where M and W represent the median annual salary (in thousands of dollars) for minorities and whites, respectively. What is the median salary range for whites that corresponds to a salary range of at least $100,000 for minorities?
(Source: *Statistical Abstract of the U.S., 1993*)

15. *Home Appraisal* A home purchased in 1996 for $190,000 was appraised at $270,000 in 2000. If the rate of increase in the value of the home is assumed to be constant,
    a. Write an equation for the value of the home as a function of the number of years, x, after 1996.
    b. If the equation in part (a) remains accurate, write an inequality that gives the range of years (until the end of 2010) when the value of the home will be greater than $400,000.

16. *Car Sales Profit* A car dealer purchases 12 new cars for $32,500 each, and sells 11 of them at a profit of 5.5%. For how much must he sell the remaining car to average a profit of at least 6% on the 12 cars?

17. *Electrical Components Profit* A company's daily profit from the production and sale of electrical components can be described by the equation $P(x) = 6.45x - 2000$ dollars, where x is the number of units produced and sold. What level of production

and sales will give a daily profit of more than $10,900?

18. *Plumbers Helper Profit* The yearly profit from the production and sale of Plumbers Helper is $P(x) = -40,255 + 9.80x$ dollars, where x is the number of Plumbers Helpers produced and sold. What level of production and sales gives a yearly profit of more than $84,355?

19. *PVC Pipe Break-even* A large hardware store's monthly profit from the sale of PVC pipe can be described by the equation $P(x) = 6.45x - 9675$ dollars, where x is the number of feet of PVC pipe sold. What level of monthly sales is necessary to avoid a loss?

20. *Plumbers Helper Break-even* The yearly profit from the production and sale of Plumbers Helper is $P(x) = -40,255 + 9.80x$ dollars, where x is the number of Plumbers Helpers produced and sold. What level of production and sales will result in a loss?

21. *Logic Board Break-even* A company is producing a new logic board for computers. The annual fixed cost for the board is $345,000 and the variable cost is $125 per board. If the logic board sells for $489, write an inequality that gives the number of logic boards that will give a profit for the product.

22. *Temperature* The temperature T (in degrees Fahrenheit) inside a concert hall m minutes after a 40-minute power outage during a summer rock concert is given by $T = 0.43m + 76.8$. Write and solve an inequality that describes when the temperature in the hall is not more than 85°F.

23. *Reading Tests* The average reading score of 17-year-olds on the National Assessment of Progress tests is given by $y = 0.155x + 244.37$ points, where x is the number of years after 1970. Assuming that this model was valid, write and solve an inequality that describes when the average 17-year-old reading score on this test was between, but not including, 245 and 248. (Your answer should be interpreted discretely.)
(Source: U.S. Department of Education)

24. *Voting* The percent of the voting population who voted in presidential elections between 1950 and 1996 is given by $p = 65.4042 - 0.3552x$, where x is the number of years after 1950. Write and solve an inequality that describes in what years the percent was
    a. less than 30.
    b. more than 75.
    c. between and including 50 and 60.

25. *Cigarette Use* The percent p of high school seniors who have tried cigarettes can be modeled by

$$p = 75.4509 - 0.706948t$$

where t is the number of years after 1975.

a. What percent does this model estimate for the year 1998?

b. This equation ceases to be an effective model when p reaches values such that $p < 0$ or if $p > 100$. Test integer values of t with the TABLE feature of your graphing utility to find the values of t for which this function fails to model the data.

c. In what years can we be certain that this equation is not an effective model for the percent?

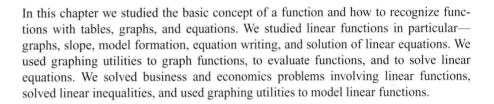

CHAPTER 1 *Summary*

In this chapter we studied the basic concept of a function and how to recognize functions with tables, graphs, and equations. We studied linear functions in particular—graphs, slope, model formation, equation writing, and solution of linear equations. We used graphing utilities to graph functions, to evaluate functions, and to solve linear equations. We solved business and economics problems involving linear functions, solved linear inequalities, and used graphing utilities to model linear functions.

Key Concepts and Formulas

1.1 Functions and Graphs

Scatter plot of data	A graph of pairs of values that represent real-world information collected in numerical form.
Function definition	A function is a rule or correspondence that determines exactly one output for each input. The function may be defined by a set of ordered pairs, a table, a graph, or an equation.
Domain and range	The set of possible inputs for a function is called its domain, and the set of possible outputs is called its range. In general, if the domain of a function is not specified, we assume that it includes all real numbers except: • Values that result in a denominator of 0, and • Values that result in an even root of a negative number
Independent variable and dependent variable	If x represents any element in the domain, then x is called the independent variable; and if y represents an output of the function from an input x, then y is called the dependent variable.
Tests for functions	We can test for a function graphically, numerically, and analytically.
Vertical line test	A set of points in a coordinate plane is the graph of a function if and only if no vertical line intersects the graph in more than one point.
Functional notation	We denote that y is a function of x using functional notation when we write $y = f(x)$.

1.2 Graphs of Functions; Mathematical Models

Point-plotting method

The point-plotting method of sketching a graph means sketching the graph of a function by plotting enough points to determine the shape of the graph and then drawing a smooth curve through the points.

Complete graph

A graph is a complete graph if it shows the basic shape of the graph and important points on the graph (including points where the graph crosses the axes and points where the graph turns), and suggests what the unseen portions of the graph will be.

Using a calculator to draw a graph

After writing the function with x representing the independent variable and y representing the dependent variable, enter the function in the equation editor of the graphing utility. Activate the graph with ZOOM or GRAPH (Set the x- and y-boundaries of the viewing window before pressing GRAPH .)

Viewing windows

The values that define the viewing window can be set manually or by using the ZOOM keys. The boundaries of a viewing window are:

x_{min}: the smallest value on the x-axis (the left boundary of the window)

y_{min}: the smallest value on the y-axis (the bottom boundary of the window)

x_{max}: the largest value on the x-axis (the right boundary of the window)

y_{max}: the largest value on the y-axis (the top boundary of the window)

Spreadsheets

Spreadsheets like Excel can be used to create accurate graphs, sometimes better-looking graphs than those created with graphing calculators.

Evaluating functions with a calculator

We can find (or estimate) the output of a function $y = f(x)$ at specific inputs with a graphing calculator by using TRACE and moving the cursor to (or close to) the value of the independent variable.

Modeling

The process of translating real-world information into a mathematical form so that it can be applied and then interpreted in the real-world setting is called modeling.

Mathematical model

A mathematical model is a functional relationship (usually in the form of an equation) that includes not only the function rule but also descriptions of all involved variables and their units of measure.

Aligning data

Using aligned inputs (input values that have been shifted horizontally to smaller numbers) rather than the actual data values results in smaller coefficients in models and less involved computations.

Graphing data points

Graphing utilities can be used to create lists of numbers and to create graphs of the data stored in the lists.

1.3 Linear Functions

Linear functions

A linear function is a function of the form $f(x) = ax + b$ where a and b are constants. The graph of a linear function is a line.

Intercepts

A point (or the x-coordinate of the point) where a graph crosses or touches the horizontal axis is called an x-intercept, and a point (or the y-coordinate of the point) where a graph crosses or touches the vertical axis is called a y-intercept. To find the x-intercept(s) of the graph of an equation, set $y = 0$ in the equation and solve for x. To find the y-intercept(s), set $x = 0$ and solve for y.

Linear model	A linear mathematical model is a linear equation that describes one quantity in terms of another quantity.
Zero of a function	Any number a for which $f(a) = 0$ is called a **zero** of the function $f(x)$. In this case, a is an x-intercept of the graph of the function.
Slope of a line	The slope of a line is a measure of its steepness and direction. The slope is defined as

$$\text{slope} = \frac{\text{vertical change}}{\text{horizontal change}} = \frac{\text{rise}}{\text{run}}$$

If (x_1, y_1) and (x_2, y_2) are two points on a line, then the slope of the line is $m = \frac{y_2 - y_1}{x_2 - x_1}$.

Slope and y-intercept of a line	The slope of the graph of the equation $y = mx + b$ is m, and the y-intercept of the graph is b. (This is called slope-intercept form.)
Rates of change	If a model is linear, the rate of change of the outputs with respect to the inputs will be constant and the rate of change equals the slope of the line that is the graph of a linear function. Thus, we can determine if a linear model fits a set of real data by determining if the rate of change remains constant for the data.

Special linear functions

• **Constant functions**	A special linear function that has the form $y = c$, where c is a real number, is called a constant function.
• **Identity function**	The identity function $y = x$ is a linear function of the form $y = mx + b$ with slope $m = 1$ and y-intercept $b = 0$.

1.4 Equations of Lines; Rates of Change

Writing equations of lines

• **General form**	The general form of the equation of a line is $ax + by = c$, where $a, b,$ and c are constants.
• **Slope-intercept form**	The slope-intercept form of the equation of a line with slope m and y-intercept b is $y = mx + b$.
• **Point-slope form**	The equation of the line with slope m and passing through a known point (x_1, y_1) is $y - y_1 = m(x - x_1)$.
• **Horizontal line**	The equation of a horizontal line is $y = b$, where b is a constant.
• **Vertical line**	The equation of a vertical line is $x = a$, where a is a constant.
Constant rate of change	We can write the equation of the linear function that fits a set of real data if the rate of change remains constant for the data.
First differences	If the changes in outputs (called *first differences*) are constant (when the changes in inputs are constant), then we can find a linear equation that models the data.
Average rate of change	The **average rate of change** of a quantity over an interval describes how a change in the output of a function describing that quantity, $f(x)$, is related to a change in the input, x, over that interval. The average rate of change of $f(x)$ with respect to x over the interval from $x = a$ to $x = b$ (where $a < b$) is calculated as

$$\text{average rate of change} = \frac{\text{change in } f(x) \text{ values}}{\text{corresponding change in } x \text{ values}} = \frac{f(b) - f(a)}{b - a}$$

Secant line	When a function is not linear, the average rate of change between two points is the slope of the line joining two points on the curve, which is called a secant line.

1.5 Algebraic and Graphical Solution of Linear Equations

Algebraic solution of linear equations	If a linear equation contains fractions, multiply both sides of the equation by a number that will remove all denominators in the equation. Next, remove any parentheses or other symbols of grouping, then perform any additions or subtractions to get all terms containing the variable on one side and all other terms on the other side of the equation. Combine like terms. Divide both sides of the equation by the coefficient of the variable. Check the solution by substitution in the original equation.
Solving real-world application problems	To solve an application problem that is set in a real-world context, use the same solution methods. However, remember to include units of measure with your answer and check that your answer makes sense in the problem situation.
Solutions, zeros, and *x*-intercepts	If a is an x-intercept of the graph of a function f, then a is a zero of the function f, and a is a solution to the equation $f(x) = 0$.
Graphical solution of linear equations	
• *x*-**Intercept method**	Rewrite the equation with 0 on one side, enter the nonzero side into the equation editor of a graphing calculator, and find the x-intercept of the graph. This is the solution to the equation.
• **Intersection method**	Enter the left side of the equation into y_1 and the right side of the equation into y_2, and find the point of intersection. The x-coordinate of the point of intersection is the solution to the equation.
Formulas; solving an equation for a specified linear variable	To solve an equation for one variable if two or more variables are in the equation, and if that variable is linear in the equation, we can solve for that variable by treating the other variables as constants and using the same steps that we used to solve a linear equation in one variable.

1.6 Fitting Lines to Data Points; Modeling Linear Functions

Fitting lines to data points	When real-world information is collected as numerical information called *data*, technology can be used to determine the pattern exhibited by the data (provided that a recognizable pattern exists). These patterns can often be described by mathematical functions.
Linear regression	We can determine the equation of the line that is the "best fit" for a set of points by using a procedure called linear regression (or the least squares method), which defines the "best fit line" as the line for which the sum of the squares of the vertical distances from the data points to the line is a minimum.
Modeling data	We can model a set of data by entering the data into a graphing utility, obtaining a scatter plot, and using the graphing utility to obtain the linear equation that is the best fit for the data. The equation and/or numerical results should be reported in a way that makes sense in the context of the problem, with the appropriate units, and with the variables identified.
Discrete versus continuous	We use the term *discrete* to describe data or a function that is presented in the form of a table or in a scatter plot. We use the term *continuous* to describe a function or graph when the inputs can be any real number or any real number between two specified values.

Goodness of fit	The goodness of fit of a linear model can be observed from a graph of the model and the data points and measured with the correlation coefficient.

1.7 Models in Business and Economics

Revenue, cost, and profit	If a company sells x units of a product for p dollars per unit, then the total revenue for this product can be modeled by the linear function $R(x) = px$.

$$\text{Cost} = \text{variable costs} + \text{fixed costs}$$

(Note: Variable costs depend on the number of units produced.)

$$\text{Profit} = \text{revenue} - \text{cost} \quad \text{or} \quad P(x) = R(x) - C(x)$$

Break-even analysis	A company is said to **break even** from the production and sale of a product if the total revenue equals the total cost; that is, if the profit for that product is zero.
Marginal cost, marginal profit, marginal revenue	Whenever the total revenue, total cost, and profit are modeled by linear functions, the slopes of their graphs give the constant rate of change of the functions. In the case of these business functions, the constant rates of change are called *marginals*.
Supply, demand, and market equilibrium	The *law of demand* states that the quantity of a product that is demanded will increase as the price decreases. The *law of supply* states that the quantity supplied for sale will increase as the price of a product increases. *Market equilibrium* is said to occur when the quantity of a commodity demanded is equal to the quantity supplied. The price at this point is called the *equilibrium price*, and the quantity at this point is called the *equilibrium quantity*.

1.8 Solutions of Linear Inequalities

Linear inequality	A linear inequality (or first-degree inequality) is an inequality that can be written in the form $ax + b > 0$ where $a \neq 0$. (The inequality symbol can be $>$, $\geq$, $<$, or $\leq$.)
Algebraically solving linear inequalities	The steps used to solve a linear inequality are the same as those used to solve linear equations, except that the inequality symbol is reversed if both sides are multiplied (or divided by) a negative number.
Graphical solution of linear inequalities	Set the left side of the inequality equal to y_1 and set the right side equal to y_2, graph the equations using a graphing utility, and find the point of intersection, with x-coordinate a. The values of x that satisfy the inequality represented by $y_1 < y_2$ are those values for which the graph of y_1 is below the graph of y_2.
x-Intercept method	To use the x-intercept method to solve an inequality, rewrite the inequality with all nonzero terms on one side and zero on the other side of the inequality and combine like terms. Graph the nonzero side of this inequality and find the x-intercept a of the graph. If the inequality to be solved is $f(x) > 0$, the solution will be the interval of x-values representing the portion of the graph above the x-axis. If the inequality to be solved is $f(x) < 0$, the solution will be the interval of x-values representing the portion of the graph below the x-axis.
Double inequalities	A double inequality represents two inequalities connected by the word "and" or "or." Double inequalities can be solved algebraically or graphically. Any operation performed on a double inequality must be performed to *all three* parts.

Chapter 1 Skills Check

Use the values in the table below in Exercises 1–4.

x	-3	-1	1	3	5	7	9	11	13
y	9	6	3	0	-3	-6	-9	-12	-15

1. Explain why the relationship shown by the table describes y as a function of x.

2. State the domain and range of the function.

3. If the function defined by the table is denoted by f, so that $y = f(x)$, what is $f(3)$?

4. Do the outputs in this table indicate that a linear function fits the data? If so, write the equation of the line.

5. If $C(s) = 16 - 2s^2$, find
   a. $C(3)$ **b.** $C(-2)$ **c.** $C(-1)$

6. Graph the function $y = 3x^2$ with a graphing utility using a standard viewing window.

7. Graph $y = -10x^2 + 400x + 10$ on a standard window and on a window with x-min = 0, x-max = 40, y-min = 0, y-max = 5000. Which window gives a better view of the graph of the function?

Use the table of data below in Exercises 8–11.

x	1	3	6	8	10
y	-9	-1	5	12	18

8. Use a graphing utility to graph the points (x, y) from the table, using a viewing window with x-min = 0, x-max = 12, y-min = -12, y-max = 20.

9. Find the linear function that is the best fit for the data in the table.

10. Use a graphing utility to graph the function found in Exercise 9 on the same set of axes as the data in the

table in Exercise 8, with x-min = 0, x-max = 12, y-min = -12, y-max = 20.

11. Do the data points in the table fit exactly on the graph of the function from Exercise 10? Should they?

12. Find the slope of the line through $(-4, 6)$ and $(8, -16)$.

13. Given the equation $2x - 3y = 12$, (a) find the x- and y-intercepts of the graph, and (b) graph the equation.

14. Find the rate of change of the function whose equation is given in Exercise 13.

15. Write the equation of the line that has slope $\dfrac{1}{3}$ and y-intercept 3.

16. Write the equation of a line that has slope $\dfrac{-3}{4}$ and passes through $(4, -6)$.

17. Write the equation of the line that passes through $(-1, 3)$ and $(2, 6)$.

In Exercises 18 and 19, solve the following equations for x (a) algebraically and (b) with a graphing utility.

18. $3x + 22 = 8x - 12$

19. $\dfrac{3(x - 2)}{5} - x = 8 - \dfrac{x}{3}$

20. For the function $y = x^2$, compute the average rate of change between $x = 0$ and $x = 3$.

21. Solve $4x - 3y = 6$ for y and graph it on a graphing utility with a standard window.

In Exercises 22–24, solve the inequalities both algebraically and graphically.

22. $3x + 8 < 4 - 2x$ 23. $3x - \dfrac{1}{2} \leq \dfrac{x}{5} + 2$

24. $18 \leq 2x + 6 < 42$

Chapter 1 Review

1. *Voters* The table gives the percent p of African American voters who have supported Democratic candidates for President for the years 1960–1996.
   a. Is the percent p a function of the year y?

   b. Let $p = f(y)$ denote that p is a function of y. Find $f(1992)$ and explain what it means.
   c. What is y if $f(y) = 94$? What does this mean?

Year	Democrat (%)	GOP (%)
1960	68	32
1964	94	6
1968	85	15
1972	87	13
1976	85	15
1980	86	12
1984	89	9
1992	82	11
1996	84	12

(Source: Joint Center for Political and Economic Studies)

2. *Voters*
   a. What is the domain of the function defined by the table in Exercise 1?
   b. Is this function defined for 1982? Why is this value not included in the table?
   c. Is this function discrete or continuous? Explain.

3. *Voters* Graph the function defined in Exercise 1 on the window [1955, 2000] by [60, 100].

4. *Voters*
   a. Find the slope of the line joining the points (1968, 85) and (1996, 84).
   b. Use the data for 1968 and 1996 from the table in Exercise 1 to find the average annual percent of African American voters who voted for the Democratic candidate for President between the years 1968 to 1996, inclusive.
   c. Is the average annual rate of change from 1968 to 1980 equal to the average annual rate of change from 1968 to 1996?
   d. Does a linear function model this data exactly?

When money is borrowed to purchase an automobile, the amount borrowed determines the monthly payment. In particular, if a dealership offers a five-year loan at 2.9% interest, then the amount borrowed for the car determines the payment according to the table below. Use the table to define the function $P = f(A)$ in Exercises 5–8.

Amount Borrowed	Monthly Payment
$10,000	$179.25
15,000	268.87
20,000	358.49
25,000	448.11
30,000	537.73

(Source: Sky Financial)

5. *Car Loans*
   a. Explain why the monthly payment P is a linear function of the amount borrowed A.
   b. Find $f(25,000)$ and interpret its meaning.
   c. If $f(A) = 358.49$, what is A?

6. *Car Loans*
   a. What are the domain and range of the function $P = f(A)$ defined by the table above?
   b. Is this function $P = f(A)$ defined for a $12,000 loan?
   c. Is the function defined by the table discrete or continuous?

7. *Car Loans*
   a. Are the first differences of the outputs in the table above constant?
   b. Is there a line on which these data points fit exactly?

8. *Car Loans*
   a. Write the equation $P = f(A)$ of the line that fits on the data points in the table above.
   b. Use the linear model from part (a) to find $P = F(28,000)$ and explain what it means.
   c. Can the function F be used to find the monthly payment for any dollar amount A of a loan if the interest rate and length of loan are unchanged?
   d. Determine the amount of a loan that will keep the payment less than or equal to $500.

9. *Life Expectancy* Figure (a) on page 126 gives the number of years the average woman is estimated to live beyond age 65 during selected years between 1950 and 2030. Let x represent the year and let y represent the expected number of years a woman will live past age 65. Write $y = f(x)$ and answer the following questions:
   a. What is $f(1960)$ and what does it mean?
   b. What is the life expectancy for the average woman in 2010?
   c. In what year is the average woman expected to live 19 years past age 65?

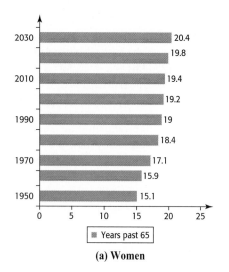

(a) Women

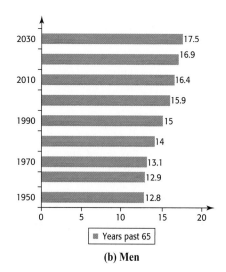

(b) Men

10. *Life Expectancy* Figure (b) in Exercise 9 gives the number of years the average man is estimated to live beyond age 65 during selected years between 1950 and 2030. Let *x* represent the year and let *y* represent the expected number of years a man will live past age 65. Write $y = g(x)$ and answer the following questions:
a. What is $g(2020)$ and what does it mean?
b. What was the life expectancy for the average man in 1950?
c. Write a functional expression that indicates that the average man in 1990 has a life expectancy of 80 years.

11. *Drug Users* The number of current users of illicit drugs increased from 12 million in 1992 to 14.5 million in 1999.

a. Find the slope of the line connecting the points (1992, 12.0) and (1999, 14.5), rounded to three decimal places.
b. Find the average annual rate of change of current users between 1992 and 1999.

12. *Fuel* The table below shows data for the number of gallons of gas purchased each day of a certain week by the 250 taxis owned by the Inner City Transportation Taxi Company. Write the equation that models this data.

Days from the first day of the week	0	1	2	3	4	5	6
Gas used by 250 taxis (gallons)	4500	4500	4500	4500	4500	4500	4500

13. *Work Hours* The average weekly hours worked by production and nonsupervisory workers on private nonfarm payrolls, seasonally adjusted, is given by the data in the table below.

Month and Year	Average Weekly Hours	Month and Year	Average Weekly Hours
June, 1998	34.6	Oct, 1998	34.6
July, 1998	34.6	Nov, 1998	34.6
Aug, 1998	34.6	Dec, 1998	34.6
Sept, 1998	34.6	Jan, 1999	34.6
		Feb, 1999	34.6

(Source: Bureau of Labor Statistics)

a. Write the equation of a function that describes the average weekly hours using an input equal to the number of months past May, 1998.
b. Is this function a constant function? Explain.

14. *Job Selection* A job candidate is given the choice of two positions, one paying $2100 per month and one paying $1000 per month plus a 5% commission on all sales made during the month.
a. How much (in dollars) must the employee sell in a month for the second position to pay as much as the first?

b. To be sure that the second position will pay more than the first, how much (in dollars) must the employee sell each month?

15. *Marketing* A car dealer purchased 12 automobiles for $24,000 each. If she sells 8 of them with an average profit of 12%, for how much must she sell the remaining 4 to obtain an average profit of 10% on all 12?

16. *Investment* A retired couple has $420,000 to invest. They chose one relatively safe investment fund that has an annual yield of 6% and another riskier investment that has a 10% annual yield. How much should they invest in each fund to earn $30,000 per year?

17. *Writing Scores* The average writing scores of eleventh-graders on the National Assessment of Educational Progress tests have changed over the years since 1984, with the average score given by $y = -0.629x + 293.871$, where x is the number of years from 1980. For what year does this model give an average score of 285?
(Source: U.S. Dept of Education)

18. *Revenue* A company has revenue given by $R(x) = 564x$ dollars and total costs given by $C(x) = 40,000 + 64x$ dollars, where x is the number of units produced and sold.
   a. What is the revenue when 120 units are produced?
   b. What is the cost when 120 units are produced?
   c. What is the marginal cost and what is the marginal revenue for this product?
   d. What is the slope of the graph of $C(x) = 40,000 + 64x$?
   e. Graph $R(x)$ and $C(x)$ on the same set of axes.

19. *Profit* A company has revenue given by $R(x) = 564x$ dollars and total costs given by $C(x) = 40,000 + 64x$ dollars, where x is the number of units produced and sold. The profit can be found by forming the function $P(x) = R(x) - C(x)$.
   a. Write the profit function.
   b. Find the profit when 120 units are produced and sold.
   c. How many units give break-even?
   d. What is the marginal profit for this product?
   e. How is the marginal profit related to the marginal revenue and the marginal cost?

20. *Depreciation* A business property can be depreciated for tax purposes by using the formula

$y + 3000x = 300,000$, where y is the value of the property x years after it was purchased.
   a. Find the y-intercept of the graph of this function. Interpret this value.
   b. Find the x-intercept. Interpret this value.

21. *Marginal Profit* A company has determined that its profit for a product can be described by a linear function. The profit from the production and sale of 150 units is $455 and the profit from 250 units is $895.
   a. What is the average rate of change of the profit for this product when between 150 and 250 units are sold?
   b. What is the slope of the graph of this profit function?
   c. Write the equation of the profit function for this product.
   d. What is the marginal profit for this product?
   e. How many units give break-even for this product?

22. *Life Expectancy*
   a. Find a linear function $y = f(x)$ that models the data shown in Figure (a) of Exercise 9 with x equal to the number of years after 1950 and y equal to the number of years the average woman is estimated to live beyond age 65.
   b. Graph the data and the model on the same set of axes.
   c. Use the model to estimate $f(104)$ and explain what it means.
   d. Determine an inequality to represent the time period (in years) for which the average woman can expect to live at least 84 years.

23. *Life Expectancy*
   a. Find a linear function $y = g(x)$ that models the data shown in Figure (b) of Problem 9 with x equal to the number of years after 1950 and y equal to the number of years the average man is estimated to live beyond age 65.
   b. Graph the data and the model on the same set of axes.
   c. Use the model to estimate $g(130)$ and explain what it means.
   d. In what year would the average man expect to live to age 90?
   e. Determine an inequality to represent the time period for which the average man could expect to live no more than 81 years.

24. *Education Spending* The figure on page 128 gives the amount (in billions of dollars) spent on education during the years 1996–2001.

a. Graph the data points *or* find the first differences to determine if a linear equation is a reasonable model for this data.
b. If it is reasonable, find the linear model that is the best fit for this data, with *x* equal to the number of years after 1990.
c. Use the unrounded model to predict the spending in 2002.

Years	Amount Spent ($billions)
1996	23
1997	26.6
1998	29.9
1999	33.5
2000	35.6
2001	40.1

25. *Population Growth* The resident population of Florida (in thousands) is given in the table below. Let *x* equal the number of years after 1980 and *y* equal the number of thousands of residents.
a. Graph the data points to determine if a linear equation is a reasonable model for this data.
b. If it is reasonable, find the linear model that is the best fit for this data, with *x* equal to the number of years after 1980.
c. What does this unrounded model predict that the population will be in 2002?

Years	1980	1985	1990	1995
Population (thousands)	9746	11,351	12,938	14,180

Years	1996	1997	1998
Population (thousands)	14,425	14,677	14,916

(Source: U.S. Census Bureau)

26. *Earnings per Share* The table below gives the earnings (in $ thousands) per share for ACS for the years 1996–2000. Let *x* equal the number of years after 1995 and *y* equal the earnings per share, in thousands of dollars.
a. Graph the data points to determine if a linear equation is a reasonable model for this data.
b. If it is reasonable, find the linear model that is the best fit for this data.
c. Graph the data points and the function on the same axes and discuss the goodness of fit.

Years	1996	1997	1998	1999	2000
Earnings per share ($ thousands)	.85	1.05	1.29	1.66	2.05

(Source: 2000 Report of ACS)

27. *Marriage Rate* According to data from the Index of Leading Cultural Indicators, the marriage rate (the number of marriages per 1000 women) can be described by $y = -0.763x + 85.284$, where *x* is the number of years after 1950. For what years does this model indicate that the rate
a. was above 48.66 per 1000 women?
b. will be below 41.03 per 1000 women?

28. *Marijuana Use* The percent *p* of high school seniors who use marijuana daily is given by the equation $30p - 19x = 1$, where *x* is the number of years after 1990. If this model is accurate, what is the range of the percent of use for the years 1996–2000?
(Source: Index of Leading Cultural Indicators)

29. *Prison Sentences* The mean time in prison, *y*, for a crime can be found as a function of the mean sentence length, *x*, using $y = 0.554x - 2.886$, where *x* and *y* are in months. If a judge sentences a convicted criminal to serve between 3 and 5 years, how many months would we expect the criminal to serve?
(Source: Index of Leading Cultural Indicators)

GROUP ACTIVITY/EXTENDED APPLICATION *1*

Body Mass Index

Obesity is a risk factor for the development of medical problems, including high blood pressure, high cholesterol, heart disease, and diabetes. Of course, how much a person can safely weigh depends on his or her height. One way of comparing weights that account for height is the **body mass index (BMI)**. The table below gives the BMI for a variety of heights and weights of people. Roche Pharmaceuticals, manufacturers of Xenical, state that a BMI of 30 or greater can create an increased risk of developing medical problems associated with obesity.

Describe how to assist a group of people in using the information. Some things you would want to include in your description follow.

a. How a person uses the table to determine his or her BMI.

b. How a person determines the weight that will put him or her at medical risk.

c. How a person whose weight or height is not in the table can determine if his or her BMI is 30. To answer this question, develop a formula to find the weight that would give a BMI of 30 for a person of a given height, including how you would:

BODY MASS INDEX FOR SPECIFIED HEIGHT (FT/IN.) AND WEIGHT (LB)

Height/Weight	120	130	140	150	160	170	180	190	200	210	220	230	240	250
5'0"	23	25	27	29	31	33	35	37	39	41	43	45	47	49
5'1"	23	25	27	28	30	32	34	36	38	40	42	44	45	47
5'2"	22	24	26	27	29	31	33	35	37	38	40	42	44	46
5'3"	21	23	25	27	28	30	32	34	36	37	39	41	43	44
5'4"	21	22	24	26	28	29	31	33	34	36	38	40	41	43
5'5"	20	22	23	25	27	28	30	32	33	35	37	38	40	42
5'6"	19	21	23	24	26	27	29	31	32	34	36	37	39	40
5'7"	19	20	23	24	25	27	28	30	31	33	35	36	38	39
5'8"	18	20	21	23	24	26	27	29	30	32	34	35	37	38
5'9"	18	19	21	22	24	25	27	28	30	31	33	34	36	37
5'10"	17	19	20	22	23	24	25	27	28	29	31	33	35	36
5'11"	17	18	20	21	22	24	25	27	28	29	31	32	34	35
6'0"	16	18	19	20	22	23	24	26	27	29	30	31	33	34
6'1"	16	17	19	20	21	22	24	25	26	28	29	30	32	33

(Source: Roche Pharmaceuticals)

1. Pick the points from the table that correspond to a BMI of 30, and create a table of these heights and weights. Change the units of measurement to simplify the data.
2. Determine whether a linear model is an exact or approximate fit for the data.
3. Create the linear model if it is reasonable.
4. Test the model with existing data.
5. Explain how to use the model to test for obesity.

GROUP ACTIVITY/EXTENDED APPLICATION 2

Research

Linear functions can be used to model many types of real data. Graphs displaying linear growth are frequently displayed in periodicals such as *Newsweek* and *Time*, in newspapers such as *USA Today* and the *Wall Street Journal*, and on numerous Web sites on the Internet. Tables of data can be also be found in these sources, especially in federal and state government Web sites, such as *www.census.gov*. (This Web site is the source of the Florida resident population data used in the Chapter Review, for example.)

Your mission is to find a company sales, stock price, biological growth, or sociological trend over a period of years (using at least four points) that is linear or "nearly" linear, and to find a linear function that is a model for this data.

A linear model will be a good fit for the data:

1. If the data is presented as a graph that is linear or "nearly linear."
2. If the data is presented in a table and the plot of the data points lies near some line.
3. If the data is presented in a table and the first differences of the outputs are nearly constant for equally spaced inputs.

After you have created the model, you should test the goodness of fit of the model to the data and discuss uses that you could make of the model.

Your completed project should include:
  a. A complete citation of the source of the data you are using.
  b. An original copy or photocopy of the data being used.
  c. A scatter plot of the data.
  d. The equation that you have created.
  e. A graph containing the scatter plot and the modeled equation.
  f. A statement about how the model could be used to make estimations or predictions about the trend you are observing.

Some helpful hints:

1. If you decide to use a relation determined by a graph you have found, read the graph very carefully to determine the data points, or contact the source of the data to get the data from which the graph was drawn.
2. Align the independent variable by letting x represent the number of years from some convenient year, then enter the data into a graphing utility and create a scatter plot.
3. Use your graphing utility to create the equation of the function that is the best fit for the data. Graph this equation and the data points on the same axes to see if the equation is reasonable.

CHAPTER

2

Quadratic and Other Nonlinear Functions

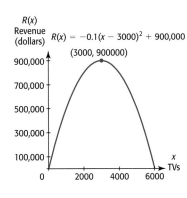

$R(x)$
Revenue (dollars)
$R(x) = -0.1(x - 3000)^2 + 900,000$
(3000, 900000)

Many applications cannot be modeled by linear functions because the rate of change is not constant. For example, revenue, cost, and profit for the production and sale of products frequently increase at rates that are not constant. The graph shows that the revenue from the sale of a product is not a linear function. In this chapter we use nonlinear functions, including quadratic and power functions, to model numerous applications in business, economics, and the life and social sciences. The main concepts and skills discussed in this chapter, along with some of their applications, follow.

	TOPICS	APPLICATIONS
2.1 Quadratic Functions; Parabolas	Graphing quadratic functions; finding vertices of parabolas; finding zeros of quadratic functions	Maximum revenue from sales, teenage drug use, height of a ball
2.2 Solving Quadratic Equations	Solving by factoring; solving graphically; combining graphs and factoring; solving with the root method; completing the square; the quadratic formula	Hospital admissions, profit, marijuana use
2.3 Power Functions and Transformations	Graphing power functions and root functions; increasing and decreasing functions; symmetry; transformations of graphs	Allometric relationships, magnitude of stimulus and response, velocity of blood
2.4 Quadratic and Power Models	Modeling with quadratic functions; comparing linear and quadratic models; modeling with power functions; comparing power and quadratic models	Cocaine use in the United Kingdom, height of a rocket, aid to dependent children, auto noise, cohabiting households, voting
2.5 Combining Functions; Reciprocal, Absolute Value, and Piecewise-Defined Functions	Adding, subtracting, multiplying and dividing functions; reciprocal functions; absolute value and piecewise-defined functions; composition of functions	Cost-benefit, average cost, residential power costs; canning orange juice
2.6 Inverse Functions	Identifying inverse functions; one-to-one functions; finding the inverse of a function; graphing inverse functions; inverse functions on limited domains	Loan balances, temperature measurement, money conversion, velocity of blood
2.7 Quadratic and Power Inequalities	Solving quadratic inequalities analytically and graphically; solving power inequalities	Height of a rocket, number of Internet users, investments

ALGEBRA *Toolbox*

In this toolbox, we discuss the properties of exponents that are used in computations with exponents, and we discuss roots and radicals. In addition, we discuss polynomials, including multiplication, factoring, and solving literal equations.

Properties of Exponents

For the real numbers a and b and positive integers m and n,

1. $a^m \cdot a^n = a^{m+n}$

2. For $a \neq 0$, $\dfrac{a^m}{a^n} = \begin{cases} a^{m-n} & \text{if } m > n \\ 1 & \text{if } m = n \\ \dfrac{1}{a^{n-m}} & \text{if } m < n \end{cases}$

3. $(ab)^m = a^m b^m$

4. $(a^m)^n = a^{mn}$

5. $\left(\dfrac{a}{b}\right)^m = \dfrac{a^m}{b^m} (b \neq 0)$

EXAMPLE 1 Using the Properties of Exponents

Use properties of exponents to simplify each of the following. Assume denominators are nonzero.

a. $\dfrac{5^6}{5^4}$ **b.** $\dfrac{y^2}{y^5}$ **c.** $(3xy)^3$ **d.** $\left(\dfrac{y}{z}\right)^4$ **e.** $3^{15-2m} \cdot 3^{2m}$

Solution

a. $\dfrac{5^6}{5^4} = 5^{6-4} = 5^2 = 25$ **b.** $\dfrac{y^2}{y^5} = \dfrac{1}{y^{5-2}} = \dfrac{1}{y^3}$ **c.** $(3xy)^3 = 3^3 x^3 y^3 = 27x^3 y^3$

d. $\left(\dfrac{y}{z}\right)^4 = \dfrac{y^4}{z^4}$ **e.** $3^{15-2m} \cdot 3^{2m} = 3^{(15-2m)+2m} = 3^{15-2m+2m} = 3^{15}$ ∎

We can also simplify and evaluate expressions involving powers of powers.

EXAMPLE 2 Powers of Powers

a. Simplify $(x^4)^5$.   **b.** Evaluate $(2^3)^5$.   **c.** Simplify $(x^2 y^3)^5$.   **d.** Evaluate 2^{3^2}.

Solution

a. $(x^4)^5 = x^{4 \cdot 5} = x^{20}$ by Property 4.

b. We can evaluate $(2^3)^5$ in two ways:

$$(2^3)^5 = 2^{15} = 32{,}768 \quad \text{or} \quad (2^3)^5 = (8)^5 = 32{,}768$$

c. $(x^2 y^3)^5 = (x^2)^5 (y^3)^5 = x^{10} y^{15}$ **d.** $2^{3^2} = 2^9 = 512$ ∎

Roots and Radicals

We know that if each side of a square is a units, then the area of the square can be written as a^2 square units, and if we know that the area of a square is S square units, then we can write the length of each edge of the square as the **square root** of S, denoted $\sqrt{S}$.

Any real number has only one odd root that is a real number. Any positive number has two even roots that are real, one that is positive and one that is negative. In this case, the positive root is called the **principal root**. The symbol $\sqrt{a}$ denotes the principal square root of a, and $\sqrt[n]{b}$ denotes the nth root of b. For example,

$$\sqrt{9} = 3, \ -\sqrt{9} = -3, \ \sqrt[3]{64} = 4$$

In $\sqrt[n]{a}$, the symbol $\sqrt[n]{}$ is called a **radical**, n is the **index of the radical**, and a is the **radicand**. If the index is 2, we omit it from the radical and write $\sqrt{a}$ to represent the square root of a. We summarize the important facts regarding roots below.

RADICAL FACTS

1. $\sqrt[n]{0} = 0$.

2. If a is a positive real number and n is even, there are two real n^{th} roots of a, $\sqrt[n]{a}$ and $-\sqrt[n]{a}$ where $\sqrt[n]{a}$ is the principal root.

3. If a is a negative real number and n is even, there is no real n^{th} root of a.

4. If a is a real number and n is odd, $\sqrt[n]{a^n} = a$.

5. If a is a real number and n is even, $\sqrt[n]{a^n} = |a|$; in particular, $\sqrt{a^2} = |a|$.

6. If $\sqrt[n]{a}$ and $\sqrt[n]{b}$ are real numbers, $\sqrt[n]{a} \cdot \sqrt[n]{b} = \sqrt[n]{ab}$.

7. If $\sqrt[n]{a}$ and $\sqrt[n]{b}$ are real numbers, $\dfrac{\sqrt[n]{a}}{\sqrt[n]{b}} = \sqrt[n]{\dfrac{a}{b}}$.

EXAMPLE 3

Find the following roots, if they are real numbers.

a. $\sqrt[4]{16}$ **b.** $\sqrt[3]{-8}$ **c.** $-\sqrt{64}$ **d.** $\sqrt{-64}$ **e.** $\sqrt[4]{(-2)^4}$

Solution

a. $\sqrt[4]{16} = 2$ because $2^4 = 16$.

b. $\sqrt[3]{-8} = -2$ because $(-2)^3 = -8$.

c. $-\sqrt{64} = -(8) = -8$.

d. $\sqrt{-64}$ is not a real number because no real number squared gives -64.

e. $\sqrt[4]{(-2)^4} = |-2| = 2$

Note also that $\sqrt[4]{(-2)^4} = \sqrt[4]{16} = 2$. ∎

We can frequently simplify radicals by removing from the radicand all factors whose indicated roots are present. For example, we can remove squares from a square root by using this procedure. Writing Radical Fact 6 ($\sqrt[n]{a} \cdot \sqrt[n]{b} = \sqrt[n]{ab}$) in the form $\sqrt[n]{ab} = \sqrt[n]{a} \cdot \sqrt[n]{b}$ is a useful step in this process.

EXAMPLE 4

Simplify the following radicals, assuming that the expressions are real.

a. $\sqrt{72}$ **b.** $\sqrt{b^2 - 4ac}$ if $a = 7, b = -10, c = -1$

Solution

a. Because 72 is divisible by the square (36), we can simplify the radical as follows:

$$\sqrt{72} = \sqrt{36 \cdot 2} = \sqrt{36}\sqrt{2} = 6\sqrt{2}$$

b. Substituting the values of a, b, and c gives $\sqrt{(-10)^2 - 4(7)(-1)} = \sqrt{128}$. We seek squares that are factors of 128 to simplify the radical.

$$\sqrt{128} = \sqrt{4 \cdot 32} = \sqrt{4} \cdot \sqrt{16 \cdot 2} = \sqrt{4}\sqrt{16}\sqrt{2} = 2 \cdot 4\sqrt{2} = 8\sqrt{2}$$

This simplification could be done more quickly by noting that the square 64 divides 128.

$$\sqrt{128} = \sqrt{64 \cdot 2} = \sqrt{64} \cdot \sqrt{2} = 8\sqrt{2}$$ ■

Radicals containing variables can be simplified in the same manner as those containing numbers.

EXAMPLE 5

Simplify the following radicals, assuming that the expressions are real.

a. $\sqrt[3]{x^6 y^5}$ **b.** $\sqrt[4]{32x^5 y^{10}}$

Solution

a. To simplify $\sqrt[3]{x^6 y^5}$, we rewrite the radicand with powers of 3.

$$\sqrt[3]{x^6 y^5} = \sqrt[3]{(x^2)^3 y^3 \cdot y^2} = \sqrt[3]{(x^2)^3} \cdot \sqrt[3]{y^3} \cdot \sqrt[3]{y^2} = x^2 y \sqrt[3]{y^2}$$

b. To simplify $\sqrt[4]{32x^5 y^{10}}$, we seek powers of 4 in the radicand.

$$\sqrt[4]{32x^5 y^{10}} = \sqrt[4]{16 \cdot 2 \cdot x^4 \cdot x \cdot y^8 \cdot y^2} = \sqrt[4]{16x^4 \cdot (y^2)^4 \cdot 2xy^2}$$
$$= \sqrt[4]{16}\sqrt[4]{x^4}\sqrt[4]{(y^2)^4}\sqrt[4]{2xy^2} = 2|x|y^2\sqrt[4]{2xy^2}$$ ■

Radicals and Rational Exponents

To use technology to graph equations involving radicals, it is sometimes easier to convert expressions involving radicals to equivalent expressions involving fractional exponents. Note that for $a \geq 0$ and $b \geq 0$

$$\sqrt{a} = b \text{ only if } a = b^2.$$

Thus

$$\left(\sqrt{a}\right)^2 = b^2 = a, \text{ so}$$
$$\left(\sqrt{a}\right)^2 = a$$

We define $a^{1/2} = \sqrt{a}$, so $(a^{1/2})^2 = a$ for $a \geq 0$

The following definitions show the connection between rational exponents and radicals.

RATIONAL EXPONENTS

If a is a real number, variable, or algebraic expression and n is a positive integer for which the principal root exists, then

$$a^{1/n} = \sqrt[n]{a}.$$

EXAMPLE 6

Write the following expressions with exponents rather than radicals.

a. $\sqrt[3]{x^2}$ **b.** $\sqrt[4]{x^3}$ **c.** $\sqrt{(3xy)^5}$ **d.** $3\sqrt{(xy)^5}$

Solution

a. $\sqrt[3]{x^2} = x^{2/3}$ **b.** $\sqrt[4]{x^3} = x^{3/4}$ **c.** $\sqrt{(3xy)^5} = (3xy)^{5/2}$ **d.** $3(xy)^{5/2}$ ∎

EXAMPLE 7

Write the following in radical form.

a. $y^{1/2}$ **b.** $(3x)^{3/7}$ **c.** $12x^{3/5}$

Solution

a. $y^{1/2} = \sqrt{y}$ **b.** $(3x)^{3/7} = \sqrt[7]{(3x)^3} = \sqrt[7]{27x^3}$ **c.** $12x^{3/5} = 12\sqrt[5]{x^3}$ ∎

Polynomials

An algebraic expression containing a finite number of additions, subtractions, and multiplications of constants and positive integer powers of variables is called a **polynomial**. A polynomial cannot contain negative powers of variables, fractional powers of variables, variables in a denominator, or variables inside a radical. The expressions $5x - 2y$ and $7z^3 + 2y$ are polynomials, but $\dfrac{3x - 5}{12 + 5y}$ and $3x^2 - 6\sqrt{x}$ are not polynomials. If the only variable in the polynomial is x, then the polynomial is called a **polynomial in** x. The general form of a polynomial in x is

$$a_n x^n + a_{n-1}x^{n-1} + \ldots + a_1 x + a_0$$

where a_0 and each coefficient $a_n, a_{n-1},\ldots$ is a real number and each exponent $n, n - 1,\ldots$ is a positive integer.

For a polynomial in the single variable x, the power of x in each term is the **degree** of that term, with the degree of a constant term equal to 0. The term that has the highest power of x is called the **leading term** of the polynomial, the coefficient of this term is the **leading coefficient**, and the degree of this term is the **degree of a polynomial**. Thus $5x^4 + 3x^2 - 6$ is a fourth degree polynomial with leading coefficient 5. If a term contains two or more variables, the **degree of the term** is the sum of the powers of the variables in that term. Thus, the degree of $8x^4y^3$ is $4 + 3 = 7$. The degree of a polynomial containing one or more variables is the degree of the term that has the highest degree in the polynomial. Polynomials with one term are called **monomials**, those with two terms are called **binomials**, and those with three terms are called **trinomials**.

EXAMPLE 8

For each polynomial, state the constant term, the degree of the highest degree term, the leading coefficient, and the degree of the polynomial.

a. $5x^2 - 8x + 2x^4 - 3$ **b.** $5xy^2 - 6x^3y + 3x^4y^2 + 7$

Solution

a. The constant term is -3; the degree of the highest degree term is 4, so the leading coefficient is 2 and the degree of the polynomial is 4.

b. The constant term is 7; the degree of the highest degree term is $4 + 2 = 6$, so the leading coefficient is 3 and the degree of the polynomial is 6. ∎

Special Products

We multiply two monomials by multiplying the coefficients and adding the exponents of the respective variables that are in both monomials. For example,

$$(3x^3y^2)(4x^2y) = 3 \cdot 4 \cdot x^3 \cdot x^2 \cdot y^2 \cdot y = 3 \cdot 4x^{3+2}y^{2+1} = 12x^5y^3$$

We can multiply more than two monomials in the same manner.

EXAMPLE 9

Find the product:

$$(-2x^4z)(4x^2y^3)(yz^5)$$

Solution

$$(-2x^4z)(4x^2y^3)(yz^5) = -2 \cdot 4x^{4+2}y^{3+1}z^{1+5} = -8x^6y^4z^6$$ ∎

We can use the **distributive property**,

$$a(b + c) = ab + ac$$

to multiply a monomial times a polynomial. For example,

$$x(3x + y) = x \cdot 3x + x \cdot y = 3x^2 + xy.$$

We can extend the property $a(b + c) = ab + ac$ to multiply a monomial times any polynomial. For example,

$$3x(2x + xy + 6) = 3x \cdot 2x + 3x \cdot xy + 3x \cdot 6 = 6x^2 + 3x^2y + 18x.$$

EXAMPLE 10

Find the following products:

a. $-2ab(3ax + 12bc - 2abx)$ **b.** $(3st^2 - 2tx + 5s^3t)6st$

Solution

a. $-2ab(3ax + 12bc - 2abx) = -2ab(3ax) + (-2ab)(12bc) - (-2ab)(2abx)$
$$= -6a^2bx - 24ab^2c + 4a^2b^2x$$

b. Multiplying by $6st$ on the right gives

$$(3st^2 - 2tx + 5s^3t)6st = 3st^2(6st) - 2tx(6st) + 5s^3t(6st)$$
$$= 18s^2t^3 - 12st^2x + 30s^4t^2. \qquad \blacksquare$$

The product of two binomials can be found by using the distributive property as follows:

$$(a + b)(c + d) = a(c + d) + b(c + d) = ac + ad + bc + bd$$

Note that this product can be remembered as the sum of the products of the **F**irst, **O**uter, **I**nner, and **L**ast terms of the binomials, and we use the word FOIL to denote this method.

EXAMPLE 11

Find the following products:

a. $(x - 4)(x - 5)$ **b.** $(2x - 3)(3x + 2)$ **c.** $(3x - 5y)(3x + 5y)$

Solution

a. $(x - 4)(x - 5) = x \cdot x + x(-5) + (-4)x + (-4)(-5)$
$$= x^2 - 5x - 4x + 20 = x^2 - 9x + 20$$

b. $(2x - 3)(3x + 2) = (2x)(3x) + (2x)2 + (-3)(3x) + (-3)2$
$$= 6x^2 + 4x - 9x - 6 = 6x^2 - 5x - 6$$

c. $(3x - 5y)(3x + 5y) = (3x)(3x) + (3x)(5y) + (-5y)(3x) + (-5y)(5y)$
$$= 9x^2 + 15xy - 15xy - 25y^2 = 9x^2 - 25y^2 \qquad \blacksquare$$

Certain products and powers involving binomials occur frequently, so the following special products should be remembered.

SPECIAL BINOMIAL PRODUCTS

I. $(x + a)(x + b) = x^2 + (a + b)x + ab$

II. $(x + a)(x - a) = x^2 - a^2$ (Difference of two squares)

III. $(x + a)^2 = x^2 + 2ax + a^2$ (Perfect square)

IV. $(x - a)^2 = x^2 - 2ax + a^2$ (Perfect square)

EXAMPLE 12

Find the following products by using the special binomial products formulas.

a. $(5x + 1)^2$ **b.** $(2x - 5)(2x + 5)$ **c.** $(3x - 4)^2$

Solution

a. $(5x + 1)^2 = (5x)^2 + 2(5x)(1) + 1^2 = 25x^2 + 10x + 1$

b. $(2x - 5)(2x + 5) = (2x)^2 - 5^2 = 4x^2 - 25$

c. $(3x - 4)^2 = (3x)^2 - 2(3x)(4) + 4^2 = 9x^2 - 24x + 16$ ■

Factoring

Factoring is the process of writing a number or an algebraic expression as the product of two or more numbers or expressions. For example, the distributive property justifies factoring of monomials from polynomials. For example,

$$5x^2 - 10x = 5x(x - 2)$$

By recognizing that a polynomial has the form of one of the special products given above, we can factor that polynomial.

EXAMPLE 13

Use knowledge of binomial products to factor the following algebraic expressions.

a. $9x^2 - 25$ **b.** $4x^2 - 12x + 9$

Solution

a. Both terms are squares, so the polynomial can be recognized as the **difference of two squares**. It will then factor as the product of the sum and the difference of the square roots (see Special Binomial Products, Formula II).

$$9x^2 - 25 = (3x + 5)(3x - 5)$$

b. Recognizing that the second-degree term and the constant term are squares leads us to investigate whether $-12x$ is twice the product of the square roots of these two terms (see Special Binomial Products, Formula IV). The answer is yes, so the polynomial is a **perfect square**, and it can be factored as follows:

$$4x^2 - 12x + 9 = (2x - 3)^2$$

This can be verified by expanding $(2x - 3)^2$.   ■

When an algebraic expression is written as the product of irreducible factors, we say the expression is **factored completely**. Applying special product formulas to factorizations will not necessarily result in the expression being factored completely. The first step in factoring is to look for common monomial factors.

EXAMPLE 14

Factor the following polynomials completely.

a. $3x^2 - 33x + 72$ **b.** $6x^2 - x - 1$

Solution

a. The number 3 can be factored from all three terms, giving

$$3x^2 - 33x + 72 = 3(x^2 - 11x + 24).$$

If the trinomial can be factored into the product of two binomials, the first term of each binomial must be x, and we seek two numbers whose product is 24 and whose sum is -11. Because -3 and -8 satisfy these requirements, we get

$$3(x^2 - 11x + 24) = 3(x - 3)(x - 8).$$

b. The four possible factorizations of $6x^2 - x - 1$ that give $6x^2$ as the product of the first terms and -1 as the product of the last terms follow:

$$(6x - 1)(x + 1) \qquad (6x + 1)(x - 1)$$
$$(2x - 1)(3x + 1) \qquad (2x + 1)(3x - 1)$$

The factorization that gives a product with middle term $-x$ is the correct factorization.

$$(2x - 1)(3x + 1) = 6x^2 - x - 1 \qquad \blacksquare$$

Some polynomials, such as $6x^2 + 9x - 8x - 12$, can be factored by **grouping**. To do this, we factor common factors from pairs of terms and then factor a common binomial expression if it exists. For example,

$$6x^2 + 9x - 8x - 12 = 3x(2x + 3) - 4(2x + 3) = (2x + 3)(3x - 4)$$

When a second-degree trinomial has a form that can be factored, but there are many possible factors to test to find the correct one, an alternate method of factoring can be used. The steps used to factor the trinomial using factoring by grouping techniques follow.

Factoring a Trinomial into the Product of Two Binomials Using Grouping

Steps	Example
To factor a quadratic trinomial in the variable x:	Factor $5x - 6 + 6x^2$
1. Arrange the trinomial with the powers of x in descending order.	1. $6x^2 + 5x - 6$
2. Form the product of the second-degree term and the constant term (first and third terms).	2. $6x^2(-6) = -36x^2$
3. Determine if there are two factors of the product in Step 2 that will sum to the middle (first-degree) term. (If there are no such factors, the trinomial will not factor into two binomials.)	3. $-36x^2 = (-4x)(9x)$ and $-4x + 9x = 5x$
4. Replace the middle term from Step 1 with the *sum* of the two factors from Step 3.	4. $6x^2 + 5x - 6 = 6x^2 - 4x + 9x - 6$
5. Factor the four-term polynomial from Step 4 by grouping.	5. $6x^2 + 5x - 6 = 2x(3x - 2) + 3(3x - 2) = (3x - 2)(2x + 3)$

EXAMPLE 15

Factor $10x^2 + 23x - 5$ using grouping.

Solution

To use this method to factor $10x^2 + 23x - 5$, we:

1. Note that the trinomial has powers $10x^2 + 23x - 5$
of x in descending order.

2. Multiply the second degree and $10x^2(-5) = -50x^2$
constant terms:

3. Factor $-50x^2$ so the sum of $-50x^2 = 25x(-2x), \; 25x + (-2x) = 23x$
factors is $23x$:

4. Rewrite the middle term of the $10x^2 + 23x - 5 = 10x^2 + 25x - 2x - 5$
expression as the sum of the
factors from Step 3.

5. Factor by grouping:

$$10x^2 + 25x - 2x - 5$$
$$= 5x(2x + 5) - (2x + 5)$$
$$= (2x + 5)(5x - 1) \quad \blacksquare$$

Solving Literal Equations

To solve a literal equation for one of two or more variables, it is sometimes necessary to use factoring. Consider the following example.

EXAMPLE 16   Solving an Equation for a Specified Variable

Solve the equation $2(2x - b) = \dfrac{5cx}{3}$ for x.

Solution

We solve the equation for x by treating the other variables as constants:

$$2(2x - b) = \frac{5cx}{3}$$

$6(2x - b) = 5cx$ Clear the equation of fractions by multiplying by the LCD, 3.

$12x - 6b = 5cx$ Multiply to remove parentheses.

$12x - 5cx = 6b$ Get all terms containing x on one side and all other terms on the other side.

$x(12 - 5c) = 6b$ Factor x from the expression. The remaining factor is the coefficient of x.

$\dfrac{x(12 - 5c)}{(12 - 5c)} = \dfrac{6b}{12 - 5c}$ Divide both sides by the coefficient of x.

$$x = \frac{6b}{12 - 5c} \quad \blacksquare$$

Toolbox EXERCISES

Use the rules of exponents to simplify.

1. **a.** $x^4 \cdot x^3$ **b.** $\dfrac{x^{12}}{x^7}$ **c.** $(4ay)^4$ **d.** $\left(\dfrac{3}{z}\right)^4$
   e. $2^3 \cdot 2^2$ **f.** $(x^4)^2$

2. Evaluate each of the following roots, if they are real numbers.
   a. $\sqrt{16}$ **b.** $-\sqrt{16}$ **c.** $\sqrt{-16}$
   d. $\sqrt[3]{27}$ **e.** $-\sqrt[5]{-32}$

3. Simplify the following radicals, assuming that the expressions are real.
   a. $\sqrt{18}$ **b.** $\sqrt{216}$
   c. $\sqrt{b^2 - 4ac}$ if $a = 2, b = -12, c = -3$
   d. $\sqrt[3]{243a^7b^4}$

4. Write each of the following expressions in simplified exponential form.
   a. $\sqrt{x^3}$ **b.** $\sqrt[4]{x^3}$ **c.** $\sqrt[5]{x^3}$ **d.** $\sqrt[6]{27y^9}$
   e. $27\sqrt[6]{y^9}$

5. Write each of the following in radical form.
   a. $a^{3/4}$ **b.** $-15x^{5/8}$ **c.** $(-15x)^{5/8}$

For each polynomial in Exercises 6–9, (a) give the degree of the polynomial and (b) give the leading coefficient, give the constant term, and decide if it is a polynomial in one or several variables.

6. $4x - 3x^2 + 8$ 7. $7x^3 - 3 + 5x^4$

8. $3x + 2y^4 - 2x^3y^4 - 119$

9. $z^4 - 15z^2 + 20z - 6$

In Exercises 10–14, find the products.

10. $(4x^2y^3)(-3a^2x^3)$

11. $2xy^3(2x^2y + 4xz - 3z^2)$

12. $(x - 7)(2x + 3)$ 13. $(k - 3)^2$

14. $(4x - 7y)(4x + 7y)$

In Exercises 15–24, factor each of the polynomials completely.

15. $3x^2 - 12x$ 16. $12x^5 - 24x^3$

17. $9x^2 - 25m^2$ 18. $x^2 - 8x + 15$

19. $x^2 - 2x - 35$ 20. $3x^2 - 5x - 2$

21. $8x^2 - 22x + 5$ 22. $6n^2 + 18 + 39n$

23. $3y^4 + 9y^2 - 12y^2 - 36$

24. $18p^2 + 12p - 3p - 2$

25. Solve $5x - 4xy = 2y$ for y.

26. Solve $2xy = \dfrac{6}{y} + \dfrac{x}{3}$ for x.

2.1 Quadratic Functions; Parabolas

Key Concepts

- Quadratic function
- $y = ax^2 + bx + c$
- $y = a(x - h)^2 + k$
- Parabola
  - Vertex
  - Axis of symmetry
  - Maxima and minima
- Zeros of quadratic functions
- *x*-Intercepts

Suppose that the monthly revenue from the sale of Carlson 32-inch televisions is given by the function

$$R(x) = -0.1x^2 + 600x \text{ dollars}$$

where x is the number of TVs sold. In this case, the revenue of this product is represented by a **second-degree**, or **quadratic**, **function**. A quadratic function is a function that can be written in the form

$$f(x) = ax^2 + bx + c \quad \text{where } a, b, \text{ and } c \text{ are real numbers with } a \neq 0.$$

The graph of the quadratic function $f(x) = ax^2 + bx + c$ is a **parabola** with a "turning point" called the **vertex**. Figure 2.1 shows the graph of the function $R(x) = -0.1x^2 + 600x$, which is a parabola that opens downward, and the vertex occurs where the function has its maximum value. Notice also that the graph of this revenue function is symmetric about a vertical line through the vertex. This vertical line is called the **axis of symmetry**.

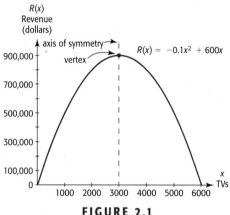

FIGURE 2.1

Knowing where the maximum value of $R(x)$ occurs can show the company how many units must be sold in order to obtain the largest revenue. Knowing the maximum output of $R(x)$ helps the company plan its sales campaign. Learning how to determine these quantities is one of the objectives of this section.

Parabolas

TABLE 2.1

x	y_1
-4	16
-3	9
-2	4
-1	1
0	0
1	1
2	4
3	9
4	16

The graph of every quadratic function has the distinctive shape known as a parabola. Three key ingredients determine the graph of a quadratic function: the location of the vertex, whether the parabola opens upward or downward, and how wide or narrow it is.

Consider the basic quadratic function $y = x^2$. Each output y is obtained by squaring an input x (see Table 2.1). The graph of $y = x^2$, shown in Figure 2.2, is a parabola that opens upward with the **vertex** (turning point) at the origin, $(0, 0)$.

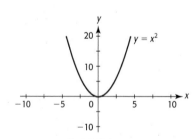

FIGURE 2.2

Observe that for $x > 0$, the graph of $y = x^2$ rises as it moves from left to right (that is, as the x-values increase), so the function $y = x^2$ is **increasing** for $x > 0$. For values of $x < 0$, the graph falls as it moves from left to right (as the x-values increase), so the function $y = x^2$ is **decreasing** for $x < 0$.

The graph of the quadratic function $y = -x^2$ (shown in Figure 2.3(a)), is a parabola that opens downward with vertex at $(0, 0)$.

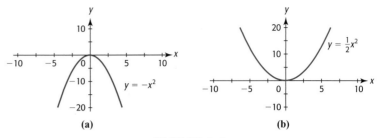

FIGURE 2.3

The function $y = \frac{1}{2}x^2$ is of the form $y = ax^2$ with $a > 0$, and its graph is a parabola opening upward (see Figure 2.3(b)). In general, the graph of a quadratic function of the form $y = ax^2$ is a parabola that opens upward (is **concave up**) if a is positive and opens downward (is **concave down**) if a is negative.* The **vertex**, which is the point where the parabola turns, is a **minimum point** if a is positive and is a **maximum point** if a is negative. The vertical line through the vertex is called the **axis of symmetry** because this line divides the graph into two halves that are reflections of each other. The vertex of $y = ax^2$ occurs at the point (0, 0) for any value of a. (See Figure 2.4.) If $a > 1$, the graph will rise more rapidly (and appear more narrow) than the graph of $y = x^2$. If $0 < a < 1$, the graph will rise more slowly (and appear more wide) than the graph of $y = x^2$. Notice that the graph of $y = \frac{1}{2}x^2$ (in Figure 2.3(b)) is more wide than that of $y = x^2$ (in Figure 2.2.)

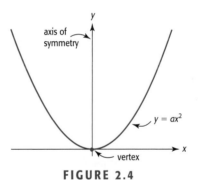

FIGURE 2.4

Standard Form of a Quadratic Function

The graph of $y = (x - 2)^2 + 3$ has the same shape as that of $y = x^2$, but it is shifted to the right 2 units and up 3 units. (See Figure 2.5(a).) In general, when a quadratic function is written in the **standard form**

$$y = a(x - h)^2 + k,$$

each point on its graph can be obtained from a point on the graph of $y = ax^2$ by shifting it h units to the right (when h is positive) or h units to the left (when h is negative) and vertically by k units (up when k is positive and down when k is negative). This means that the vertex of $y = (x - h)^2 + k$ is at (h, k). (See Figure 2.5(b).)

---

*A parabola that is concave up appears that it will "hold water" and a parabola that is concave down will appear to "shed water."

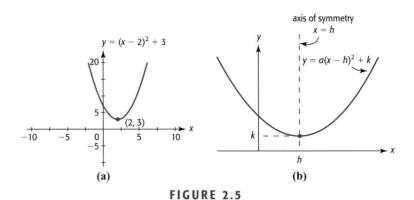

FIGURE 2.5

GRAPH OF A QUADRATIC FUNCTION IN STANDARD FORM

In general, the graph of the function

$$y = a(x - h)^2 + k$$

is a parabola with its vertex at the point (h, k).

The parabola opens upward if $a > 0$, and the vertex is a minimum.
The parabola opens downward if $a < 0$, and the vertex is a maximum.
The axis of symmetry of the parabola has equation $x = h$.

EXAMPLE 1 Vertex of a Parabola

Find the vertex of the function $R(x) = -0.1(x - 3000)^2 + 900{,}000$. Is the vertex a maximum or a minimum?

Solution

a. This function is in standard form with $h = 3000$ and $k = 900{,}000$. Thus the vertex of the function $R(x) = -0.1(x - 3000)^2 + 900{,}000$ is $(3000, 900{,}000)$.

Because $a = -0.1 < 0$, the vertex is a maximum. (The graph of this function is shown in Figure 2.6.)

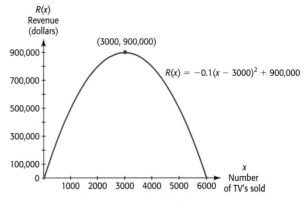

FIGURE 2.6

General Form of a Quadratic Function

Expanding the monthly Carlson revenue function $R(x) = -0.1(x - 3000)^2 + 900{,}000$ gives

$$-0.1(x - 3000)^2 + 900{,}000 = -0.1(x^2 - 6000x + 9{,}000{,}000) + 900{,}000$$
$$= -0.1x^2 + 600x,$$

so the standard form $R(x) = -0.1(x - 3000)^2 + 900{,}000$ is equivalent to $R = -0.1x^2 + 600x$. Thus these equations have the same graph, with the vertex $(3000, 900{,}000)$.

We have learned that the x-coordinate of the vertex of the graph of a quadratic function

$$y = a(x - h)^2 + k$$

occurs at $x = h$. The **general form** of a quadratic function is

$$f(x) = ax^2 + bx + c.$$

It can be shown that the x-coordinate of the vertex occurs at $x = \dfrac{-b}{2a}$. For example, we found above that the x-coordinate of the vertex of the graph of $R = -0.1x^2 + 600x$ is 3000. This could also be found by noting that $a = -0.1$ and $b = 600$ and using

$$x = \frac{-b}{2a} = \frac{-600}{2(-0.1)} = \frac{-600}{-0.2} = 3000.$$

GRAPH OF A QUADRATIC FUNCTION IN GENERAL FORM

The graph of a quadratic function in the form

$$f(x) = ax^2 + bx + c$$

is a parabola which opens upward if a is positive and downward if a is negative. The x-coordinate of the vertex is $x = \dfrac{-b}{2a}$. The y-coordinate of the vertex can be found by evaluating the function at $x = \dfrac{-b}{2a}$. That is, the coordinates of the vertex are $\left(\dfrac{-b}{2a}, f\left(\dfrac{-b}{2a}\right)\right)$. The axis of symmetry of the graph is the vertical line $x = \dfrac{-b}{2a}$. (See Figure 2.7.)

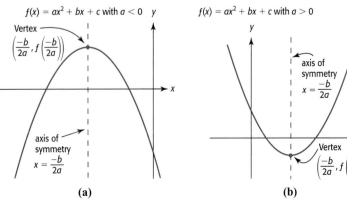

FIGURE 2.7

If we know the location of the vertex and the direction in which the parabola opens, we can make a good sketch of the graph by plotting just a few more points.

EXAMPLE 2   Vertices of Parabolas

Suppose that the profit from the sale of sweatshirts sold at a school fund-raiser is

$$P(x) = -2x^2 + 60x,$$

where $P(x)$ is the profit when each sweatshirt sells for x dollars.

a. Find the vertex and the axis of symmetry of the graph of this function.

b. Determine if the vertex represents a maximum or minimum point.

c. Interpret the vertex in the context of the application.

d. Graph this function.

Solution

a. The function is a quadratic function in general form with $a = -2$, $b = 60$, and $c = 0$. The x-coordinate of the vertex is $\dfrac{-b}{2a} = \dfrac{-60}{2(-2)} = 15$ and the axis of symmetry is the line $x = 15$. The y-coordinate of the vertex is

$$P(15) = -2(15)^2 + 60(15) = 450.$$

b. Because $a = -2$, the parabola opens downward, so the vertex is a maximum point.

c. The x-coordinate of the vertex gives the price that should be charged in order to maximize the profit, so charging $15 will result in the maximum profit. The maximum profit, $450, is the y-coordinate of the vertex.

d. A table of some values that satisfy $P(x) = -2x^2 + 60x$ and its graph are shown in Figure 2.8(a).

x	P
0	0
10	400
15	450
20	400
30	0

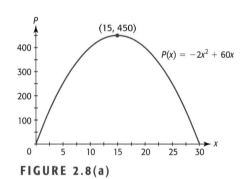

FIGURE 2.8(a)

Even when using a graphing utility to graph a quadratic function, recognizing that the graph is a parabola and locating the vertex is important. Determining which way the parabola opens and the x-coordinate of the vertex is very useful in setting the viewing window so that a complete graph (that includes the vertex and the intercepts) is shown. For example, suppose we wanted to graph the sweatshirt function with technology. Graphing this function with the standard viewing window yields a graph that appears to be a line very close to the y-axis (see Figure 2.8(b)); this graph gives no indication of the location of the vertex or which way the parabola opens. We know that the vertex of the graph of this function is (15, 450) and that it is a maximum, so we can choose a window with $x = 15$ near the center and $y = 450$ near the top. We use a window with $x_{\min} = 0$, $x_{\max} = 30$, $y_{\min} = 0$, and $y_{\max} = 460$ (see Figure 2.8(c)).

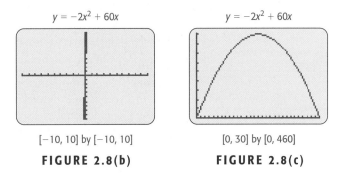

$y = -2x^2 + 60x$ $y = -2x^2 + 60x$

$[-10, 10]$ by $[-10, 10]$ $[0, 30]$ by $[0, 460]$

FIGURE 2.8(b)   **FIGURE 2.8(c)**

| EXAMPLE 3 | Teenage Drug Use |

The percent of twelfth grade students using any drug during the years 1980 to 1998 can be modeled by the equation $y = 0.181x^2 - 4.093x + 68.771$, where x is the number of years after 1980.

a. During what year does the model indicate that this usage was a minimum?

b. What is the minimum? (Source: Index of Leading Cultural Indicators)

Solution

a. This equation is in the general form $f(x) = ax^2 + bx + c$, so $a = 0.181$. Because $a > 0$, the parabola opens upward, the vertex of this parabola is a minimum, and the x-coordinate of the vertex is where the minimum usage occurs.

$$x = \frac{-b}{2a} = \frac{-(-4.093)}{2(0.181)} \approx 11.3$$

b. The minimum percent for the model is found by evaluating the function at $x = 11.3$. This value, $y = 45.6$, can be found with $\boxed{\text{TRACE}}$, $\boxed{\text{CALC}}$, $\boxed{\text{VALUE}}$, or $\boxed{\text{TABLE}}$, or with direct evaluation (see Figure 2.9(a)).

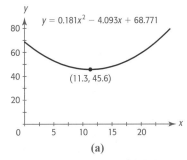

x	y_1
11	45.649
12	45.719

(b)

(a)

FIGURE 2.9

This application gives end-of-year totals, so it should be interpreted discretely. Evaluating the function at $x = 11$ gives an output of 45.649 and evaluating it at $x = 12$ gives 45.719 (See the table in Figure 2.9(b)). The smaller of these two outputs is at $x = 11$, so the answer is 1991. Twelfth grade drug use between 1980 and 1998 was least in 1991. ∎

Zeros of Quadratic Functions

As we learned in Chapter 1, the zeros of a function occur where the graph of the function crosses or touches the x-axis. We can use the graph of a quadratic function to determine the number and type of real number zeros the function has. Keep these facts in mind.

- A parabola will cross the x-axis at zero, one, or two points, which tells us that there are 0, 1, or 2 real zeros (x-intercepts) of the function. (See Figures 2.10(a), (b), and (c).) If the graph does not cross or touch the x-axis, there are no real zeros, and if the graph has its vertex on the x-axis, there is one real zero. If the graph crosses the x-axis at two points, the function has two real zeros.

- The x-intercept method can be used to find or to approximate the real zeros. Even though a decimal approximation will often suffice as an answer in an applied problem, remember that not all x-intercepts (zeros of the function) can be found exactly using a graphing utility.

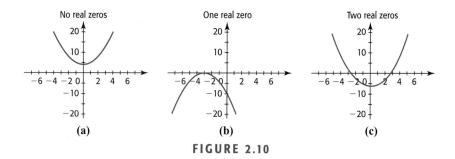

FIGURE 2.10

EXAMPLE 4   Height of a Ball

The height above ground of a ball thrown at 64 feet per second from the top of an 80-foot-high building is modeled by $S = 80 + 64t - 16t^2$ feet, where t is the number of seconds after the ball is thrown.

a. Describe the graph of the model.

b. Find the t-coordinate and S-coordinate of the vertex of the graph of this quadratic function.

c. Explain the meaning of the coordinates of the vertex for this model.

d. Over what time interval is this model increasing? What does this mean in the context of the model?

e. Over what time interval is this model decreasing, and what does this mean?

Solution

a. The function is quadratic and the coefficient of the second-degree term $(-16t^2)$ is negative, so the graph is a parabola that opens down. The graph is shown in Figure 2.11(a).

b. The height S is a function of the time t, and the t-coordinate of the vertex is

$$t = \frac{-b}{2a} = \frac{-64}{2(-16)} = 2.$$

The S-coordinate of the vertex is the value of S at $t = 2$, and $S(2) = 80 + 64(2) - 16(2^2) = 144$. Thus the vertex is (2, 144).

c. The graph is a parabola that opens down, so the vertex is the highest point on the graph and the function has its maximum there. The t-coordinate of the vertex, 2, is the time (in seconds) at which the ball reaches its maximum height, and the S-coordinate, 144, is the maximum height (in feet) that the ball reaches.

d. The function is increasing until $t = 2$, where the graph reaches its vertex. Because the model does not apply for negative values of t (the time), we consider only non-negative values of t; the ball is climbing for $0 < t < 2$; that is, for the first 2 seconds after it is thrown.

e. The function $S = 80 + 64t - 16t^2$ decreases for all $t > 2$, but the ball will only fall until it reaches the ground, so the *model* is decreasing from $t = 2$ to the time when $S = 0$. To find this value of t, we find the x-intercepts of the graph of the function. The graph of this function, with x representing the time t and y representing the height S, is shown in Figure 2.11(b) with a positive x-intercept of 5. The negative zero is not relevant to the problem, so the ball strikes the ground at $t = 5$ seconds. Thus the model is decreasing for $2 < t < 5$. This means that the ball is falling after it has reached its maximum height; this occurs between 2 seconds and 5 seconds after it is thrown.

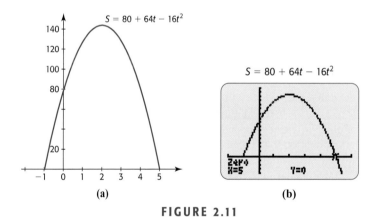

(a)   (b)

FIGURE 2.11   ■

2.1 SKILLS CHECK

In Exercises 1–6, (a) determine if the function is quadratic. If it is,
(b) Determine if the graph is concave up or concave down.
(c) Determine if the vertex of the graph is a maximum point or a minimum point.

1. $y = 2x^2 - 8x + 6$ **2.** $y = 4x - 3$

3. $y = 2x^3 - 3x^2$ **4.** $f(x) = x^2 + 4x + 4$

5. $g(x) = -5x^2 - 6x + 8$

6. $h(x) = -2x^2 - 4x + 6$

In Exercises 7–10, (a) graph each quadratic function on $[-10, 10]$ *by* $[-10, 10]$.
(b) Does this window give a complete graph?

7. $y = 2x^2 - 8x + 6$ **8.** $f(x) = x^2 + 4x + 4$

9. $g(x) = -5x^2 - 6x + 8$

10. $h(x) = -2x^2 - 4x + 6$

In Exercises 11–16, (a) give the coordinates of the vertex of the graph of each function.
(b) Graph each function on a window that includes the vertex.

11. $y = (x + 4)^2 + 3$ **12.** $y = (x - 2)^2 + 1$

13. $y = 12x - 3x^2$ **14.** $y = 3x + 18x^2$

15. $y = 3x^2 + 18x - 3$ **16.** $y = 5x^2 + 15x + 8$

Use the graph to estimate the x-intercepts of the graph of each function in Exercises 17–20.

17. $y = 2x^2 - 8x + 6$ **18.** $f(x) = x^2 + 4x + 4$

19. $g(x) = -5x^2 - 6x + 8$

20. $h(x) = -2x^2 - 4x + 6$

For Exercises 21–24, (a) Find the x-coordinate of the vertex of the graph.
(b) Set the viewing window so that the x-coordinate of the vertex is near the center of the window and the vertex is visible, and then graph the given equation.

21. $y = 2x^2 - 40x + 10$

22. $y = -3x^2 - 66x + 12$

23. $y = -0.2x^2 - 32x + 2$

24. $y = 0.3x^2 + 12x - 8$

2.1 EXERCISES

1. *Profit* The daily profit for a product is given by $P = 32x - 0.01x^2 - 1000$, where x is the number of units produced and sold.
  a. Graph this function for x between 0 and 3200.
  b. Describe what happens to the profit for this product when the number of units produced is between 1 and 1600.
  c. What happens after 1600 units are produced?

2. *Profit* The daily profit for a product is given by $P = 420x - 0.1x^2 - 4100$ dollars, where x is the number of units produced and sold.
  a. Graph this function for x between 0 and 4200.
  b. Is the graph of the function concave up or down?
  c. What is the average rate of change of the profit between $x = 400$ and $x = 800$?

3. *Revenue and Cost* The total revenue function for a certain product is given by $R = 1050x$ dollars and the total cost function for this product is $C = 10,000 + 30x + x^2$ dollars, where x is the number of units of the product that are produced and sold.
  a. Which function is quadratic and which is linear?
  b. Form the profit function for this product from these two functions.
  c. Is the profit function a linear function, a quadratic function, or neither of these?

4. *Revenue and Cost* The total revenue function for a product is given by $R = 26,600x$ dollars and the total cost function for this product is $C = 200,000 +$

$4600x + 2x^2$ dollars, where x is the number of units of the product that are produced and sold.
  a. Which function is quadratic and which is linear?
  b. Form the profit function for this product from these two functions.
  c. Is the profit function a linear function, a quadratic function, or neither of these?

5. *Profit* The profit for a product can be described by the function $P(x) = 40x - 3000 - 0.01x^2$ thousand dollars, where x is the number of units produced and sold.
  a. To maximize profit, how many units must be produced and sold?
  b. What is the maximum possible profit?

6. *Profit* The profit for a product can be described by the function $P(x) = 840x - 75.6 - 0.4x^2$ million dollars, where x is the number of units produced and sold.
  a. To maximize profit, how many units must be produced and sold?
  b. What is the maximum possible profit?

7. *Revenue and Cost* The total revenue function for a certain product is given by $R = 550x$ dollars and the total cost function for this product is $C = 10,000 + 30x + x^2$ dollars, where x is the number of units of the product that are produced and sold.
  a. Find the profit function.
  b. Find the number of units that gives maximum profit.
  c. Find the maximum possible profit.

8. *Revenue and Cost* The total revenue function for a product is given by $R = 6600x$ dollars and the total cost function for this product is $C = 2000 + 4800x + 2x^2$ thousand dollars, where x is the number of units of the product that are produced and sold.
   a. Find the profit function.
   b. Find the number of units that gives maximum profit.
   c. Find the maximum possible profit.

9. *Flight of a Ball* If a ball is thrown upward at 96 feet per second from the top of a building that is 100 feet high, the height of the ball can be modeled by $S = 100 + 96t - 16t^2$ feet, where t is the number of seconds after the ball is thrown.
   a. Describe the graph of the model.
   b. Find the t-coordinate and S-coordinate of the vertex of the graph of this quadratic function.
   c. Explain the meaning of the coordinates of the vertex for this model.

10. *Flight of a Ball* If a ball is thrown upward at 39.2 meters per second from the top of a building that is 30 meters high, the height of the ball can be modeled by $S = 30 + 39.2t - 9.8t^2$ meters, where t is the number of seconds after the ball is thrown.
    a. Describe the graph of the model.
    b. Find the t-coordinate and S-coordinate of the vertex of the graph of this quadratic function.
    c. Explain the meaning of the coordinates of the vertex for this function.
    d. Over what time interval is the function increasing? What does this mean in relation to the ball?

11. *Area* If 200 feet of fence are used to enclose a rectangular pen, the resulting area of the pen is $A = x(100 - x)$, where x is the width of the pen.
    a. Is A a quadratic function of x?
    b. What is the maximum possible area of the pen?

$(100 - x)$

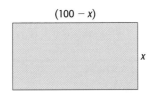

x

12. *Area* If 25,000 feet of fence are used to enclose a rectangular field, the resulting area of the field is $A = (12,500 - x)x$, where x is the width of the pen. What is the maximum possible area of the pen?

13. *Wind and Pollution* The amount of particulate pollution p in the air depends on the wind speed s, among other things, with the relationship between p and s approximated by $p = 25 - 0.01s^2$, where p is in ounces per cubic yard and s is in miles per hour.
    a. Sketch the graph of this model with s on the horizontal axis and with nonnegative values of s and p.
    b. Is the function increasing or decreasing on this domain?
    c. What is the p-intercept of the graph?
    d. What does the p-intercept mean in the context of this application?

14. *Drug Sensitivity* The sensitivity S to a drug is related to the dosage size x by $S = 1000x - x^2$.
    a. Sketch the graph of this model using a domain and range with nonnegative x and S.
    b. Is the function increasing or decreasing for x between 0 and 500?
    c. What is the positive x-intercept of the graph?
    d. Why is this x-intercept important in the context of this application?

15. *Falling Object* A tennis ball is thrown into a swimming pool from the top of a tall hotel. The height of the ball from the pool is given by $D(t) = -16t^2 - 4t + 210$ feet where t is the time, in seconds, after the ball was thrown. Graphically find the t-intercepts for this function. Interpret the value(s) that make sense in this problem context.

16. *Tourism Spending* The equation $y = 0.442x^2 + 23.126x + 181.178$, with x equal to the number of years from 1987 to 1999, models the global spending (in billions of dollars) on travel and tourism.
    a. Graph this function for $x = 0$ to $x = 12$.
    b. If $x = 0$ in 1987, find the spending projected by this model in the year 2002.
    c. Is the value in part (b) an interpolation or an extrapolation?
    (Source: *Statistical Abstract of the U.S., 1999*)

17. *World Population* A low projection scenario of world population for the years 1995 to 2150 by the United Nations is given by the function $y = -0.36x^2 + 38.52x + 5822.86$, where x is the number of years after 1990 and the world population is measured in millions of people.
    a. Graph this function for $x = 0$ to $x = 120$.
    b. What will be the world population in 2010 if the projections made using this model are accurate?
    (Source: *World Population Prospects: 1998 Revision*, United Nations, 1999)

18. *Crimes* When x represents the number of years after 1950, the number of arrests of juveniles per 100,000 juveniles for crimes is given by $f(x) = -0.027x^2 + 5.630x + 51.148$.

a. Graph this function for the years 1950 to 2010.

b. What does this model estimate the number of arrests to be in 2003?

(Source: Federal Bureau of Investigation)

19. *Bomb Threats* The number of bomb threats received against U.S. aircraft between 1980 and 1996 can be approximated by the function $B(t) = 3.268t^2 - 45.733t + 277.910$ threats, where t is the number of years after 1980.

a. What does this model estimate the number of bombing threats to be in 1997?

b. Use graphical or numerical methods with this model to estimate when the number of bomb threats will reach 1177, if this model remained valid.

(Source: *Statistical Abstract of the U.S., 1998*)

20. *Satellite Dishes* The number of U.S. schools with satellite dishes between 1992 and 1997 can be modeled by the function $S(x) = -0.877x^2 + 10.867x - 17.041$ thousand schools, where x is the number of years after 1990. Did the number of schools with satellite dishes reach a maximum or a minimum value between 1992 and 1997? If so, classify the value and give the maximum or minimum number of dishes.

(Source: *Statistical Abstract of the U.S., 1998*)

21. *Quitting Smoking* Using data from 1965 to 1990, the percent of people over 19 years of age who have ever smoked and quit is given by the equation $y = 0.010x^2 + 0.344x + 28.74$, where x is the number of years since 1960.

a. Graph this function for values of x representing 1960 to 2002.

b. Assuming that the pattern indicated by this model continues through 2002, what will be the percent in 2002?

c. When does this model indicate that the percent reached 50%?

(Source: *Substance Abuse*, Princeton N.J., 1993)

22. *U.S. Visitors* The number of international visitors to the United States, in millions, can be modeled by $y = -0.233x^2 + 7.238x - 9.167$, where x is the number of years past 1980.

a. Does the model predict that a maximum or minimum number of visitors will occur during this period? How can you tell without graphing the equation?

b. Find the input and output at the vertex.

c. Interpret the results of part (b).

(Source: *Statistical Abstract of the U.S., 1999*)

23. *Foreign-Born Population* The percent of the U.S. population that is foreign born can be modeled by the equation $y = 0.003x^2 - 0.419x + 19.858$, where x is the number of years after 1900.

a. Is the vertex of the graph of this function a maximum or a minimum?

b. Assuming that the pattern indicated by this model remains the same, find the input and output at the vertex.

c. Interpret these values.

d. Use the vertex to set the window, and graph the model.

(Source: Bureau of the Census, U.S. Department of Commerce)

24. *Abortions* Using data from 1973 to 1996, the number of abortions in the U.S. can be modeled by the function $y = -4.408x^2 + 149.930x + 383.373$, where x is the number of years after 1970 and y is measured in thousands.

a. In what year does this model indicate the number will be a maximum?

b. What is the maximum number of abortions in this year if this model is accurate?

c. For what values of x can we be sure that this model no longer applies?

(Source: Alan Guttmacher Institute)

2.2 Solving Quadratic Equations

The number of admissions at all hospitals in the United States between 1985 and 1997 can be described by the function

$$A(x) = 38.228x^2 - 1069.600x + 40{,}698.547 \text{ thousand people,}$$

where x is the number of years after 1985. Viewing and understanding the pattern indicated by this model can help the medical industry, the government, medical training facilities, and consumers to better understand current and future needs for hospital space and facilities and for hospital personnel. (Source: Based on data from the American Hospital Association: *Hospital Statistics*, 1986, 1991–1999 editions)

To find the year after 1985 that is predicted by this model to give 33,563,000 (that is, 33,563 thousand) admissions, we solve the **quadratic equation**

$$38.228x^2 - 1069.600x + 40{,}698.547 = 33{,}563, \text{ or}$$
$$38.228x^2 - 1069.600x + 7135.547 = 0$$

The values of x that satisfy this equation are called the **solutions** of the equation; they are also zeros of the function

$$y = 38.228x^2 - 1069.600x + 7135.547,$$

and they are the x-intercepts of the graph of this function. In this section, we will learn how to solve quadratic equations by using factoring methods, graphical methods, the square root property, completing the square, and the quadratic formula.

Factoring Methods

An equation that can be written in the form $ax^2 + bx + c = 0$, with $a \neq 0$, is called a **quadratic equation**. Because real-world applications are a focus of this text, we will discuss only real number solutions to quadratic equations in this chapter. Solutions to some quadratic equations can be found exactly by factoring; other quadratic equations require different types of solution methods to find or to approximate solutions.

Solution by factoring is based on the following property of real numbers.

ZERO PRODUCT PROPERTY

For real numbers a and b, the product $ab = 0$ if and only if either $a = 0$ or $b = 0$ or both a and b are zero.

To use this property to solve a quadratic equation by factoring, we must first make sure that the equation is written in a form with zero on one side. If the resulting nonzero expression is factorable, we factor it and use the zero product property to convert the equation into two linear equations that are easily solved.* Before applying this technique to real-world applications, we consider the following example.

---

*For a review of factoring methods, see the Algebra Toolbox that begins Chapter 2.

EXAMPLE 1 Solving a Quadratic Equation by Factoring

Solve the equation $3x^2 + 7x = 6$.

Solution

We first subtract 6 from both sides of the equation to rewrite the equation with 0 on one side:

$$3x^2 + 7x - 6 = 0$$

To begin factoring the trinomial $3x^2 + 7x - 6$, we seek factors of $3x^2$ (that is, $3x$ and x) as the first terms of two binomials, and factors of -6 as the last terms of the binomials. The factorization whose inner and outer products combine to $7x$ is $(3x - 2)(x + 3)$, so we have

$$(3x - 2)(x + 3) = 0.$$

Using the zero product property gives

$$3x - 2 = 0 \text{ or } x + 3 = 0.$$

Solving these linear equations gives the two solutions to the original equation.

$$x = \frac{2}{3} \text{ or } x = -3 \qquad \blacksquare$$

EXAMPLE 2 Profit

Suppose the daily profit from the production and sale of x units of a product is given by

$$P(x) = -0.01x^2 + 20x - 500 \text{ dollars.}$$

What levels of production and sales will give a daily profit of $1400?

Solution

To answer this question, note that $1400 is the amount of profit. Thus we solve the equation

$$1400 = -0.01x^2 + 20x - 500$$

to find the number of units x that gives a $1400 profit.

To find the solution by factoring, we use the following steps:

$$1400 = -0.01x^2 + 20x - 500$$
$$0 = -0.01x^2 + 20x - 1900 \qquad \text{Subtracting 1400 from both sides}$$
$$0 = -0.01(x^2 - 2000x + 190{,}000) \qquad \text{Factoring } -0.01 \text{ from the expression}$$
$$0 = -0.01(x - 1900)(x - 100) \qquad \text{Factoring the trinomial into binomial factors}$$
$$x - 1900 = 0 \text{ or } x - 100 = 0 \qquad \text{Setting the binomial factors equal to 0}$$
$$x = 1900 \text{ or } x = 100 \qquad \text{Solving for } x$$

Thus, producing and selling either 100 units or 1900 units give the daily profit of $1400. $\qquad \blacksquare$

Graphical Methods

Factoring the right-hand side of the quadratic equation in Example 2 was not especially easy. In cases where factoring $f(x)$ is difficult or impossible, graphing $y = f(x)$ after the equation is in the form $f(x) = 0$ can be helpful in finding the solution. Recall that if a is a real number, the following three statements are equivalent:

- a is a solution to the equation $f(x) = 0$

- a is a zero of the function f

- a is an x-intercept of the graph of $y = f(x)$

It is important to remember that the above three statements are equivalent, because connecting these concepts allows us to use different methods for solving equations. Sometimes graphical methods are the easiest way to find or approximate solutions to real data problems. If the x-intercepts of the graph of $y = f(x)$ are easily found, then graphical methods may be helpful in finding the solutions. Note that if the graph of $y = f(x)$ does not cross or touch the x-axis, there are no real solutions to the equation $f(x) = 0$.

We can also find solutions or decimal approximations of solutions to quadratic equations by using the intersection method with a graphing utility.

EXAMPLE 3   Profit (Revisited)

- -

Consider again the daily profit from the production and sale of x units of a product, given by

$$P(x) = -0.01x^2 + 20x - 500 \text{ dollars.}$$

a. Use a graph to find the levels of production and sales that give a daily profit of $1400.

b. Is it possible for the profit to be greater than $1400?

Solution

a. To find the level of production and sales, x, that gives a daily profit of 1400 dollars, we solve

$$1400 = -0.01x^2 + 20x - 500.$$

To solve this equation by the intersection method, we graph

$$y_1 = -0.01x^2 + 20x - 500 \text{ and } y_2 = 1400.$$

To find the appropriate window for this graph, we note that the graph of the function $y_1 = -0.01x^2 + 20x - 500$ is a parabola with the vertex at

$$x = \frac{-b}{2a} = \frac{-20}{2(-0.01)} = 1000 \text{ and } y = P(1000) = 9500.$$

We use a viewing window containing this point, with $x = 1000$ near the center, to graph the function. (See Figure 2.12). Using the intersection method, we see that (100, 1400) and (1900, 1400) are the points of intersection of the graphs of the two functions. Thus $x = 100$ and $x = 1900$ are solutions to the equation $1400 = -0.01x^2 + 20x - 500$, and the profit is $1400 when $x = 100$ units or $x = 1900$ units of the product are produced and sold. Note that this is the same solution as the one found by factoring in Example 2.

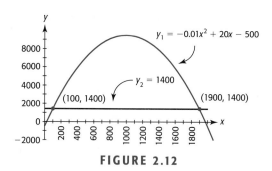

FIGURE 2.12

b. We can see from the graph in Figure 2.12 that the profit is more than $1400 for many values of x. Because the graph of this profit function is a parabola that opens down, the maximum profit occurs at the vertex of the graph. As we found in part (a), the vertex occurs at $x = 1000$ and the maximum possible profit is $P(1000) = \$9500$, which is more than $1400. ■

Combining Graphs and Factoring

Examples 2 and 3 indicate that the solutions found graphically agree with the solutions found by factoring. So if the factors of a quadratic function are not easily found, its graph may be helpful in finding the factors. Note once more the important correspondences that exist among solutions, zeros, and x-intercepts: the x-intercepts of the graph of $y = P(x)$ are the real solutions of the equation $0 = P(x)$ and the zeros of $P(x)$. In addition, the factors of $P(x)$ are also related to the zeros of $P(x)$. We saw in Example 2 that $(x - 100)$ is a factor of $-0.01x^2 + 20x - 1900$ and that $x = 100$ is a solution of $0 = -0.01x^2 + 20x - 1900$, so that $P(100) = 0$. This relationship among the factors, solutions, and zeros is true for any polynomial function f and can be generalized by the following theorem.

FACTOR THEOREM

The polynomial function f has a factor $(x - a)$ if and only if $f(a) = 0$.

Thus $(x - a)$ is factor of $f(x)$ if and only if $x = a$ is a solution to $f(x) = 0$.

This means that we can verify our factorization of f and the solutions to $0 = f(x)$ by graphing $y = f(x)$ and observing where the graph crosses the x-axis. We can also sometimes use our observation of the graph to assist us in the factorization of a quadratic function:

- If one solution can be found *exactly* from the graph, it can be used to find one of the factors of the function. The second factor can then be found easily, leading to the second solution.

EXAMPLE 4 Graphing and Factoring Methods Combined

Solve $0 = 3x^2 - x - 10$ by using the following steps.
a. Graphically find one of the x-intercepts of $y = 3x^2 - x - 10$.

b. Verify that the zero found in part (a) is an exact solution to $0 = 3x^2 - x - 10$.

c. Use the method of factoring to find the other solution to $0 = 3x^2 - x - 10$.

Solution

a. The vertex of the graph of $y = 3x^2 - x - 10$ is at $x = \dfrac{1}{6}$ and $y \approx -10.08$, and because the parabola opens up, we set a viewing window that includes this point near the bottom center of the window. Graphing the function and using the x-intercept method, we find that the graph crosses the x-axis at $x = 2$. (See Figure 2.13). This means that 2 is a zero of $f(x) = 3x^2 - x - 10$.

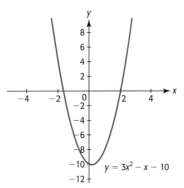
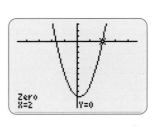

FIGURE 2.13

b. Because $x = 2$ was obtained graphically, it may be an approximation of the exact solution. To verify that it is exact, we show that $x = 2$ makes the equation $0 = 3x^2 - x - 10$ a true statement:

$$3(2)^2 - 2 - 10 = 12 - 2 - 10 = 0$$

c. Because $x = 2$ is a solution of $0 = 3x^2 - x - 10$, one factor of $3x^2 - x - 10$ is $(x - 2)$. Thus we use $(x - 2)$ as one factor and seek a second binomial factor so that the product of the two binomials is $3x^2 - x - 10$.

$$3x^2 - x - 10 = 0$$
$$(x - 2)(\quad) = 0$$
$$(x - 2)(3x + 5) = 0$$

The remaining factor is $3x + 5$, which we set equal to 0 to obtain the other solution:

$$3x = -5$$
$$x = -\frac{5}{3}$$

Thus, the two solutions to $0 = 3x^2 - x - 10$ are $x = 2$ and $x = -\dfrac{5}{3}$. ∎

Graphical and Numerical Methods

When approximate solutions to quadratic equations are sufficient, graphical and/or numerical solution methods can be used. These methods of solving are illustrated in Example 5.

Marijuana Use

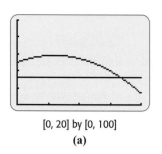

[0, 20] by [0, 100]
(a)

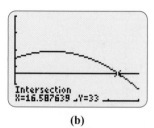

(b)

FIGURE 2.14

X	Y₁	Y₂
12	50.235	33
13	47.298	33
14	43.904	33
15	40.053	33
16	35.745	33
17	30.98	33
18	25.757	33

X=17

FIGURE 2.15

For the years from 1975 to 1991, the percent p of high school seniors who have tried marijuana can be considered as a function of time t according to the equation

$$p = -0.2286t^2 + 2.7783t + 49.8138$$

where t is the number of years past 1975. Use the indicated method to find the year after 1975 during which the percent predicted by the function reached 33%.

a. Graphical method

b. Numerical method (Source: National Institute on Drug Abuse)

Solution

Note that the output p is measured in percent, so for the output 33%, $p = 33$. Thus, to find where the percent is 33, we solve the equation

$$33 = -0.2286t^2 + 2.7783t + 49.8138$$

a. We could use the x-intercept method and find the zeros of

$$y = -0.2286t^2 + 2.7783t + 16.8138.$$

Instead, we choose to use the intersection method. Figure 2.14(a) shows the graphs of $y_1 = -0.2286t^2 + 2.7783t + 49.8138$ and $y_2 = 33$, and 2.14(b) shows the point of intersection. We find that t is approximately 16.59.

Because 16 years after 1975 is the end of 1991, the percent reaches 33 in 1992. Note that the graphs also intersect at $t \approx -4.43$, but this value would not give a year after 1975.

b. Figure 2.15 shows a table of inputs (years after 1975) and outputs (percent of high school seniors who have tried marijuana). The table shows that $p = 35.745\%$ at the end of the 16th year after 1975, or 1991, and the percent reached 33 (and fell below it) during 1992. ∎

The Square Root Method

We have solved quadratic equations by factoring and by graphical methods. Another method can be used to solve quadratic equations that are written in a particular form. In general, we can find the solutions of quadratic equations of the form $x^2 = C$ by taking the square root of both sides. For example, to solve $x^2 = 25$, we take the square root of both sides of the equation, getting $x = \pm 5$. Note that there are two solutions because $5^2 = 25$ and $(-5)^2 = 25$.

SQUARE ROOT METHOD

The solutions of the quadratic equation $x^2 = C$ are $x = \pm\sqrt{C}$.

Note that, when we take the square root of both sides, we use a $\pm$ symbol because there are both a positive and a negative value that, when squared, give C.

We can also use this method to solve quadratic equations of the form $(ax + b)^2 = C$.

EXAMPLE 6   Square Root Solution Method

Solve the following equations by using the square root method.

a. $2x^2 - 16 = 0$ **b.** $(x - 6)^2 = 18$

Solution

a. This equation can be written in the form $x^2 = C$, so the square root method can be used. We want to rewrite the equation so that the x^2-term is isolated and its coefficient is 1, and then take the square root of both sides.

$$2x^2 - 16 = 0$$
$$2x^2 = 16$$
$$x^2 = 8$$
$$x = \pm\sqrt{8} = \pm 2\sqrt{2}$$

The exact solutions to $x^2 - 8 = 0$ are $x = 2\sqrt{2}$ and $x = -2\sqrt{2}$.

b. The left side of this equation is a square, so we can take the square root of both sides to find x.

$$(x - 6)^2 = 18$$
$$x - 6 = \pm\sqrt{18}$$
$$x = 6 \pm \sqrt{18}$$
$$x = 6 \pm \sqrt{9}\sqrt{2}$$
$$x = 6 \pm 3\sqrt{2}$$

■

Completing the Square

When we convert one side of a quadratic equation to a perfect binomial square and use the square root method to solve the equation, the method used is called **completing the square**.

EXAMPLE 7   Completing the Square

Solve the equation $x^2 - 12x + 7 = 0$ by completing the square.

Solution

Because the left side of this equation is not a perfect square, we rewrite it in a form where we can more easily get a perfect square. We subtract 7 from both sides, getting

$$x^2 - 12x = -7$$

We complete the square on the left side of this equation by adding the appropriate number to both sides of the equation to make the left side a perfect square trinomial. Because $(x + a)^2 = x^2 + 2ax + a^2$, the constant term of a perfect square trinomial will be the *square of half the coefficient of* x. Using this rule with the equation $x^2 - 12x = -7$, we would need to take half the coefficient of 12 and add the square of this number to get a perfect square trinomial. Half of 12 is 6, so adding $6^2 = 36$

to $x^2 - 12x$ gives the perfect square $x^2 - 12x + 36$. We also have to add 36 to the other side of the equation (to preserve the equation), giving

$$x^2 - 12x + 36 = -7 + 36.$$

Factoring the perfect square trinomial gives

$$(x - 6)^2 = 29.$$

We can now solve the equation with the square root method.

$$(x - 6)^2 = 29$$
$$x - 6 = \pm\sqrt{29}$$
$$x = 6 \pm \sqrt{29}$$ ∎

The Quadratic Formula

We can generalize the method of completing the square to derive a general formula that gives the solution to any quadratic equation. We use this method to find the general solution to

$$ax^2 + bx + c = 0 \text{ with } a \neq 0.$$

$ax^2 + bx + c = 0$	Standard form
$ax^2 + bx = -c$	Subtracting c from both sides
$x^2 + \dfrac{b}{a}x = -\dfrac{c}{a}$	Dividing both sides by a

We would like to make the left side of the last equation a perfect square trinomial. Half the coefficient of x is $\dfrac{b}{2a}$, and squaring this gives $\dfrac{b^2}{4a^2}$. Hence adding $\dfrac{b^2}{4a^2}$ to both sides of the equation gives a perfect square trinomial on the left side, and we can continue with the solution.

$x^2 + \dfrac{b}{a}x + \dfrac{b^2}{4a^2} = \dfrac{b^2}{4a^2} - \dfrac{c}{a}$	Adding $\dfrac{b^2}{4a^2}$ to both sides of the equation
$\left(x + \dfrac{b}{2a}\right)^2 = \dfrac{b^2 - 4ac}{4a^2}$	Factoring the left side and combining the fractions on the right side
$x + \dfrac{b}{2a} = \pm\sqrt{\dfrac{b^2 - 4ac}{4a^2}}$	Taking the square root of both sides
$x = -\dfrac{b}{2a} \pm \dfrac{\sqrt{b^2 - 4ac}}{2\lvert a \rvert}$	Subtracting $\dfrac{b}{2a}$ from both sides and simplifying
$x = -\dfrac{b}{2a} \pm \dfrac{\sqrt{b^2 - 4ac}}{2a}$	$\pm 2\lvert a \rvert = \pm 2a$
$x = \dfrac{-b \pm \sqrt{b^2 - 4ac}}{2a}$	Combining the fractions

The formula we have developed is called the **quadratic formula**.

> **QUADRATIC FORMULA**
>
> The solutions of the quadratic equation $ax^2 + bx + c = 0$ with $a \neq 0$, are given by the formula
>
> $$x = \frac{-b \pm \sqrt{b^2 - 4ac}}{2a}$$
>
> Note that a is the coefficient of x^2, b is the coefficient of x, and c is the constant term.

Because of the $\pm$ sign, there are two solutions to the equation:

$$x = \frac{-b + \sqrt{b^2 - 4ac}}{2a} \quad \text{and} \quad x = \frac{-b - \sqrt{b^2 - 4ac}}{2a}$$

We can use the quadratic formula to solve all quadratic equations exactly, but it is especially useful for finding exact solutions to those equations for which factorization is difficult or impossible. For example, the solutions to $33 = -0.2286t^2 + 2.7783t + 49.8138$, were found approximately by graphical methods in Example 5. If we need to find the exact solutions, we could use the quadratic formula.

EXAMPLE 8 Solving Using the Quadratic Formula

Solve $6 - 3x^2 + 4x = 0$ by using the quadratic formula.

Solution

The equation $6 - 3x^2 + 4x = 0$ can be rewritten as $-3x^2 + 4x + 6 = 0$, so $a = -3$, $b = 4$, and $c = 6$. The two solutions to this equation are

$$x = \frac{-4 \pm \sqrt{4^2 - 4(-3)(6)}}{2(-3)} = \frac{-4 \pm \sqrt{88}}{-6} = \frac{-4 \pm 2\sqrt{22}}{-6} = \frac{2 \pm \sqrt{22}}{3}$$

Thus, the exact solutions are the irrational numbers $x = \dfrac{2 + \sqrt{22}}{3}$ and $x = \dfrac{2 - \sqrt{22}}{3}$.

Three-place decimal approximations for these solutions are $x \approx -0.897$ and $x \approx 2.230$. ∎

Decimal approximations of irrational solutions found with the quadratic formula will often suffice as answers to an applied problem. The quadratic formula is especially useful when the coefficients of a quadratic equation are decimal values that make factorization impractical. This occurs in many applied problems, such as the one discussed at the beginning of this section. If the graph of the quadratic function $y = f(x)$ does not intersect the x-axis at "nice" values of x, the solutions may be irrational numbers, and using the quadratic formula allows us to find these solutions exactly.

EXAMPLE 9 Hospital Admissions

The annual number of admissions at all hospitals in the United States between 1985 and 1997 can be described by the function $A(x) = 38.228x^2 - 1069.600x + 40,698.547$

thousand people, where x is the number of years after 1985. Find the year after 1985 that is predicted by this model to give 33,563,000 admissions. (Source: American Hospital Association: *Hospital Statistics*, 1986, 1991–1999 editions)

Solution

Note that the output of A is measured in thousands of admissions, so we must first write 33,563,000 admissions as 33,563 thousand admissions. Then, to answer the given question, we solve (using an algebraic, numerical, or graphical method) the quadratic equation

$$38.228x^2 - 1069.600x + 40{,}698.547 = 33{,}563.$$

We choose to use the quadratic formula (even though an approximate answer is all that is needed).

We first write $38.228x^2 - 1069.600x + 40{,}698.547 = 33{,}563$ with 0 on one side.

$$38.228x^2 - 1069.600x + 40{,}698.547 - 33{,}563 = 0$$

$$38.228x^2 - 1069.600x + 7135.547 = 0$$

This gives $a = 38.228$, $b = -1069.6$, and $c = 7135.547$.
Substituting in the quadratic formula gives

$$x = \frac{-(-1069.6) \pm \sqrt{(-1069.6)^2 - 4(38.228)(7135.547)}}{2(38.228)}$$

$$= \frac{1069.6 \pm \sqrt{52{,}933.39714}}{76.456}$$

$$x \approx 16.999 \text{ or } x \approx 10.981$$

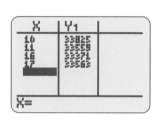

FIGURE 2.16

Thus 33,563,000 admissions occur in the 17th year from the end of 1985, in 2002, and in $1985 + 11 = 1996$. Figure 2.16 shows the outputs of

$$y_1 = 38.228x^2 - 1069.600x + 40{,}698.547$$

at integer values near the x-values that are the results from the quadratic formula, which confirms our conclusions above. ∎

Aids for Solving Quadratic Equations

The x-intercepts of the graph of the quadratic function $y = f(x)$ can be used to determine how to solve the quadratic equation $f(x) = 0$. Table 2.2 summarizes these ideas with suggested methods for solving, and the graphical representations of the solutions.

The Discriminant

We can also determine the type of solutions a quadratic equation has by looking at the expression $b^2 - 4ac$, which is called the **discriminant** of the quadratic equation $ax^2 + bx + c = 0$. The discriminant is the expression inside the radical in the quadratic formula $x = \dfrac{-b \pm \sqrt{b^2 - 4ac}}{2a}$, so it determines if the quantity inside the radical is positive, zero, or negative. Thus we have:

- If $b^2 - 4ac > 0$, there are two unique real solutions.

- If $b^2 - 4ac = 0$, there is a one real solution.

- If $b^2 - 4ac < 0$, there is no real solution.

For example, the equation $3x^2 + 4x + 2 = 0$ has no real solution because $4^2 - 4(3)(2) = -8 < 0$, and the equation $x^2 + 4x + 2 = 0$ has two unique real solutions because $4^2 - 4(1)(2) = 8 > 0$.

TABLE 2.2   Connections Between Graphs of Quadratic Functions and Solution Methods

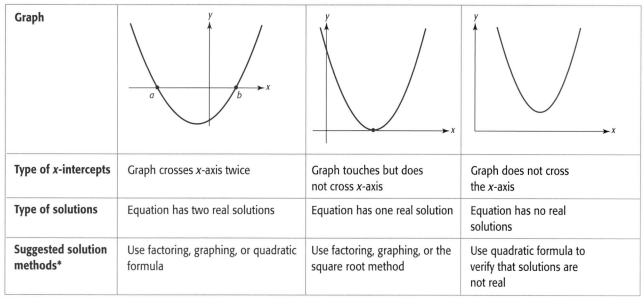

Graph			
Type of x-intercepts	Graph crosses x-axis twice	Graph touches but does not cross x-axis	Graph does not cross the x-axis
Type of solutions	Equation has two real solutions	Equation has one real solution	Equation has no real solutions
Suggested solution methods*	Use factoring, graphing, or quadratic formula	Use factoring, graphing, or the square root method	Use quadratic formula to verify that solutions are not real

*Solution methods other than the suggested methods may also be successful.

2.2 SKILLS CHECK

In Exercises 1–8, use factoring to solve the equations.

1. $x^2 - 3x - 10 = 0$ **2.** $x^2 - 9x + 18 = 0$

3. $x^2 - 11x + 24 = 0$ **4.** $x^2 + 3x - 10 = 0$

5. $2x^2 + 2x - 12 = 0$ **6.** $2s^2 + s - 6 = 0$

7. $0 = 2t^2 - 11t + 12$ **8.** $6x^2 - 13x + 6 = 0$

Use a graphing utility to find or to approximate the x-intercepts of the graph of each function in Exercises 9–12.

9. $y = 3x^2 - 8x + 4$ **10.** $y = 2x^2 + 8x - 10$

11. $y = 2x^2 + 7x - 4$ **12.** $y = 5x^2 - 17x + 6$

Use a graphing utility as an aid in factoring to solve the equations in Exercises 13 and 14.

13. $2w^2 - 5w - 3 = 0$ **14.** $3x^2 - 4x - 4 = 0$

In Exercises 15 and 16, use the square root method to solve the quadratic equations.

15. $4x^2 - 9 = 0$ **16.** $x^2 - 20 = 0$

In Exercises 17–19, complete the square to solve the quadratic equations.

17. $x^2 - 4x - 9 = 0$ **18.** $x^2 - 6x + 1 = 0$

19. $x^2 - 3x + 2$

In Exercises 20–22, use the quadratic formula to solve the equations.

20. $x^2 - 5x + 2 = 0$ **21.** $3x^2 - 6x - 12 = 0$

22. $5x + 3x^2 = 8$

In Exercises 23–28, use a graphing utility to find or approximate solutions of the equations.

23. $2x^2 + 2x - 12 = 0$ **24.** $2x^2 + x - 6 = 0$

25. $0 = 6x^2 + 5x - 6$ **26.** $10x^2 = 22x - 4$

27. $4x + 2 = 6x^2 + 3x$ **28.** $(x - 3)(x + 2) = -4$

2.2 EXERCISE

In Exercises 1–12, solve analytically and then check graphically.

1. *Flight of a Ball* If a ball is thrown upward at 96 feet per second from the top of a building that is 100 feet high, the height of the ball can be modeled by $S = 100 + 96t - 16t^2$ feet, where t is the number of seconds after the ball is thrown. How long after the ball is thrown is the height 228 feet?

2. *Falling Object* A tennis ball is thrown into a swimming pool from the top of a tall hotel. The height of the ball above the pool is modeled by $D(t) = -16t^2 - 4t + 200$ feet, where t is the time, in seconds, after the ball was thrown. How long after the ball is thrown is it 44 feet above the pool?

3. *Break-Even* The profit for a product is given by $P(x) = -12x^2 + 1320x - 21,600$, where x is the number of units produced and sold. How many units give break-even (that is, give zero profit) for this product?

4. *Break-Even* The profit for a product is given by $P(x) = -15x^2 + 180x - 405$ thousand dollars, where x is the number of tons of product produced and sold. How many tons give break-even (that is, give zero profit) for this product?

5. *Break-Even* The total revenue function for a product is given by $R = 550x$ dollars and the total cost function for this same product is given by $C = 10,000 + 30x + x^2$, where C is measured in dollars. For both functions, the input x is the number of units produced and sold.
   a. Form the profit function for this product from the two given functions.
   b. What is the profit when 18 units are produced and sold?
   c. What is the profit when 32 units are produced and sold?
   d. How many units must be sold in order to break even on this product?

6. *Break-Even* The total revenue function for a product is given by $R = 266x$ and the total cost function for this same product is $C = 2000 + 46x + 2x^2$, where R and C are each measured in thousands of dollars and x is the number of units produced and sold.
   a. Form the profit function for this product from the two given functions.

   b. What is the profit when 55 units are produced and sold?
   c. How many units must be sold in order to break even on this product?

7. *Wind and Pollution* The amount of particulate pollution p from a power plant in the air above the plant depends on the wind speed s, among other things, with the relationship between x and s approximated by $p = 25 - 0.01s^2$, with s in miles per hour.
   a. Find the value(s) of s that will make $p = 0$.
   b. What does $p = 0$ mean in this application?
   c. What solution to $0 = 25 - 0.01s^2$ makes sense in the context of this application?

8. *Drug Sensitivity* The sensitivity S to a drug is related to the dosage size x by $S = 100x - x^2$, where x is the dosage size in milliliters.
   a. What dosage(s) give 0 sensitivity?
   b. Explain what your answers in part (a) might mean.

9. *Velocity of Blood* Because of friction from the walls of an artery, the velocity of a blood corpuscle in an artery is greatest at the center of the artery and decreases as the distance r from the center increases. The velocity of the blood in the artery can be modeled by the function

$$v = k(R^2 - r^2)$$

where R is the radius of the artery and k is a constant that is determined by the pressure, viscosity of the blood, and the length of the artery. In the case where $k = 2$ and $R = 0.1$ centimeter, the velocity is $v = 2(0.01 - r^2)$ centimeters per second.
   a. What distance r would give a velocity of 0.02 cm per sec?
   b. What distance r would give a velocity of 0.015 cm per sec?
   c. What distance r would give a velocity of 0 cm per sec? Where is the blood corpuscle?

10. *Body-Heat Loss* The model for body-heat loss depends on the coefficient of convection K, which depends on wind speed s according to the equation $K^2 = 16s + 4$, where s is in miles per hour. Find the positive coefficient of convection when the wind speed is:
   a. 20 mph.   b. 60 mph.
   c. What is the change in K for a change in speed from 20 mph to 60 mph?

11. *Market Equilibrium* Suppose that the demand for artificial Christmas trees is given by the function

$$p = 109.70 - 0.10q$$

and that the supply of these trees is given by

$$p = 0.01q^2 + 5.91$$

where p is the price of a tree in dollars and q is the quantity of trees that are demanded/supplied in hundreds. Find the price that gives market equilibrium price and the number of trees that will be sold/bought at this price.

12. *Market Equilibrium* The demand for a product is given by $p = 7000 - 2x$ dollars and the supply for this product is given by $p = 0.01x^2 + 2x + 1000$ dollars, where x is the number of units demanded and supplied when the price per unit is p dollars. Find the equilibrium quantity and equilibrium price.

13. *Foreign-born Population* The percent of the U.S. population that is foreign born can be modeled by the equation $y = 0.003x^2 - 0.42x + 19.8$, where x is the number of years after 1900.
    a. Verify graphically that one solution to $7.8 = 0.003x^2 - 0.42x + 19.8$ is $x = 40$.
    b. Use this information and factoring to find the year or years when the percent of the U.S. population that is foreign born is 7.8.
    (Source: U.S. Census Bureau)

14. *Smoking Cessation* Using data from 1965 to 1990, the percent of people over 19 years of age who have ever smoked and quit is given by the equation $y = 0.010x^2 + 0.344x + 28.74$, where x is the number of years after 1960.
    a. One solution to the equation $58.5 = 0.010x^2 + 0.344x + 28.74$ is $x = 40$. What does that mean?
    b. To find when the percent of people over 19 years of age who have ever smoked and quit is 58.5, do we need to find the second solution to this equation? Why or why not?
    c. Graphically verify that $x = 40$ is a solution to $58.5 = 0.010x^2 + 0.344x + 28.74$.
    (Source: *Substance Abuse*, Princeton, N.J., 1993)

15. *Baggage Complaints* The number of consumer complaints about baggage against U.S. airlines between 1989 and 1997 is given by

$$C(x) = 41.018x^2 - 428.194x + 1662.14$$

where x is the number of years after 1989.
(Source: *Statistical Abstract of the U.S., 1999*)

a. Describe how to set the viewing window so that you can see the important points for this model for the years 1989–1997. Graph the function $y = C(x)$.
b. When did the minimum number of baggage complaints against U.S. airlines occur?
c. Graphically determine the year after 1990 when the number of baggage complaints reached 1482.
d. Is the result in part (*c*) an interpolation or an extrapolation?

16. *Internet Use* An equation that models the number of users of the Internet is

$$y = 11.786x^2 - 142.214x + 493 \text{ million users}$$

where x is the number of years after 1990.
a. If the pattern indicated by the model is based on data collected between 1990 and 1999, when does this model predict there will be 812 million users?
b. Of what significance in this real-world situation is the answer to part (*a*)?
(Source: *CyberAtlas*, 1999)

17. *Tourism Spending* The global spending on travel and tourism (in billions of dollars) can be described by the equation $y = 0.442x^2 + 23.126x + 181.178$, where x equals the number of years past 1987. Graphically find the year in which spending is projected to reach \$820.5 billion.
(Source: *Statistical Abstract of the U.S.*, 1999)

18. *Prison Population* The average daily number of inmates in the Beaufort County, South Carolina prison can be modeled by

$$I(x) = 0.161x^2 + 7.582x + 82.157$$

where x is the number of years after 1991.
a. How would you set the viewing window if you wanted a complete graph of the function? Graph the function.
b. Graphically estimate when this model predicts that the average daily inmate population will be 185.
(Source: *The Island Packet*, November 1999)

19. *Hospital Admissions* The number of admissions to all types of hospitals in the U.S. between 1985 and 1997 can be described by the function

$$A(t) = 38.228t^2 - 1069.60t + 40{,}698.547 \\ \text{thousand people,}$$

where t is the number of years after 1980.
a. By how much did the number of hospital admissions change between 1990 and 1997?

b. Use this model and graphical methods to find in what year after 1980 the number of admissions was equal to 33,217,000.

c. Use the model to predict the number of hospital admissions in 2005. What assumptions are you making when you make this prediction?
(Source: American Hospital Association, *Hospital Statistics*)

20. *Smoking Cessation* Using data from 1965 to 1990, the percent of people over 19 years of age who have ever smoked and quit is given by the equation $y = 0.010x^2 + 0.344x + 28.74$, where x is the number of years after 1960. In what year will this model predict y to be over 100 percent, making the model invalid?
(Source: *Substance Abuse*, Princeton, N.J., 1993)

21. *Student Aid* Federal funds providing for student financial assistance through programs funded by congressional appropriations between 1980 and 2000 can be described by the model

$$f(x) = 9.032x^2 + 99.970x + 3645.90 \text{ million dollars,}$$

where x is the number of years after 1980.

a. Use the model to compare the federal aid available to students in 1990 and 2000.

b. When did the available federal aid reach and exceed $8,000,000,000?
(Source: U.S. Department of Education, National Center for Educational Statistics)

2.3 Power Functions and Transformations

The data in Table 2.3 gives the total U.S. cigarette advertising and promotional expenditures, in millions of dollars, for selected years from 1975 to 1993. These data can be modeled by the **power function** $y = 33.036x^{1.623}$, where x is the number of years after 1970 and y is in millions of dollars. The data points and the function are graphed in Figure 2.17.

TABLE 2.3

Year	Cigarette Advertising (in millions of dollars)	Year	Cigarette Advertising (in millions of dollars)
1975	491.3	1991	4650.1
1980	1242.3	1992	5231.9
1985	2476.4	1993	6035.4
1990	3992.0		

(Source: *Federal Trade Commission Report to Congress* for 1990–1996)

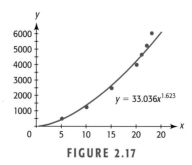

FIGURE 2.17

Power functions are used in applications in the life, social, and physical sciences as well as in business. Power functions are one of a group of special functions that are the basic building blocks for many mathematical applications. In this section we see how to graph, evaluate, and apply power functions. We will also see how power functions and other functions can be transformed by shifts, stretches, and compressions.

Power Functions

We define power functions as follows.

> ### POWER FUNCTIONS
>
> A **power function** is a function of the form $y = ax^b$ where a and b are real numbers, $b > 0$.

We know that the area of a rectangle is $A = lw$ square units, where l and w are the length and the width, respectively, of the rectangle. Because $l = w$ when the rectangle is a square, the area of a square can be found with the formula $A = x^2$ if each side is x units long. Table 2.4 gives the area of a square for the selected lengths of sides.

TABLE 2.4

Side Length x (units)	Area of Square $A = x^2$ (square units)
1	1
2	4
3	9
4	16
5	25
6	36
7	49

The graph of these data points is shown in Figure 2.18. Because the formula for area applies for any (positive) length of a side, we can graph the model for the area of a square for any nonnegative value of x (see Figure 2.19). This is the function $A = x^2$ for $x \geq 0$. A related function is $y = x^2$, defined on the set of all real numbers; this is a power function with power 2. This function is also a basic quadratic function, and its complete graph is a parabola (see Figure 2.20).

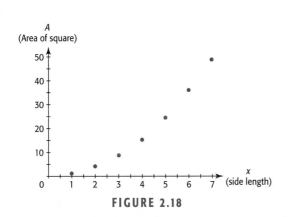

FIGURE 2.18

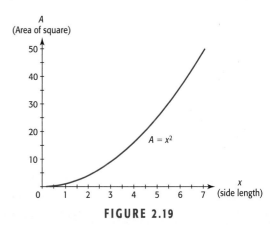

FIGURE 2.19

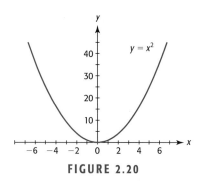

FIGURE 2.20

Symmetry

Notice that the complete graph of $y = x^2$ in Figure 2.20 is symmetric (that is, it is a reflection of itself) about the y-axis. This will occur for any function f if for every point (x, y) on the graph, the point $(-x, y)$ is also on the graph.

SYMMETRY WITH RESPECT TO THE *Y*-AXIS

The graph of $y = f(x)$ is symmetric with respect to the y-axis if

$$f(-x) = f(x)$$

for all x in its domain. Such a function is called an **even function**.

All power functions with even integer exponents will have graphs that are symmetric about the y-axis. This is why all functions satisfying this condition are called even functions, although functions other than power functions exist that are even functions.

If we fill a larger cube with cubes that are 1 unit on each edge and have a volume of 1 cubic unit, the number of these small cubes that fit gives the volume of the larger cube. Table 2.5 contains the results of measuring the edges and the volumes of cubes of different sizes, and shows the relationship between the edge length and the volume for cubes. It is easily observed that if the length of an edge of a cube is x units, its volume is x^3 cubic units.

TABLE 2.5

Edge Length x (units)	Volume of Cube $V = x^3$ (cubic units)
1	1
2	8
3	27
4	64
5	125
6	216

The graph of these data points is shown in Figure 2.21(a) and the graph of $y = x^3$ for any nonnegative value of x is shown in Figure 2.21(b). The related function $y = x^3$, defined for all real numbers x, is called the **cubing function**. A complete graph of this function is shown in Figure 2.22(a).

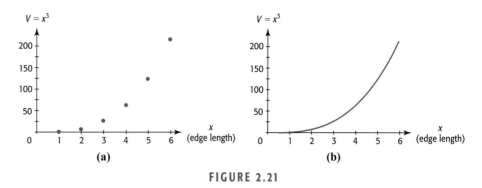

FIGURE 2.21

Notice in Figure 2.22(b) that if we draw a line through the origin and any point on the graph, it will intersect the graph at a second point, and the distances from the two points to the origin will be equal.

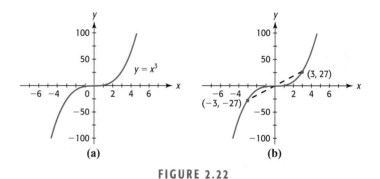

FIGURE 2.22

For example, the point (3, 27) is on the graph of $y = x^3$, and the line through (3, 27) and the origin intersects the graph at the point $(-3, -27)$, which is the same distance from the origin as (3, 27). In general, the graph of a function is called symmetric with respect to the origin if for every point (x, y) on the graph, the point $(-x, -y)$ is also on the graph.

SYMMETRY WITH RESPECT TO THE ORIGIN

The graph of $y = f(x)$ is symmetric with respect to the origin if

$$f(-x) = -f(x)$$

for all x in its domain. Such a function is called an **odd function**.

Because all power functions with odd integer powers will have graphs that are symmetric about the origin, we say that any function whose graph is symmetric about the origin is an odd function. However, there are other functions that are not power functions that are odd functions.

Additional Power Functions; Root Functions

The functions $y = x$, $y = x^2$, and $y = x^3$ are examples of power functions with positive integer powers. Additional examples of power functions include functions with noninteger powers of x, like the function in the following example.

EXAMPLE 1   Allometric Relationships

For fiddler crabs, data shows that the relationship between the weight C of the claw and the weight W of the body is given by

$$C = 0.11W^{1.54},$$

with W and C in grams. (Source: d'Arcy Thompson, *On Growth and Form*, Cambridge University Press, 1961)

a. Graph this function.

b. Use this function to compute the weight of a claw if the weight of the crab is 5 grams.

Solution

a. It is appropriate to graph this model only in the first quadrant because $C > 0$ and $W > 0$ in the problem context. The graph is shown in Figure 2.23.

b. We find the weight of the claw by evaluating the function with an input of 5, giving an output of $C = 0.11(5)^{1.54} \approx 1.312$. Thus, the weight of the claw is approximately 1.3 grams.

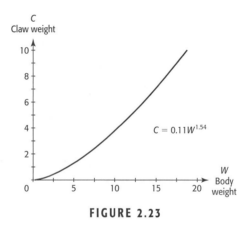

FIGURE 2.23   ∎

Some functions such as $y = x^{1/2} = \sqrt{x}$, and $y = x^{1/3} = \sqrt[3]{x}$, are power functions with fractional powers. The functions $y = \sqrt{x}$ and $y = \sqrt[3]{x}$ are also called **root functions**.

ROOT FUNCTIONS

A **root function** is a function of the form $y = ax^{1/n}$, or $y = a\sqrt[n]{x}$, where n is an integer, $n \geq 2$.

The graphs of $y = \sqrt{x}$ and $y = \sqrt[3]{x}$ are shown in Figures 2.24 and 2.25. Note that the domain of $y = \sqrt{x}$ is $x \geq 0$, because $\sqrt{x}$ is undefined for negative values of x. Note also that these root functions can be written as power functions with powers less than 1. The graphs of these functions increase in the first quadrant, but not as fast as $y = x$ does for values of x greater than 1, so we say that they are *concave down* in the first quadrant.

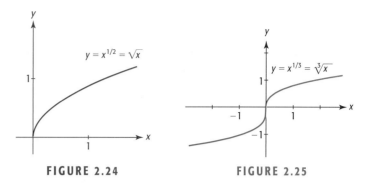

FIGURE 2.24   **FIGURE 2.25**

In general, if $a > 0$ and $x > 0$, the graph of $y = ax^b$ is concave up if $b > 1$ and concave down if $0 < b < 1$. Figures 2.26(a) and 2.26(b) show the first-quadrant portion of typical graphs of $y = ax^b$ for $a > 0$ and different values of b.

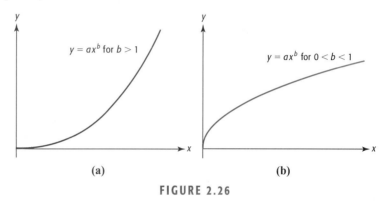

(a)   **(b)**

FIGURE 2.26

Transformations of Graphs

Knowledge of shifting, stretching, compressing, and reflecting the graph of a basic function is useful in obtaining graphs of additional functions and also in finding windows in which to graph them. Some examples of the shifts of the power function $f(x) = x^2$ are shown in Figure 2.27.

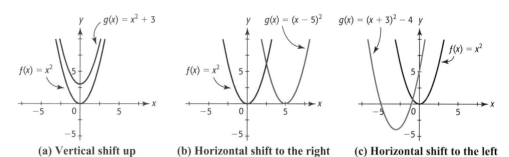

(a) Vertical shift up   **(b) Horizontal shift to the right**   **(c) Horizontal shift to the left and vertical shift down**

FIGURE 2.27

In Figure 2.27(a), we see that the graph of $g(x) = x^2 + 3$, with minimum at $(0, 3)$, can be obtained by shifting the graph of $f(x) = x^2$, the parabola with minimum at the origin, upward 3 units. The equation of the shifted graph is therefore $g(x) = f(x) + 3 = x^2 + 3$. In general, we have the following.

VERTICAL SHIFTS OF GRAPHS

If k is a positive real number:

The graph of $g(x) = f(x) + k$ can be obtained by shifting the graph of $f(x)$ upward k units.

The graph of $g(x) = f(x) - k$ can be obtained by shifting the graph of $f(x)$ downward k units.

In Figure 2.27(b), the graph of $g(x) = (x - 5)^2$, the parabola with minimum at $(5, 0)$, can be obtained by shifting the graph of $f(x) = x^2$ to the right 5 units. The equation of the shifted graph is therefore

$$g(x) = f(x - 5), \text{ or } g(x) = (x - 5)^2.$$

The general rule is the following.

HORIZONTAL SHIFTS OF GRAPHS

If h is a positive real number:

The graph of $g(x) = f(x - h)$ can be obtained by shifting the graph of $f(x)$ to the right h units.

The graph of $g(x) = f(x + h)$ can be obtained by shifting the graph of $f(x)$ to the left h units.

Note that a positive value of h indicates a shift to the right and that a negative value of h indicates a shift to the left when the function is in the form $g(x) = f(x - h)$. In Figure 2.27(c), the graph of $g(x) = (x + 3)^2 - 4$ is obtained by shifting the graph of $f(x)$ to the left 3 units and down 4 units.

EXAMPLE 2   Stimulus-Response

One of the early results in psychology relates the magnitude of a stimulus x to the magnitude of the response y with the model $y = kx^2$, where k is an experimental constant.

a. Sketch the graph of this function for $k = 1$ and $k = 2$.

b. How are these graphs related?

Solution

a. The graphs of these two functions are shown in Figure 2.28(a).

b. The graph of $y = 2x^2$ has y-coordinates that are twice as large as those on the graph of $y = x^2$. It is called a **stretch** of the graph of $y = x^2$. ∎

The graph of $y = \dfrac{1}{2}x^2$, shown in Figure 2.28(b), is called a **compression** of $y = x^2$, because each y-value is divided by 2.

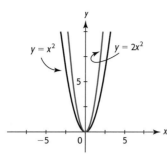 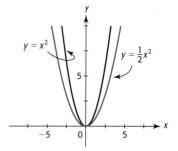

(a) Stretch by a factor of 2
(Graph rises more quickly)

(b) Compression by a factor of 2
(Graph rises more slowly)

FIGURE 2.28

STRETCHING AND COMPRESSING GRAPHS

The graph of $y = af(x)$ is obtained by stretching the graph of $f(x)$ by a factor of a if $a > 1$, and compressing the graph of $f(x)$ by a factor of a if $0 < a < 1$.

In Figure 2.29(a), we see that the graph of $y = -x^2$ is a parabola that *opens down*. It can be obtained by reflecting the graph of $f(x) = x^2$, which *opens up*, across the *x*-axis. We can compare the *y*-coordinates of the graphs of these two functions by looking at Table 2.6 and Figure 2.29(a). Notice that for a given value of *x*, the *y*-coordinates of $y = x^2$ and $y = -x^2$ are negatives of each other. (See Figure 2.29(b).)

TABLE 2.6

x	$y = x^2$	$y = -x^2$
−2	4	−4
−1	1	−1
0	0	0
1	1	−1
2	4	−4

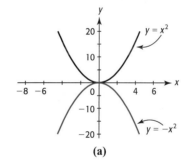 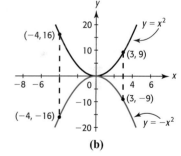

(a)

(b)

FIGURE 2.29

We can also see that the graph of $y = (-x)^3$ is a reflection of the graph of $y = x^3$ across the *y*-axis by looking at Table 2.7 and Figure 2.30.

TABLE 2.7

x	$y = x^3$	$y = (-x)^3$
−2	−8	8
−1	−1	1
0	0	0
1	1	−1
2	8	−8

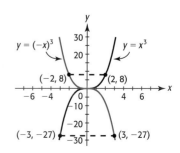

FIGURE 2.30

REFLECTIONS OF GRAPHS ACROSS THE COORDINATE AXES

1. The graph of $y = -f(x)$ can be obtained by reflecting the graph of $y = f(x)$ across the x-axis.

2. The graph of $y = f(-x)$ can be obtained by reflecting the graph of $y = f(x)$ across the y-axis.

EXAMPLE 3 Transformations

Graph each function together on the same axes with the graph of $f(x) = x^3$, and describe the shift, reflection, stretch, or compression involved to obtain the graph of g from the graph of f.

a. $g(x) = -x^3 + 4$ **b.** $g(x) = (x - 4)^3 + 2$ **c.** $g(x) = 3x^3 - 5$

Solution

a. The graph of $g(x) = -x^3 + 4$ is obtained by reflecting the graph of $f(x) = x^3$ across the x-axis and shifting this graph up 4 units. (See Figure 2.31(a).)

b. The graph of $g(x) = (x - 4)^3 + 2$ is obtained by shifting the graph of $f(x) = x^3$ to the right 4 units and up 2 units. (See Figure 2.31(b).)

c. The graph of $g(x) = 3x^3 - 5$ is obtained by stretching the graph of $f(x) = x^3$ by a factor of 3, then shifting this graph down 5 units. (See Figure 2.31(c).)

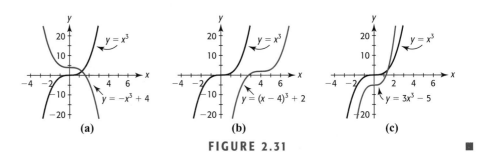

(a) **(b)** **(c)**

FIGURE 2.31

A summary of the transformations of a graph follows.

GRAPH TRANSFORMATIONS

In general, the graph of a function in the form

$$g(x) = a \cdot f(x - h) + k$$

is the graph of $y = f(x)$ shifted h units horizontally, k units vertically, and stretched or compressed by a factor of a. The horizontal shift is to the right when $h > 0$ and to the left when $h < 0$. The vertical shift is up when $k > 0$ and down when $k < 0$. The graph of f is compressed by a factor of a when $0 < a < 1$ and stretched by a factor of a when $a > 1$. If a is negative, the graph is reflected across the x-axis.

Knowing the shape of the graph and the shifts of the function is very useful in determining a viewing window that will give a complete graph.

EXAMPLE 4   Setting Windows

Use the graph of $y = -x^2$ and the appropriate transformation to set the window and graph the profit function

$$P = -(x - 10)^2 + 30.$$

Solution

The graph of the function $P = -(x - 10)^2 + 30$ is the graph of $y = -x^2$ shifted 10 units to the right and up 30 units (see Figure 2.32(a)). This means that the vertex $(0, 0)$ on the graph of $y = -x^2$ will be shifted to the point $(10, 30)$ on the graph of $P = -(x - 10)^2 + 30$. If the window used to graph $y = -x^2$ is set so that $(0, 0)$ is near the center of the viewing window, then the window used to graph $P = -(x - 10)^2 + 30$ should be set so that $(10, 30)$ is near the center. Such a window might have x-min $= 0$, x-max $= 20$, y-min $= 0$, y-max $= 60$. The graph with this window is shown in Figure 2-32(b). ■

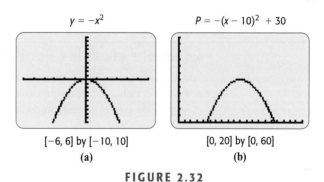

$y = -x^2$	$P = -(x - 10)^2 + 30$
$[-6, 6]$ by $[-10, 10]$	$[0, 20]$ by $[0, 60]$
(a)	(b)

FIGURE 2.32

EXAMPLE 5   Velocity of Blood

Because of friction from the walls of an artery, the velocity of blood is greatest at the center of the artery and decreases as the distance r from the center increases. The velocity of the blood in the artery can be modeled by the function

$$v = k(R^2 - r^2)$$

where R is the radius of the artery and k is a constant that is determined by the pressure, viscosity of the blood, and the length of the artery. In the case where $k = 2$ and $R = 0.1$ cm, the velocity is

$$v = 2(.01 - r^2) \text{ cm per second.}$$

a. Graph this function and the functions $v = r^2$ and $v = -2r^2$, on the viewing window $[-0.1, 0.1]$ by $[-0.05, 0.05]$.

b. How is the function $v = 2(.01 - r^2)$ related to the second-degree power function $v = r^2$?

Solution

a. The graph of the function $v = 2(.01 - r^2)$ is shown in Figure 2.33(a), the graph of $v = r^2$ is shown in Figure 2.33(b), and the graph of $v = -2r^2$ is shown in Figure 2.33(c).

b. The function $v = 2(.01 - r^2)$ can also be written in the form $v = -2r^2 + .02$, so it should be no surprise that the graph is a reflected, shifted, and stretched form of the graph of $v = r^2$. If the graph of $v = r^2$ is reflected about the r-axis, has each r-value doubled, and then is shifted upward .02 unit, the graph of $v = -2r^2 + .02$ results.

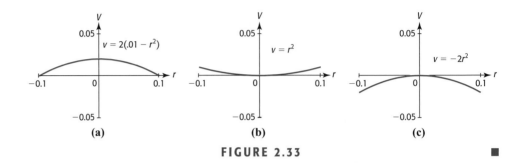

FIGURE 2.33 ■

Equations Involving Roots; Radical Equations

An equation containing a radical can frequently be converted to an equation that does not contain radicals by raising both sides of the equation to an appropriate power. For example, an equation containing a square root radical can usually be converted by squaring both sides of the equation. It is possible, however, that raising both sides of an equation to a power may produce an extraneous solution (a solution that does not satisfy the original solution). For this reason, we must check solutions when using this technique. To solve an equation containing a radical, we use the following steps.

SOLVING RADICAL EQUATIONS

1. Isolate a single radical on one side of the equation.

2. Square both sides of the equation. (Or, raise both sides to a power that is equal to the index of the radical.)

3. If a radical remains, repeat steps 1 and 2.

4. Solve the resulting equation.

5. All solutions must be checked in the original equation, and only those that satisfy the original solution are actual solutions.

EXAMPLE 6 Radical Equation

Solve $\sqrt{x + 5} + 1 = x$.

Solution

In order to eliminate the radical, we isolate it on one side and square both sides.

$$\sqrt{x + 5} = x - 1$$
$$x + 5 = x^2 - 2x + 1$$
$$0 = x^2 - 3x - 4$$
$$0 = (x - 4)(x + 1)$$
$$x = 4 \text{ or } x = -1$$

We can check graphically by the intersection method. Graphing $y = \sqrt{x + 5} + 1$ and $y = x$, we find the point of intersection to be $(4, 4)$. (See Figure 2.34.) Note that there is *not* a point of intersection at $x = -1$.

Thus, the solution to the equation $\sqrt{x + 5} + 1 = x$ is $x = 4$.   ■

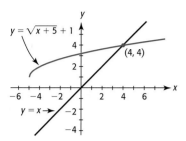

FIGURE 2.34

2.3 SKILLS CHECK

In Exercises 1–6, sketch the graph of each pair of functions using a standard window.

1. $y = x^3$, $y = x^3 + 5$ **2.** $y = x^2$, $y = x^2 + 3$

3. $y = \sqrt{x}$, $y = \sqrt{x - 4}$

4. $y = x^2$, $y = (x + 2)^2$

5. $y = x^3$, $y = (x - 5)^3 - 3$

6. $y = \sqrt[3]{x}$, $y = \sqrt[3]{x + 2} - 1$

For each of the functions in Exercises 7–10, find the value of (a) f(−1) and (b) f(3), if possible.

7. $f(x) = x^2 + 3$ **8.** $f(x) = \sqrt{x - 2}$

9. $f(x) = (x - 2)^3$ **10.** $f(x) = \sqrt[3]{x^2 - 2}$

11. Find the average rate of change of the function $y = x^2$:
   a. From $x = -2$ to $x = 2$.
   b. From $x = 0$ to $x = 2$.

12. Determine if the function $y = -3x^2$ is increasing or decreasing: **a.** for $x < 0$.   **b.** for $x > 0$.

13. Determine if the function $y = 4x^3$ is increasing or decreasing: **a.** for $x < 0$.   **b.** for $x > 0$.

For each of the functions in Exercises 14 and 15, determine if the function is concave up or concave down in the first quadrant.

14. $y = x^{3/2}$ **15.** $y = x^{1/2}$

16. Suppose that the graph of $y = x^{3/2}$ is shifted up 5 units. What is the equation that gives the new graph?

17. Suppose that the graph of $y = x^{3/2}$ is shifted to the left 4 units. What is the equation that gives the new graph?

18. Suppose that the graph of $y = x^{3/2}$ is shifted down 5 units and to the right 4 units. What is the equation that gives the new graph?

19. How is the graph of $y = (x - 2)^2 + 3$ transformed from the graph of $y = x^2$?

20. How is the graph of $y = (x + 4)^3 - 2$ transformed from the graph of $y = x^3$?

Solve the following equations and check graphically.

21. $\sqrt{2x^2 - 1} - x = 0$

22. $\sqrt{x^2 + 4} = 2x - 1$

23. $\sqrt{3x - 2} + 2 = x$

24. $\sqrt{x - 2} + 2 = x$

25. $\sqrt[3]{4x + 5} = \sqrt[3]{x^2 - 7}$

26. $\sqrt{4x - 8} - 1 = \sqrt{2x - 5}$

2.3 EXERCISES

----

1. *Suicides* The number of suicides can be modeled by the function $y = 20,000x^{0.11}$, where x is the number of years after 1960.
   a. What type of function is this?
   b. What is $f(5)$? What does this mean?
   c. How many suicides does this model estimate for 1982?
   (Source: National Center for Health Statistics, for selected years 1962–1996)

2. *Single Parent Families* The percent of all families that were single parent families between 1960 and 1998 was found to be modeled by $y = 1.053x^{0.888}$, with $x = 0$ in 1950.
   a. What is $f(45)$? What does this mean?
   b. Does this model indicate that the percent of single parent families increased or decreased during the period from 1960 to 2000?

3. *Taxi Miles* The Inner City Taxi Company estimated, on the basis of collected data, that the number of taxi miles driven each day by their cabs can be modeled by the function $Q = 489L^{0.6}$, when they employ L drivers per day.
   a. Graph this function for $0 \le L \le 35$.
   b. How many taxi miles are driven each day if there are 32 drivers employed?
   c. Does this model indicate that the number of taxi miles increases or decreases as the number of drivers increases? Is this reasonable?

4. *Single Parent Families* The percent of all families that were single parent families between 1960 and 1998 can be modeled by $y = 1.053x^{0.888}$, with $x = 0$ in 1950.
   a. According to the model, what percent of all families were estimated to have been single parent families in 1975?
   b. What percent of all families does the model predict will be single parent families in 2005? Comment on the reliability of your answer.
   c. Is this function concave up or concave down during this period of time?
   d. Use numerical or graphical methods to find when the model predicts that the percent will be 40%.
   (Source: U.S. Census Bureau)

5. *Mutual Funds* The household assets in mutual funds, measured in billions of dollars, for the years 1995 to 1999 can be modeled by $f(x) = 105.095x^{1.5307}$ billion dollars, where x is the number of years after 1990.

   a. Is this function increasing or decreasing from 1995 to 1999?
   b. Is this function concave up or concave down during this period of time?
   c. Use numerical or graphical methods to find when the model predicts that the assets will reach $4,000,000,000,000.
   (Source: Federal Reserve, *Time*, April 3, 2000)

6. *U.S. Population* U.S. population can be modeled by the function $y = 165.6x^{1.345}$, where y is in thousands and x is the number of years after 1800.
   a. What was the population in 1960, according to this model?
   b. Is the graph of this function concave up or concave down?
   c. Use numerical or graphical methods to find when the model estimates the population to be 92,370,000.

7. *Ballistics* Ballistic experts are able to identify the weapon that fired a certain bullet by studying the markings on the bullet. If a test is conducted by firing the bullet into a water tank, the distance that the bullet will travel is given by $s = 27 - (3 - 10t)^3$ inches, for $0 \le t \le 0.3$, t in seconds.
   a. Graph this function for $0 \le t \le 0.3$.
   b. How far does the bullet travel during this time period?

8. *Production Output* The monthly output of a product (in units) is given by $P = 1200x^{5/2}$, where x is the capital investment in thousands of dollars.
   a. Graph this function for x from 0 to 10 and P from 0 to 200,000.
   b. Is the graph concave up or concave down?

9. *Supply and Demand* The supply function for a product is given by $p = q^2 + 500$, and the demand function for this product is $p = -40q + 1124$, where p is the price in dollars and x is the number of hundreds of units.
   a. Is the supply function a shifted power function or a linear function?
   b. Graph the supply and demand functions on the same axes.
   c. Is there a point where the graphs intersect? What does this point of intersection represent?
   d. Use numerical or graphical methods to find this point.

10. *Supply and Demand* The supply function for a product is given by $p = q + 578$, and the demand function for this product is $p = 4000 - q^2$, where p is the price in dollars and q is the quantity supplied or demanded at that price.
    a. Is the demand function a shifted power function or a linear function?
    b. Graph the supply and demand functions on the same axes.
    c. Is there a point where the graphs intersect? What does this point of intersection represent?
    d. Use numerical or graphical methods to find this point.

11. *Break-Even* The weekly total cost function for a piece of medical equipment is $C(x) = 3x^2 + 1228$ dollars and the weekly revenue for the equipment is $R(x) = 177.50x$ dollars, where x is the number of thousands of units.
    a. Which function is linear and which function is a shifted power function?
    b. Graph the weekly total cost and revenue functions on the same axes.
    c. Use numerical or graphical methods to find where break-even occurs.

12. *Break-Even* The weekly total cost function for a piece of medical equipment is $C(x) = 2x^2 + 3402$ dollars and the weekly revenue for the equipment is $R(x) = 500x$ dollars, where x is the number of thousands of units.
    a. Which function is linear and which function is a transformed power function?
    b. Graph the weekly total cost and revenue functions on the same axes.
    c. Use numerical or graphical methods to find where break-even occurs.

13. *Mutual Funds* The household assets in mutual funds, measured in billions of dollars, for the years 1995 to 1999 was found to be modeled by $f(x) = 105.095x^{1.5307}$ billion dollars, where x is the number of years from 1990. Rewrite the model with x equal to the number of years from 1995. Hint: Shift the x-value by the difference in years.

14. *Single Parent Families* The percent of all families that were single parent families between 1960 and 1998 was found to be modeled by $y = 1.053x^{0.888}$, with $x = 0$ in 1950. Rewrite the model with x equal to the number of years from 1960.

15. *Prison Population* The average number of inmates in the Beaufort County, South Carolina jail can be modeled by $y = 0.161x^2 + 7.582x + 82.157$, where x is the number of years after 1991.
    a. Graph this function from $x = 0$ to $x = 20$.
    b. Does this model indicate that the population will increase or decrease in years 1991 through 2011?
    c. How would this information be useful to government officials planning for the future?
    (Source: The *Island Packet*, November 1999)

16. *World Population* A low projection scenario of world population for 1995 to 2150 by the United Nations is given by the function $y = -0.36x^2 + 38.52x + 5822.86$, where x is the number of years after 1990 and the world population is measured in millions of people.
    a. Find the input and output at the vertex of the graph of this model.
    b. Interpret these values.
    c. For what years after 1995 does this model predict that the population will increase?
    (Source: *World Population Prospects: 1998 Revision*, United Nations, 1999)

17. *Voter Turnout* The function $y = 92.239x^{-0.1585}$ gives the percent of voter turnout during presidential election years, with x representing the number of years after 1950.
    a. Graph the function for $x > 0$.
    b. Does this model indicate that the percent of voter turnout is increasing or decreasing?
    c. Use the graph to determine the percent in 1996.
    d. What percent voter turnout does this model estimate for the 2000 election?
    e. Did the 2000 turnout agree with the prediction from the model?
    (Source: Federal Election Commission)

18. *Trust in the Government* The percent of people who say they trust the government in Washington always or most of the time is given by $y = 154.131x^{-0.492}$, with x equal to the number of years after 1960.
    a. Graph the function.
    b. Does this model indicate that trust in the government is increasing or decreasing?
    c. Use the graph to estimate the percent of people who trust the government in 1998.
    (Source: Pew Research Center)

2.4 Quadratic and Power Models

Key Concepts

- Modeling quadratic functions
- Modeling power functions
- Comparison of models; first and second differences
- Modeling with power functions
- Comparison of quadratic and power models

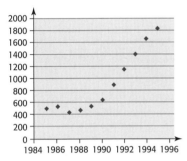

FIGURE 2.35

Table 2.8 shows that the number of registered cocaine addicts rocketed in the United Kingdom for the years 1985 to 1995. The scatter plot of this data, shown in Figure 2.35, shows that a line would not fit these data points well, so a linear function is not a good model for this data. The use of graphing utilities permits us to model nonlinear data with other types of functions by using steps similar to those that we used to model data with linear functions. In this section we will model sets of data with quadratic and power functions.

TABLE 2.8 Cocaine Use in the United Kingdom

Year	Number of Registered Cocaine Addicts	Year	Number of Registered Cocaine Addicts
1985	490	1991	882
1986	520	1992	1131
1987	431	1993	1375
1988	462	1994	1636
1989	527	1995	1809
1990	633		

(Source: www.ukcia.org/lib/stats/ukstats.htm, 4/23/2001)

Modeling with Quadratic Functions

If the graph of a set of data has a pattern that approximates the shape of a parabola or part of a parabola, a quadratic function may be appropriate to model the data. The scatter plot in Figure 2.35 has this shape.

EXAMPLE 1 Cocaine Use in the United Kingdom

Table 2.8 gives the number of registered cocaine addicts in the United Kingdom for the years 1985 to 1995.

a. Create a quadratic function that models the data, using the number of years after 1985 as the input x.

b. Graph the aligned data and the quadratic function on the same axes. Does this model seem like a reasonable fit?

c. Use the model to estimate the number of registered addicts in 1991. Is this estimate close to the actual number?

d. Use the model to estimate the number in 2000. Discuss the reliability of this estimate.

Solution

a. Figure 2.35 shows the scatter plot of the data. Enter the aligned inputs 0 through 10 in a graphing utility, and enter the corresponding output values from the second column of

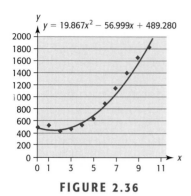

FIGURE 2.36

the table. Using quadratic regression in a graphing utility gives a quadratic function that models the data. The function, with coefficients rounded to three decimal places, is

$$y = 19.867x^2 - 56.999x + 489.280$$

where x is the number of years from 1985. Note that using a rounded model in calculations may result in slight calculation errors.

b. The scatter plot of the aligned data and the graph of the function are shown in Figure 2.36. The function appears to be a good fit to the data.

c. Evaluating the reported function at $x = 6$ gives the number of registered drug addicts as 862 in 1991. This is a reasonable estimate of 882, the actual number.

d. Evaluating the reported function at $x = 15$ gives the number of registered drug addicts as 4104 in 2000. This estimate is an extrapolation that is 5 years after the last data, and many factors may have changed the pattern that predicts this number. Thus the number is not very reliable. Finding and including data points for more recent years would give a new model that would be a better predictor for 2000. ∎

Comparison of Linear and Quadratic Models

Recall that when the changes in inputs are constant and the (first) differences of the outputs are constant or nearly constant, a linear model will give a good fit for the data. In a similar manner, we can compare the differences of the first differences, which are called the **second differences**. If the second differences are constant for equally spaced inputs, the data can be modeled by a quadratic function.

Consider the data in Table 2.9, which gives the measured height y of a toy rocket x seconds after it has been shot into the air from the ground.

TABLE 2.9 Height of a Rocket

Time x (seconds)	Height (meters)
1	68.6
2	117.6
3	147
4	156.8
5	147
6	117.6
7	68.6

Table 2.10 gives the first differences and second differences for the equally spaced inputs for the rocket height data in Table 2.9.

TABLE 2.10

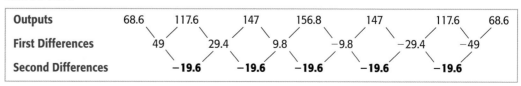

Outputs	68.6		117.6		147		156.8		147		117.6		68.6
First Differences		49		29.4		9.8		−9.8		−29.4		−49	
Second Differences			−19.6		−19.6		−19.6		−19.6		−19.6		

The first differences are not constant, but each of the second differences is -19.6, which indicates that the data can be fit exactly by a quadratic formula. In Exercise 2.4, we will see that the quadratic model for the height of the toy rocket is $y = 78.4x - 9.8x^2$ meters, where x is the time in seconds.

EXAMPLE 2 Aid to Dependent Children

Table 2.11 gives the number of families (in thousands) who were recipients under the Federal Aid to Families with Dependent Children program for the years 1990 to 1996.

a. Find the first and second differences for the data to justify that a quadratic function will be a better model for the data than a linear function.

b. Align the data so that the input is the number of years from 1990, and create a quadratic function that models the data.

c. Graph the aligned data and the quadratic function on the same axes.

d. What does the model give as the maximum number of families who were recipients of federal aid during the period 1990 to 1996?

TABLE 2.11

Years	1990	1991	1992	1993	1994	1995	1996
Recipient Families (in thousands)	4218	4708	4936	5050	4979	4641	4166

(Source: U.S. Census Bureau, *Statistical Abstract of the U.S.*)

Solution

a. The inputs are uniform, and the first and second differences for the data are shown in Table 2.12. The second differences are closer to being equal than the first differences, so a quadratic model will be a better fit for the data.

TABLE 2.12

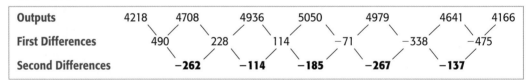

Outputs	4218	4708	4936	5050	4979	4641	4166
First Differences		490	228	114	−71	−338	−475
Second Differences			−262	−114	−185	−267	−137

b. Enter the aligned inputs 0, 1, 2, 3, 4, 5, and 6 in a graphing utility, and enter the corresponding output values from the second row of Table 2.11. Using quadratic regression in a graphing utility gives a quadratic function that models the data. The function, with the coefficients rounded to four decimal places, is

$$y = -95.5357x^2 + 564.3929x + 4219.9286.$$

c. The scatter plot of the aligned data and the graph of the function are shown in Figure 2.37(a). We see that a quadratic model fits the data well even though the second differences are not really constant.

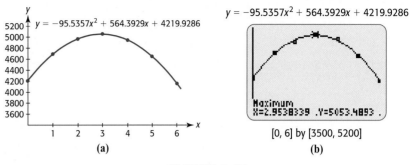

FIGURE 2.37

d. The maximum value of the rounded function occurs at the vertex of the parabola shown in Figure 2.37(a). The x-coordinate of this vertex is $x = -\dfrac{-564.3929}{2(-95.5357)} = 2.9538 \approx 3$, and the y-coordinate is $y \approx 5053$. Thus the model gives 5053 thousand, or 5,053,000 families as the maximum number of families who received federal aid under this program in $1990 + 3 = 1993$. This can be verified by finding the maximum point on the graph with technology. (See Figure 2.37(b).) ∎

Spreadsheet Solution

We can use graphing calculators, software programs, and spreadsheets to find the quadratic function that is the best fit for data. Table 2.13 shows the Excel spreadsheet for the aligned data of Example 2. Selecting the cells containing the data, using Chart Wizard to get the scatter plot of the data, and selecting ADD TRENDLINE and picking POLYNOMIAL with order 2 gives the equation of the quadratic function that is the best fit for the data, along with the scatter plot and the graph of the best-fitting parabola. The equation and graph are shown in Figure 2.38.

TABLE 2.13

	A	B
1	Year x	Recipient Families (thousands) y
2	0	4218
3	1	4708
4	2	4936
5	3	5050
6	4	4979
7	5	4641
8	6	4166

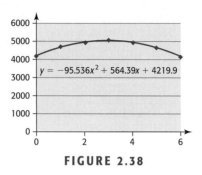

Modeling with Power Functions

We can model some experimental data by observing patterns. For example, by observing how the area of each of the (square) faces of a cube is found and that a cube has six faces, we can deduce that the surface area of a cube that is x units on each edge is

$$S = 6x^2 \text{ square units.}$$

We can also measure and record the surface areas for cubes of different sizes to investigate the relationship between the edge length and the surface areas for cubes. Table 2.14 contains selected measures of edges and the resulting surface areas.

TABLE 2.14

Edge length x (units)	Surface Area of Cube (square units)
1	6
2	24
3	54
4	96
5	150

We can enter the lengths from the table as the independent (x) variable and the corresponding surface areas as the dependent (y) variable in a graphing utility, and then have the utility create the **power function** that is the best model for the data. (See Figure 2.39.) This model also has the equation $y = 6x^2$ square units.

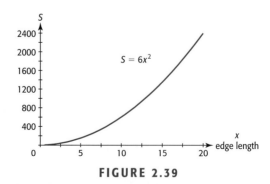

FIGURE 2.39

EXAMPLE 3   Auto Noise

The noise level of a Vauxhall VX220 increases as the speed of the car increases. Table 2.15 gives the noise, in decibels, at different speeds.

TABLE 2.15

Speed (miles per hour)	Noise Level (decibels)
10	50
30	68
50	75
70	79
100	84

(Source: *Auto Car Magazine*: November 2001, p. 63)

a. Fit a power function model to the data.

b. Graph the data points and the model on the same axes.

c. Use the result from part (a) to estimate the noise level at 80 miles per hour.

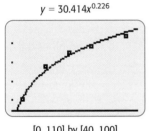

$y = 30.414x^{0.226}$

[0, 110] by [40, 100]

FIGURE 2.40

Solution

a. Entering the data in a graphing utility and creating the power regression model with the utility gives the equation. The model is

$$y = 30.414x^{0.226},$$

where x is in mph, and y is in decibels.

b. The graph of the data points and the model is shown in Figure 2.40.

c. Evaluating the model at $x = 80$ gives 81.88, so the noise level at 80 mph is 81.88 decibels. ∎

EXAMPLE 4   Cohabiting Households

The data in Table 2.16 gives the number of cohabiting (without marriage) households (in thousands) for selected years from 1960 to 1998. The scatter plot of this data, with x representing the number of years from 1950 and y representing thousands of households, is shown in Figure 2.41.

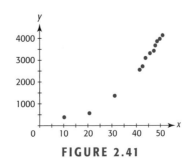

FIGURE 2.41

TABLE 2.16 Cohabiting Households

Year	Cohabiting Households (in thousands)	Year	Cohabiting Households (in thousands)
1960	439	1993	3510
1970	523	1994	3661
1980	1589	1995	3668
1985	1983	1996	3958
1990	2856	1997	4130
1991	3039	1998	4236
1992	3308		

(Source: Index of Leading Cultural Indicators)

a. Find the power function that models this data.

b. Graph the data and the model on the same axes.

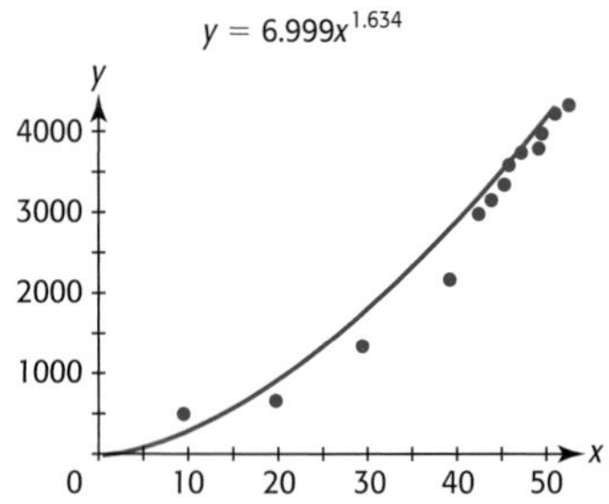

Solution

Using power regression in a graphing utility gives the power function that models this data. The power function that is the best fit is

$$y = 6.999x^{1.634},$$

where y is in thousands and x is the number of years from 1950. The graph of the data and the power function that models it is shown in Figure 2.42. ■

$y = 6.999x^{1.634}$

FIGURE 2.42

Spreadsheet Solution

Power models can also be found with Excel. The Excel spreadsheet in Table 2.17 shows the household assets in mutual funds, measured in billions of dollars, for 1995 to 1999, listed as years after 1990. Selecting the cells containing the data, using Chart Wizard to get the scatter plot of the data, and selecting Add Trendline and picking Power gives the equation of the power function that is the best fit for the data, along with the scatter plot and the graph of the best-fitting power function. The equation and graph of the function are shown in Figure 2.43.

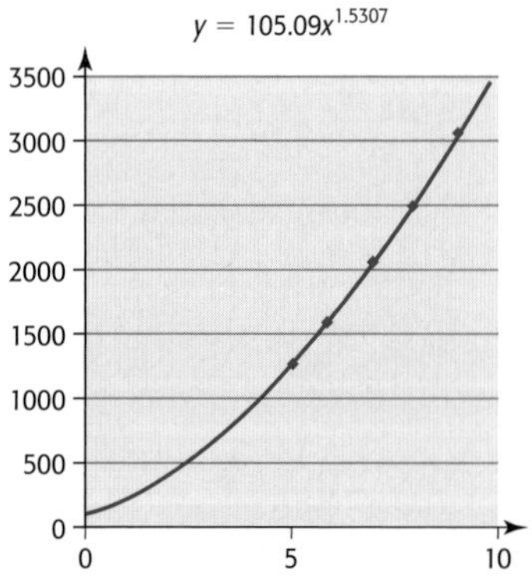

TABLE 2.17

$y = 105.09x^{1.5307}$

	A	B
1	Year x	Assets in mutual funds y ($ dollars)
2	5	1265
3	6	1586
4	7	2057
5	8	2501
6	9	3104

FIGURE 2.43

(Source: "Federal Reserve," *Time*, April 3, 2000)

Comparison of Power and Quadratic Models

We found a power function that is a good fit for the data in Table 2.17, but a linear function or a quadratic function may also be a good fit for this data. A quadratic function may fit data points even if there is no obvious "turning point" in the graph of the data points. If the data points appear to rise (or fall) more rapidly than a line, then a quadratic model or a power model may fit the data well. In some cases it may be necessary to find both models to determine which is the better fit for the data.

EXAMPLE 5   Voting

Table 2.18 shows the percent of registered voters who voted in presidential elections for the years 1960 through 2000.

TABLE 2.18 Percent Voting in Presidential Elections

Years	Percent	Years	Percent
1960	63.1	1984	53.1
1964	61.9	1988	50.1
1968	60.8	1992	55.1
1972	55.2	1996	49.1
1976	53.6	2000	51.0
1980	52.6		

(Source: Federal Election Commission)

a. Graph the data points, representing the years from 1950 as x and the percent as y.

b. Find the quadratic model that is the best fit for the data.

c. Find the power model that is the best fit for the data.

d. Discuss the use of the two models to predict the percent voting after 2000.

Solution

a. The scatter plot of the data is shown in Figure 2.44(a). It has no obvious "turning point," so either a quadratic or a power model may possibly be a good fit.

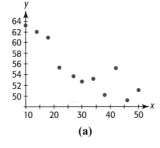

(a)

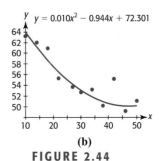

$y = 0.010x^2 - 0.944x + 72.301$

(b)

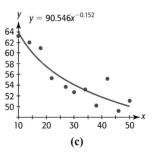

$y = 90.546x^{-0.152}$

(c)

FIGURE 2.44

b. The quadratic model that is the best fit for this data is

$$y = 0.010x^2 - 0.944x + 72.301$$

The graph of this model and the data points is shown in Figure 2.44(b).

c. The power model that is the best fit for this data is

$$y = 90.546x^{-0.152}$$

The graph of this model and the data points is shown in Figure 2.44(c).

d. The graphs of the two models appear to be very similar until $x = 50$, which represents 2000. The quadratic model looks like it is beginning to rise after 1996 ($x = 46$) and could be used for prediction if the percent voting is expected to increase. The graph of the power model could be used for prediction if the percent voting is expected to continue to fall. ■

Statistical measurements exist that can be used to measure the "goodness of fit" of models to the data, but care must be used in interpreting these measures. For example, the statistical measurement for quadratic functions is different from the statistical measurement for linear functions.

2.4 SKILLS CHECK

1. Find the quadratic function that models the data in the table below.

x	-2	-1	0	1	2	3	4
y	16	5	0	1	8	21	40

x	5	6	7	8	9	10
y	65	96	133	176	225	280

2. The following table has the inputs, x, and the outputs for three functions, f, g, and h. Use second differences to determine which function is exactly quadratic, which is approximately quadratic, and which is not quadratic.

x	$f(x)$	$g(x)$	$h(x)$
0	0	2	0
2	399	0.8	110
4	1601	1.2	161
6	3600	3.2	195
8	6402	6.8	230
10	9998	12	290

3. As you can verify, the following data points give constant second differences, but the points do not fit on the graph of a quadratic function. How can this be?

x	1	2	4	8	16	32	64
y	1	6	15	28	45	66	91

4. **a.** Make a scatter plot of the data in the table below.
   b. Does it appear that a linear model or a power model is the best fit for the data?

x	y
1	4
2	9
3	11
4	21
5	32
6	45

5. Find the quadratic function that is the best fit for $f(x)$ defined by the table in Exercise 2.

6. Find the quadratic function that is the best fit for $g(x)$ defined by the table in Exercise 2.

7. **a.** Find a power function that models the data in the table in Exercise 4.
   b. Find a linear function that models the data.
   c. Visually determine which model is the better fit for the data.

8. **a.** Make a scatter plot of the data in the table below.
   b. Does it appear that a linear model or a power model is the better fit for the data?

x	y
3	4.7
5	8.6
7	13
9	17

9. **a.** Find a power function that models the data in the table in Exercise 8.
   b. Find a linear function that models the data.
   c. Visually determine if each model is a good fit.

10. Find the quadratic function that models the data in the table below.

x	−2	−1	0	1	2	3
y	15	5	2	1	3	10

x	4	5	6	7	8
y	20	35	55	75	176

11. Find the power function that models the data in the table below.

x	1	2	3	4	5	6	7	8
y	3	4.5	5.8	7	8	9	10	10.5

2.4 EXERCISES

Calculate numerical results with the unrounded models, unless otherwise instructed.

1. *Income* The median annual income of male householders in the U.S. with children under 18 years of age for selected years between 1969 and 1996 are shown in the table below.

Year	Median Annual Income (dollars)	Year	Median Annual Income (dollars)
1969	33,749	1984	36,002
1972	36,323	1987	34,747
1975	33,549	1990	33,769
1978	37,575	1993	29,320
1981	33,337	1996	31,020

(Source: U.S. Bureau of the Census)

   a. Find the quadratic function that models the median income as a function of x, the number of years after 1960.
   b. Use the model from part (a) to estimate the median annual income of male householders with children under 18 years of age in 1997 and in 2002.

   c. Do you feel that these estimates are reliable? Explain.

2. *Aircraft Accidents* The table below shows data recorded on December 31 of the indicated years for the number of U.S. air carrier accidents, for all military services.
   a. Plot the data in a scatter plot, with x equal to the number of years after 1970.
   b. Fit a quadratic function to these data.
   c. Use the model from part (b) to estimate the number of aircraft accidents in 1992 and in 2001.
   Do you feel that these estimates are reliable? Explain.

Year	1975	1980	1985	1990
Accidents	37	19	21	24

Year	1995	1996	1997
Accidents	36	38	49

(Source: National Traffic Safety Council)

3. *Unemployment* Total unemployment in South Carolina for the years 1982–1992 is given by the data in the table below.
   a. Create a scatter plot for the data, with x equal to the number of years after 1980.

b. Does it appear that a quadratic model will fit the data? If so, find the best-fitting quadratic model.

c. Does the y-intercept of the function in part (b) have meaning in the context of this problem? If so, interpret the value.

Year	Total Unemployment (thousands)	Year	Total Unemployment (thousands)
1982	162	1988	76
1983	148	1989	80
1984	105	1990	81
1985	107	1991	108
1986	99	1992	111
1987	91		

(Source: South Carolina Employment Security Commission, September 1993)

4. *Unemployment* The unemployment as a percent of the labor force in South Carolina for the years 1982–1992 is given by the table below. The best-fitting quadratic model, with x representing the number of years after 1980, is $y = 0.144x^2 - 2.48x + 15.47$. In what year past 1990 does this model indicate that the unemployment percent is 5.6 percent?

Year	Total Unemployment as a Percent of Labor Force
1982	10.8
1983	10.1
1984	7.1
1985	6.9
1986	5.6
1987	5.6
1988	4.6
1989	4.7
1990	4.7
1991	6.2
1992	6.2

(Source: South Carolina Employment Security Commission, September 1993)

5. *World Population* One projection of the world population by the United Nations (a low projection sce-

nario) is given in the table below. This data can be modeled by $y = -0.36x^2 + 38.52x + 5822.86$ million people, where x is the number of years from 1990. In what year past 1990 does this model predict the world population will first reach 6,702,000,000?

Year	Projected Population (in millions)
1995	5666
2000	6028
2025	7275
2050	7343
2075	6402
2100	5153
2125	4074
2150	3236

(Source: *World Population Prospects: 1998 Revision*, United Nations, 1999)

6. *U.S. Population* The data in the table below gives the U.S. population for selected years from 1790–2000.

a. Create a quadratic function that models this data, with y equal to the population in millions and x equal to the number of years from 1700.

b. During what year past 1700 does this model indicate that the population was 140.2 million?

Year	Population
1790	3,929,214
1870	38,558,371
1930	123,202,624
1990	248,790,925
2000	281,421,906

(Source: *Newsweek*, January 8, 2001)

7. *Educational Funding* Data that gives the amount of federal on-budget funds for research programs at universities and related institutions appears in the table below.

Year	1965	1970	1975	1980	1985
Federal funds (in billions of dollars)	1.82	2.28	3.42	5.80	8.84

Year	1990	1995	1998	1999	2000
Federal funds (in billions of dollars)	12.61	15.68	18.48	20.24	21.02

(Source: U.S. Department of Education, National Center for Educational Statistics)

   a. Using an input equal to the number of years after 1960, find a quadratic function that models the data.

   b. Using an input equal to the number of years after 1965, find a quadratic function that models the data.

   c. Use the model in part (a) to find when (between 1965 and 2000) federal funds for research first exceeded $10 billion.

   d. If you had used the model in part (b) instead of the one in part (a) to answer part (c), how would your results have differed?

8. *Medicare Trust Fund Balance* The year 1994 marked the 30th anniversary of Medicare. A 1994 pamphlet by Representative Lindsey O. Graham gave projections for the Medicare Trust Fund Balance as shown in the table below.

   a. Representative Graham stated, "The fact of the matter is that Medicare is going broke." Use the data in the table to find when he predicted that this would happen.

   b. Find a quadratic model to fit the data, with *x* equal to the number of years after 1990. When does this model predict that the Medicare Trust Fund balance will be 0?

   c. The pamphlet reported the Medicare trustees as saying: ". . . the present financing schedule for the hospital insurance program is sufficient to ensure the payment of benefits only over the next 7 years." Does the function in part (b) confirm or refute the value 7 in this statement?

   d. Find and interpret the vertex of the quadratic function in part (b).

Year	Medicare Trust Fund Balance (billions of dollars)	Year	Medicare Trust Fund Balance (billions of dollars)
1993	128	1999	98
1994	133	2000	72
1995	136	2001	37
1996	135	2002	−7
1997	129	2003	−61
1998	117	2004	−126

9. *Unemployment* The unemployment as a percent of the labor force in South Carolina for the years 1982–1992 is given by the table below.

   a. Create a scatter plot for the data, with *x* equal to the number of years from 1980.

   b. Does it appear that a quadratic model will fit the data? If so, find the best-fitting quadratic model.

   c. Does the *y*-intercept of the function in part (b) have meaning in the context of this problem? If so, interpret the value.

Year	Total Unemployment as a Percent of Labor Force	Year	Total Unemployment as a Percent of Labor Force
1982	10.8	1988	4.6
1983	10.1	1989	4.7
1984	7.1	1990	4.7
1985	6.9	1991	6.2
1986	5.6	1992	6.2
1987	5.6		

(Source: South Carolina Employment Security Commission, September 1993)

10. *Shrimp-Baiting* The table below gives the number of shrimp-baiting permits in South Carolina in each of the years from 1994 through 1998.

   a. Create a quadratic model for the data, using *x* as the number of years from 1990.

   b. Use the model to predict the number of permits in 1999.

   c. Because of a busy hurricane season, the number of permits issued in 1999 was 15,894. How does this compare with the prediction from the model, and what does it tell us about the dangers of extrapolating from models?

Year	1994	1995	1996	1997	1998
Number of permits	13,366	13,919	14,150	15,488	17,467

(Source: South Carolina Department of Natural Resources)

11. *Volume* The measured volume of a pyramid with each edge of the base equal to *x* units and with its altitude (height) equal to *x* units is given in the table below.

   a. Determine if the second differences of the outputs are constant.

b. If the answer is yes, find the quadratic model that is the best fit for the data. Otherwise, find the power function that is the best fit.

Edge Length x (units)	Volume of Pyramid (cubic units)
1	1/3
2	8/3
3	9
4	64/3
5	125/3
6	72

12. *Age and Income* The median annual income of men for certain average ages is given in the table below.
a. Find the quadratic function that models the annual income as a function of the average age.
b. Does the model appear to be a good fit?
c. Find and interpret the vertex of the graph of the *rounded* function.

Average Age (years)	Median Income (dollars)	Average Age (years)	Median Income (dollars)
19.5	6960	49.5	36,526
29.5	25,179	59.5	29,526
39.5	32,167	69.5	16,684

(Source: *Statistical Abstract of the U.S., 1998*)

13. *World Population* One projection of the world population by the United Nations for selected years (a low projection scenario) is given in the table below.

Year	Projected Population (in millions)	Year	Projected Population (in millions)
1995	5666	2075	6402
2000	6028	2100	5153
2025	7275	2125	4074
2050	7343	2150	3236

(Source: *World Population Prospects: 1998 Revision*, United Nations, 1999)

a. Find a quadratic function that fits these data, using the number of years after 1990 as the input.

b. Find the positive x-intercept of this graph, to the nearest year.
c. When can we be certain that this model no longer applies?

14. *Personal Savings* The table below gives Americans' personal-savings rate for selected years from 1960–1998.
a. Find a quadratic function that models the personal-savings rate as a function of the number of years from 1960.
b. Find and interpret the vertex of the graph of the *rounded* model from part (a).
c. Find the positive x-intercept of this graph, to the nearest tenth.
d. When can we be certain that this model no longer applies?

Year	Personal-Savings Rate (percent)	Year	Personal-Savings Rate (percent)
1960	6.6	1985	6.9
1965	7.8	1990	5.1
1970	8.5	1995	3.4
1975	9.3	1998	0.5
1980	8.5		

(Source: Index of Leading Cultural Indicators)

15. *Mortgages* The balance owed, y, on a $50,000 mortgage after x monthly payments is shown in the table below. Graph the data points with each of the equations below to determine which of them is the better model for the data if x is the number of months that payments have been made.
a. $y = 338{,}111.278x^{-0.676}$
b. $y = 4700\sqrt{110 - x}$

Monthly Payments	Balance Owed (in dollars)
12	47,243
24	44,136
48	36,693
72	27,241
96	15,239
108	8074

16. *International Visitors* The number of international visitors to the U.S. for each of the years 1986–1997 is given in the table below.

Year	U.S. Visitors (millions)	Year	U.S. Visitors (millions)
1986	26	1992	47.3
1987	29.5	1994	45.5
1988	34.1	1995	44
1989	36.6	1996	46.3
1990	39.5	1997	48.9
1991	43		

(Source: *1998 World Almanac*)

a. Using an input equal to the number of years past 1980, graph the aligned data points and both of the given equations to determine which equation is the better model for the aligned data.
   (i) $y = 12\sqrt{x}$
   (ii) $y = -.2327x^2 + 7.237x - 9.167$
b. If you had to pick one of these models to predict the number of international visitors in the year 2002, which model would be the more reasonable choice?

17. *Trust in the Government* The percent of people who say they trust the government in Washington always or most of the time has decreased over the years.

Year	Believe government always or most of time (percent)	Year	Believe government always or most of time (percent)
1964	76	1982	33
1966	65	1984	44
1968	61	1986	38
1970	54	1988	41
1972	53	1990	28
1974	36	1992	29
1976	34	1994	21
1978	29	1996	27
1980	25		

(Source: Pew Research Center)

a. Find a power function that models the data in the table using an input equal to the number of years from 1960.
b. According to the model, what percent of people said they trust the government always or most of the time in 2000?

18. *Box-office Revenues* The data in the table below gives the box-office revenues, in billions of dollars, for movies released in selected years between 1980 and 1998.
a. Find the power function that best fits the revenues as a function of the number of years from 1970.
b. What does the unrounded model predict as the revenue in 2005?
c. Discuss the reliability of this prediction.

Year	Revenue (in billions of dollars)	Year	Revenue (in billions of dollars)
1980	2.75	1994	5.39
1985	3.75	1995	5.49
1990	5.02	1996	5.91
1991	4.80	1997	6.37
1992	4.87	1998	6.95
1993	5.15		

(Source: Index of Leading Cultural Indicators)

19. *Classroom Size* The data in the table below gives the number of students per teacher for selected years between 1960 and 1998.

Year	Students per Teacher	Year	Students per Teacher
1960	25.8	1992	17.4
1965	24.7	1993	17.4
1970	22.3	1994	17.3
1975	20.4	1995	17.3
1980	18.7	1996	17.1
1985	17.9	1997	17.0
1990	17.2	1998	17.2
1991	17.3		

(Source: U.S. Department of Education)

a. Find the power function that is the best fit for the data using as input the number of years after 1950.
b. According to the unrounded model, how many students per teacher were there in 2000?
c. Is this function increasing or decreasing during this time period?
d. What does the model predict will happen to the number of students per teacher as time goes on?

20. *Classroom Teachers* The percent of all full-time school staff that are classroom teachers in the U.S. has decreased over the years since 1960, as seen in the table below.

Year	Classroom Teachers (percent)	Year	Classroom Teachers (percent)
1960	64.8	1992	52.2
1970	60.0	1993	52.1
1980	52.4	1994	52.0
1990	53.4	1995	52.0
1991	53.3	1996	52.1

(Source: U.S. Department of Education)

a. Find a power function that models the data using an input equal to the number of years from 1950.
b. What does the unrounded model predict as the percent of all full-time school staff that are classroom teachers in 2008?
c. Do we have enough data to be confident that the 2008 prediction is accurate?
d. Use graphical or numerical methods to predict when the model output will fall below 50%.

21. *Cigarette Advertising* The data in the table below gives the total U.S. cigarette advertising and promotional expenditures, in millions of dollars, for the years 1975–1993.
a. Using an input equal to the number of years after 1970, find a power function that models the data.
b. Use the model to predict the expenditure for cigarette advertising and promotion in 2000.
c. Use the model to estimate the year when the expenditure was $3.962 billion.
d. Cigarette smoking has decreased during this period. Has advertising expenditure increased or decreased?

Year	Cigarette Advertising (in millions of dollars)	Year	Cigarette Advertising (in millions of dollars)
1975	491.3	1985	2476.4
1976	639.1	1986	2382.4
1977	779.5	1987	2580.5
1978	875.0	1988	3274.9
1979	1083.4	1989	3617.0
1980	1242.3	1990	3992.0
1981	1547.7	1991	4650.1
1982	1800.4	1992	5231.9
1983	1901.5	1993	6035.4
1984	2095.2		

(Source: Federal Trade Commission Report to Congress for 1990–1996)

22. *Travel and Tourism Spending* The global spending on travel and tourism (in billions of dollars) for the years 1988–1997 is given in the table below.
a. Write the equation of a power function that models the data, letting your input represent the number of years after 1980.
b. Use the model to predict the global spending for 1999.
c. When did the global spending reach $300 billion, according to this model?

Year	Spending (billions of dollars)	Year	Spending (billions of dollars)
1988	204.7	1993	322.3
1989	221.0	1994	353.5
1990	269.2	1995	403.6
1991	277.6	1996	438.8
1992	315.5	1997	447.7

(Source: *1998 World Almanac*)

23. *Insurance Rates* The table gives the monthly insurance rates for a $100,000 life insurance policy for smokers 35 to 50 years of age.
a. Create a scatter plot for the data.
b. Does it appear that a quadratic function can be used to model the data? If so, find the best-fitting quadratic model.

c. Find the power model that is the best fit for the data.
d. Compare the two models by graphing each model on the same graph with the data points. Which model appears to be the better fit?

Age (years)	Monthly Insurance Rate (dollars)	Age (years)	Monthly Insurance Rate (dollars)
35	17.32	43	23.71
36	17.67	44	25.11
37	18.02	45	26.60
38	18.46	46	28.00
39	19.07	47	29.40
40	19.95	48	30.80
41	21.00	49	32.55
42	22.22	50	34.47

(Source: American General Life Insurance Company, November 1999)

24. *U.S. Gross Domestic Product* The graph in the figure below gives the U.S. gross domestic product (in billions of dollars) for selected years from 1940 through 1995.
   a. Create a scatter plot for the data, with *x* equal to the number of years from 1930 and *y* equal to the gross domestic product in billions of dollars.
   b. Does it appear that a quadratic function can be used to model these data? If so, find the best fitting quadratic model.
   c. Find the power model that is the best fit for the data.
   d. Compare the two models by graphing each model on the same graph with the data points. Which model appears to be the better fit?

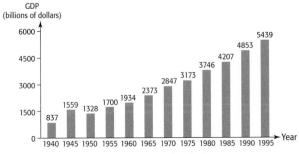

GDP (billions of dollars)

(Source: U.S. Department of Management and Budget)

25. *Hospital Discharges* The number of discharges per 1000 people in nonfederal short-stay hospitals for the

indicated year from 1980 to 1996 is given in the table below.
   a. Align the data as the number of years after 1970 and find both quadratic and power models for the data in the table. Discuss the fit of each of the models.
   b. Even though you know the dangers of extrapolation, use the data in the table to make a quick estimate of the discharge rate in 2005. Discuss how your prediction for the year 2005 compares with the 1996 value.
   c. Use both of the functions in part (a) to predict the number of discharges per 1000 people in 2005. Compare the values with what you expected from the discussion in part (b).

Year	Discharges per 1000 People
1980	158.5
1985	137.7
1988	117.6
1990	113.0
1992	110.5
1994	106.5
1996	102.3

(Source: National Hospital Discharge Survey)

26. *Abortions* The total number of abortions, in thousands, in the U.S. for selected years from 1973 through 1996 are given in the table below.
   a. Use the data to find a quadratic function that models the number of abortions as a function of years from 1970.
   b. Does the function reported in part (a) yield a maximum or minimum? When is this value obtained, and what does the value represent?

Year	Number of Abortions (thousands)	Year	Number of Abortions (thousands)
1973	744.6	1992	1528.9
1975	1034.2	1993	1500.0
1980	1553.9	1994	1431.0
1985	1588.6	1995	1363.7
1990	1608.6	1996	1365.7
1991	1556.5		

(Source: Alan Guttmacher Institute)

2.5 Combining Functions; Reciprocal, Absolute Value, and Piecewise-Defined Functions

If the daily total cost to produce x units of a product is

$$C(x) = 360 + 40x + 0.1x^2 \text{ thousand dollars,}$$

and the daily revenue from the sale of x units of this product is

$$R(x) = 60x \text{ thousand dollars,}$$

we can model the profit function as $P(x) = R(x) - C(x)$, giving

$$P(x) = 60x - (360 + 40x + 0.1x^2)$$

or

$$P(x) = -0.1x^2 + 20x - 360.$$

In a manner similar to the one that formed this function, we can construct new functions by performing algebraic operations with two or more functions. For example, we can build an average cost function by finding the quotient of two functions. In addition to using arithmetic operations with functions, we can create new functions using function composition.

Operations with Functions

New functions that are the sum, difference, product, and quotient of two functions are defined as follows:

Operation	Formula	*Example with* $f(x) = \sqrt{x}$ and $g(x) = x^3$
Sum	$(f + g)(x) = f(x) + g(x)$	$(f + g)(x) = \sqrt{x} + x^3$
Difference	$(f - g)(x) = f(x) - g(x)$	$(f - g)(x) = \sqrt{x} - x^3$
Product	$(f \cdot g)(x) = f(x) \cdot g(x)$	$(f \cdot g)(x) = \sqrt{x} \cdot x^3 = x^3 \sqrt{x}$
Quotient	$\left(\dfrac{f}{g}\right)(x) = \dfrac{f(x)}{g(x)}, \quad g(x) \neq 0$	$\left(\dfrac{f}{g}\right)(x) = \dfrac{\sqrt{x}}{x^3}, \quad x \neq 0$

The domain of the sum, difference, and product of f and g consists of all real numbers of the input variable for which f and g are defined. The domain of the quotient function consists of all real numbers for which f and g are defined and $g \neq 0$.

EXAMPLE 1   Operations with Functions

If $f(x) = x^3$ and $g(x) = x - 1$, find the following functions and give their domains:

a. $(f + g)(x)$ **b.** $(f - g)(x)$ **c.** $(f \cdot g)(x)$ **d.** $\left(\dfrac{f}{g}\right)(x)$

Solution

a. $(f + g)(x) = f(x) + g(x) = x^3 + x - 1$; all real numbers

b. $(f - g)(x) = f(x) - g(x) = x^3 - (x - 1) = x^3 - x + 1$; all real numbers

c. $(f \cdot g)(x) = f(x) \cdot g(x) = x^3(x - 1) = x^4 - x^3$; all real numbers

d. $\left(\dfrac{f}{g}\right)(x) = \dfrac{f(x)}{g(x)} = \dfrac{x^3}{x - 1}, x \neq 1$ ■

Reciprocal Function

The function formed by the quotient of $f(x) = 1$ and the identity function $g(x) = x$ is the **reciprocal function** with equation

$$y = \frac{1}{x}.$$

This graph of the reciprocal function, shown in Figure 2.45, is called a **rectangular hyperbola**. Note that x cannot equal 0, so 0 is not in the domain of this function. As the graph of $y = \dfrac{1}{x}$ shows, as x gets close to 0, $|y|$ gets large and the graph approaches, but does not touch, the y-axis. We say that the y-axis is a **vertical asymptote** of this graph.

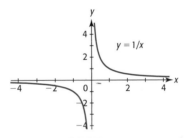

FIGURE 2.45

VERTICAL ASYMPTOTE

If $|y|$ gets large without bound as x approaches a, then $x = a$ is a vertical asymptote of the graph of $y = f(x)$. We can denote this by

$$f(x) \rightarrow \infty \text{ or } f(x) \rightarrow -\infty \text{ as } x \rightarrow a \text{ from the left or right.}$$

Note also that the values of $y = \dfrac{1}{x}$ get very small as $|x|$ gets large, and the graph approaches the x-axis. We say that the x-axis is a **horizontal asymptote** for this graph.

HORIZONTAL ASYMPTOTE

If the values of y approach some finite number b as $|x|$ becomes very large, we say that the graph of $y = f(x)$ has a horizontal asymptote at $y = b$. We can denote this as

$$f(x) \rightarrow b \text{ as } x \rightarrow \infty \text{ or } x \rightarrow -\infty.$$

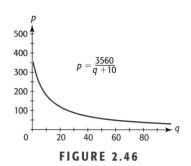

FIGURE 2.46

Suppose the number of units q of a product that is demanded is related to the price p by the function $p = \dfrac{3560}{q + 10}$, which is a stretched and shifted reciprocal function. The graph of this function is the graph of the reciprocal function $p = \dfrac{1}{q}$ with each point shifted 10 units to the left and each p-coordinate multiplied by 3560. (See Figure 2.46.) Because q is the number of units demanded and p is the price, only nonnegative inputs and outputs make sense, and the graph is restricted to the first quadrant.

EXAMPLE 2   Cost-Benefit

Suppose that for a certain city the cost C of obtaining drinking water that contains p percent impurities (by volume) is given by

$$C = \frac{120{,}000}{p} - 1200$$

a. Determine the domain of this function and graph the function without concern for the context of the problem.

b. Use knowledge of the context of the problem to graph the function in a window that applies to the application.

Solution

a. All values of p except $p = 0$ result in real values for the function. Thus the domain of C is all real numbers except 0. The graph of this function is a rectangular hyperbola, shifted downward 1200 units (and stretched by a factor of 120,000). The viewing window should have its center near $p = 0$ (horizontally) and $C = -1200$ (vertically). We increase the vertical view to allow for the large stretching factor, using the viewing window $[-100, 100]$ by $[-20{,}000, 20{,}000]$. The graph, shown in Figure 2.47(a), has the same shape as the graph of $y = \dfrac{1}{x}$ in Figure 2.45.

b. Because p represents the percent of impurities, the viewing window is set for values of p from 0 to 100. (Recall from part (a) that p cannot be 0.) The cost of reducing the impurities cannot be negative, so the vertical view is set from 0 to 20,000. The graph is shown in Figure 2.47(b).

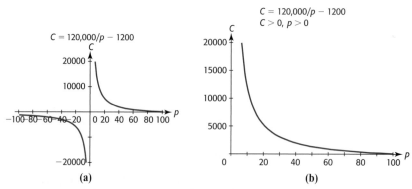

(a)   (b)

FIGURE 2.47

The quotient of the function $C(x)$, which gives total production cost, and the identity function, which gives the number of units produced, gives a new function, the **average cost** function, $\overline{C}$.

$$\overline{C}(x) = \frac{C(x)}{x}$$

EXAMPLE 3   Average Cost

----

a. Form the average cost function if the total cost function for the production of x units of a product is

$$C(x) = 50x + 5000.$$

b. For which input values is $\overline{C}$ defined? Give a real-world explanation of this answer.

Solution

a. The average cost function is the quotient of the cost function $C(x) = 50x + 5000$ and the identity function $I(x) = x$.

$$\overline{C}(x) = \frac{50x + 5000}{x}$$

b. The average cost function is defined for all real numbers such that $x > 0$, because producing negative units is not possible and the function is undefined for $x = 0$. (This is reasonable, because if nothing is produced, it does not make sense to discuss average cost per unit of product). ■

Absolute Value and Piecewise-Defined Functions

We can also construct a new function by combining two or more functions into a **piecewise-defined function**. For example, we can write a function with the form

$$f(x) = \begin{cases} x & \text{if } x \geq 0 \\ -x & \text{if } x < 0 \end{cases}$$

Because $f(x)$ is constructed by using two basic functions ($y_1 = x$ and $y_2 = -x$) that define this function for different parts of its domain, we call this function a **piecewise-defined function**.

This function is called the **absolute value function**, which is denoted by $f(x) = |x|$ and is derived from the definition of the absolute value of a number. Recall that the definition of the absolute value of a number is

$$|x| = \begin{cases} x & \text{if } x \geq 0 \\ -x & \text{if } x < 0 \end{cases}$$

To graph $f(x) = |x|$, we graph the portion of the line $y_1 = x$ for $x \geq 0$ (see Figure 2.48(a)) and the portion of the line $y_2 = -x$ for $x < 0$ (see Figure 2.48(b)). When these pieces are joined on the same graph (see Figure 2.48(c)), we have the graph of $y = |x|$.

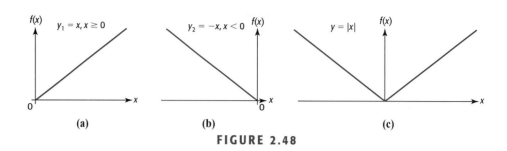

FIGURE 2.48

Note that the graphs of $y = |x + 3|$ and $y = |x + 3| + 2$ are shifted graphs with the same shape as the graph of $y = |x|$. (See Figure 2.49.)

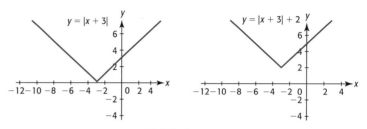

FIGURE 2.49

Other piecewise-defined functions can be viewed by graphing their pieces on the same set of axes. For example, we can use $H(x) = 5x + 2$ and $K(x) = x^3$ to construct the piecewise-defined function

$$f(x) = \begin{cases} 5x + 2 & \text{if } 0 \le x < 3 \\ x^3 & \text{if } 3 \le x \le 5 \end{cases}$$

TABLE 2.19(a)

x	0	1	2	2.99
$f(x)$	2	7	12	16.95

TABLE 2.19(b)

x	3	4	5
$f(x)$	27	64	125

To graph this function, we can construct tables of values for each of the pieces. Table 2.19(a) gives outputs of the function for some sample inputs x on the interval [0, 3]; on this interval $f(x)$ is defined by $f(x) = 5x + 2$. Table 2.19(b) gives outputs for some sample inputs on the interval [3, 5]; on this interval $f(x)$ is defined by $f(x) = x^3$.

Plotting the points from Table 2.19(a) and connecting them with a smooth curve gives the graph of $y = f(x)$ on the x-interval [0, 3). (See Figure 2.50.) The open circle on this piece of the graph indicates that this piece of the function is not defined for $x = 3$. Plotting the points from Table 2.19(b) and connecting them with a smooth curve gives the graph of $y = f(x)$ on the x-interval [3, 5]. (See Figure 2.50.) The closed circle on this piece of the graph indicates that this piece of the function is defined for $x = 3$. The graph of $y = f(x)$, shown in Figure 2.50, consists of these two pieces.

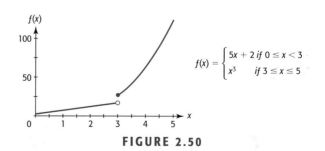

FIGURE 2.50

<table>
<tr><td>**E X A M P L E 4**</td><td>Residential Power Costs</td></tr>
</table>

Excluding fuel adjustment costs and taxes, Georgia Power Company charges its residential power customers for electricity during the months of June through September according to Table 2.20.

TABLE 2.20

Monthly Kilowatt-Hours (kWh)	Monthly Charge
0 to 650	$7.50 plus $0.04783 per kWh
More than 650, up to 1000	$38.59 plus $0.07948 per kWh above 650
More than 1000	$66.41 plus $0.08184 per kWh above 1000

a. Write the piecewise-defined function C that gives the monthly charge for residential electricity, with input x equal to the monthly number of kilowatt-hours.

b. Find $C(950)$ and explain what it means.

c. Find the charge for using 1560 kWh in a month.

Solution

a. The monthly charge is given by the function

$$C(x) = \begin{cases} 7.50 + 0.04783x & \text{if } 0 \leq x \leq 650 \\ 38.59 + 0.07948(x - 650) & \text{if } 650 < x \leq 1000 \\ 66.41 + 0.08184(x - 1000) & \text{if } x > 1000 \end{cases}$$

where $C(x)$ is the charge in dollars for x kilowatt-hours of electricity.

b. To evaluate $C(950)$, we must determine which "piece" defines the function when $x = 950$. Because 950 is between 650 and 1000, we use the "middle piece" of the function:

$$C(950) = 38.59 + 0.07948(950 - 650) = 62.434$$

Companies regularly round charges *up* to the next cent if any part of a cent is due. This means that if 950 kWh are used in a month, the bill is $62.44.

c. To find the charge for 1560 kWh, we evaluate $C(1560)$, which uses the "bottom piece" of the function because $1560 > 1000$. Evaluating this gives

$$C(1560) = 66.41 + 0.08184(1560 - 1000) = 112.2404,$$

so the charge for the month is $112.25. ∎

Solving Absolute Value Equations

Recall that $|x|$ is defined as follows:

$$|x| = \begin{cases} x & \text{if } x \geq 0 \\ -x & \text{if } x < 0 \end{cases}$$

We now consider the solution of equations that contain absolute value symbols, called **absolute value equations.** Consider the equation $|x| = 5$. To solve this equation, we know that $|5| = 5$ and $|-5| = 5$. Therefore, the solution to $|x| = 5$ is $x = 5$ or $x = -5$. Also, if $|x| = 0$, then x must be 0. Finally, because the absolute value of a number is never negative, we cannot solve $|x| = a$ if a is negative. We generalize this as follows.

ABSOLUTE VALUE EQUATION

If $|x| = a$ and $a > 0$, then $x = a$ or $x = -a$.

There is no solution to $|x| = a$ if $a < 0$; $|x| = 0$ has solution $x = 0$.

If one side of an equation is a function contained in an absolute value and the other side is a nonnegative constant, we can solve the equation by using the method above.

EXAMPLE 5   Absolute Value Equations

Solve the following equations.

a. $|x - 3| = 9$ **b.** $|2x - 4| = 8$ **c.** $|x^2 - 5x| = 6$

Solution

a. If $|x - 3| = 9$, then $x - 3 = 9$ or $x - 3 = -9$.
  Thus the solution is $x = 12$ or $x = -6$.

b. If $|2x - 4| = 8$, then

$$2x - 4 = 8 \quad \text{or} \quad 2x - 4 = -8$$
$$2x = 12 \quad \bigg| \quad 2x = -4$$
$$x = 6 \quad \bigg| \quad x = -2$$

Thus the solution is $x = 6$ or $x = -2$.

c. If $|x^2 - 5x| = 6$, then

$$x^2 - 5x = 6 \quad \text{or} \quad x^2 - 5x = -6$$
$$x^2 - 5x - 6 = 0 \quad \bigg| \quad x^2 - 5x + 6 = 0$$
$$(x - 6)(x + 1) = 0 \quad \bigg| \quad (x - 3)(x - 2) = 0$$
$$x = 6 \text{ or } x = -1 \quad \bigg| \quad x = 3 \text{ or } x = 2$$

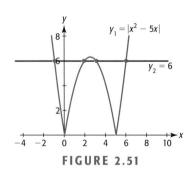

FIGURE 2.51

Thus there are four values of x that satisfy the equation. We can check these solutions by graphing or by substitution. Using the intersection method, we graph $y_1 = |x^2 - 5x|$ and $y_2 = 6$ and find the points of intersection. The solutions to the equation, $x = 6, -1, 3$, and 2, found above, are also the x-values of the points of intersection. (See Figure 2.51.)   ∎

Composition of Functions

In the 1980s, several U.S. oil companies attempted to convert to the metric system by replacing gasoline pumps with new pumps that measured gasoline in liters rather than gallons. The plan was eventually aborted because consumers did not know how to convert liters to gallons and they distrusted the oil companies' prices when they couldn't judge the price per gallon. If a gasoline pump registered 40 liters, how many gallons did the consumer receive? What is the conversion formula from liters to gallons?

The equations below give the conversions from liters to quarts and from quarts to gallons.

$$1 \text{ liter } = 1.0567 \text{ quarts}$$

$$1 \text{ quart } = .25 \text{ gallon}$$

We can use this table to convert 40 liters to quarts, and then we can convert the quarts to gallons.

$$40 \text{ liters } = 40(1.0567 \text{ quarts}) = 42.268 \text{ quarts}$$

$$42.268 \text{ quarts } = 42.268(.25 \text{ gallons}) = 10.567 \text{ gallons}$$

Rather than use two steps every time we want to make this conversion, we can create a new function that gives the formula for converting liters to gallons. If we write the formula for conversion from liters to quarts as the function

$$Q = f(L) = 1.0567L$$

and we write the formula for conversion from quarts to gallons as the function

$$G = g(Q) = .25Q,$$

then we can replace Q with $f(L)$ in $G = g(Q) = .25Q$ to write the conversion formula from liters to gallons as

$$g(f(L)) = .25(1.0567L) = 0.264175L.$$

Thus if we have x liters of gasoline, it is equivalent to $g(f(x)) = 0.264175x$ gallons.

This example illustrates the concept of **composition of functions.** The notation $g(f(x))$ is called a **composite** function. We read $g(f(x))$ as "g of f of x," and denote it by $(g \circ f)(x)$. In general, we define a composite function as follows.

COMPOSITE FUNCTION

The composite function, f of g, is denoted by $f \circ g$ and defined by

$$(f \circ g)(x) = f(g(x))$$

The domain of $f \circ g$ is the subset of the domain of g for which $f \circ g$ is defined. The composite function $g \circ f$ is defined by $(g \circ f)(x) = g(f(x))$. The domain of $g \circ f$ is the subset of the domain of f for which $g \circ f$ is defined.

When computing the composite function $f \circ g$, keep in mind that the output of g becomes the input for f, and that the rule for f is applied to this new input.

A composite function machine can be thought of as a machine within a machine. Figure 2.52 shows a composite "function machine" in which the input is denoted by x, the inside function is g, the outside function is f, the composite function rule is denoted by $f \circ g$, and the composite function output is symbolized by $f(g(x))$. Note that for most functions f and g, $f \circ g \neq g \circ f$.

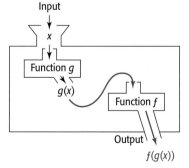

FIGURE 2.52

EXAMPLE 6   Function Composition and Orange Juice

There are two machines in a room; one of the machines squeezes oranges to make orange juice and the other is a canning machine that puts the orange juice into cans.

Thinking of these two processes as functions we name the squeezing function g and the canning function f.

a. Describe the composite function $(f \circ g)(\text{orange}) = f(g(\text{orange}))$.

b. Describe the composite function $(g \circ f)(\text{orange}) = g(f(\text{orange}))$.

c. Which function, $f(g(\text{orange}))$, $g(f(\text{orange}))$, neither, or both, makes sense in context?

Solution

a. The process by which this output $f(g(\text{orange})$ is obtained is easiest to understand thinking "from the inside out." Think of the input as an orange. The orange first goes into the g machine, so it is squeezed. The output of g, liquid orange juice, is then put into the f machine, which puts it into a can. The result is a can containing orange juice. (See Figure 2.53.)

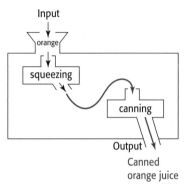

FIGURE 2.53

b. Again thinking from the inside out, the output $g(f(\text{orange}))$ is obtained by first putting an orange into the f machine, which puts it into a can. The output of f, the canned orange, is then put into the squeezing machine, g. The result is a compacted can, containing an orange! (See Figure 2.54.)

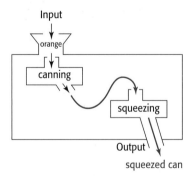

FIGURE 2.54

c. As seen from the results to part (a) and part (b), the order in which the functions appear in the composite function symbol does make a difference. The process that makes sense is the one in part (a), $f(g(\text{orange}))$. ∎

| **EXAMPLE 7** | **Finding Composite Function Outputs** |

Find the following composite function outputs, using $f(x) = 2x - 5$, $g(x) = 6 - x^2$, and $h(x) = \dfrac{1}{x}$. Give the domain of each new function formed.

a. $(h \circ f)(x) = h(f(x))$ **b.** $(f \circ g)(x) = f(g(x))$ **c.** $(g \circ f)(x) = g(f(x))$

Solution

a. First, the output of f becomes the input for h.

$$(h \circ f)(x) = h(f(x)) = h(2x - 5) = \frac{1}{2x - 5}$$

Next, the rule for h is applied to this input.

The domain of this function is all $x \neq \dfrac{5}{2}$ because $\dfrac{1}{2x - 5}$ is undefined if $x = \dfrac{5}{2}$.

b. $(f \circ g)(x) = f(g(x)) = f(6 - x^2) = 2(6 - x^2) - 5 = 12 - 2x^2 - 5 = -2x^2 + 7$

The domain of this function is the set of all real numbers.

c. $(g \circ f)(x) = g(f(x)) = g(2x - 5) = 6 - (2x - 5)^2 = 6 - (4x^2 - 20x + 25)$

$$= 6 - 4x^2 + 20x - 25 = -4x^2 + 20x - 19$$

The domain of this function is the set of all real numbers.   ■

2.5 SKILLS CHECK

1. a. Graph $f(x) = |x|$. **b.** Find $f(-2)$ and $f(5)$.
  c. State the domain of the function.

2. a. Graph $f(x) = |x - 4|$. **b.** Find $f(-2)$ and $f(5)$.
  c. State the domain of the function.

3. a. Graph $f(x) = \dfrac{1}{x}$. **b.** Find $f(-2)$ and $f(5)$.
  c. State the domain of the function.

4. a. Graph $f(x) = \dfrac{1}{x} - 3$. **b.** Find $f(-3)$ and $f(1)$.
  c. State the domain of the function.

5. a. Graph $f(x) = \dfrac{-2}{x - 1} + 1$. **b.** Find $f(-2)$ and $f(5)$.
  c. State the domain of the function.

6. a. Graph $f(x) = \begin{cases} 3 - x & \text{if } x \leq 2 \\ x^2 & \text{if } x > 2 \end{cases}$ **b.** Find $f(2)$
  and $f(3)$. **c.** State the domain of the function.

7. a. Graph $f(x) = \begin{cases} 4x - 3 & \text{if } x \leq 3 \\ x^2 & \text{if } x > 3 \end{cases}$ **b.** Find $f(2)$
  and $f(4)$. **c.** State the domain of the function.

In Exercises 8–18, using the functions $f(x) = \dfrac{1}{x}$, $g(x) = 4x + 1$, $h(x) = \sqrt{x}$, (a) create the listed functions, and (b) give the domains of the new functions.

8. $(f + g)(x)$ **9.** $(h - f)(x)$

10. $(g \cdot h)(x)$ **11.** $(f \cdot g)(x)$

12. $(f/g)(x)$ **13.** $(g/h)(x)$

14. $f(h(x))$ **15.** $g(h(x))$

16. $(f \circ g)(x)$ **17.** $(g \circ f)(x)$

18. $(h \circ g)(x)$

In Exercises 19–24, solve the equations and check graphically.

19. $|2x - 5| = 3$

20. $\left|x - \dfrac{1}{2}\right| = 3$

21. $|x| = x^2 + 4x$

22. $|x^2 - 4| = 0$

23. $|3x - 1| = 4$

24. $|x - 5| = x^2 - 5x$

2.5 EXERCISES

----

1. *Fixed and Variable Cost* The daily total cost function for Smores, $C(x) = 0.20x + 1234$, is the sum of two functions, the fixed cost $F(x)$ and the variable cost $V(x)$.
   a. Which function in $C(x)$ is the fixed cost and which is the variable cost?
   b. Which of the two functions is a constant function?

2. *Cost-benefit* Suppose that for a certain city the cost C of obtaining drinking water that contains p percent impurities (by volume) is given by

   $$C = \frac{120{,}000}{p} - 1200.$$

   a. This function can be considered as the difference of what two functions?
   b. What is the cost of drinking water that is 100% impure?

3. *Average Cost* If the monthly total cost of producing 27-inch television sets is given by $C(x) = 50{,}000 + 105x$, where x is the number of sets produced per month, then the *average cost* per unit is given by

   $$\overline{C}(x) = \frac{50{,}000 + 105x}{x}.$$

   a. Explain how $C(x)$ and another function can be combined to obtain the average cost function.
   b. What is the average cost per set if 3000 sets are produced?

4. *T-Shirt Sales* Let $T(c)$ be the number of T-shirts that are sold when the shirts have c colors and $P(c)$ be the price, in dollars, of a T-shirt that has c colors. Write a sentence explaining the meaning of the function $(T \cdot P)(c)$.

5. *Football Tickets* At a certain school, the number of student tickets sold for a home football game can be modeled by $S(p) = 62p + 8500$, where p is the winning percentage of the home team. The number of nonstudent tickets sold for these home games is given by $N(p) = 0.5p^2 + 16p + 4400$ tickets.

   a. Write an equation for the total number of tickets sold for a home football game at this school as a function of the winning percent p.
   b. What is the domain for the function in part (a) in this context?
   c. Assuming that the football stadium is filled to capacity when the team wins 90% or more of its home games, what is the capacity of the school's stadium?

6. *Harvesting* A farmer's main cash crop is tomatoes, and the tomato harvest begins in the month of May. The number of bushels of tomatoes harvested on the xth day of May is given by the equation $B(x) = 6(x + 1)^{3/2}$. The market price in dollars of 1 bushel of tomatoes on the xth day of May is given by the formula $P(x) = 8.5 - 0.12x$.

   a. How many bushels did the farmer harvest on May 8?
   b. What was the market price of tomatoes on May 8?
   c. How much was the farmer's tomato harvest worth on May 8?
   d. Write a model for the worth W of the tomato harvest on the xth day of May.

7. *Printers* The weekly total cost function for producing a dot matrix printer is $C(x) = 3000 + 72x$, where x is the number of printers produced per week.

   a. Form the weekly average cost function for this product.
   b. Find the average cost for the production of 100 printers.

8. *Electronic Components* The monthly cost of producing x electronic components is $C(x) = 2.15x + 2350$.

   a. Find the monthly average cost function.
   b. Find the average cost for the production of 100 components.

9. *Population of Children* The table gives the estimated population (in millions) of U.S. boys age 5 and under and the estimated U.S. population (in millions) of girls age 5 and under in selected years.

Year	1995	2000	2005	2010
Boys	10.02	9.71	9.79	10.24
Girls	9.57	9.27	9.43	9.77

(Source: U.S. Department of Commerce)

A function that models the population (in millions) of U.S. boys age 5 and under t years after 1990 is $B(t) = 0.0076t^2 - 0.1752t + 10.705$, and a function that models the population (in millions) of U.S. girls age 5 and under t years after 1990 is $G(t) = 0.0064t^2 - 0.1448t + 10.12$.

a. Find the equation of a function that models the estimated U.S. population (in millions) of children age 5 and under t years after 1990.

b. Use the result of part (a) to estimate the U.S. population of children age 5 and under in 2003.

10. *Educational Attainment* Data collected in March of the indicated years given in the table below shows the percentage of 25- to 29-year-old male and female high school graduates who also have a bachelor's degree or higher.

Year	1975	1980	1985	1990
Percent of males	29.7	28.1	26.9	28.0
Percent of females	22.9	24.5	24.6	26.2

Year	1993	1995	1997	1999
Percent of males	27.2	28.4	30.7	31.2
Percent of females	27.4	28.5	32.9	33.0

(Source: U.S. Census Bureau)

a. Will adding the percents for males and females create a new function that gives the total percent of all 25- to 29-year-old high school graduates who also have a bachelor's degree or higher?

b. A function that approximately models the total percent of all 25- to 29-year-old high school graduates who also have a bachelor's degree or higher is $h(t) = 0.024t^2 - 0.595t + 29.121$ where t is the number of years after 1970. Use this model to estimate the total percent in 1990 and 1999.

c. Add the percent of males and the percent of females for the 1990 data and the 1999 data in the table and divide by 2. Do these values conflict with your answers in part (b)?

11. *First-Class Postage* The postage charged for first-class mail is a function of its weight. The U.S. Postal Service uses the following table to describe the rates for 2003.

Weight Increment x (ounces)	First Class (cents) Postage $P(x)$
First ounce or fraction of an ounce	37
Each additional ounce or fraction	23

(Source: pe.usps.gov/text)

a. Convert this table to a piecewise-defined function that represents first-class postage for letters weighing up to 4 oz, using x as the weight in ounces and P the postage in cents.

b. Find $P(1.2)$ and explain what it means.

c. Give the domain of P as it is defined above.

d. Find $P(2)$ and $P(2.01)$.

e. Find the postage for a 2-oz letter and for a 2.01-oz letter.

12. *Federal Support for Education* The federal on-budget funds for all educational programs (in millions of constant 2000 dollars) between 1965 and 2000 can be modeled by the function

$$P(t) = \begin{cases} 1.965t - 5.65 & \text{when } 5 \le t \le 20 \\ 0.095t^2 - 2.925t + 54.429 & \text{when } 20 < t \le 40 \end{cases}$$

where t is the number of years after 1960.

a. Graph the function P for $5 \le t \le 40$. Describe how the funding for education programs varied between 1965 and 2000.

b. What was the amount of funding for educational programs in 1980?

c. How much federal funding was allotted for educational programs in 1998?

(Source: U.S. Department of Education, National Center for Educational Statistics)

13. *Income Tax* The 2000 U.S. federal income tax owed by a married couple filing jointly can be found from Schedule Y-1.

Schedule Y-1—Use if your filing status is **Married filing jointly** or **Qualifying widow(er)**

If the amount on Form 1040, line 39, is: Over—	But not over—	Enter on Form 1040, line 40	of the amount over—
$0	$43,850	----- 15%	$0
43,850	105,950	$6,577.50 + 28%	43,850
105,950	161,450	23,965.50 + 31%	105,950
161,450	288,350	41,170.50 + 36%	161,450
288,350	-----	86,854.50 + 39.6%	288,350

(Source: Internal Revenue Service, 2000, Form 1040)

a. Write the piecewise-defined function T with input x that models the federal tax dollars owed as a function of x, the taxable income dollars earned, with $0 < x \leq 105{,}950$.

b. Use the function to find $T(42{,}000)$.

c. Find the tax owed on a taxable income of $55{,}000.

d. A friend tells Jack Waddell not to earn any money over $43{,}850 because it would raise his tax rate to 28% on all of his taxable income. Test this statement by finding the tax on $43{,}850 and $ 43{,}850 + $ 1. What do you conclude?

14. *Federal Support for Education* The federal on-budget funds for all educational agencies between 1965 and 2000 can be modeled by the function

$$f(t) = \begin{cases} 0.011t^3 - 0.696t^2 + 14.380t - 29.370 \\ \qquad\qquad \text{when } 5 \leq t < 35 \\ 0.931t^2 - 67.374t + 1295.785 \\ \qquad\qquad \text{when } 35 \leq t \leq 40 \end{cases}$$

where t is the number of years after 1960 and $f(t)$ is in millions of constant 2000 dollars.

a. Graph the function f for $5 \leq t \leq 40$.

b. What was the amount of funding given to educational agencies in 1989?

c. How much federal funding was given to educational agencies in 1997?

(Source: U.S. Department of Education, National Center for Educational Statistics)

15. *Mob Behavior* In a study of lynchings between 1899 and 1946, psychologist Brian Mullin concluded that the size of the mob relative to the number of victims predicted the level of brutality. The formula he developed gives y, the other-total ratio that predicts the level of self-attentiveness of people in a crowd of size x with one victim. Mullin's formula is $y = \dfrac{1}{x + 1}$, and the lower the value of y, the more likely an individual is to be influenced by "mob psychology."

a. This function is a shifted version of what basic function, and how is it shifted?

b. Graph this function for $x > 0$.

c. Will the amount of self-attentiveness of a person increase or decrease as the crowd size becomes larger?

16. *Concentration of Body Substances* The concentration C of a substance in the body depends on the quantity of substance Q and the volume V through which it is distributed. For a static substance, the concentration is given by

$$C = \frac{Q}{V}$$

a. For 1000 ml of a substance, graph the concentration as a function of the volume on the interval from $V = 1000$ ml to $V = 5000$ ml.

b. For a fixed quantity of a substance, does the concentration of the substance in the body increase or decrease as the volume through which it is distributed increases?

17. *Population Growth* Suppose that the population of a certain microorganism at time t (in minutes) is given by

$$P = -1000\left(\frac{1}{t + 10} - 1\right)$$

a. Describe the transformations needed to obtain this function from the basic function $f(t) = \dfrac{1}{t}$.

b. Graph this function for values of t representing 0 to 100 minutes.

18. *Supply and Demand* The supply function for a commodity is given by $p = 58 + \dfrac{q}{2}$, and the demand function for this commodity is given by $p = \dfrac{2555}{q + 5}$.

a. Is the supply function a linear function or a shifted reciprocal function?

b. Is the demand function a shifted power function or a shifted reciprocal function? Describe the transformations needed to obtain the specific function from the basic function.

19. *Function Composition* Think of each of the following processes as a function designated by the indicated letter: f, placing in a styrofoam container; g, grinding. Describe each of the functions in parts (a) to (e), then answer part (f).

a. $f(\text{meat})$ b. $g(\text{meat})$ c. $g \circ g(\text{meat}))$
d. $f(g(\text{meat}))$ e. $g(f(\text{meat}))$
f. Which of the functions in part (d) and part (e) gives a sensible operation?

20. *Function Composition* Think of each of the following processes as a function designated by the indicated letter: f, putting a sock on; g, taking a sock off. Describe each of the functions in parts (a) to (c).

a. $f(\text{left foot})$ b. $f(f(\text{left foot}))$
c. $(g \circ f)(\text{right foot})$

21. *Shoe Sizes* A woman's shoe that is size x in Japan is size $s(x)$ in the United States, where $s(x) = x - 17$. A woman's shoe that is size x in the United States is size $p(x)$ in Britain, where $p(x) = x - 1.5$. Find a function that will convert Japanese shoe size to British shoe size.
(Source: Kuru International Exchange Association)

22. *Shoe Sizes* A man's shoe that is size x in Britain is size $d(x)$ in the United States, where $d(x) = x + 0.5$. A man's shoe that is size x in the United States is size $t(x)$ in Continental size, where $t(x) = x + 34.5$. Find a function that will convert British shoe size to Continental shoe size.
(Source: Kuru International Exchange Association)

23. *Exchange Rates* On January 19, 2001, each Austrian schilling was worth 1.987376 Russian rubles and each Chilean pesos was worth 0.025202 Austrian schillings. Find the value of 1000 Chilean pesos in Russian rubles on January 19, 2001. Round the answer to two decimal places.
(Source: Expedia.com)

24. *Exchange Rates* On January 19, 2001, each British pound sterling was worth 1.495701 US dollars and each French franc was worth 0.096115 British pound sterling. Find the value of 100 French francs in US dollars.

25. *AOL* If $f(x)$ represents the number of AOL subscribers x years after 1992 and $g(x)$ represents the number of Internet users x years after 1992, what function represents the percent of Internet users who are AOL subscribers x years after 1992?

26. *Home Computers* If $f(x)$ represents the percent of American homes with computers and $g(x)$ represents the number of American homes, with x equal to the

number of years after 1990, then what function represents the number of American homes with computers, with x equal to the number of years after 1990?

27. *Education* If the function $f(x)$ gives the number of female PhDs produced by American universities x years after 1990 and the function $g(x)$ gives the number of male PhDs produced by American universities x years after 1990, what function gives the total number of PhDs produced by American universities x years after 1990?

28. *Pollution* The daily cost C (in dollars) of removing pollution from the smokestack of a coal-fired electric power plant is related to the percent of pollution p being removed according to the equation $100C - Cp = 10,500$.
  a. Solve this equation for C to write the daily cost as a function of p, the percent of pollution removed.
  b. What is the daily cost of removing 50 percent of the pollution?
  c. Why would this company probably resist removing 99 percent of the pollution?

29. *Discount Prices* Half-Price Books has a sale with an additional 20% off the regular price of their books. What percent of the retail price is charged during this sale?
(Source: Half-Price Books, Cleveland, Ohio)

30. *Wind Chill* If the air temperature is 25° F, the wind chill / temperature C is given by $C = 59.914 - 235s - 20.14\sqrt{s}$, where s is the wind speed in miles per hour.
  a. State two functions whose difference gives this function.
  b. Graph this function for $3 \le s \le 12$.
  c. Is the function increasing or decreasing on this domain?

2.6 Inverse Functions

Key Concepts

- Inverse functions
  One-to-one functions
  Horizontal line test

- Finding inverse functions
  Graphs of inverse functions

- Inverse functions on
  limited domains

A business property is purchased with a promise to pay off a $60,000 loan plus the $16,500 interest on this loan by making 60 monthly payments of $1275. The amount of money remaining to be paid on the loan plus interest is given by the function

$$f(x) = 76,500 - 1275x,$$

where x is the number of months remaining for payments to be made.

If we know how much remains to be paid and want to find how many months remain to make payments, we can find the **inverse** of the above function.

In this section we investigate how to determine if two functions are inverses of each other, how to determine if a function has an inverse, and how to find the inverse of a function, and we solve applied problems by using inverse functions.

Inverse Functions

The following example shows that the composition of some pairs of functions is the identity function, $y = x$. Functions that have this relationship are called **inverse functions**.

EXAMPLE 1   Composite Function Outputs and Identity Functions

If $f(x) = \sqrt[3]{x}$ and $g(x) = x^3$, find $f(g(x))$ and $g(f(x))$.

Solution

To find $f(g(x))$, remember that the output for g becomes the input for f, and then the rule for f is applied to the new input. That is,

$$f(g(x)) = f(x^3) = \sqrt[3]{x^3} = x$$

Similarly, to find $g(f(x))$, the output for f becomes the input for g, and then the rule for g is applied to the new input. That is,

$$g(f(x)) = g(\sqrt[3]{x}) = (\sqrt[3]{x})^3 = x$$ ■

As Example 1 illustrates for the functions $f(x) = \sqrt[3]{x}$ and $g(x) = x^3$, $f(g(x)) = x$ and $g(f(x)) = x$. This means that both of these composite functions act as identity functions because their outputs are the same as their inputs. This happens because the second function "undoes" the operation performed by the first function. Because of this, the functions that define these models are called **inverse functions**.

> ### INVERSE FUNCTIONS
>
> Functions f and g for which $f(g(x)) = x$ and $g(f(x)) = x$ for all x in the domains of f and g are called **inverse functions**.
> In this case, we denote g by f^{-1}, read "f inverse."

EXAMPLE 2   Temperature Measurement

To see how the two conversion formulas for temperature measurement are related,

a. Convert the Fahrenheit temperature 32° to the equivalent Celsius temperature by using the formula

$$C(F) = \frac{5F - 160}{9}.$$

b. Convert the Celsius temperature found in part (a) to Fahrenheit temperature by using the formula

$$F(C) = \frac{9}{5}C + 32.$$

How does the Fahrenheit temperature found in part (b) compare with the original Fahrenheit temperature given?

c. Find $C(F(x))$ and $F(C(x))$ and determine if the functions are inverse functions.

Solution

a. Evaluating $C(F) = \dfrac{5F - 160}{9}$ for $F = 32$ gives

$$C(32) = \frac{5(32) - 160}{9} = 0.$$

b. Evaluating $F(C) = \dfrac{9}{5}C + 32$ for $C = 0$ gives

$$F(0) = \frac{9}{5}(0) + 32 = 32.$$

The resulting Fahrenheit temperature is the same as the original Fahrenheit temperature.

c. $C(F(x)) = \dfrac{5(F(x)) - 160}{9} = \dfrac{5\left(\dfrac{9}{5}x + 32\right) - 160}{9} = \dfrac{9x + 160 - 160}{9} = x$

and

$$F(C(x)) = \frac{9}{5}(C(x)) + 32 = \frac{9}{5} \cdot \frac{5x - 160}{9} + 32$$

$$= \frac{5x - 160}{5} + 32 = x - 32 + 32 = x$$

Because $C(F(x)) = x$ and $F(C(x)) = x$, the two functions are inverses. ■

Consider a function that doubles each input. Its inverse takes half of each input. For example, the discrete function f with equation $f(x) = 2x$ and domain $\{1, 4, 5, 7\}$ has the range $\{2, 8, 10, 14\}$. The inverse of this function has the equation $g(x) = \dfrac{x}{2}$. The domain of the inverse function is the set of outputs of the original function, $\{2, 8, 10, 14\}$. The outputs of the inverse function form the set $\{1, 4, 5, 7\}$, which is the domain of the original function. Figure 2.55 illustrates the relationship between the domains and ranges of the function f and its inverse f^{-1}. In this case and in every case, the domain of the inverse function is the range of the original function, and the range of the inverse function is the domain of the original function.

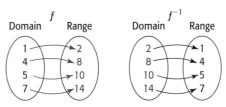

FIGURE 2.55

We summarize the information about inverse functions as follows.

> ### INVERSE FUNCTIONS
>
> The functions f and g are inverse functions if, whenever the pair (a, b) satisfies $y = f(x)$, the pair (b, a) satisfies $y = g(x)$. Note that when this happens,
>
> $$f(g(x)) = x \quad \text{and} \quad g(f(x)) = x.$$

It is very important to note that the "$^{-1}$" used in f^{-1} and g^{-1} is *not* an exponent, but rather a symbol used to denote the inverse of the function. The expression f^{-1} *always* refers to the inverse function of f and *never* to the reciprocal $\dfrac{1}{f}$ of f; that is, $f^{-1}(x) \neq \dfrac{1}{f(x)}$.

In general, we can show that a function has an inverse if it is a **one-to-one function**. We define a one-to-one function as follows.

> ### ONE-TO-ONE FUNCTION
>
> A one-to-one function has exactly one output for each input and exactly one input for each output. This means there is a one-to-one correspondence between the independent and dependent variables defining the function. Every one-to-one function has an inverse function.

The statement above means that for a one-to-one function f, $f(a) \neq f(b)$ if $a \neq b$. Recall that no vertical line can intersect the graph of a function in more than one point. The definition of a one-to-one function means that if a function is one-to-one, a horizontal line will intersect its graph in at most one point.

> ### HORIZONTAL LINE TEST
>
> A function is one-to-one if no horizontal line can intersect the graph of the function in more than one point.

We can see that $y = 3x^4$ is not a one-to-one function because we can observe that the graph of this function does not pass the horizontal line test. (See Figure 2.56(a).) Note that when $x = 2$, $y = 48$, and when $x = -2$, $y = 48$. Therefore, this function does not have an inverse function.

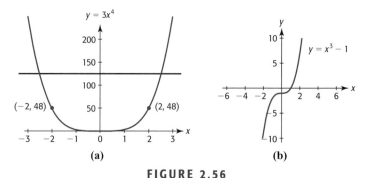

(a)   **(b)**

FIGURE 2.56

On the other hand, each different value of x results in a unique value for $y = x^3 - 1$, so it is a one-to-one function. The graph in Figure 2.56(b) and the horizontal line test also indicate that this function is one-to-one.

EXAMPLE 3 Inverse Functions

---

a. Determine if $f(x) = x^3 - 1$ has an inverse function.

b. Verify that $g(x) = \sqrt[3]{x + 1}$ is the inverse function of $f(x) = x^3 - 1$.

c. Find the domain and range of each function.

Solution

a. We used the horizontal line test in Figure 2.56(b) to show that the function $y = x^3 - 1$ is one-to-one. Thus it has an inverse function.

b. To verify that the functions $f(x) = x^3 - 1$ and $g(x) = \sqrt[3]{x + 1}$ are inverse functions, we show that the compositions $f(g(x))$ and $g(f(x))$ are each equal to the identity function.

$$f(g(x)) = f(\sqrt[3]{x + 1}) = (\sqrt[3]{x + 1})^3 - 1 = (x + 1) - 1 = x \text{ and}$$
$$g(f(x)) = g(x^3 - 1) = \sqrt[3]{(x^3 - 1) + 1} = \sqrt[3]{x^3} = x$$

Thus, we have verified that these functions are inverses.

c. The cube of any real number decreased by 1 is a real number, so the domain of f is the set of all real numbers. The cube root of any real number plus 1 is also a real number, so the domain of g is the set of all real numbers. The domain of the function f is the range of its inverse, g, and the domain of g is the range of f. Thus the ranges of f and g are the set of real numbers. ∎

By using the definition of inverse functions, we can find the equation for the inverse function of f by interchanging x and y in the equation $y = f(x)$ and solving the new equation for y.

FINDING THE INVERSE OF A FUNCTION

To find the inverse of the function f that is defined by the equation $y = f(x)$:

1. Rewrite the equation replacing $f(x)$ with y.

2. Interchange x and y in the equation defining the function.

3. Solve the new equation for y. If this equation cannot be solved uniquely for y, the original function has no inverse function.

4. Replace y with $f^{-1}(x)$.

EXAMPLE 4 Finding an Inverse Function

---

a. Find the inverse function of $f(x) = \dfrac{2x - 1}{3}$.

b. Graph $f(x) = \dfrac{2x - 1}{3}$ and its inverse function on the same axes.

Solution

a. Using the steps for finding the inverse of a function, we have:

1. $y = \dfrac{2x - 1}{3}$ Replacing $f(x)$ with y.

2. $x = \dfrac{2y - 1}{3}$ Interchanging x and y.

3. $3x = 2y - 1$ Solving for y

$3x + 1 = 2y$

$\dfrac{3x + 1}{2} = y$

4. $f^{-1}(x) = \dfrac{3x + 1}{2}$ Replacing y with $f^{-1}(x)$.

b. The graph of $f(x) = \dfrac{2x - 1}{3}$ and of its inverse $f^{-1}(x) = \dfrac{3x + 1}{2}$ are shown in Figure 2.57.

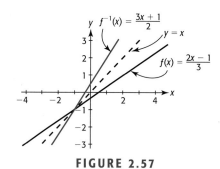

FIGURE 2.57 ■

Notice that the graphs of $y = f(x)$ and $y = f^{-1}(x)$ in Figure 2.57 appear to be reflections of each other about the line $y = x$. This occurs because if you were to choose any point (a,b) on the graph of the function f and interchange the x- and y-coordinates, the new point (b,a) will be on the graph of the inverse function f^{-1}. This should make sense, because the inverse function is formed by interchanging x and y in the equation defining the original function. In fact, this relationship occurs for every function and its inverse.

GRAPHS OF INVERSE FUNCTIONS

The graphs of a function and its inverse are symmetric with respect to the line $y = x$.

We again illustrate the symmetry of graphs of inverse functions in Figure 2.58, which shows the graphs of the inverse functions $f(x) = x^3 - 2$ and $f^{-1}(x) = \sqrt[3]{x + 2}$. Note that the point $(0, -2)$ is on the graph of $f(x) = x^3 - 2$ and the point $(-2, 0)$ is on the graph of $f^{-1}(x) = \sqrt[3]{x + 2}$.

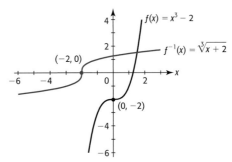

FIGURE 2.58

EXAMPLE 5   Loan Repayment

A business property is purchased with a promise to pay off a $60,000 loan plus the $16,500 interest on this loan by making 60 monthly payments of $1275. The amount of money remaining to be paid on the loan plus interest is given by the function

$$f(x) = 76{,}500 - 1275x,$$

where x is the number of months remaining for payments to be made.

a. Find the inverse of this function.

b. Use the inverse to determine how many months remain to make payments if $35,700 remains to be paid.

Solution

a. 1. Replacing $f(x)$ with y gives $y = 76{,}500 - 1275x$.

   2. Interchanging x and y gives the equation $x = 76{,}500 - 1275y$.

   3. Solving this new equation for y gives $y = \dfrac{76{,}500 - x}{1275}$.

   4. Replacing y with $f^{-1}(x)$ gives the inverse function $f^{-1}(x) = \dfrac{76{,}500 - x}{1275}$.

b. The inverse function gives the number of months remaining to make payments if x dollars remain to be paid.

$$f^{-1}(35{,}700) = \frac{76{,}500 - 35{,}700}{1275} = 32,$$

so 32 months remain to make payments.   ■

Inverse Functions on Limited Domains

As we have stated, a function cannot have an inverse function if it is not a one-to-one function. However, if there is a limited domain over which such a function is a one-to-one function, then it has an inverse function for this domain. Consider the function $f(x) = x^2$. The horizontal line test on the graph of this function (shown in Figure 2.59(a)) indicates that this function is not a one-to-one function and that the function f

does not have an inverse function. To see why, consider the attempt to find the inverse function:

$$y = x^2$$
$$x = y^2$$
$$\pm\sqrt{x} = y$$

This equation is not the inverse function because it is not a function (many values of x give two values for y). Because we cannot solve uniquely for y, $f(x) = x^2$ does not have an inverse function.

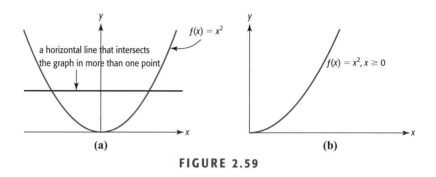

FIGURE 2.59

However, if we limit the domain of the function to $x \geq 0$, no horizontal line intersects the graph more than once, so the function is one-to-one on this limited domain (see Figure 2.59(b)), and the function has an inverse. If we restrict the domain of the original function by requiring that $x \geq 0$, then the range of the inverse function is restricted to $y \geq 0$, and the equation $y = \sqrt{x}$ or $f^{-1}(x) = \sqrt{x}$ defines the inverse function. Figure 2.60 shows the graph of $f(x) = x^2$ for $x \geq 0$ and its inverse $f^{-1}(x) = \sqrt{x}$. Note that the graphs are symmetric about $y = x$.

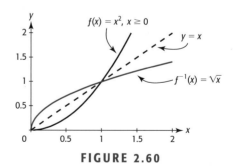

FIGURE 2.60

EXAMPLE 6 Velocity of Blood

Because of friction from the walls of an artery, the velocity of blood is greatest at the center of the artery and decreases as the distance x from the center increases. The velocity in centimeters per second of a blood corpuscle in the artery can be modeled by the function

$$v = k(R^2 - x^2),$$

where R is the radius of the artery and k is a constant that is determined by the pressure, viscosity of the blood, and the length of the artery. In the case where $k = 2$ and $R = 0.1$ cm, the velocity can be written as a function of the distance x from the center as

$$f(x) = 2(0.01 - x^2) \text{ cm per second.}$$

a. Does the inverse of this function exist in the context of the application?

b. What is the inverse function?

c. What does the inverse function mean in the context of this application?

Solution

a. Because x represents distance, it is nonnegative in this application. Hence the function is one-to-one and we can find its inverse.

b. We find the inverse function as follows.

$$f(x) = 2(0.01 - x^2)$$

$$y = 2(0.01 - x^2) \qquad \text{Replacing } f(x) \text{ with } y$$

$$x = 2(0.01 - y^2) \qquad \text{Interchanging } y \text{ and } x$$

$$x = 0.02 - 2y^2 \qquad \text{Solving for } y$$

$$2y^2 = 0.02 - x$$

$$y^2 = \frac{0.02 - x}{2}$$

$$y = \sqrt{\frac{0.02 - x}{2}} \qquad \text{Only nonnegative values of } y \text{ are possible.}$$

$$f^{-1}(x) = \sqrt{\frac{0.02 - x}{2}} \qquad \text{Replacing } y \text{ with } f^{-1}(x).$$

c. The inverse function gives the distance from the center of the artery as a function of the velocity of a blood corpuscle. ∎

2.6 SKILLS CHECK

In each of Exercises 1 and 2, determine if the function f defined by the arrow diagram has an inverse. If it does, create an arrow diagram that defines the inverse.

1.

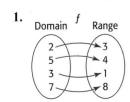

2.

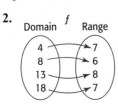

3. If $f(x) = 3x$ and $g(x) = \dfrac{x}{3}$

    a. What are $f(g(x))$ and $g(f(x))$?

    b. Are $f(x)$ and $g(x)$ inverse functions?

4. If $f(x) = 4x - 1$ and $g(x) = \dfrac{x + 1}{4}$

    a. What are $f(g(x))$ and $g(f(x))$?

    b. Are $f(x)$ and $g(x)$ inverse functions?

5. If $f(x) = x^3 + 1$ and $g(x) = \sqrt[3]{x - 1}$, are $f(x)$ and $g(x)$ inverse functions?

6. If $f(x) = (x - 2)^3$ and $g(x) = \sqrt[3]{x + 2}$, are $f(x)$ and $g(x)$ inverse functions?

7. For the function f defined by $f(x) = 3x - 4$, complete the tables below for f and f^{-1}.

x	$f(x)$
-1	-7
0	
1	
2	
3	

x	$f^{-1}(x)$
-7	-1
-4	
-1	
2	
5	

8. a. Write the inverse of $f(x) = 3x - 4$.
   b. Do the values for f^{-1} in the table in Exercise 7 fit the equation for f^{-1}?

9. If the graph of $y = f(x)$ has the point (a, b) on its graph, name a point on the graph of $y = f^{-1}(x)$.

10. Find the inverse of $f(x) = \dfrac{1}{x}$.

11. Find the inverse of $g(x) = 4x + 1$.

12. Find the inverse of $f(x) = 4x^2$ for $x \geq 0$.

13. Graph $g(x) = \sqrt{x}$ and its inverse $g^{-1}(x)$ for $x \geq 0$ on the same axes.

14. $f(x) = (x - 2)^2$ and $g(x) = \sqrt{x} + 2$ are inverse functions for what values of x?

15. Is the function $f(x) = 2x^3 + 1$ a one-to-one function? Does it have an inverse?

16. Is the function $f(x) = 3x^2 + 1$ a one-to-one function? Does it have an inverse?

17. Is the function defined by $\{(1, 3), (2, 2), (3, 4)\}$ a one-to-one function?

18. Sketch the graph of $y = f^{-1}(x)$ on the axes with the graph of $y = f(x)$, shown below.

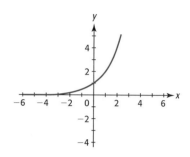

19. Is the function with the graph below one-to-one?

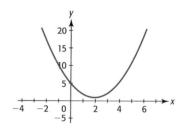

2.6 EXERCISES

1. *Shoe Sizes* A man's shoe that is size x in Britain is equivalent to size $d(x)$ in the United States, where $d(x) = x + 0.5$.
   a. Find the inverse of the function.
   b. Use the inverse function to find the British size of a shoe if it is U.S. size $8\frac{1}{2}$.
   (Source: Kuru International Exchange Association)

2. *Shoe Sizes* A man's shoe that is size x in the United States is size $t(x)$ in Continental size, where $t(x) = x + 34.5$.
   a. Find a function that will convert Continental shoe size to U.S. shoe size.
   b. Use the inverse function to find the U.S. size if the Continental size of a shoe is 43.
   (Source: Kuru International Exchange Association)

3. *Cigarettes* For the years 1975–1991, the percent of high school seniors who have tried cigarettes is given by $f(t) = 75.451 - 0.707t$, where t is the number of years after 1975. Find the inverse of this function and use it to find the year in which the percent fell below 65%.
   (Source: National Institute on Drug Abuse)

4. *Investments* If x is invested at 10% for 6 years, the future value of the investment is given by $S(x) = x + 0.6x$.
   a. Find the inverse of this function.
   b. What are the outputs of the inverse function?
   c. Use this function to find the amount of money that must be invested for 6 years at 10% to have a future value of 24,000.

5. *Ritalin* The function that models the consumption of Ritalin over the years 1990–1996 is $f(x) = 225.304x + 493.432$, where x is the number of years after 1990 and y is measured in grams per 100,000 people.
  a. Find the inverse of this function. What are the outputs of the inverse function?
  b. Use the inverse function to find when consumption equals 1170 grams per 100,000 people.

6. *Apparent Temperature* If the outside temperature is 90° F, the apparent temperature is given by $A(x) = 82.35 + 29.3x$, where x is the humidity written as a decimal. Find the inverse of this function and use it to find the percent humidity that will give an apparent temperature of 97° if the temperature is 90°F.
  (Source: "Temperature-Humidity Indices," *The UMAP Journal*, Fall 1989)

7. *Body-Heat Loss* The model for body-heat loss depends on the coefficient of convection $K = f(x)$, which depends on wind speed x according to the equation $f(x) = 4\sqrt{4x + 1}$.
  a. What are the domain and range of this function without regard to the context of this application?
  b. Find the inverse of this function.
  c. What are the domain and range of the inverse function?
  d. In the context of the application, what are the domain and range of the inverse function?

8. *Algorithmic Relationship* For many species of fish, the weight W is a function of the length x, given by $W = kx^3$, where k is a constant depending on the species. Suppose that $k = .002$, W is in pounds, and x is in inches, so that the weight is $w(x) = .002x^3$.
  a. Find the inverse function of this function.
  b. What does the inverse function give?
  c. Use the inverse function to find the length of a fish that weighs 2 lb.
  d. In the context of the application, what are the domain and range of the inverse function?

9. *Decoding Messages* If we assign numbers to the letters of the alphabet as follows below and assign 27 to a blank space, we can convert a message to a numerical sequence. We can "encode" a message by adding 3 to each number that represents a letter in a message.

A	B	C	D	E	F	G	H	I	J	K	L	M
1	2	3	4	5	6	7	8	9	10	11	12	13

N	O	P	Q	R	S	T	U	V	W	X	Y	Z
14	15	16	17	18	19	20	21	22	23	24	25	26

Thus the message "Go for it" can be encoded by using the numbers to represent the letters, and further encode it by using the function $C(x) = x + 3$. The coded message would be 10 18 30 9 18 21 30 12 23. Find the inverse of the function and use it to decode 23 11 8 30 21 8 4 15 30 23 11 12 17 10.

10. *Decoding Messages* Use the numerical representation from Exercise 9 and the inverse of the encoding function $C(x) = 3x + 2$ to decode 41 5 35 17 83 41 77 83 14 5 77.

11. *Social Security Numbers and Income Taxes* Consider the function that assigns each person who pays federal income tax his or her Social Security number. Is this a one-to-one function? Explain.

12. *Checkbook Balance* Consider the function with the check number in your checkbook as input and the dollar amount of the check as the output. Is this a one-to-one function? Explain.

13. *Volume of a Cube* The volume of a cube is $f(x) = x^3$ cubic inches, where x is the length of the edge of the cube in inches.
  a. Is this function one-to-one?
  b. Find the inverse of this function.
  c. What is the domain and range of this inverse function in the context of the application?
  d. How could this inverse function be used?

14. *Volume of a Sphere* The volume of a sphere is $f(x) = \frac{4}{3}\pi x^3$ cubic inches, where x is the radius of the sphere in inches.
  a. Is this function one-to-one?
  b. Find the inverse of this function.
  c. What are the domain and range of this inverse function?
  d. How could this inverse function be used?
  e. What is the radius of a sphere if its volume is 65,450 cubic inches?

15. *Currency Conversion* The function that converts Canadian dollars to US dollars according to the January 19, 2001 values is $f(x) = 0.66832x$, where x is the number of Canadian dollars and $f(x)$ is the number of US dollars.
  a. Find the inverse function for f and interpret its meaning.
  b. Use f and f^{-1} to determine the money you will have if you take 500 US dollars to Canada, convert it to Canadian dollars, don't spend any, and then convert it back to US dollars. (Assume that there is no fee for conversion and the conversion rate remains the same.)
  (Source: Expedia.com)

16. *Surface Area* The surface area of a cube is $y = 6x^2$ cm^2, where x is the length of the edge of the cube in centimeters.

   a. For what values of x does this model make sense? Is the model a one-to-one function for these values of x?

   b. What is the inverse of this function on this interval?

   c. How could the inverse function be used?

17. *Illumination* The intensity of illumination of a light is a function of the distance from the light. For a given light, the intensity is given by $I(x) = \dfrac{300,000}{x^2}$ candle power, where x is the distance in feet from the light.

   a. Is this function a one-to-one function?

   b. What is the domain of this function in the context of the application?

   c. Is the function one-to-one for the domain in part (b)?

   d. Find the inverse of this function and use it to find the distance at which the intensity of the light is 75,000 candlepower.

18. *Supply* The supply function for a product is $p(x) = \dfrac{1}{4}x^2 + 20$, where x is the number of thousands of units a manufacturer will supply if the price is $p(x)$ dollars.

   a. Is this function a one-to-one function?

   b. What is the domain of this function in the context of the application?

   c. Is the function one-to-one for this domain?

   d. Find the inverse of this function and use it to find how many units the manufacturer is willing to supply if the price is $101.

19. Suppose that the function that converts United Kingdom (UK) pounds to US dollars is $f(x) = 1.7655x$, where x is the number of pounds and $f(x)$ is the number of US dollars.

   a. Find the inverse function for f and interpret its meaning.

   b. Use f and f^{-1} to determine the money you will have if you take $1000 US dollars to the U.K., convert it to pounds, don't spend any, and then convert it back to US dollars. (Assume that there is no fee for conversion and the conversion rate remains the same.)

   (Source: International Monetary Fund)

20. *First Class Postage* The postage charged for first-class mail is a function of its weight. The U.S. Postal Service uses the following table to describe the rates for 2003.

Weight Increment, w	Postal Rate
First ounce or fraction of an ounce	0.370
Each additional ounce or fraction	0.230

(Source: pe.usps.gov/text)

   a. Convert this table to a piecewise-defined function $P(x)$ that represents postage for letters weighing more than 0 and no more than 3 oz, using w as the weight in ounces and $P(x)$ as the postage in cents.

   b. Does P have an inverse function? Why or why not?

21. *Path of a Ball* If a ball is thrown into the air at a velocity of 96 feet per second from a building that is 256 feet high, the height of the ball after x seconds is $f(x) = 256 + 96x - 16x^2$ feet.

   a. For how many seconds will the ball be in the air?

   b. Is this function a one-to-one function over this time interval?

   c. Give an interval over which the function is one-to-one.

   d. Find the inverse of the function over the interval $0 \le x \le 3$. What does it give?

2.7 Quadratic and Power Inequalities

Key Concepts

- Quadratic inequalities
- Solving analytically
- Solving graphically
- Critical values
- Sign diagrams
- Power inequalities

The daily profit from the production and sale of x units of a product is given by

$$P(x) = -0.01x^2 + 20.25x - 500.$$

Because a profit occurs when $P(x)$ is positive, there is a profit for those values of x that make $P(x) > 0$. Thus the values of x that give a profit are the solutions to

$$-0.01x^2 + 20.25x - 500 > 0$$

In this section we solve quadratic inequalities by using both analytical and graphical methods.

Solution of Quadratic Inequalities

A **quadratic inequality** is an inequality that can be written in the form

$$ax^2 + bx + c > 0, \text{ where } a, b, \text{ and } c \text{ are real numbers and } a \neq 0.$$

(or with $>$ replaced by $<$, $\geq$, or $\leq$).

Recall that we solved a linear inequality $f(x) > 0$ or $f(x) < 0$ graphically by graphing the related equation $f(x) = 0$ and observing where the graph is above or below the x-axis, that is, where $f(x)$ is positive or negative.

Similarly, we can solve a quadratic inequality $f(x) > 0$ or $f(x) < 0$ graphically by graphing the related equation $f(x) = 0$ and observing the x-values of the intervals where $f(x)$ is positive or negative.

For example, we can graphically solve the inequality

$$x^2 - 3x - 4 \leq 0$$

by graphing

$$y_1 = x^2 - 3x - 4$$

and observing the part of the graph that is below or on the x-axis. Using the x-intercept method, we see that the graph is below or on the x-axis for values of x satisfying $-1 \leq x \leq 4$. (See Figure 2.61.) Thus the solution to the inequality is $-1 \leq x \leq 4$.

To solve a quadratic inequality $f(x) > 0$ or $f(x) < 0$ analytically, we first need to find the zeros of $f(x)$. The zeros can be found analytically by factoring or the quadratic formula. If the quadratic function intersects the x-axis in two points, the function has two real zeros. These two zeros divide the real number line into three intervals. Within each interval, the value of the quadratic function is either always positive or always negative, so we can use one value in each interval to test the function on the interval. We can use the following steps, which summarize these ideas, to solve quadratic inequalities.

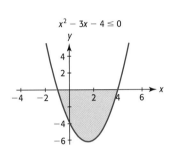

$x^2 - 3x - 4 \leq 0$

FIGURE 2.61

TO SOLVE A QUADRATIC INEQUALITY ANALYTICALLY:

1. Write an equivalent inequality with 0 on one side and with the function $f(x)$ on the other side.

2. Solve $f(x) = 0$ analytically.

3. Create a sign diagram that uses the solutions from step 2 to divide the number line into intervals. Pick a test value in each interval to create a sign diagram. Determine whether $f(x)$ is positive or negative in that interval.*

4. Identify the intervals that satisfy the inequality in step 1. The values of x that define these intervals are solutions to the original inequality.

EXAMPLE 1

Solve the inequality $x^2 - 3x > 3 - 5x$.

Solution

Rewriting the inequality with 0 on the right side of the inequality gives $f(x) > 0$ with $f(x) = x^2 + 2x - 3$:

$$x^2 + 2x - 3 > 0$$

---

*The numerical feature of your graphing utility can be used to test the x-values.

Writing the equation $f(x) = 0$ and solving for x gives

$$x^2 + 2x - 3 = 0$$

$$(x + 3)(x - 1) = 0$$

$$x = -3 \quad \text{or} \quad x = 1$$

The two values of x, -3 and 1, divide the number line into three intervals, the numbers less than -3, the numbers between -3 and 1, and the numbers greater than 1. We need only find the sign of $(x + 3)$ and $(x - 1)$ in each interval and then find the sign of their product to find the solution to the original inequality. Testing a value in each interval determines the sign of each factor in each interval. (See the sign diagram in Figure 2.62.)

sign of $(x + 3)(x - 1)$ ++++++++++++   – – – – – – – – –   ++++++++++

sign of $(x - 1)$ – – – – – – – – – –   – – – – – – – – –   ++++++++++

sign of $(x + 3)$ – – – – – – – – – –   ++++++++++++   ++++++++++

```
                          |                    |
                         -3                    1
```

FIGURE 2.62

The function $f(x)$ is positive on the intervals $(-\infty, -3)$ and $(1, \infty)$, so the solution to $x^2 + 2x - 3 > 0$ and the original inequality is

$$x < -3 \quad \text{or} \quad x > 1. \qquad \blacksquare$$

Table 2.21 shows the possible graphs of quadratic functions and the solutions to the related inequalities. Note that if the graph of $y = f(x)$ lies entirely above the x-axis, the solution to $f(x) > 0$ is the set of all real numbers and there is no solution to $f(x) < 0$. (Other possible orientations of the graph of $y = f(x)$ exist, of course.)

TABLE 2.21

Orientation of Graph of $y = f(x)$	Inequality to Be Solved	Part of Graph That Satisfies the Inequality	Solution to Inequality
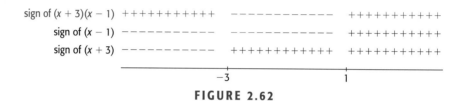	$f(x) > 0$	where the graph is above the x-axis	$x < a$ or $x > b$
	$f(x) < 0$	where the graph is below the x-axis	$a < x < b$
	$f(x) > 0$	where the graph is above the x-axis	$a < x < b$
	$f(x) < 0$	where the graph is below the x-axis	$x < a$ or $x > b$

TABLE 2.21 *(continued)*

Orientation of Graph of $y = f(x)$	Inequality to Be Solved	Part of Graph That Satisfies the Inequality	Solution to Inequality
(upward parabola above x-axis)	$f(x) > 0$	the entire graph	all real numbers
	$f(x) < 0$	none of the graph	no solution
(downward parabola below x-axis)	$f(x) > 0$	none of the graph	no solution
	$f(x) < 0$	the entire graph	all real numbers

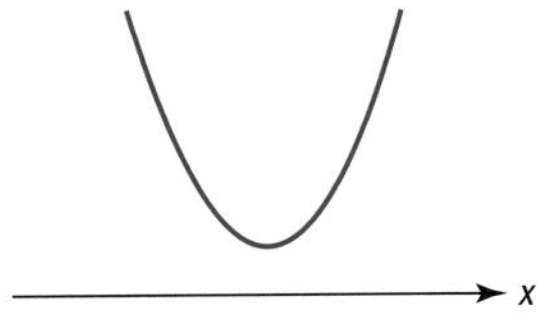

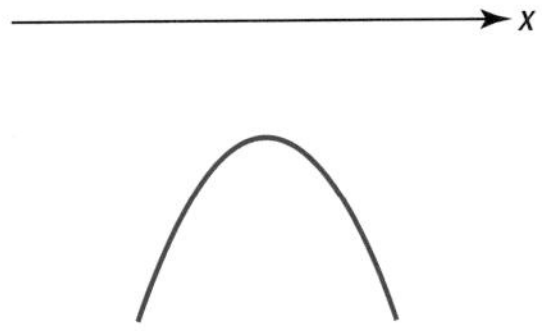

EXAMPLE 2   Height of a Model Rocket

A model rocket is projected straight upward from ground level according to the equation

$$h = -16t^2 + 192t, \; t \geq 0,$$

where h is the height in feet and t is the time in seconds. During what time interval will the height of the rocket exceed 320 feet?

Analytical Solution

To find the time interval when the rocket is higher than 320 feet, we will find the values of t for which

$$-16t^2 + 192t > 320.$$

Getting 0 on the right side of the inequality, we have

$$-16t^2 + 192t - 320 > 0.$$

Writing the equation $h = 0$ and solving gives

$$-16t^2 + 192t - 320 = 0$$
$$-16(t^2 - 12t + 20) = 0$$
$$-16(t - 2)(t - 10) = 0$$
$$t = 2 \text{ or } t = 10$$

The values of $t = 2$ and $t = 10$ divide the number line into three intervals—the numbers less than 2, the numbers between 2 and 10, and the numbers greater than 10. (See Figure 2.63.) To solve the inequality, we find the sign of the product of -16, $(t - 2)$, and $(t - 10)$ in each interval to find the solution. We do this by testing a value in each interval.

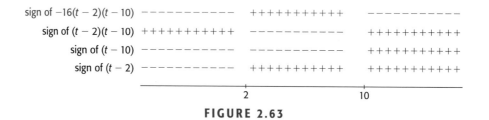

FIGURE 2.63

This shows that the product $-16(t - 2)(t - 10)$ is positive only for $2 < t < 10$, so the solution to $-16t^2 + 192t - 320 > 0$ is $2 < t < 10$. This solution is also a solution to the original inequality.

Note that if the inequality problem is applied, we must check that the solution makes sense in the context of the problem.

Graphical Solution

To find the value of t where the height exceeds 320 feet, we solve $-16t^2 + 192t > 320$ using a graphing utility. To use the intercept method, we rewrite the inequality with 0 on the right side:

$$-16t^2 + 192t - 320 > 0.$$

Using the variable x in place of the variable t, we enter $y_1 = -16x^2 + 192x - 320$ and graph the function. (See Figure 2.64.) The x-intercepts of the graph are $x = 2$ and $x = 10$.

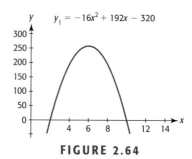

FIGURE 2.64

The solution to $-16t^2 + 192t - 320 > 0$ is the interval where the graph is above the x-axis, which is

$$2 < x < 10.$$

Thus, the height of the rocket exceeds 320 feet between 2 and 10 seconds. ∎

| **EXAMPLE 3** | Internet Use |

According to the 1999 *CyberAtlas*, the number of Internet users, in thousands, can be predicted by the equation $y = 11.786x^2 - 142.214x + 493$, where x is the number of years after 1990. If this model is accurate, for what years between 1990 and 2020 will the number of Internet users be over one million?

Solution

To find the years when the number of Internet users is over one million, we solve the inequality

$$11.786x^2 - 142.214x + 493 > 1000.$$

We will begin solving this inequality by graphing $y = 11.786x^2 - 142.214x + 493$. We set the viewing window from $x = 0$ (1990) to $x = 30$ (2020). The graph of this function is shown in Figure 2.65.

We seek the point(s) where the graph of $y = 1000$ intersects the graph of $y = 11.786x^2 - 142.214x + 493$. Graphing the line $y = 1000$, and intersecting it with the graph of the quadratic function gives the point of intersection $(14.945, 1000)$ as shown in Figure 2.66. This graph shows that the number of users is above 1000 thousand for $x \geq 15$. Because x corresponds to the number of years after 1990, the number of Internet users will be over one million between the years 2005 and 2020.

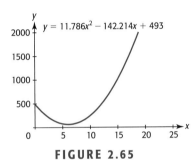

FIGURE 2.65

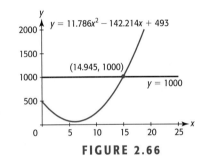

FIGURE 2.66 ■

Power Inequalities

To solve a power inequality, we will use a combination of analytical and graphical methods.

POWER INEQUALITIES

To solve a power inequality, first solve the related equation analytically by raising each side of the related equation to the reciprocal of the power. Then use graphical methods to find the values of the variable that satisfy the inequality.

EXAMPLE 4   Investment

The future value of $3000 invested for 3 years at rate r, compounded annually, is given by $S = 3000(1 + r)^3$. What interest rate will give a future value of at least $3630?

Solution

To solve this problem, we solve the inequality

$$3000(1 + r)^3 \geq 3630.$$

We begin by solving the related equation by using the root method.

$$3000(1 + r)^3 = 3630$$

$$(1 + r)^3 = 1.21 \qquad \text{Dividing both sides by 3000}$$

$$1 + r = \sqrt[3]{1.21} \qquad \text{Taking the cube root (1/3 power) of both sides}$$

$$1 + r \approx 1.0656$$

$$r \approx 0.0656$$

This tells us that the investment will have a future value of $3630 in 3 years if the interest rate is approximately 6.56%, and helps us set the window to solve the inequality graphically. Graphing $y_1 = 3000(1 + r)^3$ and $y_2 = 3630$ on the same axes shows that $3000(1 + r)^3 \geq 3630$ if $r \geq .0656$. (See Figure 2.67.) Thus the investment will result in more than $3630 if the interest rate is higher than 6.56%. Note that in the context of this problem, only values of r from 0 (0%) to 1 (100%) make sense.

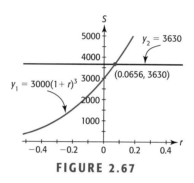

FIGURE 2.67

2.7 SKILLS CHECK

In Exercises 1–10, use analytical methods to solve the inequalities.

1. $x^2 + 4x < 0$

2. $x^2 - 25x < 0$

3. $9 - x^2 \geq 0$

4. $x > x^2$

5. $-x^2 + 9x - 20 > 0$

6. $2x^2 - 8x < 0$

7. $2x^2 - 8x \geq 24$

8. $t^2 + 17t \leq 8t - 14$

9. $x^2 - 6x < 7$

10. $4x^2 - 4x + 1 > 0$

Use the quadratic formula and graphical methods to solve the inequalities in Exercises 11–14.

11. $2x^2 - 7x + 2 \geq 0$

12. $w^2 - 5w + 4 > 0$

13. $5x^2 \geq 2x + 6$

14. $2x^2 \leq 5x + 6$

Use graphical methods to solve the inequalities in Exercises 15–18. Give the answers correct to three decimal places.

15. $-2x^2 + 9x - 4 \geq 0$

16. $3x^2 + 5x - 1 \geq 0$

17. $x^2 - 3x - 5 < 0$

18. $-x^2 - 3x - 5 < 0$

2.7 EXERCISES

Use factoring and graphical methods to solve Exercises 1–6.

1. *Profit* The monthly profit from producing and selling x units of a product is given by

$$P(x) = -0.3x^2 + 1230x - 120,000.$$

Producing and selling how many units will result in a profit for this product?

2. *Profit* The monthly profit from producing and selling x units of a product is given by

$$P(x) = -0.01x^2 + 62x - 12,000.$$

Producing and selling how many units will result in a profit for this product?

3. *Profit* The revenue from sales of x units of a product is given by $R(x) = 200x - 0.01x^2$ and the cost of producing and selling the product is $C(x) = 38x + .01x^2 + 16,000$. Producing and selling how many units will result in a profit?

4. *Profit* The revenue from sales of x units of a product is given by $R(x) = 300x - 0.01x^2$ and the cost of producing and selling the product is $C(x) = 40x + 0.02x^2 + 28,237$. Producing and selling how many units will result in a profit?

5. *Projectiles* Two projectiles are fired into the air over a lake, with the height of the first projectile given by $y = 100 + 130t - 16t^2$ and the height of the second projectile given by $y = -16t^2 + 180t$, where y is in feet and t is in seconds. Over what time interval, before the lower one hits the lake, is the second projectile above the first?

6. *Projectile* A rocket shot into the air has height $s = 128t - 16t^2$ feet, where t is the number of seconds after the rocket is shot. During what time after the rocket is shot is it at least 240 feet high?

Use the quadratic formula to solve Exercises 7–10.

7. *Tobacco Sales* The domestic sales of tobacco (in millions of kilograms) in Canada is given by $y = -0.084x^2 + 1.124x + 4.028$, where x is the number of years after 1986. During what years from 1986–1990 does this model indicate that the sales will be at least 5,940,000 kilograms?
(Source: Canadian Council on Smoking and Health)

8. *World Population* The low long-range world population numbers and projections for the years 1995–2150 are given by the equation $y = -0.00036x^2 + 0.0385x + 5.823$, where x is the number of years after 1990 and y is in billions. During what years from 1990 through 2000 does this model estimate that the population was above 6 billion?
(Source: UN Department of Economic and Social Affairs)

9. *Airplane Crashes* The number of all airplane crashes (in thousands) is given by the equation $y = 0.0057x^2 - 0.197x + 3.613$, where x is the number of years after 1980. During what years does the model indicate the number of crashes as being below 3000?

10. *Gross Domestic Product* The U.S. gross domestic product (in billions of constant dollars) for the

years 1940–1995 can be modeled by the equation $y = 0.875x^2 + 23.406x + 886.82$, where x is the number of years after 1935. During what years prior to 2010 is the gross domestic product greater than $4 trillion?
(Source: Microsoft Encarta 98)

Use graphical and/or numerical methods to solve Exercises 11–16.

11. *Homicide Rate* The homicide rate (number per 100,000 people) is given by the function $r = -0.0128x^2 + 0.586x + 2.986$, where x is the number of years after 1960. During what years does this model indicate that the rate will be above 8 per 100,000 people?
(Source: Bureau of Justice Statistics)

12. *Hotel Room Supply* The percentage change p (from the previous year) in the hotel room supply is given by $p = 0.025x^2 - 0.645x + 4.005$, where x is the number of years after 1998. If this model is accurate, during what years after 1998 will the percentage change be less than 3 percent?
(Source: Hotel and Motel Management, September 6, 1999)

13. *Unemployment* For the years 1983–1992, the unemployment rate in South Carolina can be modeled by $y = 0.144x^2 - 2.482x + 15.467$, where y is the percent and x is the number of years after 1980. During what years in this period does the model indicate that the unemployment rate fell below 6 percent?
(Source: *South Carolina Statistical Abstract*)

14. *Tobacco Sales* The domestic sales of tobacco (in millions of kilograms) in Canada is given by $y = -0.084x^2 + 1.124x + 4.028$, where x is the number of years after 1986. During what years from 1986 does this model indicate that the sales will exceed 6,643,000 kilograms?
(Source: Canadian Council on Smoking and Health)

15. *Foreign born Population* The percent of the U.S. population that is foreign born is given by $y = 0.0019x^2 - 0.305x + 18.885$, where x is the number of years after 1900. During what years does this model indicate that the percent is less than 8 percent?
(Source: U.S. Bureau of the Census)

16. *Marriage Age* Based on data between 1960 and 1997 appearing in the *1999 World Almanac*, the median age at first marriage for women in the U.S. can be described by $A = 0.0024x^2 + 0.0418x + 20.24$ years of age, where x is the number of years after 1960. During what years does this model indicate that the median age at first marriage for women is at least 23?

CHAPTER 2 *Summary*

In this chapter we presented the building blocks for function construction. We began with an in-depth discussion of quadratic functions, including realistic applications that involve the vertex and x-intercepts of parabolas and solving quadratic equations. We then studied power functions and how to transform them to obtain other functions. Real-world data is provided throughout the chapter, and we learned how to fit power and quadratic functions to some of these data. We finished the chapter by learning how to combine two or more functions to create new functions, how to find the inverse of a function, and how to solve quadratic and power inequalities.

Key Concepts and Formulas

2.1 Quadratic Functions; Parabolas

Quadratic function

Also called a *second-degree function*, this function can be written in the form $f(x) = ax^2 + bx + c$ where $a \neq 0$.

Parabola

A parabola is the graph of a quadratic equation.

Vertex

The high or low point on a parabola is the vertex.

Maximum point

If a quadratic function has $a < 0$, the parabola has a high point called the maximum point and the parabola opens downward.

Minimum point

If a quadratic function has $a > 0$, the parabola has a low point called the minimum point and the parabola opens upward.

Axis of symmetry

The axis of symmetry is a vertical line through the vertex of the parabola.

Forms of quadratic functions

Standard Form	General Form
$y = a(x - h)^2 + k$	$y = ax^2 + bx + c$
Vertex at (h, k)	x-coordinate of vertex at $x = -\dfrac{b}{2a}$
Axis of symmetry is the line $x = h$.	Axis of symmetry is the line $x = -\dfrac{b}{2a}$
Parabola opens up if a is positive and down if a is negative.	Parabola opens up if a is positive and down if a is negative.

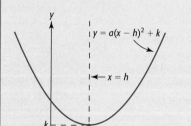

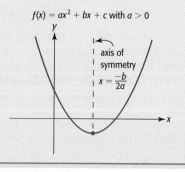

x-Intercepts of the graph of a quadratic function	Occur where the graph of the function crosses or touches the x-axis. The parabola will cross the x-axis at 0, 1, or 2 points.
Zeros of a quadratic function	Zeros are values of x that make the function 0. The zeros of a function are also the x-intercepts of the graph of the function.

2.2 Solving Quadratic Equations

Solving quadratic equations	An equation that can be written in the form $ax^2 + bx + c = 0$ is called a *quadratic equation*.
Zero product property	For real numbers a and b, the product $ab = 0$ if and only if either $a = 0$ or $b = 0$ or both a and b are zero.
Solving by factoring	To solve a quadratic equation by factoring, write the equation in a form with 0 on one side. Then factor the nonzero side of the equation, if possible, and use the zero product property to convert the equation into two linear equations that are easily solved.
Solving and checking graphically	Solutions or decimal approximations of solutions to quadratic equations can be found by using TRACE, ZERO, or INTERSECT with a graphing utility.
Factor Theorem	The factorization of a polynomial function $f(x)$ and the solutions to $f(x) = 0$ can be verified by graphing $y = f(x)$ and observing where the graph crosses the x-axis. If $x = a$ is a solution to $f(x) = 0$, then $x - a$ is a factor of f.
Solving using the square root method	When a quadratic equation has the simplified form $x^2 = C$, the solutions are $x = \pm\sqrt{C}$. This method can also be used to solve equations of the form $(ax + b)^2 = C$.
Completing the square	A quadratic equation can be solved by converting one side to a perfect binomial square and taking the square root of both sides to solve the equation.
Solving using the quadratic formula	The solutions of the quadratic equation $ax^2 + bx + c = 0$ with $a \neq 0$, are given by the formula $$x = \frac{-b \pm \sqrt{b^2 - 4ac}}{2a}$$
Solutions, zeros, x-intercepts, and factors	If a is a real number, the following three statements are equivalent: • a is a solution to the equation $f(x) = 0$. • a is a zero of the function $f(x)$. • a is an x-intercept of the graph of $y = f(x)$.

2.3 Power Functions and Transformations

Power functions	A power function is a function of the form $y = ax^b$ with $b > 1$ where a and b are real numbers.
Root functions	A root function is a function of the form $y = ax^{1/n}$, or $y = a\sqrt[n]{x}$, where n is an integer, $n \geq 2$.
Radical equations	An equation containing radicals can frequently be converted to an equation that does not contain radicals by raising both sides of the equation to a power that is equal to the index of the radical. The solutions must be checked when using this technique.
Transformations of graphs	The graph of a function can be obtained by shifting (horizontally or vertically), compressing, stretching, or reflecting another graph. In particular, the graph of $f(x) = a \cdot f(x - h) + k$ is the graph of $y = f(x)$ shifted h units horizontally and k units vertically, and stretched or compressed by a factor of a. If $a < 0$, the graph is reflected across the x-axis.
Symmetry with Respect to the Origin	The graph of $y = f(x)$ is symmetric with respect to the origin if $f(-x) = -f(x)$.

Symmetry with Respect to the y-axis	The graph of $y = f(x)$ is symmetric with respect to the y-axis if $f(-x) = f(x)$.

2.4 *Quadratic and Power Models*

Quadratic modeling	The use of graphing utilities permits us to fit quadratic functions to data by using technology.
Second differences	If the second differences of data are constant for equally spaced inputs, a quadratic function is the exact fit for the data.
Power modeling	The use of graphing utilities permits us to fit power functions to nonlinear data by using technology.

2.5 *Combining Functions; Reciprocal, Absolute Value, and Piecewise-Defined Functions*

Operations with Functions	*Sum*: $(f + g)(x) = f(x) + g(x)$

Difference: $(f - g)(x) = f(x) - g(x)$

Product: $(f \cdot g)(x) = f(x) \cdot g(x)$

Quotient: $\left(\dfrac{f}{g}\right)(x) = \dfrac{f(x)}{g(x)}, g(x) \neq 0$

The domain of the sum, difference, and product of f and g consists of all real numbers of the input variable for which f and g are defined. The domain of the quotient function consists of all real numbers for which f and g are defined and $g \neq 0$.

Reciprocal function	$$f(x) = \frac{1}{x}$$

This function is the quotient of a constant function and the identity function.

Piecewise-defined functions	This is a function that is created by combining two or more functions. Piecewise-defined functions can be graphed by graphing their pieces on the same axes. A familiar example of a piecewise-defined function is the *absolute value function*.
Absolute value function	$$\lvert x \rvert = \begin{cases} x & \text{if } x \geq 0 \\ -x & \text{if } x < 0 \end{cases}$$
Absolute value equations	The solution of the absolute value equation $\lvert x \rvert = a$ is $x = a$ or $x = -a$ when $a \geq 0$. There is no solution to $\lvert x \rvert = a$ if $a < 0$.
Composite functions	The notation for the composite function "f of g" is $(f \circ g)(x) = f(g(x))$. The domain of $f \circ g$ is the subset of the domain of g for which $f \circ g$ is defined.

2.6 *Inverse Functions*

Inverse functions	If $f(g(x)) = x$ and $g(f(x)) = x$, f and g are inverse functions. In addition, the functions f and g are inverse functions if, whenever the pair (a, b) satisfies $y = f(x)$, the pair (b, a) satisfies $y = g(x)$. We denote g by f^{-1}, read "f inverse".
One-to-one functions	In order for a function f to have an inverse, f must be one-to-one. A function f is one-to-one if $f(a) = f(b)$ implies that $a = b$.
Horizontal line test	If no horizontal line can intersect the graph of a function in more than one point, then the function is one-to-one.

Finding the equation of the inverse function	To find the inverse of the function f that is defined by the equation $y = f(x)$:

Finding the equation of the inverse function

To find the inverse of the function f that is defined by the equation $y = f(x)$:
1. Rewrite the equation with y replacing $f(x)$.
2. Interchange x and y in the equation defining the function.
3. Solve the new equation for y. If this equation cannot be solved uniquely for y, the function has no inverse.
4. Replace y by $f^{-1}(x)$.

Graphs of inverse functions

The graphs of a function and its inverse are symmetric with respect to the line $y = x$.

Inverse functions on limited domains

If the function is one-to-one on a limited domain, then it has an inverse function on that domain.

2.7 Quadratic and Power Inequalities

Quadratic inequality

A quadratic inequality is an inequality that can be written in the form $ax^2 + bx + c > 0$, where a, b, c are real numbers and $a \neq 0$. (or with $>$ replaced by $<, \geq,$ or $\leq$).

Solving a quadratic inequality graphically

To solve a quadratic inequality $f(x) < 0$ or $f(x) > 0$ graphically, we can graph the related equation $y = f(x)$ and observe the x-values of the intervals where the graph of $f(x)$ is below or above the x-axis.

Solving a quadratic inequality analytically

To solve a quadratic inequality $f(x) < 0$ or $f(x) > 0$ analytically, solve $f(x) = 0$, and use the solutions to divide the number line into intervals. Pick a test value in each interval to determine whether $f(x)$ is positive or negative in that interval and identify the intervals that satisfy the original inequality.

Power inequalities

To solve a power inequality, first solve the related equation analytically by raising each side of the related equation to the reciprocal of the power. Then use graphical methods to find the values of the variable that satisfy the inequality.

Chapter 2 Skills Check

---

1. Find the coordinates of the vertex of the graph of $f(x) = 3x^2 - 6x - 24$.

2. Graph $f(x) = 3x^2 - 6x - 24$.

3. Use graphical and algebraic methods to find the x-intercepts of the graph of $f(x) = 3x^2 - 6x - 24$.

4. Find the solutions to $f(x) = 0$ if $f(x) = 3x^2 - 6x - 24$.

In Exercises 5 and 6, use factoring to solve the equations.

5. $x^2 - 5x + 4 = 0$ **6.** $6x^2 + x - 2 = 0$

Use a graphing utility as an aid in factoring to solve the equations in Exercises 7 and 8.

7. $5x^2 - x - 4 = 0$ **8.** $3x^2 + 4x - 4 = 0$

In Exercises 9 and 10, use the quadratic formula to solve the equations.

9. $x^2 - 4x + 3 = 0$ **10.** $4x^2 + 4x - 3 = 0$

11. a. Graph the functions $f(x) = \sqrt{x}$ and $g(x) = \sqrt{x + 2} - 3$.
    b. How are the graphs related?

12. What is the domain of the function $g(x) = \sqrt{x + 2} - 3$?

13. Determine if the function $y = -3x^2$ is increasing or decreasing
    a. for $x < 0$.   **b.** for $x > 0$.

14. For each of the functions, determine if the function is concave up or concave down.
    a. $y = x^{3/2}$ **b.** $y = x^{1/2}$

15. Find the inverse of $f(x) = 3x - 2$.

16. Find the inverse of $g(x) = \sqrt[3]{x - 1}$.

17. Graph $f(x) = (x + 1)^2$ and its inverse $f^{-1}(x)$ on the domain $[-1, 10]$.

18. Graph $f(x) = \begin{cases} 4 - x & \text{if } x \le 3 \\ x^2 - 5 & \text{if } x > 3 \end{cases}$

19. Is the function $f(x) = \dfrac{x}{x-1}$ a one-to-one function?

20. Find a power function that models the data below.

x	1	2	3	4	5	6
y	4	9	11	21	32	45

21. Find a quadratic function that models the data below.

x	1	3	4	6	8
y	2	8	15	37	63

For Exercises 22–24, match each graph with the correct equation.

a. $y = |x| + 2$

b. $y = x^3$

c. $y = \dfrac{3}{x-1} + 1$

d. $y = \dfrac{-3}{x+1} + 1$

e. $y = 2x^3$

f. $y = |x + 2|$

22.

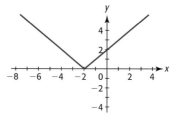

23.

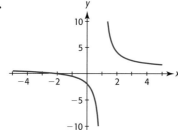

24.

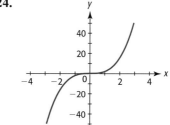

25. Solve the inequality $x^2 - 7x \le 18$.

26. Solve the inequality $2x^2 + 5x \ge 3$.

Chapter 2 Review

1. *Height of a Ball* If a ball is thrown into the air at 64 feet per second from a height of 192 feet, its height (in feet) is given by $S = 192 + 64t - 16t^2$, where t is in seconds.
   a. Find the values of t that make $S = 0$.
   b. Do both of these values of t have meaning in the context of this application?
   c. When will the ball strike the ground?

2. *Falling Ball* If a ball is dropped from the top of a 400-foot-high building, its height S in feet is given by $S = 400 - 16t^2$, where t is in seconds. In how many seconds will it hit the ground?

3. *Break-Even* The profit for a product is given by $P = 3600 - 150x + x^2$, where x is the number of units produced and sold. How many units will give break-even (that is, return a profit of 0)?

4. *Profit* The profit from producing and selling x units of a product is given by the function $P(x) = -0.3x^2 + 1230x - 120{,}000$. Producing and selling how many units will result in a profit of \$324,000 for this product?

5. *Firefighter Injuries* The number of on-duty injuries to firefighters during the years 1988 to 1998 is given by $y = 161{,}488.4931x^{-0.4506}$, with $x = 0$ in 1980.
   a. What does this model indicate the number of injuries to be in 1998?
   b. Graph the function and use the graph to determine in what year the model indicates that the number of injuries was 57,219.

6. *Taxes* The tax burden has grown rapidly in the U.S. during the 20th century. Using data from 1910 to 1999, the per capita (per person) tax paid can be modeled by the power function $S = 11.23t^{1.44}$, where t is the number of years from 1900.
   a. Does this model indicate that the function is increasing?
   b. Graph the function on the interval from $t = 0$ to $t = 100$.
   c. Is the graph concave up or concave down?
   d. What does the model estimate the per capita tax paid to be in 1995?

7. *Maximizing Profit* The monthly profit from producing and selling x units of a product is given by the function $P(x) = -0.01x^2 + 62x - 12{,}000$.
   a. Producing and selling how many units will result in the maximum profit for this product?
   b. What is the maximum possible profit for the product?

8. *Unemployment* For the years 1983–1992, the unemployment percent in South Carolina can be modeled by $y = 0.144x^2 - 2.482x + 15.467$, where x is the number of years from 1980.
   a. Use numerical methods to find the year when the unemployment percent is a minimum.
   b. What is the unemployment rate percent for the year in (*a*)?
   c. During what year after 1988 does the model indicate that the unemployment was 5.047 percent?
   (Source: *South Carolina Statistical Abstract, 1994*)

9. *Profit* The revenue from sales of x units of a product is given by $R(x) = 200x - 0.01x^2$, and the cost of producing and selling the product can be described by $C(x) = 38x + .01x^2 + 16{,}000$. Producing and selling how many units will give break-even?

10. *Tobacco Sales* The domestic sales of tobacco (in millions of kilograms) in Canada can be described by $y = -0.084x^2 + 1.124x + 4.028$, where x is the number of years after 1986. During what years does this model indicate that the sales will equal 6,745,000 kilograms?
    (Source: Canadian Council on Smoking and Health)

11. *Homicide Rate* The homicide rate (the number of homicides per 100,000 people) is given by the function $r = -0.0128x^2 + 0.586x + 2.986$, where x is the number of years after 1960. Use a graphing utility to find during what years this model indicates that the rate will be above 8 per 100,000 people.
    (Source: Bureau of Justice Statistics)

12. *Hotel Room Supply* The percentage change p (from the previous year) in the hotel room supply is given by $p = 0.025x^2 - 0.645x + 4.005$, where x is the number of years after 1998. If this model is accurate, use a graphing utility to find during what years between 1998 and 2010 the percentage change will fall below 3 percent.
    (Source: Hotel and Motel Management, September 6, 1999)

13. *Airplane Crashes* The number of all airplane crashes (in thousands) is given by the equation $y = 0.0057x^2 - 0.197x + 3.613$, where x is the number of years after 1980.
    a. During what year does the model indicate that the number of crashes was at a minimum?
    b. Use a graphing utility to determine the years during which the model indicates that the annual number of crashes is below 3000.

14. *Average Cost* Suppose the total cost function for a product is determined to be $C(x) = 30x + 3150$, where x is the number of units produced and sold.
    a. Write the average cost function.
    b. Graph this function, which is the quotient of two functions.

15. *Supply and Demand* The price per unit of a product is \$$p$ and the number of units of the product is denoted by q. Suppose the supply function for a product is given by $p = \dfrac{180 + q}{6}$ and that the demand for the product is given by $p = \dfrac{30{,}000}{q} - 20$.
    a. Which of these functions is a shifted reciprocal function?
    b. For what values of q will the supply and demand prices be equal?
    c. What will the supply and demand prices be at market equilibrium?

16. *Supply and Demand* The supply function for a commodity is given by $p = 58 + \dfrac{q}{2}$ and the demand function for this commodity is given by $p = \dfrac{2555}{q + 5}$.

    a. Which of these functions is a shifted reciprocal function?

    b. Graph the supply and demand functions on the same graph.

    c. What is the point in the first quadrant where the graphs intersect? What occurs at this point?

17. *Indiana Population* The population of Indiana (in thousands) between 1980 and 1997 can be described by

    $$P(x) = \begin{cases} 2.320x^2 - 389x + 21762 \\ \text{where } 80 \le x \le 90 \\ 46.0x + 1411.667 \\ \text{where } x > 90 \end{cases}$$

    and x is the number of years after 1900.

    a. What was the Indiana population in 1990?

    b. Use P to predict when the population of Indiana reached 6 million people. What assumptions are made if this prediction is to be considered valid?
    (Source: *Statistical Abstract of the U.S., 1998*)

18. *Marijuana Use* The percent of persons 12 years of age and over in the U.S. who said that they had used marijuana at least once within the month prior to being asked during selected years is described by the function

    $$M(x) = \frac{-1}{12}\left(x - \frac{789}{10}\right)^2 + \frac{15{,}541}{1200}$$

    when $79 \le x \le 88$, where x is the number of years after 1900.
    (Source: National Household Survey on Drug Abuse, Series H6 and H7)

    a. Describe the function M in the context of the problem.

    b. Find and interpret $M(79)$, $M(80)$, and $M(88)$.

    c. Using only the information in part (a) and part (b) (and not your calculator), sketch a graph of M.

    d. The survey on which M is based was taken once during each of the selected years. State the domain of the related discrete function.

    e. At what value of x does the maximum of the function M occur? If you put this result in the context of this problem, do you feel that it might be valid?

19. *U. S. Foreign Trade* The *trade balance* is defined as the value of exports minus the value of imports.

Between 1880 and 1980, the value of exports from the U.S. exceeded the value of imports into the U.S., resulting in a positive trade balance. However, this has not been the case since the early 1980s. The value of U.S. imports between 1990 and 1999 is modeled by $I(x) = 61487.24x + 435{,}606.84$ million dollars, and the value of U.S. exports between 1990 and 1999 can be modeled by the function $E(x) = -191.73x^2 + 39{,}882.69x + 377{,}849.85$ million dollars. For both functions, x is the number of years past 1990.

   a. Use the functions I and E to write a model for the U.S. trade balance between 1990 and 1999.

   b. Estimate the year 2000 trade deficit.
   (Source: Office of Trade and Economic Analysis, U.S. Department of Commerce)

20. *Southwest Airlines* The number of employees of Southwest Airlines Co. between 1990 and 1998 can be modeled by the function $E(t) = 0.017t^2 + 2.164t + 0.861$ thousand employees, where t is the number of years after 1990. Southwest Airlines' revenue during the same time period is given by the model $R(E) = 0.165E - 0.226$ billion dollars when there are E thousand employees of the company.
   (Source: Hoover's Online Capsules)

   a. Find and interpret the meaning of the function $R(E(t))$.

   b. Use the result of part *a* to find $R(E(3))$. Interpret this result.

   c. How many employees worked for Southwest Airlines in 1997?

   d. What was the 1997 revenue for this airline?

21. *Prison Sentences* The mean time in prison, y, for certain crimes can be found as a function of the mean sentence length, x, using $f(x) = 0.554x - 2.886$, where x and y are measured in months.

   a. Find the inverse of this function.

   b. Interpret the inverse function from part (a).
   (Source: Index of Leading Cultural Indicators)

22. *Milk Cows* The number of cows and heifers kept for milk production in the U.S. between 1990 and 1997 can be approximated by the function $C(x) = -0.093x + 9.929$ million head, where x is the number of years after 1990.

   a. Find a formula for the inverse of the function C.

   b. Interpret your answer to part (a).

   c. If the inverse function is called C^{-1}, find $C^{-1}(C(8))$. Is the result what you expected? Explain.
   (Source: *Statistical Abstract of the U.S., 1998*)

23. *Internet Usage* The worldwide Internet usage from 1997 through 2002 (projected) is shown in the table below.
  a. Find the quadratic function that models this data, with *x* equal to the number of years after 1990.
  b. Graph the data and the function that models the data, for $6 \leq x \leq 12$.
  c. Does this model appear to be a good fit for the data?
  d. Use the result of part (a) to estimate when the number of Internet users is 550 billion.

Year	Internet Users (in billions)	Year	Internet Users (in billions)
1997	75	2000	255
1998	105	2001	340
1999	175	2002	490

(Source: *CyberAtlas*, 1999)

24. *Personal Income* The income received by persons from all sources minus their personal contributions for social security insurance is called *personal income*. The table below lists the personal income, in billions of dollars, of persons living in the U.S. for the indicated years.

Year	Personal Income (in billions of dollars)
1990	4802.4
1991	4981.6
1992	5277.2
1993	5519.2
1994	5791.8
1995	6150.8
1996	6495.2
1997	6873.9

(Source: *Statistical Abstract of the U.S., 1998*)

  a. Using an input equal to the number of years after 1990, find a quadratic function that models this data.
  b. Use your unrounded model to estimate when the personal income was $5500 billion.

c. If the pattern indicated by the model remains valid, when will the personal income be double its 1990 value?

25. *Resident Population* The resident population of the U.S. from 15 to 19 years of age and the percent of this resident population that are males is given for selected years in the table below. Fill in the blank column of the table and write the quadratic equation (to two decimal places) that models the 15- to 19-year-old male population as a function of the years from 1980. Include the description of this population and its units of measure.

Year	Age 15–19 Population (in thousands)	Male population (%)	Age 15–19 Male Population (in thousands)
1980	21,168	48.6	
1985	18,727	48.7	
1990	17,890	48.7	
1995	18,152	48.9	
1997	19,068	49.0	

(Source: U.S. Bureau of the Census, *Current Population Reports*)

26. *Insurance Premiums* The table below gives the annual premiums required for a $250,000 term life insurance policy on a 35-year-old female nonsmoker for different guaranteed term periods.
  a. Find a quadratic function that models the monthly premium as a function of the length of term for a 36-year-old female nonsmoking policyholder.
  b. Assuming that the domain of the function contains integer values between 10 years and 30 years, what term in years could a 35-year-old nonsmoking female purchase for $130 a month?

Term Period (years)	Monthly Premium for 35-Year-Old Female (dollars)
10	103
15	125
20	145
25	183
30	205

(Source: Quotesmith.com, December 1999)

27. *Computer Usage* The number of students per computer in U. S. public schools from the 1983–84 through the 1997–98 school years is shown in the table below.

School Year	Students per Computer
1983–84	125
1985–86	50
1987–88	32
1989–90	22
1991–92	18
1993–94	14
1995–96	10
1997–98	6.1

(Source: *CyberAtlas*, 1999)

a. Align the input data as the number of years after the beginning of the 1980–81 school year, and find a power function f to fit the data. Graph the data and the function $y = f(t)$ that models the data. Does this model appear to be a good fit?

b. If we assume that the function in part (a) is valid for all school years after the 1997–98 one, will the function in part (a) ever indicate that there will be one student per computer? Explain.

c. Another function that might be used to model this data is

$$C(t) = \frac{380}{t + 0.3} - 15 \text{ students per computer,}$$

where t is the number of years after the beginning of the 1980–81 school year. What is the basic function that can be transformed to obtain C? Describe the transformation.

d. Do you feel that the function $y = f(t)$ in part (a) or $y = C(t)$ better fits the data? Why?

e. There were 5.7 students per computer in 1998–99. Which function, $y = f(t)$ or $y = C(t)$, comes closer to the actual value? Does this result agree with your thoughts in part (d)?

28. *Hawaii Population* The data in the table below gives the 1990 through 1997 population of Hawaii for selected years.

Year	Population (in thousands)	Year	Population (in thousands)
1990	1108	1994	1173
1991	1131	1995	1179
1992	1150	1996	1183
1993	1160	1997	1187

(Source: *Statistical Abstract of the U.S.*, 1998)

a. Align the data with $x =$ the number of years after 1985 and find a power function to fit the aligned data.

b. If the pattern indicated by the model remains valid, estimate in what year Hawaii's population will rise to 1.3 million people.

29. *Hospital Utilization* The average length of stay in noninstitutional, short-stay hospitals (exclusive of federal hospitals and not counting newborn infants) for selected years is given in the table below. Align the years to be the number of years after 1980.

Year	Average Stay (days)	Year	Average Stay (days)
1980	7.3	1993	6.0
1985	6.5	1994	5.7
1990	6.4	1995	5.4
1991	6.4	1996	5.2
1992	6.2		

a. Fit a quadratic model to the 1980–1991 aligned data.

b. Fit a linear model to the 1992–1996 aligned data.

c. Combine the results of part (a) and part (b) to form a piecewise model for the 1980–1996 data. The function should be defined for all years between 1980 and 1991.

d. Use the piecewise-defined model to answer the following:

    i. Find and interpret the output in 1987.

    ii. When was the average stay 6.1 days?

    iii. What was the average stay in 1991?

30. *Consumer Price Index* Prices as measured by the U.S. Consumer Price Index have risen steadily since World War II. The data in the table gives the CPI for selected years between 1975 and 1995. The CPI in this table has 1967 as a reference year; that is, what cost $1 in 1967 cost about $1.61 in 1975 and $4.57 in 1995.

Year	CPI
1975	161.2
1980	248.8
1985	322.2
1990	391.4
1995	456.5

(Source: Congressional Budget Office, *Investor's Business Daily*, July 28, 1999)

Align the input data as the number of years after 1975, and report any models with two decimal places.

a. Find a linear model for the data. Discuss the fit to the data.

b. Find a quadratic model for the data. How good is the fit?

c. The CPI for 2000 was predicted to be 511.5. Evaluate the unrounded linear and quadratic functions at the input corresponding to 2000. Which does a better job of predicting this value?

d. Use the model you chose in part (c) to estimate when, before 2010, the CPI is 550.

31. *Profit* The monthly profit from producing and selling x units of a product is given by

$$P(x) = -0.01x^2 + 62x - 12{,}000.$$

Producing and selling how many units will result in profit for this product ($P(x) > 0$)?

32. *Unemployment* For the years 1983–1992, the unemployment rate (as a percent) in South Carolina can be modeled by $y = 0.144x^2 - 2.482x + 15.467$, where y is the percent and x is the number of years from 1980. Use graphical or numerical methods to find the years during which the model indicates that the unemployment rate is above 5.5 percent. (Source: *South Carolina Statistical Abstract*)

GROUP ACTIVITY/EXTENDED APPLICATION *1*

The following table gives the weekly revenue and cost, respectively, for a selected number of units of production and sale of a product by the Quest Manufacturing Company.

Number of Units	Revenue (in dollars)	Number of Units	Cost (in dollars)
100	6800	100	32,900
300	20,400	300	39,300
500	34,000	500	46,500
900	61,200	900	63,300
1400	95,200	1400	88,800
1800	122,400	1800	112,800
2500	170,000	2500	162,500

I. Provide the information requested, and answer the questions.

1. Use technology to determine the equations that model revenue and cost functions for this product, using x as the number of units produced and sold.

2. a. Combine the revenue and cost functions with the correct operation to create the profit function for this product.

b. Use the profit function to complete the following table:

x (number of units)	Profit, P(x) (dollars)
0	
100	
600	
1600	
2000	
2500	

3. Find the number of units of this product that must be produced and sold to break even.
4. Find the maximum possible profit and the number of units that gives the maximum profit.
5. **a.** Use operations with functions to create the average cost function for the product.
   b. Complete the following table.

x (number of units)	Average Cost, C(x) (dollars per unit)
1	
100	
300	
1400	
2000	
2500	

c. Graph this function using the viewing window [0, 2500] by [0, 400].
6. Graph the average cost function using the viewing window [0, 4000] by [0, 100] . Determine the number of units that should be produced to minimize the average cost and the minimum average cost.
7. Compare the number of units that produced the minimum average cost with the number of units that produced the maximum profit. Are they the same number of units? Discuss which of these values is more important to the manufacturer and why.

GROUP ACTIVITY/EXTENDED APPLICATION 2

Graphs displaying linear and nonlinear growth are frequently displayed in periodicals such as *Newsweek* and *Time*, in newspapers such as *USA Today* and the *Wall Street Journal*, and on numerous Web sites on the Internet. Tables of data can also be found in these sources, especially in federal and state government Web sites, such as www.census.gov.

Your mission is to find a company sales, stock price, biological growth, or sociological trend over a period of years that is nonlinear, to determine which type of nonlinear function is the best fit for the data, and to find a nonlinear function that is a model for this data.

A quadratic model will be a good fit for the data:

1. If the data is presented as a graph that resembles a parabola or part of a parabola.
2. If the data is presented in a table and the plot of the data points lie near some parabola or part of a parabola.
3. If the data is presented in a table and the second differences of the outputs are nearly constant for equally spaced inputs.

A power model may be a good fit if the shape of the graph or graph of the data points resembles part of a parabola, but a parabola is not a good fit for the data.

After you have created the model, you should test the goodness of fit of the model to the data and discuss uses that you could make of the model.

Your completed project should include:
a. A complete citation of the source of the data you are using.
b. An original copy or photocopy of the data being used.
c. A scatter plot of the data.
d. The equation that you have created.
e. A graph containing the scatter plot and the modeled equation.
f. A statement about how the model could be used to make estimations or predictions about the trend you are observing.
Some helpful hints:

1. If you decide to use a relation determined by a graph you have found, read the graph very carefully to determine the data points, or contact the source of the data to get the data from which the graph was drawn.
2. Align the independent variable by letting *x* represent the number of years from some convenient year, then enter the data into a graphing utility and create a scatter plot.
3. Use your graphing utility to create the equation of the function that is the best fit for the data. Graph this equation and the data points on the same axes to see if the equation is reasonable.

CHAPTER

3

Exponential and Logarithmic Functions

Many applications in life science, business, and economics involve rapid growth or decay, so they can be modeled by exponential equations. For example, growth of money at compound interest and growth of bacterial populations can be modeled by exponential growth functions, and radioactive decay and sales decay can be modeled by exponential decay functions. Logarithmic functions are used in many applications, including the pH of substances, the stellar magnitude system for measuring the brightness of stars, and the Richter scale for measuring the intensity of earthquakes.

The main concepts and skills discussed in this chapter, along with some of their applications, follow.

		TOPICS	APPLICATIONS
3.1	Exponential Functions	Graphing exponential functions; exponential growth and decay; transformations of graphs; spreadsheets; and the number e	Growth of paramecia, Head Start program growth, sales decay, carbon-14 dating, investments
3.2	Logarithmic Functions	Graphing and evaluating logarithmic functions; exponential and logarithmic forms; common and natural logarithms; change of base	Richter scale, doubling time for investments, mothers in the workforce
3.3	Solving Exponential Equations; Properties of Logarithms	Solving exponential equations; Logarithmic properties; solving logarithmic equations; exponential inequalities	Carbon-14 dating, measuring earthquake intensity, sales decay
3.4	Exponential and Logarithmic Models	Modeling with exponential functions, spreadsheets; constant rates of change; logarithmic models; comparison of models	Insurance premiums, sales decay, carbon-14 dating, smokers, inflation, mothers in the workforce, international visitors to the United States
3.5	Exponential Functions and Investing	Compound interest; continuous compounding; the number e; present value; investment models	Future value of an account, continuous compounding, present value of an investment, mutual fund growth
3.6	Annuities; Loan Repayment	Future value of an annuity; present value of an annuity; amortization	Future value of an ordinary annuity, present value of an ordinary annuity, home mortgage, loan repayment
3.7	Logistic and Gompertz Functions	Logistic growth functions; logistic decay functions; Gompertz functions	DVD players, expected life span, deer populations, company growth

ALGEBRA *Toolbox*

In this chapter, we discuss zero and negative exponents and properties of integer exponents. We also discuss scientific notation, which uses positive and negative powers of 10. To prepare us for the discussion of other financial functions, we also discuss the computation of simple interest.

Zero and Negative Exponents

Several applications in this chapter require the use of zero and negative exponents. To extend the rules for positive exponents to include the **zero exponent**, consider the following. For $a \neq 0$, we know that

$$\frac{a^n}{a^n} = 1 \quad \text{and} \quad \frac{a^n}{a^n} = a^{n-n}$$

so we define $a^{n-n} = a^0$ and $a^0 = 1$ for $a \neq 0$. (0^0 is undefined.)

To extend the rules to **negative exponents**, we know that for $a \neq 0$ and n a positive integer

$$\frac{a^m}{a^{m+n}} = \frac{1}{a^{m+n-m}} = \frac{1}{a^n} \quad \text{and} \quad \frac{a^m}{a^{m+n}} = a^{m-(m+n)}.$$

So we define

$$a^{m-(m+n)} = a^{-n} \quad \text{and} \quad a^{-n} = \frac{1}{a^n} \quad \text{for} \ a \neq 0.$$

Special cases of this definition are

$$a^{-1} = \frac{1}{a} \quad \text{for} \ a \neq 0 \quad \text{and} \quad \left(\frac{a}{b}\right)^{-n} = \frac{1}{\left(\frac{a}{b}\right)^n} = \left(\frac{b}{a}\right)^n.$$

These definitions are summarized below.

1. $a^0 = 1 \quad (a \neq 0)$ 2. $a^{-1} = \dfrac{1}{a} \quad (a \neq 0)$

3. $a^{-n} = \dfrac{1}{a^n} \quad (a \neq 0)$ 4. $\left(\dfrac{a}{b}\right)^{-n} = \left(\dfrac{b}{a}\right)^n \quad (a, b \neq 0)$

The rules of exponents that we discussed for integer exponents are also true for zero and negative integer exponents. In addition, these rules apply if the exponents are rational numbers.

EXAMPLE 1 Zero and Negative Exponents

Simplify the following expressions by removing all zero and negative exponents, for nonzero a, b, and c.

a. $(4c)^0$ **b.** $4c^0$ **c.** $(5b)^{-1}$ **d.** $5b^{-1}$ **e.** $(6a)^{-3}$ **f.** $6a^{-3}$

Solution

a. $(4c)^0 = 1$ **b.** $4c^0 = 4(1) = 4$ **c.** $(5b)^{-1} = \dfrac{1}{(5b)} = \dfrac{1}{5b}$

d. $5b^{-1} = 5 \cdot \dfrac{1}{b} = \dfrac{5}{b}$ **e.** $(6a)^{-3} = \dfrac{1}{(6a)^3} = \dfrac{1}{6^3 a^3} = \dfrac{1}{216a^3}$

f. $6a^{-3} = 6 \cdot \dfrac{1}{a^3} = \dfrac{6}{a^3}$ ∎

EXAMPLE 2 Applying the Rules for Exponents

If a, b, c and d are nonzero real numbers, write $\dfrac{5a^3b^{-2}}{(4c)^0 d^{-1}}$ without zero or negative exponents, and simplify the expression.

Solution

We write $\dfrac{5a^3b^{-2}}{(4c)^0 d^{-1}}$ without zero or negative exponents, and then simplify the expression, as follows:

$$\frac{5a^3b^{-2}}{(4c)^0 d^{-1}} = \frac{5a^3 \cdot \dfrac{1}{b^2}}{1 \cdot \dfrac{1}{d}} = \frac{\dfrac{5a^3}{b^2}}{\dfrac{1}{d}} = \frac{5a^3}{b^2} \cdot \frac{d}{1} = \frac{5a^3 d}{b^2}$$

∎

Multiplications and divisions with algebraic expressions involving zero and negative exponents are performed by using the properties of exponents.

EXAMPLE 3 Applying Exponent Rules to Products and Quotients

Compute the following products and quotients, and write the answer with positive exponents.

a. $(-2x^{-2}y)(5x^{-2}y^{-3})$ **b.** $\dfrac{8xy^{-2}}{2x^4 y^{-6}}$ **c.** $\dfrac{\dfrac{2x^{-1}y}{3a}}{\dfrac{6xy^{-2}}{5a}}$ **d.** $(-3x^{1/2})^3(2x^{-1/3})$

Solution

a. $(-2x^{-2}y)(5x^{-2}y^{-3}) = -10x^{-2+(-2)}y^{1+(-3)} = -10x^{-4}y^{-2}$

$$= -10 \cdot \frac{1}{x^4} \cdot \frac{1}{y^2} = \frac{-10}{x^4 y^2}$$

b. $\dfrac{8xy^{-2}}{2x^4y^{-6}} = 4x^{1-4}y^{-2-(-6)} = 4x^{-3}y^4 = 4 \cdot \dfrac{1}{x^3} \cdot y^4 = \dfrac{4y^4}{x^3}$

c. To perform this division, we invert and multiply.

$$\dfrac{\dfrac{2x^{-1}y}{3a}}{\dfrac{6xy^{-2}}{5a}} = \dfrac{2x^{-1}y}{3a} \cdot \dfrac{5a}{6xy^{-2}} = \dfrac{10ax^{-1}y}{18axy^{-2}} = \dfrac{5x^{-1-1}y^{1-(-2)}}{9} = \dfrac{5x^{-2}y^3}{9} = \dfrac{5y^3}{9x^2}$$

d. $(-3x^{1/2})^3(2x^{-1/3}) = \left((-3)^3x^{(1/2)(3)}\right)(2x^{-1/3}) = (-27x^{3/2})(2x^{-1/3}) = -54x^{7/6}$ ∎

Scientific Notation

A convenient way to write very large (positive or negative) numbers or numbers close to zero is to express them with exponents. This can be done by converting the numbers to **scientific notation**, which has the form

$$N \times 10^p \quad \text{where } 1 \le N < 10 \text{ and } p \text{ is an integer.}$$

For instance, the scientific notation form of 2,654,000 is 2.654×10^6.

Calculators often automatically express numbers by using scientific notation when the numbers are very large (positive or negative) numbers or when the numbers are close to zero. Most calculators display a number in scientific notation as a number N, the symbol E, and the power of 10. Figure 3.1 shows two numbers expressed in standard notation and in scientific notation, and the calculator display of scientific notation.

Standard Notation	Scientific Notation	Calculator Display of Scientific Notation
62256	6.2256×10^4	62256 6.2256E4
0.000235	2.35×10^{-4}	0.000235 2.35E -4

FIGURE 3.1

A number written in scientific notation can be converted to standard notation by multiplying (when the exponent on 10 is positive) or by dividing (when the exponent on 10 is negative.) For example, we convert 7.983×10^5 to standard notation by multiplying 7.983 by $10^5 = 100,000$:

$$7.983 \cdot 100,000 = 798,300$$

and we convert 4.563×10^{-7} to standard notation by dividing 4.563 by $10^7 = 10,000,000$:

$$4.563/10,000,000 = 0.0000004563.$$

Multiplying two numbers in scientific notation involves adding the powers of 10 and dividing them involves subtracting the powers.

EXAMPLE 4   Scientific Notation

Compute the following and write the answers in scientific notation.

a. $(7.983 \times 10^5)(4.563 \times 10^{-7})$ **b.** $(7.983 \cdot 10^5)/(4.563 \times 10^{-7})$

Solution

a. $(7.983 \times 10^5)(4.563 \times 10^{-7}) = 36.426429 \times 10^{5+(-7)}$
$= 36.426429 \times 10^{-2} = (3.6426429 \times 10^1) \times 10^{-2} = 3.6426429 \times 10^{-1}$

The calculation using technology is shown in Figure 3.2(a).

b. $(7.983 \times 10^5)/(4.563 \times 10^{-7}) = 1.749506903 \times 10^{5-(-7)}$
$= 1.749056903 \times 10^{12}$. Figure 3.2(b) shows the calculation using technology.

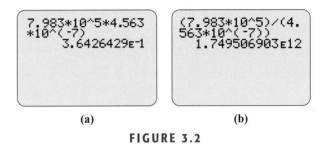

(a)   (b)

FIGURE 3.2   ∎

Simple Interest

The formula for the simple interest I earned on an investment of P dollars at rate r per year for t years is

$$I = Prt.$$

The sum of the original investment and the earned interest is called the **future value** of the investment. The formula for calculating the future value S of this investment is

$$S = P + I = P + Prt = P(1 + rt).$$

EXAMPLE 5   Simple Interest

a. Find the interest on an investment of $2000 earning interest at a rate of 10% per year for 3 years. Also find the future value of this investment.

b. Find the interest on an investment of $2000 at 12% per year for 3 months, and find the future value of this investment.

Solution

a. The interest on the investment of $2000 at 10% per year for 3 years is

$$I = Prt = (\$2000)(0.10)(3) = \$600.$$

The future value of this investment is

$$S = P + I = \$2000 + \$600 = \$2600.$$

b. The interest on an investment of $2000 at 12% per year for 3 months (which is $\frac{3}{12}$ of a year) is

$$I = Prt = (\$2000)(.12)\left(\frac{3}{12}\right) = \$60.$$

The future value of this investment is

$$S = P + I = \$2000 + \$60 = \$2060. \qquad \blacksquare$$

Toolbox EXERCISES

In Exercises 1–6, use the rules of exponents to simplify the following expressions and remove all zero and negative exponents. Assume that all variables are nonzero.

1. $x^{-4} \cdot x^{-3}$

2. $\dfrac{a^{-4}}{a^{-5}}$

3. $(c^{-6})^3$

4. $(x^{-1/2})(x^{2/3})$

5. $\left(\dfrac{2x^{-3}}{x^2}\right)^{-2}$

6. $(3a^{-3}b^2)(2a^2b^{-4})$

7. Simplify: **a.** $(a^3)^4$ **b.** $(x^{-2})^4$

8. Evaluate: **a.** $(3^2)^4$ **b.** $(2^{-1})^3$

9. Evaluate: **a.** 10^{5^0} **b.** 4^{2^2}

10. Simplify: **a.** $(x^2)^3$ **b.** x^{2^3}

Write the following numbers correct to 3 decimal places in scientific notation.

11. 3344563

12. 0.0012334

In Exercises 13 and 14, write each of the following numbers in standard form.

13. 4.372×10^5

14. 5.6294×10^{-4}

Multiply or divide, as indicated, and write the result correct to 3 decimal places in scientific notation.

15. $(6.25 \times 10^7)(5.933 \times 10^{-2})$

16. $\dfrac{2.961 \times 10^{-2}}{4.583 \times 10^{-4}}$

17. $2^5 \cdot 3^6$

18. $\dfrac{2^5}{3^{12}}$

In Exercises 19–21, compute the simple interest and future value of each of the following investments.

19. $2000 invested for 5 years at 6% per year.

20. $3500 invested for 6 months at 1% per month.

21. $5600 invested for 6 months at 6% per year.

3.1 Exponential Functions

The primitive single-cell animal called a paramecium reproduces by splitting into two pieces (called binary fission) so that the population doubles each time there is a split. If we assume that the population begins with one paramecium and doubles each hour, then there will be 2 paramecia after 1 hour, 4 paramecia after 2 hours, 8 after 3 hours, and so on. If we let y represent the number of paramecia in the population after x hours have passed, the points (x, y) that satisfy this function for the first 9 hours of growth are described by the data in Table 3.1 and the plot in Figure 3.3a. We can show that each of these points and the data points for $x = 10$, 11, and so forth lie on the graph of the function $y = 2^x$ with $x \geq 0$. (See Figure 3.3(b).)

TABLE 3.1

x (hours)	0	1	2	3	4	5	6	7	8	9
y (paramecia)	1	2	4	8	16	32	64	128	256	512

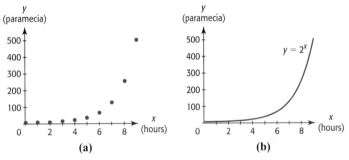

FIGURE 3.3

Note that not every point on the graph of $y = 2^x$ in Figure 3.3b describes the number of paramecia. For example, the point $(7.1, 2^{7.1})$ is on this graph, but $2^{7.1} \approx 137.187$ does not represent a number of paramecia because it is not a whole number. However, every point describing the number of paramecia is on the graph of $y = 2^x$, so the function $y = 2^x$ with $x \geq 0$ is a continuous model for this discrete growth function. Functions like $y = 2^x$, which have a constant base raised to a variable power, are called **exponential functions**. We discuss exponential functions and their applications in this section.

Exponential Functions

The function that models the growth of paramecia was restricted to $x \geq 0$ because of the physical setting, but if the domain of the mathematical function $y = 2^x$ is not restricted, it is the set of real numbers. Some values satisfying this function are shown in Table 3.2, and the graph of the function is shown in Figure 3.4. Note that the range of this function is the set of all positive real numbers because there is no power of 2 that results in a value of 0 or a negative number.

TABLE 3.2

x	y
-3	$2^{-3} = 0.125$
-1.5	$2^{-1.5} \approx 0.35$
-0.5	$2^{-0.5} \approx 0.71$
0	$2^0 = 1$
0.5	$2^{0.5} \approx 1.41$
1.5	$2^{1.5} \approx 2.83$
3	$2^3 = 8$
5	$2^5 = 32$

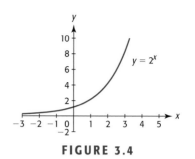

FIGURE 3.4

The graph in Figure 3.4 approaches but never touches the x-axis as x becomes more negative (approaching $-\infty$), so the x-axis is a **horizontal asymptote** for the graph. As previously mentioned, the function $y = 2^x$ is an example of a special class of functions called **exponential functions**. In general, we define an exponential function as follows.

EXPONENTIAL FUNCTION

If b is a positive real number, $b \neq 1$, then the function $f(x) = b^x$ is an exponential function. The constant b is called the *base* of the function and the variable x is the *exponent*.

If we graph another exponential function $y = b^x$ that has a base b greater than 1, the graph will have the same basic shape as the graph of $y = 2^x$. (See the graphs of $y = 1.5^x$ and $y = 12^x$ in Figure 3.5.)

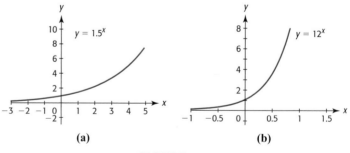

(a) (b)

FIGURE 3.5

Transformations of Graphs of Exponential Functions

Exponential functions, like other types of functions, can be shifted, reflected, or stretched. The concept of transformations introduced in Chapter 2 can be applied to graphs of exponential functions.

EXAMPLE 1   Transformations of Graphs of Exponential Functions

a. Explain how the graph of $y = 2 + 3^{x-4}$ compares to the graph of $y = 3^x$ and sketch the graphs.

b. Compare the graph of $y = 5(3^x)$ with the graph of $y = 3^x$.

Solution

a. The graph of $y = 2 + 3^{x-4}$ has the same shape as $y = 3^x$, but it is shifted 4 units to the right and 2 units up. The graph of $y = 3^x$ is shown in Figure 3.6(a) and the graph of $y = 2 + 3^{x-4}$ is shown in Figure 3.6(b).

b. As we saw in Chapter 2, multiplication of a function by a constant greater than 1 stretches the graph of the function. Each of the y-values of $y = 5(3^x)$ is 5 times the corresponding y-value of $y = 3^x$. The graph of $y = 5(3^x)$ is shown in Figure 3.6(c).

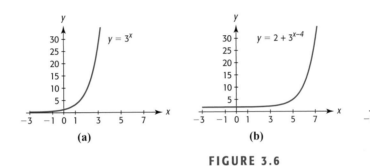

(a)   (b)   (c)

FIGURE 3.6   ■

Exponential Growth

Exponential functions can be used to model growth in many different applications. Whenever the base b of $y = a(b^x)$ is greater than 1 and $a > 0$, the exponential function is increasing and can be used to model growth.

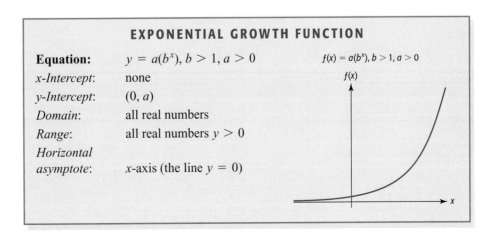

EXPONENTIAL GROWTH FUNCTION

Equation:	$y = a(b^x), b > 1, a > 0$
x-Intercept:	none
y-Intercept:	$(0, a)$
Domain:	all real numbers
Range:	all real numbers $y > 0$
Horizontal asymptote:	*x*-axis (the line $y = 0$)

$f(x) = a(b^x), b > 1, a > 0$

$f(x)$

EXAMPLE 2 ## Head Start Program Growth

Federal funds for the Head Start Program in the U.S. Department of Heath and Human Services increased dramatically between 1975 and 2000. The on-budget (federal) funds that were allocated for Head Start during this time period can be described by the model

$$f(x) = 400(1.107)^x \text{ million dollars,}$$

where x is the number of years after 1975.

a. Graph this function. Find and interpret the vertical-axis intercept.

b. According to the model, how much was allocated for Head Start in 2000?

(Source: *Federal Support for Education, Fiscal Years 1980–2000*, National Center for Educational Statistics, U.S. Department of Education, September 2000.)

Solution

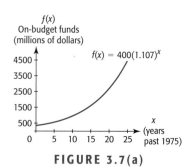

FIGURE 3.7(a)

a. The graph is shown in Figure 3.7(a).

The y-intercept is found by evaluating the function at $x = 0$, obtaining

$$f(0) = 400(1.107)^0 = 400.$$

Because our data is aligned so that $x = 0$ at the end of 1975, one possible interpretation is this: $400 million in federal funds were allocated for the Head Start Program in 1975.

b. The year 2000 is 25 years after 1975, so we find the allocation for 2000 by evaluating $f(25)$:

$$f(25) = 400(1.107)^{25} \approx 5078.68 \text{ million dollars,}$$

so federal funds allocated for Head Start in 2000 amounted to about $5.079 billion. Note that we could find the value of the function at $x = 25$ graphically or numerically by using the table feature of a graphing utility.   ■

Spreadsheet Solution

As with linear and quadratic functions, computer software of several types can be used to create exponential graphs, and spreadsheets like Excel can also be used. Figure 3.7(b) shows the output of $y = 400(1.107)^x$ at several values of x and the graph of this function on Excel.

	A	B
1	0	400
2	5	664.9639
3	10	1105.443
4	15	1837.699
5	20	3055.008
6	25	5078.676

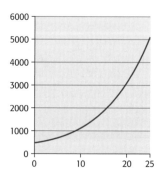

FIGURE 3.7(b)

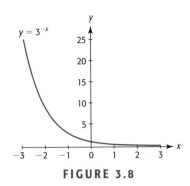

$y = 3^{-x}$

FIGURE 3.8

Exponential Decay

Replacing x with $-x$ in $y = f(x)$ gives a function whose graph is a **reflection** of the graph of $y = f(x)$ about the y-axis. For example, the graph of $y = 3^{-x}$ is the reflection of the graph of $y = 3^x$, and the function $y = 3^{-x}$ decreases rather than increases. (See Figure 3.8.)

We can also write the function $y = 3^{-x}$ in the form $y = \left(\dfrac{1}{3}\right)^x$ because

$$3^{-x} = (3^{-1})^x = \left(\dfrac{1}{3}\right)^x.$$

In general, a function of the form $f(x) = a(b^{-x})$ with $b > 1$ can also be written in the form $f(x) = a(c^x)$ with $0 < c < 1$ and $c = 1/b$. Functions with equations of this form can be used to model **exponential decay**.

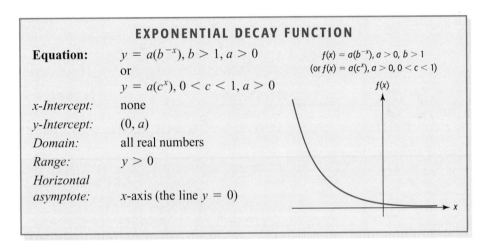

EXPONENTIAL DECAY FUNCTION

Equation:	$y = a(b^{-x}), b > 1, a > 0$
	or
	$y = a(c^x), 0 < c < 1, a > 0$
x-Intercept:	none
y-Intercept:	$(0, a)$
Domain:	all real numbers
Range:	$y > 0$
Horizontal asymptote:	x-axis (the line $y = 0$)

$f(x) = a(b^{-x}), a > 0, b > 1$
(or $f(x) = a(c^x), a > 0, 0 < c < 1$)

EXAMPLE 3   Sales Decay

It pays to advertise, and it is frequently true that weekly sales will drop rapidly for certain products after an advertising campaign ends. This decline in sales is called *sales decay*. Suppose that the decay in the sales of a product is given by

$$S = 1000(2^{-0.5x}) \text{ dollars,}$$

where x is the number of weeks after the end of a sales campaign. Use this function to answer the following.

a. What is the level of sales when the advertising campaign ends?

b. What is the level of sales 1 week after the end of the campaign?

c. Use a graph of the function to estimate the week in which sales equal $500.

d. According to this model, will sales ever fall to zero?

Solution

a. The campaign ends when $x = 0$, so $S = 1000(2^{-0.5(0)}) = 1000(2^0) = 1000(1) = \$1000$.

b. At 1 week after the end of the campaign, $x = 1$. Thus, $S = 1000(2^{-0.5(1)}) = \$707.11$.

c. The graph of sales decay is shown in Figure 3.9(a). One way to find the x-value for which $S = 500$ is to graph $y_1 = 1000(2^{-0.5x})$ and $y_2 = 500$, and find the point of intersection of the two graphs. See Figure 3.9(b), which shows that $y = 500$ when $x = 2$.

Thus, sales fall to half their original amount after about 2 weeks.

d. The graph of this sales decay function approaches the positive x-axis as x gets large, but it never reaches the x-axis. Thus sales will never reach a value of $0.

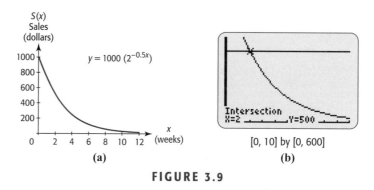

(a)

$[0, 10]$ by $[0, 600]$

(b)

FIGURE 3.9

The equations of growth and decay functions are related as follows.

GROWTH AND DECAY

For initial amount $y_0 > 0$, the equation $y = y_0 b^{kx}$ for $b > 1$ defines a growth function if $k > 0$ and a decay function if $k < 0$.

The Number *e*

Many real applications involve exponential functions with the base e. The number e is an irrational number with decimal approximation 2.7182818 (to seven decimal places).

$$e \approx 2.7182818$$

The exponential function with base e occurs frequently in biology and in finance. We discuss the derivation of e and how it is used in finance in Section 3.5. Because e is close in value to 3, the graph of the exponential function $y = e^x$ has the same basic shape as $y = 3^x$. (See Figure 3.10, which compares the graphs of $y = e^x$, $y = 3^x$, and $y = 2^x$.)

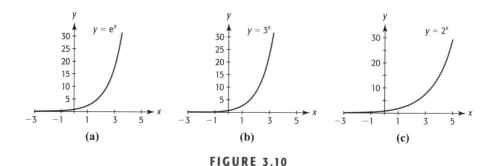

(a)

(b)

(c)

FIGURE 3.10

EXAMPLE 4   Comparing Exponential Functions

Compare the functions $f(x) = 5e^{2x}$ and $g(x) = 5e^{-2x}$ algebraically and graphically.

Solution

To compare these functions algebraically, we see that the base in each function is e, and that the function g can be created by replacing x with $-x$ in $f(x) = 5e^{2x}$. Thus the graph of $g(x) = 5e^{-2x}$ is a reflection of the graph of $f(x) = 5e^{2x}$ about the y-axis. The function $f(x) = 5e^{2x}$ is an exponential growth function and the function $g(x) = 5e^{-2x}$ is an exponential decay function.

The graphs of $y = f(x)$ in Figure 3.11(a) and $y = g(x)$ in Figure 3.11(b) confirm the results of the algebraic investigation. In addition, both graphs have $(0, 5)$ as the y-intercept and the x-axis as a horizontal asymptote.

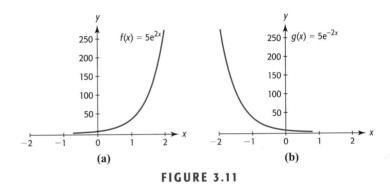

(a)   (b)

FIGURE 3.11   ■

Note in Example 4 that the exponent of $f(x) = 5e^{2x}$ is $2x$. To see the effect of the 2 in this exponent, we can use Property 4 of exponents* to rewrite $f(x) = 5e^{2x}$ as $f(x) = 5(e^x)^2$, so multiplying the exponent by 2 has the effect of squaring the exponential e^x. In general,

$$y = e^{rx}$$

will increase more rapidly than $y = e^x$ if $r > 1$, will increase more slowly than $y = e^x$ if $0 < r < 1$, and will decrease if $r < 0$.

An exponential decay function can be used to model the amount of a radioactive material that remains after a period of time. Carbon-14 decays over time, with the amount remaining after t years given by

$$f(t) = y_0 e^{-0.00012378t}.$$

where y_0 is the original amount. This formula is used by archeologists to date fossils and artifacts from ancient civilizations.

EXAMPLE 5   Carbon-14 Dating

a. If the original amount of carbon-14 present in an artifact is 100 grams, how much remains after 2000 years?

b. What percent of the original amount of carbon-14 remains after 4500 years?

---

*See the Algebra Toolbox at the beginning of Chapter 3.

Solution

a. The initial amount y_0 is 100 grams and the time t is 2000 years, so we substitute these values in $f(t) = y_0 e^{-0.00012378t}$ and calculate:

$$f(2000) = 100e^{-0.00012378(2000)} \approx 78.07$$

The amount of carbon-14 present after 2000 years is approximately 78 grams.

b. If y_0 is the original amount, the amount that will remain after 4500 years is

$$f(4500) = y_0 e^{-0.00012378(4500)} = 0.57y_0.$$

Thus approximately 57% of the original amount will remain after 4500 years. (Note that you do not need to know the original amount of carbon-14 in order to answer this question.) ∎

3.1 SKILLS CHECK

1. Which of the following functions are exponential functions?
   a. $y = 3x + 3$ **b.** $y = x^3$ **c.** $y = 3^x$
   d. $y = x^2 - 3x$ **e.** $y = e^{x^2 - 3x}$ **f.** $y = x^e$

2. **a.** Graph the function $f(x) = 5^x$ on the window $[-5, 5]$ by $[-60, 100]$.
   b. Find $f(1)$, $f(3)$, and $f(-2)$.
   c. What is the horizontal asymptote of the graph?
   d. What is the y-intercept?

3. **a.** Graph the function $f(x) = e^x$ on $[0, 5]$ by $[-2, 100]$.
   b. Find $f(1)$, $f(-1)$, and $f(4)$, rounded to three decimal places.
   c. What is the horizontal asymptote of the graph?
   d. What is the y-intercept?

4. Graph the function $y = 3^x$ on $[0, 5]$ by $[-2, 100]$.

5. Graph the function $y = 3^{(x-2)} - 4$ on $[0, 7]$ by $[-6, 100]$.

6. Compare the graphs of the functions in Exercises 4 and 5.

In Exercises 7–12, use your knowledge of transformations to compare the graph of the function with the graph of $f(x) = 4^x$. *Then graph each function for* $-4 \le x \le 4$.

7. $y = 4^x + 2$ 8. $y = 4^{x-1}$

9. $y = 4^{-x}$ 10. $y = -4^x$

11. $y = 3(4^x)$ 12. $y = 3 \cdot 4^{(x-2)} - 3$

13. How does the graph of Exercise 12 compare with the graph of Exercise 11?

14. Which of the functions in Exercises 7–12 are increasing functions and which are decreasing functions?

15. **a.** Graph the function $f(x) = 12e^{-0.2x}$ for $-5 \le x \le 15$.
    b. Find $f(10)$ and $f(-10)$.
    c. Does this function represent growth or decay?

16. If $y = 200(2^{-0.01x})$,
    a. find the value of y (to two decimal places) when $x = 20$.
    b. Use graphical or numerical methods to determine the value of x that gives $y = 100$.

3.1 EXERCISES

1. *Sales Decay* At the end of an advertising campaign, weekly sales declined according to the equation $y = 2000(2^{-0.1x})$ dollars, where x is the number of weeks after the end of the campaign.
   a. Determine the sales at the end of the ad campaign.
   b. Determine the sales 6 weeks after the end of the campaign.

   c. Does this model indicate that sales will eventually reach $0?

2. *Sales Decay* At the end of an advertising campaign, weekly sales declined according to the equation $y = 40,000(3^{-0.1x})$ dollars, where x is the number of weeks after the campaign ended.

a. Determine the sales at the end of the ad campaign.
b. Determine the sales 8 weeks after the end of the campaign.
c. How do we know, by inspecting the equation, that this function is decreasing?

3. *Sales Decay* At the end of an advertising campaign, weekly sales at an electronics store declined according to the equation $y = 2000(2^{-0.1x})$ dollars, where x is the number of weeks after the end of the campaign.
a. Graph this function for $0 \le x \le 60$.
b. Use the graph to find the weekly sales 10 weeks after the campaign ended.
c. Comment on "It pays to advertise" for this store.

4. *Sales Decay* At the end of an advertising campaign, weekly retail sales of a product declined according to the equation $y = 40,000(3^{-0.1x})$ dollars, where x is the number of weeks after the campaign ended.
a. Graph this function for $0 \le x \le 50$.
b. Find the weekly sales 10 weeks after the campaign ended.
c. Should the retailers consider another advertising campaign even if it costs $5000?

5. *Purchasing Power* The purchasing power (real value of money) decreases if inflation is present in the economy. For example, the purchasing power of R after t years of 5% inflation is given by the model
$$P = R(0.95^t) \text{ dollars}$$
a. What will be the purchasing power of $40,000 after 20 years of 5% inflation?
b. How does this impact people planning to retire at age 50?
(Source: *Viewpoints*, VALIC, 1993)

6. *Purchasing Power* If a retired couple has a fixed income of $60,000 per year, the purchasing power (adjusted value of the income) after t years of 5% inflation is given by the equation $P = 60,000(0.95^t)$ dollars.
a. What is the purchasing power of their income after 4 years?
b. Test numerical values in this function to determine the years when their purchasing power will be less than $30,000.

7. *Real Estate Inflation* During a 5-year period of constant inflation, the value of a $100,000 property will increase according to the equation $v = 100,000e^{0.05t}$.
a. What will be the value of this property in 4 years?
b. Use a table or graph to estimate when this property will double in value.

8. *Inflation* An antique table increases in value according to the function $v(x) = 850(1.04^x)$ dollars, where x is the number of years past 1990.
a. How much was the table worth in 1990?
b. If the pattern indicated by the function remains valid, what will be the value of the table in 2005?
c. Use a table or graph to estimate the year when this table will reach double its 1990 value.

9. *Radioactive Decay* The amount of radioactive isotope thorium-234 present at time t is given by $A(t) = 500e^{-0.02828t}$ grams, where t is the time in years that the isotope decays. The initial amount present is 500 grams.
a. How many grams remain after 10 years?
b. Graph this function for $0 \le t \le 100$.
c. If the half-life is the time it takes for half of the initial amount to decay, use graphical methods to estimate the half-life of this isotope.

10. *Radioactive Decay* A breeder reactor converts stable uranium-238 into the isotope plutonium-239. The decay of this isotope is given by $A(t) = 100e^{-0.00002876t}$, where $A(t)$ is the amount of the isotope at time t (in years) and 100 is the original amount.
a. How many grams remain after 100 years?
b. Graph this function for $0 \le t \le 50,000$.
c. The half-life is the time it takes for half of the initial amount to decay; use graphical methods to estimate the half-life of this isotope.

11. *Carbon-14 Dating* An exponential decay function can be used to model the number of atoms of a radioactive material that remain after a period of time. Carbon-14 decays over time, with the amount remaining after t years given by
$$y = 100e^{-0.00012378t}$$
if 100 grams is the original amount.
a. How much remains after 1000 years?
b. Use graphical methods to estimate the number of years until 10 grams of carbon-14 remains.

12. *Drugs in the Bloodstream* If a drug is injected into the bloodstream, the percent of the maximum dosage that is present at time t is given by $y = 100(1 - e^{-0.35(10-t)})$, where t is in hours, with $0 \le t \le 10$.
a. What percent of the drug is present after 2 hours?
b. Graph this function.
c. When is the drug totally gone from the bloodstream?

13. *Population* The population in a certain city was 53,000 in 2000, and its future size is predicted to be

$P(t) = 53,000e^{0.015t}$ people, where t is the number of years after 2000.

a. Does this model indicate that the population is increasing or decreasing?

b. Use this function to predict the population of the city in 2005.

c. Use this function to predict the population of the city in 2010.

d. What is the average rate of growth between 2000 and 2010?

14. *Population* The population in a certain city was 800,000 in 2003 and its future size is predicted to be $P = 800,000e^{-0.020t}$ people, where t is the number of years after 2003.

a. Does this model indicate that the population is increasing or decreasing?

b. Use this model to predict the population of the city in 2010.

c. Use this model to predict the population of the city in 2020.

d. What is the average rate of change in population between 2003 and 2010?

15. *Normal Curve* The "curve" on which many students like to be graded is the bell-shaped normal curve. The equation $y = \dfrac{1}{\sqrt{2\pi}}e^{-(x-50)^2/2}$ describes the normal curve for a standardized test, where x is the test score before curving.

a. Graph this function for x between 47 and 53 and for y between 0 and 0.5.

b. The average score for the test is the score that gives the largest output y. Use the graph to find the average score.

16. *I.Q. Measure* An I.Q. measure follows a bell-shaped normal curve with equation $y = \dfrac{1}{\sqrt{2\pi}}e^{-(x-100)^2/20}$. Graph this equation for x between 80 and 120 and for y between 0 and 0.4.

3.2 Logarithmic Functions

Key Concepts

- Logarithmic functions
  Base
  Logarithm
  Exponent
- Logarithmic versus exponential form
- Common logarithms
- Natural logarithms
  Doubling time
- Change of base formula

Suppose that $2500 is invested in an account earning 10% annual interest compounded continuously. How long will it take for the amount to grow to $5000 (that is, to double its value)? To find the time t in which this investment will double, we can solve

$$2 = e^{0.10t}.$$

We can solve this equation graphically as we did in the last section. (See Figure 3.12.) In this section, we learn how to rewrite this equation in a new form, called the **logarithmic form** of the equation, in order to find the value of t analytically. We will introduce logarithmic functions, discuss the relationship between exponential and logarithmic functions, and apply logarithmic functions in real situations.

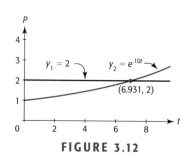

FIGURE 3.12

Logarithmic Functions

Recall that we discussed inverse functions in Section 2.6. The inverse of a function is a second function that "undoes" what the original function does. For example, if the original function doubles each input, its inverse takes half of each input. If a function is defined by $y = f(x)$, the inverse of this function can be found by interchanging x and y in $y = f(x)$ and solving the new equation for y. As we saw in Section 2.6, a function has an inverse if it is a **one-to-one function** (that is, there is a one-to-one correspondence between the independent and dependent variables defining the function).

Every exponential function of the form $y = b^x$ is one-to-one, so every exponential function of this form has an inverse function. The inverse function of $y = b^x$ is found by interchanging x and y and then solving the new equation for y. Interchanging x and y in $y = b^x$ gives $x = b^y$, and the function that results from solving this equation for y is denoted as

$$y = \log_b x.$$

This inverse function is called a **logarithmic function** with base b.

LOGARITHMIC FUNCTION

For $x > 0$, $b > 0$, and $b \neq 1$, the logarithmic function to the base b is

$$y = \log_b x$$

which is defined by $x = b^y$.

This function is the inverse function of the exponential function $y = b^x$.

Note that, according to the definition of the logarithmic function, $y = \log_b x$ and $x = b^y$ are two different forms of the same equation. We call $y = \log_b x$ the *logarithmic form* of the equation and $x = b^y$ the *exponential form* of the equation. The number b is called the **base** in both $y = \log_b x$ and $x = b^y$, and y is the **logarithm** in $y = \log_b x$ and the **exponent** in $x = b^y$. Thus, a logarithm is an exponent.

EXAMPLE 1   Exponential and Logarithmic Forms

a. Write the equation $y = \log_3 x$ in exponential form.

b. Write $x = 3^y$ in logarithmic form.

Solution

a. The base of $y = \log_3 x$ is 3, and y, which equals the logarithm, is the exponent. Thus the exponential form is $x = 3^y$.

b. To write $x = 3^y$ in logarithmic form, note that the base of the logarithm will be 3, the same as the base of the exponential expression. Also note that y is the exponent in $x = 3^y$, so y is the value of the logarithm. That is, the exponential form of $x = 3^y$ is $y = \log_3 x$. ∎

Table 3.3 shows some logarithmic equations and their equivalent exponential forms.

TABLE 3.3

Logarithmic Form	Exponential Form
$\log_2 x = y$	$2^y = x$
$\log_{10} 100 = 2$	$10^2 = 100$
$\log_a 1 = 0 \ (a > 0)$	$a^0 = 1$
$\log_a a = 1 \ (a > 0)$	$a^1 = a$

We can sometimes more easily evaluate logarithmic functions for different inputs by changing the function from logarithmic form to exponential form.

EXAMPLE 2 Evaluating Logarithms

a. Write $y = \log_2 x$ in exponential form.

b. Use the exponential form from part (a) to find y when $x = 8$.

c. Evaluate $\log_2 8$.

Solution

a. To rewrite $y = \log_2 x$ in exponential form, note that the base of the logarithm, 2, becomes the base of the exponential expression; and y equals the exponent. That is, the exponential form is

$$2^y = x.$$

b. To find y when x is 8, we solve $2^y = 8$. We can easily see that $y = 3$ because $2^3 = 8$.

c. To evaluate $\log_2 8$, we write $\log_2 8 = y$, which is equivalent to $2^y = 8$ in exponential form. From part (b), $y = 3$. Thus $\log_2 8 = 3$. ∎

Figure 3.13 shows a graph of the logarithmic function $y = \log_b x$ for the base $b > 0$.

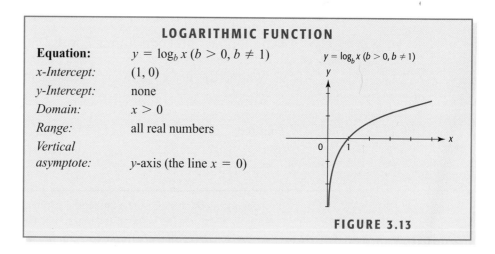

LOGARITHMIC FUNCTION

Equation:	$y = \log_b x \ (b > 0, b \neq 1)$
x-Intercept:	$(1, 0)$
y-Intercept:	none
Domain:	$x > 0$
Range:	all real numbers
Vertical asymptote:	y-axis (the line $x = 0$)

$y = \log_b x \ (b > 0, b \neq 1)$

FIGURE 3.13

Recall that exponential and logarithmic functions are inverse functions. This relationship is shown in the following example.

EXAMPLE 3 Logarithmic and Exponential Functions

a. Write the inverse of $y = 2^x$ in logarithmic form.

b. Use the exponential form of the inverse function from part (a) to create a table of values for x and y.

c. Use the table of values to graph this inverse function.

d. Graph $y = 2^x$ and its inverse on the same set of axes.

Solution

a. We find the inverse of the function $y = 2^x$ as follows:

$$x = 2^y \qquad \text{Interchanging } x \text{ and } y$$

$$y = \log_2 x \qquad \text{Writing in logarithmic form}$$

b. The exponential form of the inverse function is $x = 2^y$. Each entry in the table of values (see Table 3.4) can be found by entering a value for y in $x = 2^y$ and evaluating the function to obtain x.

TABLE 3.4

y	-2	-1	0	1	2	3
x	$2^{-2} = \dfrac{1}{4}$	$2^{-1} = \dfrac{1}{2}$	$2^0 = 1$	$2^1 = 2$	$2^2 = 4$	$2^3 = 8$

c. Plotting the points (x, y) and connecting them gives the graph of $y = \log_2 x$, shown in Figure 3.14(a).

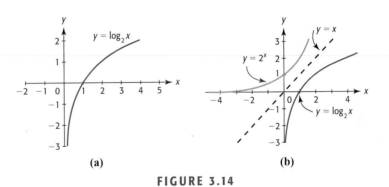

(a)   (b)

FIGURE 3.14

d. The graphs $y = 2^x$ and its inverse $y = \log_2 x$ are shown in Figure 3.14(b). Because the logarithmic function $y = \log_2 x$ is the inverse of the exponential function $y = 2^x$, the graphs of these functions are symmetric about the graph of the line $y = x$. Note that the graph of the exponential function has the x-axis as a horizontal asymptote and that the graph of the logarithmic function has the y-axis as a vertical asymptote. ■

Common Logarithms

We can use technology to evaluate logarithmic functions if the base of the logarithm is 10 or e. Because our number system uses base 10, logarithms with a base of 10 are called **common logarithms**, and, by convention, $\log_{10} x$ is simply denoted $\log x$. Graphing utilities and spreadsheets can be used to evaluate $f(x) = \log x$, using the key or command $\boxed{\log\ x}$.

EXAMPLE 4 Evaluating Base 10 Logarithms

Find $f(10,000)$ if $f(x) = \log x$ using each of the indicated methods.

a. Write the equation $y = \log x$ in exponential form and find y when $x = 10,000$.

b. Use technology to evaluate log 10,000.

Solution

a. The exponential form of $y = \log x = \log_{10} x$ is $10^y = x$. Substituting 10,000 for x in this equation gives $10^y = 10,000$. Because $10^4 = 10,000$, we see that $y = 4$. Thus $f(10,000) = 4$.

b. The value of log 10,000 is shown to be 4 on the calculator screen in Figure 3.15, so $f(10,000) = 4$.

FIGURE 3.15   ∎

Graphing utilities and spreadsheets can also be used to graph $y = \log x$. Figure 3.16 shows the graph of $y = \log x$ on a graphing calculator and on a spreadsheet. The graph of a logarithm with any base can be created with a spreadsheet such as Excel by entering the formula $\boxed{= \log\ (x,\ \text{base})}$.

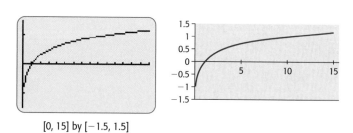

[0, 15] by [−1.5, 1.5]

FIGURE 3.16

EXAMPLE 5 Richter Scale

Because the intensities of earthquakes are so large, the Richter scale was developed to provide a "scaled down" measuring system that permits earthquakes to be more easily measured and compared. The Richter scale gives the magnitude R of an earthquake using the formula

$$R = \log\left(\frac{I}{I_0}\right),$$

where I is the intensity of the earthquake and I_0 is a certain minimum intensity used for comparison.

a. If an earthquake has an intensity of 10,000 times I_0, what is the magnitude of the earthquake?

b. An earthquake that measured 6.5 on the Richter scale occurred in Pakistan in February 1991. Express this intensity in terms of I_0.

c. How many times more is the intensity of the Pakistani earthquake than one with a Richter scale measurement of 5.5?

Solution

a. If the intensity of the earthquake is $10,000I_0$, we substitute $10,000I_0$ for I in the equation $R = \log\left(\dfrac{I}{I_0}\right)$, getting $R = \log\dfrac{10,000I_0}{I^0} = \log 10,000 = 4$. Thus, the magnitude of the earthquake is 4.

b. Substituting 6.5 for R in the equation $R = \log\left(\dfrac{I}{I_0}\right)$, gives $6.5 = \log\dfrac{I}{I_0}$. Changing this equation to exponential form gives the equivalent equation

$$10^{6.5} = \frac{I}{I_0}, \text{ so } 3,162,280 \approx \frac{I}{I_0}, \text{ or } I \approx 3,162,280I_0.$$

Thus, the Pakistani earthquake had an intensity of approximately 3,162,280 times I_0 (rounded to six significant digits).

c. To find the intensity of an earthquake with Richter scale measurement of 5.5, we substitute 5.5 for R in the equation, which gives

$$5.5 = \log\frac{I}{I_0}.$$

Changing this equation to exponential form gives the equivalent equation

$$10^{5.5} = \frac{I}{I_0}, \text{ so } 316,228 \approx \frac{I}{I_0}, \text{ or } I \approx 316,228I_0.$$

Thus, the earthquake had an intensity of approximately 316,228 times I_0 (rounded to six significant digits).

The ratio of the intensities of these two earthquakes is $\dfrac{3,162,280I_0}{316,228I_0} = \dfrac{10}{1}$. ∎

Note that in Example 5, the Richter scale measurement of the Pakistani earthquake is 1 larger than the one with Richter scale measurement of 5.5, and that the ratio of the intensities is 10:1. Thus the Pakistani earthquake is 10 times more intense. In general, we have the following.

> If the difference of the Richter scale measurements of two earthquakes is the positive number d, the intensity of the larger earthquake is 10^d times more than that of the smaller earthquake.

Natural Logarithms

The logarithm with base e is $y = \log_e x$. It is defined by the exponential equation $x = e^y$. Because logarithms with base e are used frequently, especially in science, they are called **natural logarithms**. By convention, we use the special notation $\ln x$ to denote $\log_e x$. Most graphing utilities and spreadsheets have the key or command $\boxed{\ln\ x}$ to use in evaluating or graphing functions involving natural logarithms.

We have previously used graphical methods to find the time it would take for an investment to double. We can now develop a formula to find the *doubling time* when the interest is compounded continuously.

If P dollars are invested for t years at an annual interest rate r, compounded continuously (at every instant), then the investment will grow to a future value S given by the exponential function

$$S = Pe^{rt}$$

and that the investment will be doubled when $S = 2P$, giving

$$2P = Pe^{rt} \quad \text{or} \quad 2 = e^{rt}.$$

Converting the equation to its equivalent logarithmic form gives $\log_e 2 = rt$ and solving this equation for t gives the **doubling time**, $t = \dfrac{\ln 2}{r}$.

DOUBLING TIME

The time it takes for an investment to grow to double its value is

$$t = \frac{\ln 2}{r}$$

if the interest rate is r, compounded continuously.

| EXAMPLE 6 | Time for an Investment to Double |

Suppose that $2500 is invested in an account earning 6% annual interest compounded continuously. How long will it take for the amount to grow to $5000?

Solution

We seek the time it takes to double the investment, and the doubling time is given by

$$t = \frac{\ln 2}{.06} \approx 11.5525.$$

Thus it will take approximately 11.6 years for an investment of $2500 to grow to $5000 if it is invested at an annual rate of 6% compounded continuously. Note that this is the same time it would take any investment at 6%, compounded continuously, to double. We can confirm this solution graphically or numerically. (See Figure 3.17.)

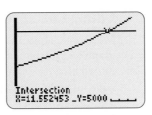 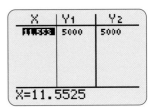

[0, 15] by [0, 6000]

FIGURE 3.17   ■

EXAMPLE 7   Mothers in the Workforce
- -

Between the years 1976 and 1998, the percent of moms who returned to the workforce within 1 year after having a child is given by $w(x) = 0.815 + 17.062 \ln x$, where x is the number of years after 1970. (Source: Associated Press)

a. Graph this function.

b. What does this model estimate the percent of mothers returning to the workforce to be for 1994?

c. Use the graph drawn in part (a) to estimate the year in which the percent reached 40.

Solution

a. We are given that the model is valid between 1976 and 1998, so an appropriate domain is $6 \le x \le 28$. This logarithmic function increases as the input becomes larger, so a reasonable range includes the interval $w(6) \le y \le w(28)$, or approximately $31.386 \le y \le 57.669$. We choose to use the viewing window [0, 28] by [0, 60], but there are many other appropriate windows. The graph of $w(x) = 0.815 + 17.062 \ln x$ appears in Figure 3.18(a).

b. The year 1994 is 24 years past 1970, so we find

$$w(24) = 0.815 + 17.062 \ln 24 \approx 55.04.$$

So the model estimates that in 1994 about 55% of moms returned to the workforce within 1 year after having a child.

c. To find the x-value for which $y = 40$, we graph $y_1 = 0.815 + 17.062 \ln x$ and $y_2 = 40$ on the same window and find the point of intersection of graphs. (See Figure 3.18(b)). We see that the input of this point is $x \approx 10$, which corresponds to the year 1980. So we conclude that approximately 40% of mothers returned to the workforce within 1 year after having a child in 1980.

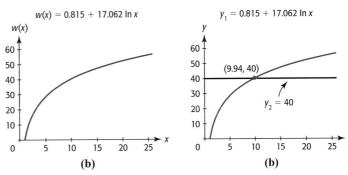

FIGURE 3.18   ■

Change of Base

Most calculators and other graphing utilities can be used to evaluate logarithms and graph logarithmic functions if the base is 10 or e. Graphing logarithmic functions with other bases on a graphing utility may require a special formula called the **change of base formula**. This formula allows us to rewrite logarithms so that the base is 10 or e. The general change of base formula is summarized below.

CHANGE OF BASE FORMULA

If $b > 0$, $b \neq 1$, $c > 0$, $c \neq 1$, and $x > 0$, then

$$\log_c x = \frac{\log_b x}{\log_b c}$$

In particular, for base 10 and base e,

$$\log_c x = \frac{\log x}{\log c} \quad \text{and} \quad \log_c x = \frac{\ln x}{\ln c}$$

EXAMPLE 8   Applying the Change of Base Formula

Graph each function below by changing the logarithm to a common logarithm and then by changing the logarithm to a natural logarithm.

a. $y = \log_3 x$

b. $y = \log_2(-3x)$

Solution

a. Changing to base 10 gives $\log_3 x = \dfrac{\log x}{\log 3}$, so we graph $y = \dfrac{\log x}{\log 3}$ (shown in Figure 3.19(a)). Changing to base e gives $\log_3 x = \dfrac{\ln x}{\ln 3}$, so we graph $y = \dfrac{\ln x}{\ln 3}$ (shown in Figure 3.19(b)). Note that the graphs appear to be identical (as they should be, since they both represent $y = \log_3 x$).

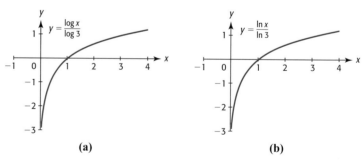

(a)   (b)

FIGURE 3.19

b. Changing to base 10 gives $\log_2(-3x) = \dfrac{\log(-3x)}{\log 2}$, so we graph $y = \dfrac{\log(-3x)}{\log 2}$ (shown in Figure 3.20(a)). Changing to base e gives $\log_2(-3x) = \dfrac{\ln(-3x)}{\ln 2}$, so we graph

$y = \dfrac{\ln(-3x)}{\ln 2}$ (shown in Figure 3.20(b)). Note that the graphs are identical. Also notice that because logarithms are defined only for positive inputs, x must be negative so that $-3x$ is positive. Thus, the domain of $y = \log_2(-3x)$ is the interval $(-\infty, 0)$.

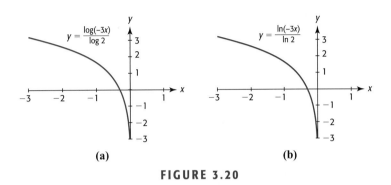

(a)   (b)

FIGURE 3.20

In Example 6 we found the doubling time for money invested with continuous compounding. If the interest is compounded *annually* on an investment, we can find the doubling time by writing $2 = (1 + r)^t$ in logarithmic form, getting

$$t = \log_{(1+r)} 2,$$

where r is the annual interest rate.

EXAMPLE 9   Investment

Find the number of years that it takes for an investment to double if it is invested at 10%, compounded annually.

Solution

The number of years that it takes for an investment to double if it is invested at 10%, compounded annually, can be found by evaluating

$$t = \log_{(1+0.1)} 2 = \log_{1.1} 2.$$

We can evaluate the expression $\log_{1.1} 2$ by using the change of base formula:

$$\log_{1.1} 2 = \frac{\log 2}{\log 1.1} \approx 7.273.$$

Thus, it takes approximately 7.3 years for an investment to double if it is invested at 10% compounded annually. ∎

3.2 SKILLS CHECK

In Exercises 1–4, write the logarithmic equations in exponential form.

1. $y = \log_3 x$

2. $2y = \log_5 x$

3. $y = \ln(2x)$

4. $y = \log(-x)$

In Exercises 5–8, write the exponential equations in logarithmic form.

5. $x = 4^y$

6. $m = 3^p$

7. $32 = 2^5$

8. $9^{2x} = y$

In Exercises 9–14, evaluate the logarithms, if possible. Round each answer to three decimal places.

9. log 7   **10.** log 456

11. ln 86   **12.** log (-12)

13. log 63,980   **14.** ln 10

In Exercises 15–22, find the value of the logarithms without using a calculator.

15. $\log_2 32$ **16.** $\log_9 81$

17. $\ln (e^3)$ **18.** log 100

19. $\log_3 \dfrac{1}{27}$ **20.** ln 1

21. ln e **22.** log 0.0001

In Exercises 23–25, use a change of base formula to evaluate each of the logarithms. Give your answers rounded to four decimal places.

23. $\log_6(18)$ **24.** $\log_7(215)$

25. $\log_8\left(\sqrt{2}\right)$

In Exercises 26–29, graph the functions with technology.

26. $y = 2 \ln x$ for $0 \le x \le 10$

27. $y = \ln (x - 2) - 3$ for $0 \le x \le 10$

28. $y = \ln x^2$ for $-8 \le x \le 8$

29. $y = \log_4 x$ for $-2 \le x \le 12$

30. Write the inverse of $y = 4^x$ in logarithmic form.

31. Graph $y = 4^x$ and its inverse, and discuss the symmetry of their graphs.

32. Write $\log_a 1 = x$ in exponential form and find x to evaluate $\log_a 1$ for any $a > 0, a \ne 1$.

33. Write $\log_a a = x$ in exponential form and find x to evaluate $\log_a a$ for any $a > 0, a \ne 1$.

3.2 EXERCISES

----

1. *Life Span* On the basis of data for the years 1910 through 1998, the expected life span of people in the U.S. can be described by the function $f(x) =$ 12.734 ln $x + 17.875$ years, where x is the number of years from 1900 to the person's birth year.
  a. What does this model estimate the life span to be for people born in 1925? in 1996? (Give each answer to the nearest year.)
  b. Explain why these numbers are so different.
  (Source: National Center for Health Statistics)

2. *Japan's Population* The population of Japan for the years 1984 to 1999 is approximated by the logarithmic function $y = 114.198 + 4.175 \ln x$ million people, with x equal to the number of years after 1980. According to the model, what is the estimated population in 1986? in 2000?
  (Source: www.jinjapan.org/stat/)

3. *Poverty Threshold* A single person is considered as living in poverty if his or her income falls below the federal government's official poverty level, which is adjusted every year for inflation. The function that models the poverty threshold for the years 1987–1999

is $f(x) = 282.1666 + 2771.0125 \ln x$ dollars, where x is the number of years after 1980.
  a. What does this model give as the poverty threshold in 1990 and in 1999?
  b. Is this function increasing or decreasing?
  c. What factors have led to this change in the threshold?
  (Source: U.S. Bureau of the Census)

4. *Out-of-Wedlock Births* For the years 1960–1997, the percent of births among blacks that were out-of-wedlock is given by the model $y = -61.0399 +$ 34.326 ln x, where x is the number of years after 1950.
  a. What does this model give as the percent in 1960 and in 1995?
  b. Is the percent of out-of-wedlock births increasing or decreasing?
  (Source: National Center for Health Statistics)

5. *Doubling Time* If \$4000 is invested in an account earning 10% annual interest compounded continuously, then the number of years that it takes for the amount to grow to \$8000 is $n = \dfrac{\ln 2}{0.10}$. Find the number of years.

6. *Doubling Time* If $5400 is invested in an account earning 7% annual interest compounded continuously, then the number of years that it takes for the amount to grow to $10,800 is $n = \dfrac{\ln 2}{0.07}$. Find the number of years.

7. *Doubling Time* The number of quarters needed to double an investment when a lump sum is invested at 8%, compounded quarterly, is given by $n = \log_{1.02} 2$.
   a. Use the change of base formula to find n.
   b. In how many years will the investment double?

8. *Doubling Time* The number of periods needed to double an investment when a lump sum is invested at 12%, compounded semiannually, is given by $n = \log_{1.06} 2$.
   a. Use the change of base formula to find n.
   b. How many years pass before the investment doubles in value?

9. *Annuities* If $2000 is invested at the end of each year in an annuity that pays 5%, compounded annually, the number of years it takes for the future value to amount to $40,000 is given by $t = \log_{1.05} 2$. Use the change of base formula to find the number of years until the future value is $40,000.

10. *Annuities* If $1000 is invested at the end of each year in an annuity that pays 8%, compounded annually, the number of years it takes for the future value to amount to $30,000 is given by $t = \log_{1.08} 3.4$. Use the change of base formula to find the number of years until the future value is $30,000.

11. *Cost* Suppose that the weekly cost for the production of x units of a product is given by $C(x) = 3452 + 50 \ln(x + 1)$ dollars. Use graphical methods to estimate the number of units produced if the total cost is $3556.

12. *Supply* Suppose that the daily supply function for a product is $p = 31 + \ln(x + 2)$, where p is in dollars and x is the number of units supplied. Use graphical methods to estimate the number of units that will be supplied if the price is $35.70.

13. *Life Span* Based on data for the years 1910–1998, the expected life span of people in the U.S. can be described by the function $f(x) = 12.734 \ln x + 17.875$, where x is the number of years from 1900 to the person's birth year.
    (Source: National Center for Health Statistics)
    a. Replace $f(x)$ with y and solve the equation for $\ln x$.
    b. Write the exponential form of the result of part (a).
    c. Use the exponential function from part (b) to estimate the birth year for which the expected life span is 75 years.

d. Use graphical methods to determine the birth year for which the expected life span is 75 years. Does this agree with the solution in part (c)?

14. *Mothers in Workforce* Between the years 1976 and 1998, the percent of moms who returned to the workforce within one year after they had a child is given by $w(x) = 1.11 + 16.94 \ln x$, where x is the number of years past 1970.
    a. Replace $w(x)$ with y and solve the equation for $\ln x$.
    b. Write the exponential form of the result of part (a).
    c. Use the exponential function from part (b) to estimate the year in which the percent is 50.
    d. Use graphical methods to determine the birth year for which the percent is 50. Does this agree with the solution in part (c)?
    (Source: Associated Press)

15. *Earthquakes* If an earthquake has an intensity of 25,000 times I_0, what is the magnitude of the earthquake?

16. *Earthquakes*
    a. If an earthquake has an intensity of 250,000 times I_0, what is the magnitude of the earthquake?
    b. Compare the magnitudes of two earthquakes when the intensity of one is 10 times the intensity of the other.

17. *Earthquakes*
    a. Write the exponential form of the function $R = \log\left(\dfrac{I}{I_0}\right)$ used for the Richter scale.
    b. The Richter scale reading for the San Francisco earthquake of 1989 was 7.1. Use the result of part (a) to find the intensity I of this earthquake as a multiple of I_0.

18. *Earthquakes* In 1906, an earthquake measuring 8.25 on the Richter scale occurred in San Francisco. Write the formula $R = \log\left(\dfrac{I}{I_0}\right)$ in exponential form and find the intensity I of this earthquake as a multiple of I_0.

19. *Earthquakes* The largest earthquake ever to strike San Francisco (in 1906) measured 8.25 on the Richter scale and the second largest (in 1989) measured 7.1. How many times more intense was the 1906 earthquake than the 1989 earthquake?

20. *Earthquakes* The largest earthquake ever recorded measured 8.9 on the Richter scale, and an earthquake measuring 8.25 occurred in Japan in 1983. Calculate the ratio of the intensities of these earthquakes.

Use the following information to answer the questions in Exercises 21–24. To quantify the intensity of sound, the decibel scale (named for Alexander Graham Bell) was developed. The formula for loudness L *on the decibel scale is* $L = 10 \log\left(\dfrac{I}{I_0}\right)$, *where* I_0 *is intensity of sound just below the threshold of hearing, which is approximately* 10^{-16} *watt per cm^2.*

21. *Decibel Scale* Find the decibel reading for a sound with intensity 20,000 times I_0.

22. *Decibel Scale* Compare the decibel readings of two sounds if the intensity of the first sound is 100 times more than the intensity of the second sound.

23. *Decibel Scale* Use the exponential form of the function $L = 10 \log\left(\dfrac{I}{I_0}\right)$ to find the intensity of a sound if its decibel reading is 40.

24. *Decibel Scale* Use the exponential form of $L = 10 \log\left(\dfrac{I}{I_0}\right)$ to find the intensity of a sound if its decibel reading is 140. (Sound at this level causes pain.)

Use the following information to answer the questions in Exercises 25–27. To simplify the measurement of the acidity or basicity of a solution, the pH (hydrogen potential) measure was developed. The pH is given by the formula pH $= -\log[\text{H}^+]$, *where* $[\text{H}^+]$ *is the concentration of hydrogen ions in moles per liter in the solution.*

25. *pH Levels* Find the pH value of beer for which $[\text{H}^+] = 0.0000631$.

26. *pH Levels* The pH value of eggs is 7.79. Find the hydrogen ion concentration for eggs.

27. *pH Levels* The most common solutions have a pH range between 1 and 14. Use an exponential form of the pH formula to find the values of $[\text{H}^+]$ associated with these pH levels.

3.3 Solving Exponential Equations; Properties of Logarithms

Key Concepts

- Solving an exponential equation by writing it in logarithmic form
- Properties of logarithms
  - Special cases of logarithm properties
- Solving an exponential equation by using properties of logarithms
- Exponential inequalities

Carbon-14 dating is a process by which scientists can tell the age of many fossils. To find the age of a fossil that originally contained 1000 grams of carbon-14 and now contains 1 gram, we solve the equation

$$1 = 1000e^{-0.00012378t}$$

An approximate solution of this equation can be found graphically, but sometimes a window that shows the solution is hard to find. If this is the case, it may be easier to solve an exponential equation like this analytically. We consider two analytical methods of solving exponential equations. The first method involves converting the equation to logarithmic form and then solving the logarithmic equation for the variable. The second method involves taking the logarithm of both sides of the equation and using the properties of logarithms to solve for the variable.

Solving Exponential Equations Using Logarithmic Forms

When we wish to solve an equation for a variable that is contained in an exponent, we can remove the variable from the exponent by converting the equation to its logarithmic form. The steps in this solution method follow.

SOLVING EXPONENTIAL EQUATIONS USING LOGARITHMIC FORMS

To solve an exponential equation using logarithmic forms:

1. Rewrite the equation with the term containing the exponent by itself on one side.

2. Divide both sides by the coefficient of the term containing the exponent.

3. Change the new equation to logarithmic form.

4. Solve for the variable.

This method is illustrated in the following example.

EXAMPLE 1   Solving a Base 10 Exponential Equation

a. Solve the equation $3000 = 150(10^{4t})$ for t analytically, by converting it to logarithmic form.

b. Solve the equation graphically to confirm the solution.

Solution

a. **1.** The term containing the exponent is by itself on one side of the equation.

   2. We divide both sides of the equation by 150.

$$3000 = 150(10^{4t})$$
$$20 = 10^{4t}$$

   3. We rewrite this equation in logarithmic form.

$$4t = \log_{10} 20 \quad \text{or} \quad 4t = \log 20$$

   4. Solving for t gives the solution:

$$t = \frac{\log 20}{4} \approx 0.32526$$

b. We solve the equation graphically by using the intersection method. Entering $y_1 = 3000$ and $y_2 = 150(10^{4t})$, we graph using the window $[0, 0.5]$ by $[0, 3500]$ and find the point of intersection to be about $(0.32526, 3000)$. (See Figure 3.21.) Thus, the solution to the equation $3000 = 150(10^{4t})$ is $t \approx 0.32526$, as we found analytically in part (a).

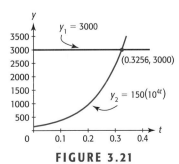

FIGURE 3.21

EXAMPLE 2 Carbon-14 Dating

Radioactive carbon-14 decays according to the equation

$$y = y_0 e^{-0.00012378t},$$

where y_0 is the original amount and y is the amount of carbon-14 at time t years. To find the age of a fossil if the original amount of carbon-14 was 1000 grams and the present amount is 1 gram, we solve the equation

$$1 = 1000 e^{-0.00012378t}.$$

a. Find the age of the fossil analytically, by converting it to logarithmic form.

b. Find the age of the fossil using graphical methods.

Solution

a. To write the equation so that the coefficient of the exponential term is 1, we divide both sides by 1000, obtaining

$$0.001 = e^{-0.00012378t}.$$

Writing this equation in logarithmic form gives the equation

$$-0.00012378t = \log_e 0.001 \text{ or } -0.00012378t = \ln 0.001.$$

Solving the equation for t gives

$$t = \frac{\ln 0.001}{-0.00012378} \approx 55{,}806.7.$$

Thus the age of the fossil is calculated to be about 55,807 years.

We can confirm that this is the solution to $1 = 1000 e^{-0.00012378t}$ by using a table of values. The result of entering 55,806.7 for x is shown in Figure 3.22.

b. We can solve the equation graphically by using the x-intercept method. Entering

$$y_1 = 1000 e^{-0.00012378t} - 1$$

and graphing using the window $[0, 80000]$ by $[-3, 3]$, we find the x-intercept to be 55806.716. (See Figure 3.23). Thus, as we found in part (a), the age of the fossil is about 55,807 years.

While reading this solution from a graph is easy, actually finding a window that contains the intersection is not. It may be easier to solve the equation analytically, as we did in part (a), and then use graphical or numerical methods to confirm the solution. ∎

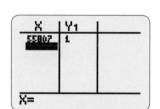

FIGURE 3.22

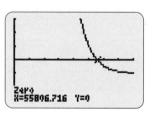

FIGURE 3.23

Logarithmic Properties

We have learned that logarithms are exponents and that a logarithmic function with base a is the inverse of an exponential function with base a. Thus the properties of logarithms can be derived from the properties of exponents. These properties are useful in solving equations involving exponents and logarithms.

The first properties of logarithms are easy to derive from the definition of a logarithm.

BASIC PROPERTIES OF LOGARITHMS

For $b > 0, b \neq 1$

1. $\log_b b = 1$
2. $\log_b 1 = 0$
3. $\log_b b^x = x$
4. $b^{\log_b x} = x$

These properties are easily justified by rewriting the logarithmic form as an exponential form or vice versa.

LOGARITHMIC VERSUS EXPONENTIAL FORM OF LOGARITHMIC PROPERTIES

Logarithmic Form	Exponential Form
$\log_b b = 1$	$b^1 = b$
$\log_b 1 = 0$	$b^0 = 1$
$\log_b b^x = x$	$b^x = b^x$
$\log_b x = \log_b x$	$b^{\log_b x} = x$

We can write these properties using base 10 and base e to obtain the following special cases.

LOGARITHMIC PROPERTIES USING BASE 10 AND BASE e

Property	Base 10	Base e
1.	$\log 10 = 1$	$\ln e = 1$
2.	$\log 1 = 0$	$\ln 1 = 0$
3.	$\log 10^x = x$	$\ln e^x = x$
4.	$10^{\log x} = x$	$e^{\ln x} = x$

EXAMPLE 3   Logarithmic Properties

Use the basic properties of logarithms to simplify the following:

a. $\log_5 5^{10}$ **b.** $\log_4 4$ **c.** $\log_4 1$ **d.** $\log 10^7$ **e.** $\ln e^3$ **f.** $\log\left(\dfrac{1}{10^3}\right)$

Solution

a. $\log_5 5^{10} = 10$ **b.** $\log_4 4 = 1$ **c.** $\log_4 1 = 0$

d. $\log 10^7 = 7$ **e.** $\ln e^3 = 3$ **f.** $\log\left(\dfrac{1}{10^3}\right) = \log 10^{-3} = -3$ ∎

Other logarithmic properties are helpful in simplifying logarithmic expressions, which in turn will help us solve some logarithmic equations. These properties are also directly related to the properties of exponents.

ADDITIONAL LOGARITHMIC PROPERTIES

For $b > 0$, $b \neq 1$, k a real number, and M and N positive real numbers,

Product Property   5. $\log_b(MN) = \log_b M + \log_b N$

Quotient Property   6. $\log_b\left(\dfrac{M}{N}\right) = \log_b M - \log_b N$

Power Property   7. $\log_b M^k = k \log_b M$

EXAMPLE 4   Rewriting Logarithms

Rewrite the following expressions as the sum, difference, or product of logarithms, and simplify if possible.

a. $\log_4 5(x - 7)$ **b.** $\ln(e^2(e + 3))$ **c.** $\log\left(\dfrac{x - 8}{x}\right)$

d. $\log_4 y^6$ **e.** $\ln\left(\dfrac{1}{x^5}\right)$

Solution

a. By Property 5, $\log_4 5(x - 7) = \log_4 5 + \log_4(x - 7)$.

b. By Property 5, $\ln(e^2(e + 3)) = \ln e^2 + \ln(e + 3)$.
By Property 3, this equals $2 + \ln(e + 3)$.

c. By Property 6, $\log\left(\dfrac{x - 8}{x}\right) = \log(x - 8) - \log x$.

d. By Property 7, $\log_4 y^6 = 6 \log_4 y$.

e. By Property 6, Property 2, and Property 7,

$$\ln\left(\frac{1}{x^5}\right) = \ln 1 - \ln x^5 = 0 - 5 \ln x = -5 \ln x$$ ∎

Caution: There is no property of logarithms to rewrite a logarithm of a sum or difference. That is why, in part (a) of Example 4, the expression $\log_4(x - 7)$ was **not** written as $\log_4 x - \log_4 7$.

EXAMPLE 5   Rewriting Logarithms

Rewrite the following expressions as a single logarithm.

a. $\log_3 x + 4 \log_3 y$

b. $\dfrac{1}{2} \log a - 3 \log b$

c. $\ln(5x) - 3 \ln z$

Solution

a. $\log_3 x + 4 \log_3 y = \log_3 x + \log_3 y^4$ Using Logarithmic Property 7

$= \log_3 xy^4$ Using Logarithmic Property 5

b. $\dfrac{1}{2} \log a - 3 \log b = \log a^{1/2} - \log b^3$ Using Logarithmic Property 7

$= \log\left(\dfrac{a^{1/2}}{b^3}\right)$ Using Logarithmic Property 6

c. $\ln(5x) - 3 \ln z = \ln(5x) - \ln z^3$ Using Logarithmic Property 7

$= \ln\left(\dfrac{5x}{z^3}\right)$ Using Logarithmic Property 6   ■

EXAMPLE 6   Applying the Properties of Logarithms

a. Use logarithmic properties to estimate $\ln(5e)$ if $\ln 5 \approx 1.61$. Check your answer with technology.

b. Find $\log_a\left(\dfrac{15}{8}\right)$ if $\log_a 15 = 2.71$ and $\log_a 8 = 2.079$.

c. Find $\log_a\left(\dfrac{15^2}{\sqrt{8}}\right)^3$ if $\log_a 15 = 2.71$ and $\log_a 8 = 2.079$.

Solution

a. $\ln(5e) = \ln 5 + \ln e \approx 1.61 + 1 = 2.61$

b. $\log_a\left(\dfrac{15}{8}\right) = \log_a 15 - \log_a 8 = 2.71 - 2.079 = 0.631$

c. $\log_a\left(\dfrac{15^2}{\sqrt{8}}\right)^3 = 3 \log_a\left(\dfrac{15^2}{\sqrt{8}}\right) = 3\left[\log_a 15^2 - \log_a 8^{1/2}\right] = 3\left[2 \log_a 15 - \dfrac{1}{2} \log_a 8\right]$

$= 3\left[2(2.71) - \dfrac{1}{2}(2.079)\right] = 13.1415$ ■

EXAMPLE 7   Earthquakes

If one earthquake has intensity 320,000 times I_0 and a second has intensity 3,200,000 times I_0, what is the difference in their Richter scale measurements?

Solution

Recall that the Richter scale measurement is given by $R = \log\left(\dfrac{I}{I_0}\right)$. The difference of the Richter scale measurements is

$$R_2 - R_1 = \log\frac{3,200,000 I_0}{I_0} - \log\frac{320,000 I_0}{I_0} = \log 3,200,000 - \log 320,000.$$

Using Property 6 gives us

$$\log 3,200,000 - \log 320,000 = \log\frac{3,200,000}{320,000} = \log 10 = 1.$$

Note that the intensity of the second quake is 10 times the intensity of the first one and that the difference in Richter scale measurements is 1. This is true in general as well as in the case of these two earthquakes. ∎

Solving Exponential Equations Using Logarithmic Properties

Now that we have the properties of logarithms at our disposal, we consider a second method for solving exponential equations, which may be easier to use. This method of solving an exponential equation involves taking the logarithm of both sides of the equation and then using properties of logarithms to write the equation in a form that we can solve.

SOLVING EXPONENTIAL EQUATIONS USING LOGARITHMIC PROPERTIES

To solve an exponential equation using logarithmic properties:

1. Rewrite the equation with a base raised to a power on one side.
2. Take the logarithm, base e or 10, of both sides of the equation.
3. Use a logarithmic property to remove the variable from the exponent.
4. Solve for the variable.

This method is illustrated in the following example.

EXAMPLE 8 Solution of Exponential Equations

Solve the following exponential equations:

a. $4096 = 8^{2x}$ **b.** $500 = 10e^{0.1x}$

Solution

a. 1. This equation has the base 8 raised to a variable power on one side.

   2. Taking the logarithm, base 10, of both sides of the equation $4096 = 8^{2x}$ gives

$$\log 4096 = \log 8^{2x}.$$

   3. Using Logarithmic Property 7 removes the variable x from the exponent:

$$\log 4096 = 2x \log 8$$

  4. Solving for x gives the solution.

$$\frac{\log 4096}{2 \log 8} = x$$

$$x = 2$$

b. To get the equation in a form with a base raised to a power on one side of the equation, we divide both sides by 10.

$$500 = 10e^{0.1x}$$

$$50 = e^{0.1x}$$

Taking the logarithm, base e, of both sides of this equation gives

$$\ln 50 = \ln e^{0.1x}.$$

By Logarithmic Property 3 for base e, $\ln e^{0.1x} = 0.1x$, so

$$\ln 50 = 0.1x$$

$$\frac{\ln 50}{0.1} = x$$

$$x \approx 39.120. \qquad \blacksquare$$

The following example revisits Example 2.

EXAMPLE 9   Carbon-14 Dating

Carbon-14 decays according to the equation

$$y = y_0 e^{-0.00012378t},$$

where y_0 is the original amount and y is the amount of carbon-14 at time t years. To find the age of a fossil if the original amount of carbon-14 was 1000 grams and the present amount is 1 gram, we solve the equation

$$1 = 1000e^{-0.00012378t}.$$

Solution

To solve $1 = 1000e^{-0.00012378t}$, we begin by dividing both sides by 1000:

$$0.001 = e^{-0.00012378t}$$

We take the logarithm, base e, of both sides of the equation to obtain

$$\ln 0.001 = \ln e^{-0.00012378t}.$$

By Logarithmic Property 3 for base e, $\ln e^{-0.00012378t} = -0.00012378t$, so we have

$$\ln 0.001 = -0.00012378t.$$

Dividing both sides of the equation by the coefficient of t gives the solution:

$$t = \frac{\ln 0.001}{-0.00012378} \approx 55{,}806.716$$

Thus the fossil is approximately 55,807 years old. $\qquad \blacksquare$

Solution of Logarithmic Equations

Some logarithmic equations can be solved analytically by using the properties that relate exponential and logarithmic forms of an equation. In some cases, it may be necessary to use more than one of the properties of logarithms to write the equation in a form suitable for solving. Graphical solutions using technology do not depend on the particular type of equation being solved, so those methods also apply to logarithmic equations.

EXAMPLE 10   ## Solving a Logarithmic Equation

a. Solve $6 + 3 \ln x = 12$ by writing the equation in exponential form.

b. Solve the equation graphically.

Solution

a. We first solve the equation for $\ln x$; then we write the equation in exponential form, which gives the solution.

$$6 + 3 \ln x = 12$$
$$3 \ln x = 6$$
$$\ln x = 2$$
$$x = e^2$$

b. We solve $6 + 3 \ln x = 12$ graphically by the intersection method. Entering $y_1 = 6 + 3 \ln x$ and $y_2 = 12$, we graph using the window $[-2, 10]$ by $[-10, 15]$ and find the point of intersection to be about $(7.38906, 12)$. (See Figure 3.24.) Thus, the solution to the equation $6 + 3 \ln x = 12$ is $x \approx 7.38906$. Because $e^2 \approx 7.38906$, the solutions agree.

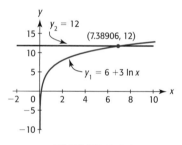

FIGURE 3.24   ■

Some exponential and logarithmic equations are difficult or impossible to solve analytically, and finding approximate solutions to real data problems is frequently easier with graphical methods.

Exponential Inequalities

Inequalities involving exponential functions can be solved by solving the related equation analytically and then investigating the inequality graphically. Consider the following example.

EXAMPLE 11   ## Sales Decay

After the end of an advertising campaign, the daily sales of Genapet fell rapidly, with daily sales given by $S = 3200e^{-0.08x}$ dollars, where x is the number of days from the end of the campaign. For how many days after the campaign ended were sales greater than \$1980?

Solution

To solve this problem, we find the solution to the inequality $3200e^{-0.08x} > 1980$. We begin our solution by solving the related equation.

$$3200e^{-0.08x} = 1980$$

$e^{-0.08x} = .61875$ Dividing both sides by 3200

$\ln e^{-0.08x} = \ln.61875$ Taking the logarithm, base e, of both sides

$-0.08x \approx -.4801$ Using Logarithmic Property 3

$x \approx 6$

We now investigate the inequality graphically.

To solve this inequality graphically, we graph $y_1 = 3200e^{-0.08x}$ and $y_2 = 1980$ on the same axes, with nonnegative x-values since x represents the number of days. The graph shows that $y_1 = 3200e^{-0.08x}$ is above $y_2 = 1980$ for $x \leq 6$. (See Figure 3.25.) Thus the daily revenue is greater than \$1980 for each of the first 6 days after the end of the sales campaign.

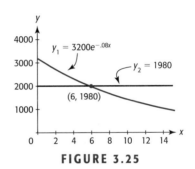

FIGURE 3.25

3.3 SKILLS CHECK

In Exercises 1–8, solve the following exponential equations, accurate to three decimal places.

1. $2500 = e^x$ **2.** $8900 = e^{5x}$

3. $4000 = 200e^{8x}$ **4.** $5200 = 13e^{12x}$

5. $1600 = 10^x$ **6.** $8000 = 500(10^x)$

7. $18000 = 30(2^{12x})$ **8.** $5880 = 21(2^{3x})$

In Exercises 9 and 10, solve for x.

9. $2P_0 = P_0e^x$ **10.** $0.5y_0 = y_0(2^x)$

In Exercises 11–17, use the properties of logarithms to evaluate the following expressions.

11. $\log 10^{14}$ **12.** $\ln e^5$

13. $\log_a(100)$, if $\log_a(20) = 1.4406$ and $\log_a(5) = 0.7740$

14. $\log_a(4)$, if $\log_a(20) = 1.4406$ and $\log_a(5) = 0.7740$

15. $\log_b(mn)$, if $\log_b(m) = p$ and $\log_b(n) = q$

16. $\log_a\left(\dfrac{5^3}{\sqrt{20}}\right)$, if $\log_a(20) = 1.4406$ and $\log_a(5) = 0.7740$

17. $\log_b\left(\dfrac{m^2}{n^3}\right)$, if $\log_b(m) = p$ and $\log_b(n) = q$

In Exercises 18–20, rewrite the following expressions as the sum, difference, or product of logarithms, and simplify if possible.

18. $\ln\dfrac{3x - 2}{x + 1}$ **19.** $\log[x^3(3x - 4)^5]$

20. $\log_3\dfrac{\sqrt[4]{4x + 1}}{4x^2}$

In Exercises 21-23, rewrite the expressions as a single logarithm.

21. $3 \log_2 x + \log_2 y$

22. $\log x - \dfrac{1}{3} \log y$

23. $4 \ln(2a) - \ln b$

24. Solve $12 + 4 \ln x = 36$.

25. By definition, $y = \log_b x$ means that $x = b^y$, so $\log_a x = \log_a b^y$. Use this to verify the change of base formula, $\log_b x = \dfrac{\log_a x}{\log_a b}$.

3.3 EXERCISES

----

1. *Supply* The supply function for a certain size boat is given by $p = 340(2^q)$ boats, where p dollars is the price per boat and q is the quantity of boats supplied at that price. What quantity will be supplied if the price is $10,880 per boat?

2. *Demand* The demand function for a dining room table is given by $p = 4000(3^{-q})$ dollars per table, where p is the price and q is the quantity, in thousands of tables, demanded at that price. What quantity will be demanded if the price per table is $256.60?

3. *Sales Decay* After a television advertising campaign ended, the weekly sales of Korbel champagne fell rapidly. Weekly sales in a city were given by $S = 25{,}000e^{-0.072x}$ dollars, where x is the number of weeks after the campaign ended.
   a. Write the logarithmic form of this function.
   b. Use the logarithmic form of this function to find the number of weeks after the end of the campaign before weekly sales fell to $16,230.

4. *Sales Decay* After a television advertising campaign ended, sales of Genapet fell rapidly with daily sales given by $S = 3200e^{-0.08x}$ dollars, where x is the number of days after the campaign ended.
   a. Write the logarithmic form of this function.
   b. Use the logarithmic form of this function to find the number of days after the end of the campaign that passed before daily sales fell to $2145.

5. *Sales Decay* After the end of an advertising campaign, the daily sales of Genapet fell rapidly, with daily sales given by $S = 3200e^{-0.08x}$ dollars, where x is the number of days from the end of the campaign.
   a. What were daily sales when the campaign ended?
   b. How many days passed after the campaign ended before daily sales were below half of what they were at the end of the campaign?

6. *Sales Decay* After the end of a television advertising campaign, the weekly sales of Korbel champagne fell rapidly, with weekly sales given by $S = 25{,}000e^{-0.072x}$ dollars, where x is the number of weeks from the end of the campaign.
   a. What were daily sales when the campaign ended?
   b. How many weeks passed after the campaign ended before weekly sales were below half of what they were at the end of the campaign?

7. *Snapple Beverage Revenues* Prior to the November 1994 $1.7 billion takeover proposal by Quaker Oats, Snapple Beverage Corporation's revenues were given by the function $B(t) = 1.337e^{0.718t}$ million dollars, where t is the number of years after 1985.
   a. According to the model, what was Snapple's 1995 revenue?
   b. If the revenue continued to increase as described by this model, when did it reach $3599 million?
   (Source: *U.S. News and World Report*, November 14, 1994)

8. *Population Growth* The population in a certain city was 53,000 in 2000 and its future size is predicted to be $P = 53{,}000e^{0.015t}$ people, where t is the number of years after 2000. Determine when the population will reach 60,000.

9. *Purchasing Power* The purchasing power (real value of money) decreases if inflation is present in the economy. For example, the purchasing power of $40,000 after t years of 5% inflation is given by the model

$$P = 40{,}000e^{-0.05t} \text{ dollars}$$

How long will it take for the value of a $40,000 pension to have a purchasing power of $20,000 under 5% inflation?
   (Source: *Viewpoints*, VALIC, 1993)

10. *Purchasing Power* If a retired couple has a fixed income of $60,000 per year, the purchasing power (adjusted value of the money) after t years of 5%

inflation is given by the equation $P = 60,000e^{-0.05t}$. In how many years will the purchasing power of their income be half of their current income?

11. *Real Estate Inflation* During a 5-year period of constant inflation, the value of a $100,000 property increases according to the equation $v = 100,000e^{0.03t}$ dollars. In how many years will the value of this building be double its current value?

12. *Real Estate Inflation* During a 10-year period of constant inflation, the value of a $200,000 property is given by the equation $v = 200,000e^{0.05t}$ dollars. In how many years will the value of this building be $254,250?

13. *Doubling Time* If $P is invested at an annual interest rate r, compounded continuously, for t years, the future value of the investment is given by $S = Pe^{rt}$. Find a formula for the number of years it will take to double the initial investment.

14. *Doubling Time* If $P is invested at 10% APR compounded continuously for t years, the future value of the investment is given by $S = Pe^{0.10t}$. Find the number of years (to the nearest whole year) it will take to double the investment.

15. *Investment* At the end of t years, the future value of an investment of $10,000 in an account that pays 8% APR compounded monthly is $S = 10,000\left(1 + \dfrac{0.08}{12}\right)^{12t}$ dollars. Assuming no withdrawals or additional deposits, how long will it take for the investment to amount to $40,000?

16. *Investment* At the end of t years, the future value of an investment of $25,000 in an account that pays 12% APR compounded quarterly is $S = 25,000\left(1 + \dfrac{0.12}{4}\right)^{4t}$ dollars. In how many years will the investment amount to $60,000?

17. *Radioactive Decay* The amount of radioactive isotope thorium-234 present in a certain sample at time t is given by $A(t) = 500e^{-0.02828t}$ grams, where t years is the time since the initial amount was measured.
   a. Find the initial amount of the isotope that is present in the sample.
   b. Find the half-life of this isotope. That is, find the number of years until half of the original amount of the isotope remains.

18. *Drugs in a Bloodstream* The concentration of a drug in the bloodstream from the time the drug is injected until 8 hours later is given by

$$y = 100(1 - e^{-0.312(8-t)}) \text{ percent,}$$

where the drug is administered at time $t = 0$. In how many hours will the drug concentration be 79% of the initial dose?

19. *Drugs in the Bloodstream* If a drug is injected into the bloodstream, the percent of the maximum dosage that is present at time t is given by $y = 100(1 - e^{-0.35(10-t)})$, where t is in hours, with $0 \le t \le 10$. In how many hours will the percent reach 65%?

20. *Radioactive Decay* The amount of radioactive isotope thorium-234 present in a certain sample at time t is given by $A(t) = 500e^{-0.02828t}$ grams, where t years is the time since the initial amount was measured. How long will it take for the amount of isotope to equal 318 grams?

21. *Radioactive Decay* A breeder reactor converts stable uranium-238 into the isotope plutonium-239. The decay of this isotope is given by $A(t) = A_0e^{-0.00002876t}$, where $A(t)$ is the amount of the isotope at time t (in years) and A_0 is the original amount. If the original amount is 100 lb, find the half-life of this isotope.

22. *Recording Media Sales* On the basis of data collected between 1982 and 1997, manufacturer's shipments of vinyl record albums (LPs and EPs) declined according to the function $A(t) = 482.794e^{-0.351t}$ million albums, where t is the number of years after 1980.
   a. Graph the function A for $0 \le t \le 20$.
   b. If album sales continue according to the pattern indicated in the graph, what sales level are vinyl record albums approaching?
   c. According to the function, when did vinyl record album sales fall to 5 million?
   d. Why are the sales decreasing?
   (Source: *Statistical Abstract of the U.S., 1998*)

23. *Carbon-14 Dating* An exponential decay function can be used to model the number of atoms of a radioactive material that remain after a period of time. Carbon-14 decays over time, with the amount remaining after t years given by $y = y_0e^{-0.00012378t}$, where y_0 is the original amount. If the original amount of carbon-14 is 200 grams, find the number of years until 155.6 grams of carbon-14 remains.

24. *Sales Decay* After a television advertising campaign ended, the weekly sales of Korbel champagne fell

rapidly. Weekly sales in a city were given by $S = 25{,}000e^{-0.072x}$ dollars, where x is the number of weeks after the campaign ended.

a. Use the logarithmic form of this function to find the number of weeks after the end of the campaign that passed before weekly sales fell below $16,230.

b. Check your solution by graphical or numerical methods.

25. *Sales Decay* After a television advertising campaign ended, the weekly sales of Turtledove bars fell rapidly. Weekly sales are given by $S = 600e^{-0.05x}$ thousand dollars, where x is the number of days after the campaign ended.

a. Use the logarithmic form of this function to find the number of weeks after the end of the campaign before weekly sales fell below $269.60.

b. Check your solution by graphical or numerical methods.

26. *Investment* At the end of t years, the future value of an investment of $10,000 in an account that pays 8%,

compounded monthly, is $S = 10{,}000\left(1 + \dfrac{0.08}{12}\right)^{12t}$ dollars. Assuming no withdrawals or additional deposits, for what time period (in years) will the future value of the investment be less than $40,000?

27. *Investment* At the end of t years, the future value of an investment of $25,000 in an account that pays 12%, compounded quarterly, is $S = 25{,}000\left(1 + \dfrac{.12}{4}\right)^{4t}$ dollars. For what time period (in years) will the future value of the investment be less than $60,000?

28. *Drugs in the Bloodstream* The concentration of a drug in the bloodstream from the time the drug is administered until 8 hours later is given by $y = 100\left(1 - e^{-0.312t}\right)$ percent, where the drug is administered at time $t = 0$. For what time period is the amount of drug present less than 79%?

3.4 Exponential and Logarithmic Models

Key Concepts

- Modeling with exponential functions

- Constant percent change

- Comparing quadratic and exponential models

- Logarithmic modeling

- Exponents, logarithms, and linear regression

We saw in Section 3.1 that the decay of carbon-14 can be modeled by the exponential function

$$y = y_0e^{-0.00012378t},$$

where y_0 is the original amount and t is the number of years, and that the federal funds that have been allocated for the Head Start program can be modeled with

$$f(x) = 400.004(1.107)^x \text{ million dollars,}$$

where x is the number of years after 1975. Many other different sources provide data that can be modeled by exponential growth and decay functions, and technology can be used to find exponential functions that model data. In this section we model real data with exponential functions and determine when this model is appropriate. We also create exponential functions that model phenomena characterized by constant rates of change, and we model real data with logarithmic functions when appropriate.

Modeling with Exponential Functions

The exponential functions discussed above were used to model real data because the data described rapid growth or decay. When the scatter plot of data shows a very rapid increase or decrease, it is possible that an exponential function can be used to model the data. Consider the following examples.

EXAMPLE 1   Insurance Premiums

The monthly premiums for $250,000 in term life insurance over a 10-year term period increase with the age of the men purchasing the insurance. The monthly premiums for nonsmoking males are shown in Table 3.5.

TABLE 3.5

Age (years)	Monthly Premium for 10-Year Term Insurance (dollars)	Age (years)	Monthly Premium for 10-Year Term Insurance (dollars)
35	123	60	783
40	148	65	1330
45	225	70	2448
50	338	75	4400
55	500		

(Source: Quotesmith.com)

a. Graph the data in the table with x as age and y in dollars.

b. Create an exponential function that models these premiums as a function of age.

c. Graph the data and the exponential function that models the data on the same axes.

Solution

a. The scatter plot of the data is shown in Figure 3.26(a). It shows that the outputs rise rapidly as the inputs increase.

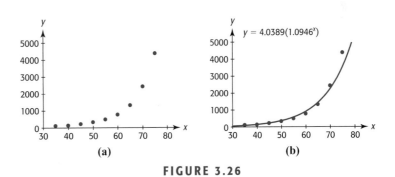

(a)   (b)

FIGURE 3.26

b. Using technology gives the exponential model, rounded to four decimal places, as

$$y = 4.0389(1.0946^x) \text{ dollars}$$

as the monthly premium where x is the age in years.

c. Figure 3.26(b) shows a scatter plot of the data and a graph of the (unrounded) exponential equation used to model it. The graph of the model appears to be a good, but not perfect, fit for the data. ∎

EXAMPLE 2 Smokers

The percent of U.S. males 18 years of age and older who smoked in selected years from 1965 to 1995 is given in Table 3.6.

TABLE 3.6

Year	Percent	Year	Percent
1965	51.6	1990	28
1974	42.9	1991	27.5
1979	37.2	1992	28.2
1983	34.7	1993	27.5
1985	32.1	1994	27.8
1987	31	1995	26.7

(Source: Centers for Disease Control, www.cdc.gov/nchs/datawh/statabl/)

a. Graph the data, with x equal to the number of years after 1960.

b. Find an exponential function that models the data, using as input the number of years after 1960.

c. Graph the data and the exponential function that models the data on the same axes.

d. Use the model to estimate the percent of U.S. male smokers age 18 and over in 2000.

Solution

a. The graph of this data is shown in Figure 3.27(a).

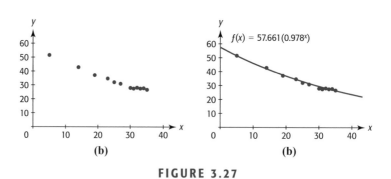

FIGURE 3.27

b. Exponential regression gives the function $f(x) = 57.661(0.978^x)$, with three-decimal place accuracy.

c. The graph of the (unrounded) model and the data is shown in Figure 3.27(b).

d. Evaluating the function at $x = 40$ gives $f(40) \approx 23.3$, so the estimated percent in 2000 is 23%. ∎

Spreadsheet Solution

We can use software programs and spreadsheets to find the exponential function that is the best fit for the data. Table 3.7 shows the Excel spreadsheet for the aligned data of Example 2. Selecting the cells containing the data, using Chart Wizard to get the scatter plot of the data, selecting ADD TRENDLINE , and picking EXPONENTIAL gives the equation of the exponential function that is the best fit for the data, along with the scatter plot and the graph of the best-fitting function. The equation and graph are shown in Figure 3.28.

TABLE 3.7

	A		B	
1	Year	x	Percent	y
2	5		51.6	
3	14		42.9	
4	19		37.2	
5	23		34.7	
6	25		32.1	
7	27		31	
8	30		28	
9	31		27.5	
10	32		28.2	
11	33		27.5	
12	34		27.8	
13	35		26.7	

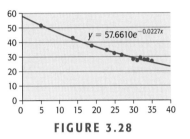

$$y = 57.6610e^{-0.0227x}$$

FIGURE 3.28

The function created by Excel,

$$y = 57.661(e^{-0.0227x}),$$

is in a different form than the model found in Example 2, which is

$$y = 57.661(0.978^{x}).$$

However, they are equivalent because $e^{-0.0227} \approx 0.978$ which means that

$$57.661(e^{-0.0227})^{x} \approx 57.661(0.978)^{x}.$$

Constant Percent Change in Exponential Models

How can we determine if a set of data can be modeled by an exponential function? To investigate this, we look at Table 3.8, which gives the growth of paramecia for each hour up to 9 hours.

TABLE 3.8

x (hours)	0	1	2	3	4	5	6	7	8	9
y (paramecia)	1	2	4	8	16	32	64	128	256	512

If we look at the first differences in the output, we see that they are not constant. (See Table 3.9.) But if we calculate the percent change of the outputs for uniform inputs, we see that they are constant. For example, from hour 2 to hour 3, the population grew by 4 units; this represents a 100% increase from the hour-2 population. From hour 3 to hour 4, the population grew by 8, which is a 100% increase over the hour-3 population. In fact, this population increases by 100% each hour. This means that the population 1 hour from now will be the present population plus 100% of the present population.

TABLE 3.9

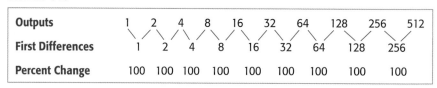

Outputs	1	2	4	8	16	32	64	128	256	512
First Differences		1	2	4	8	16	32	64	128	256
Percent Change		100	100	100	100	100	100	100	100	100

Because the percent change in the outputs is constant for uniform inputs in this example, an exponential model ($y = 2^x$) fits the data perfectly.

CONSTANT PERCENT CHANGES

If the percent change of the outputs of a set of data is constant for uniform inputs, an exponential function will be a perfect fit for the data.

If the percent change of the outputs is approximately constant for uniform inputs, an exponential function will be an approximate fit for the data.

EXAMPLE 3   Sales Decay

Suppose a company develops a product that is released with great expectations and extensive advertising, but sales suffered due to bad word of mouth from dissatisfied customers.

a. Use the monthly sales data shown in Table 3.10 to determine the percent change for each of the months given.

TABLE 3.10

Month	1	2	3	4	5	6	7	8
Sales (in $1000)	780	608	475	370	289	225	176	137

b. Find the exponential function that models the data.

c. Graph the data and the model on the same axes.

Solution

a. The inputs are uniform; the differences of outputs and the percent changes are shown in Table 3.11.

TABLE 3.11

Outputs	780	608	475	370	289	225	176	137
First Differences		−172	−133	−105	−81	−64	−49	−39
Percent of Change		−22.1	−21.9	−22.1	−21.9	−22.1	−21.8	−22.2

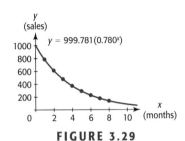

FIGURE 3.29

The percent change of the outputs is approximately -22%. This means that the sales 1 month from now will be approximately 22% less than the sales now.

b. Because the percent change is nearly constant, an exponential function should fit this data well. Technology gives the model

$$y = 999.781(0.780^x).$$

where x is the month and y is the sales in $1000.

c. The graphs of the data and the (unrounded) model are shown in Figure 3.29.   ∎

Exponential Models

Most graphing utilities give the best-fitting exponential model in the form

$$y = a \cdot b^x.$$

It can be shown that the constant percent change of this function is

$$(b - 1) \cdot 100\%.$$

Thus if $b > 1$, the function is increasing (growing) and if $0 < b < 1$, the function is decreasing (decaying).

For the function $y = 2^x$, $b = 2$, so the constant percent change is $(2 - 1) \cdot 100\% = 100\%$ (which we found in Table 3.6), and for the function $y = 1000(0.78^x)$, the base is $b = 0.78$, and the percent change is $(0.78 - 1) \cdot 100\% = -22\%$.

Because $a \cdot b^0 = a \cdot 1 = a$, the value of $y = a \cdot b^x$ is a when $x = 0$, so a is the y-intercept of the graph of the function. Because $y = a$ when $x = 0$, we say that the **initial value** of the function is a. If the constant percent change (as a decimal) is $r = b - 1$, then $b = 1 + r$, and we can write $y = a \cdot b^x$ as

$$y = a(1 + r)^x.$$

EXPONENTIAL MODEL

If a set of data has initial value a and a constant percent change r (written as a decimal) for each equally spaced input x, the data can be modeled by the exponential function

$$y = a(1 + r)^x.$$

To illustrate this, suppose that we know that the present population of a city is 100,000 and that it will grow at a constant annual percent of 10% per year. Then we know that the future population can be modeled by an exponential function with initial value $a = 100{,}000$ and constant percent change $r = 10\% = 0.10$ per year. That is, the population can be modeled by

$$y = 100{,}000(1 + .10)^x = 100{,}000(1.10)^x,$$

where x is the number of years from the present.

EXAMPLE 4   Inflation

Suppose that inflation averages 4% per year for each year from 2000 to 2010. This means that an item that costs $1 one year will cost $1.04 one year later. In the second year, the $1.04 cost will increase by a factor of 1.04, to $(1.04)(1.04) = (1.04)^2$.

a. Write an expression that gives the cost t years after 2000 of an item costing $1 in 2000.

b. Write an exponential function that models the cost of an item t years from 2000 if its cost was $100 in 2000.

c. Use the model to find the cost of the item from part (b) in 2005.

Solution

a. The cost of an item is $1 in 2000 and the cost increases at a constant percent of $4\% = 0.04$ per year, so the cost after t years will be

$$1(1 + .04)^t = (1.04)^t \text{ dollars.}$$

b. The inflation rate is $4\% = 0.04$. Thus if an item costs $100 in 2000, the function that gives the cost after t years is

$$f(t) = 100(1 + .04)^t = 100(1.04)^t \text{ dollars.}$$

c. The year 2005 is 5 years after 2000, so the cost of an item costing $100 in 2000 is

$$f(5) = 100(1.04)^5 = 121.67 \text{ dollars.} \quad \blacksquare$$

Comparison of Models

Sometimes it is hard to determine whether a linear model or an exponential model is the better fit for a set of data. When a scatter plot exhibits a single curvature without a visible high or low point, it is sometimes difficult to determine whether a power, quadratic, or exponential function should be used to model the data. Suppose that after graphing the data points we are unsure which of these models is more appropriate. Then we can find the three models, graph them on the same axes as the data points, and inspect the graphs to see which is a better visual fit.*

Consider the exponential model for insurance premiums as a function of age that we found in Example 1. The curvature indicated by the scatter plot suggests that either a power, quadratic, or exponential model may fit the data. To compare the goodness of fit of the models, we find each model and compare its graph with the scatter plot. (See Figure 3.30.)

---

*Statistical measures of goodness of fit exist, but caution must be used in applying these measures. For example, the correlation coefficient r found for linear fits to data is different from the coefficient of determination R^2 found for quadratic fits to data.

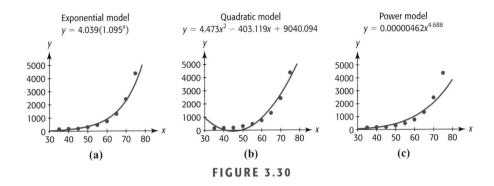

FIGURE 3.30

The exponential function appears to provide a better fit than the quadratic function because the data points seem to be approaching the *x*-axis as the input becomes closer to zero from the right. (Recall that the *x*-axis is a horizontal asymptote for the basic exponential function.) The exponential function also appears to be a better fit than the power function, especially for larger values of *x*.

When fitting exponential functions to data using technology, it is important to know that problems arise if the input values are large. For instance, when the input values are years (like 1995 and 1998), a technology-determined model may appear to be of the form $y = 0 \cdot b^x$, a function that certainly does not make sense as a model for exponential data. At other times when large inputs are used, your calculator or computer software may give an error message and not even produce an equation. To avoid these problems, it is helpful to align the inputs by converting years to the number of years from a specified year (for example, inputting years from 1990 rather than all possible years reduces the size of the inputs).

TECHNOLOGY NOTE

When using technology to fit an exponential model to data, you should align the inputs to reasonably small values.

Logarithmic Models

As with other functions we have studied, we can create models involving logarithms. Data that exhibits an initial rapid increase and then has a slow rate of growth can often be described by the function

$$f(x) = a + b \ln x \text{ for } b > 0.$$

Note that the parameter *a* is the vertical shift of the graph of $y = \ln x$ and the parameter *b* affects how much the graph of $y = \ln x$ is stretched. If $b < 0$, the graph of $f(x) = a + b \ln x$ is reflected about the line $y = a$ and decreases rather than increases. Most graphing utilities will create logarithmic models in the form $y = a + b \ln x$.

EXAMPLE 5   ## Mothers in the Workforce

Table 3.12 gives the percent of moms who returned to the workforce within 1 year after having a child (for the years 1976 through 1998). Find a logarithmic model for this data, with *x* equal to the number of years after 1970.

TABLE 3.12

Year	Returning Mothers (percent)	Year	Returning Mothers (percent)
1976	31	1988	51
1978	36	1990	52
1980	39	1992	53
1982	43	1994	52
1984	48	1995	55
1986	50	1998	59

(Source: Associated Press)

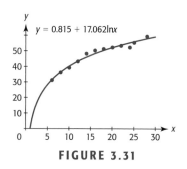

FIGURE 3.31

Solution

Entering the aligned input data (the number of years after 1970) as the values of x and the percents as the values of y, we use logarithmic regression on a graphing utility to find the function that models the data. This function, rounded to three decimal places, is

$$y = 0.815 + 17.062 \ln x,$$

where x is the number of years after 1970. The graphs of this function and the data are shown in Figure 3.31. ∎

Note that the input data in Example 5 was not aligned as the number of years after 1976 because the first aligned input value would be 0, and the logarithm of 0 does not exist.

TECHNOLOGY NOTE

When using technology to fit a logarithmic model to data, you must align the data so that all input values are positive.

EXAMPLE 6 International Visitors to the United States

The number of international visitors to the United States for each of the years 1986 through 1997 is given in Table 3.13.

a. Find a logarithmic function that models this data. Align the input to be the number of years after 1980.

b. Graph the equation and the aligned data points. Comment on how the model fits the data.

c. Assuming that the unrounded model $y = f(x)$ is valid in 2002, use it to estimate the number of international visitors to the United States in 2002.

d. Why might this model not apply for 2002?

TABLE 3.13

Years	International Visitors (in millions)	Years	International Visitors (in millions)
1986	26	1992	47.3
1987	29.5	1994	45.5
1988	34.1	1995	44
1989	36.6	1996	46.3
1990	39.5	1997	48.9
1991	43		

(Source: *1998 World Almanac*)

Solution

a. A logarithmic function that models this data is

$$y = -9.644 + 20.907 \ln x$$

million international visitors, where x is the number of years from 1980.

b. The scatter plot of the data and the graph of the logarithmic function that models the data are shown in Figure 3.32. The fit is fairly good except for 1991 and 1992. The number of international visitors during those two years was higher than what is given by the model.

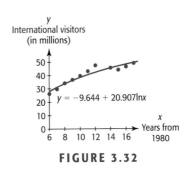

FIGURE 3.32

c. The number of international visitors in 2002 is estimated to be $f(22) = 55.98$, or approximately 56 million.

d. Immigration has been restricted since 9/11 of 2001.   ∎

Spreadsheet Solution

We can use software programs and spreadsheets to find the logarithmic function that is the best fit for data. Table 3.14 shows the Excel spreadsheet for the data of Example 6. Selecting the cells containing the data, using Chart Wizard to get the scatter plot of the data, selecting ADD TRENDLINE, and picking LOGARITHMIC gives the equation of the logarithmic function that is the best fit for the data, along with the scatter plot and the graph of the best-fitting function. The equation and graph are shown in Figure 3.33.

TABLE 3.14

	A	B
1	Year x	International Visitors (millions) y
2	6	26
3	7	29.5
4	8	34.1
5	9	36.6
6	10	39.5
7	11	43
8	12	47.3
9	14	45.5
10	15	44
11	16	46.3
12	17	48.9

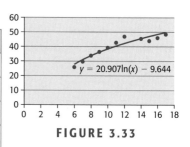

FIGURE 3.33

Exponents, Logarithms, and Linear Regression

Now that we have knowledge of the properties of logarithms and how exponential functions are related to logarithmic functions, we can show that linear regression can be used to create exponential models and logarithmic models that fit data. (The development of linear regression is discussed in Section 1.6.)

Logarithmic Property 3 states that $\log_b b^x = x$ and in particular that $\ln e^x = x$, so if we have data that can be approximated by an exponential function, we can convert the data to a linear form by taking the logarithm, base e, of the outputs. We can then use linear regression to find the linear function that is the best fit for the converted data and then use the linear function as the exponent of e, which gives the exponential function that is the best fit for the original data. Consider Table 3.15, which gives the number y of paramecia in a population after x hours. We have shown that the population can be modeled by $y = 2^x$ in Section 3.1.

TABLE 3.15

x (hours)	0	1	2	3	4	5	6	7	8	9
y (paramecia)	1	2	4	8	16	32	64	128	256	512
$\ln y$	0	.6931	1.3863	2.0794	2.7726	3.4657	4.1589	4.8520	5.5452	6.2383

The third row of Table 3.15 has the logarithms, base e, of the numbers (y-values) in the second row, and the relationship between x and $\ln y$ is linear. The first differences of the $\ln y$ values are constant, with the difference approximately equal to 0.6931. Using linear regression on the x and ln y values (to 10 decimal places) gives the equation of the linear model*:

$$\ln y = 0.6931471806x$$

Because we seek the equation solved for y, we can write this equation in its exponential form, getting

$$y = e^{0.6931471806x}.$$

To show that this model for the data is equal to $y = 2^x$, which we found in Section 3.1, we use properties of exponents:

$$y = e^{0.6931471806x} = \left(e^{0.6931471806}\right)^x \approx 2^x$$

Fortunately most graphing utilities have combined these steps to give the exponential model directly from the data.

3.4 SKILLS CHECK

1. Find the exponential function that models the data in the table below.

x	-2	-1	0	1	2	3	4	5
y	2/9	2/3	2	6	18	54	162	486

2. The following table has the inputs, x, and the outputs for three functions, f, g, and h. Test the percent change of the outputs to determine which function is exactly exponential, which is approximately exponential, and which is not exponential.

x	$f(x)$	$g(x)$	$h(x)$
1	4	2.5	1.5
2	16	6	2.25
3	64	8.5	3.8
4	256	10	5
5	1024	8	11
6	4096	6	17

3. Find the exponential function that is the best fit for $f(x)$ defined by the table in Exercise 2.

4. Find the exponential function that is the best fit for $h(x)$ defined by the table in Exercise 2.

5. **a.** Make a scatter plot of the data in the table below.
   b. Does it appear that a linear model or an exponential model is the better fit for the data?

x	1	2	3	4	5	6
y	2	3.1	4.3	5.4	6.5	7.6

6. Find the exponential function that models the data in the table below. Round the model with 3 decimal place accuracy.

x	-2	-1	0	1	2	3
y	5	30	150	1000	4000	20,000

7. Compare the first differences and the percent change of the outputs to determine if the data in the table below should be modeled by a linear or an exponential function.

x	1	2	3	4	5
y	2	6	14	34	81

*This equation could be found directly, in the form ln $y = ax + b$, where $b = 0$ and a equals the constant difference. This is because the exponential function is a perfect fit for the data in this application.

8. Use a scatter plot to determine if a linear or exponential function is the better fit for the data in Exercise 7.

9. Find the linear *or* exponential function that is the better fit for the data in Exercise 7. Round the model with three-decimal-point accuracy.

10. Find the logarithmic function that models the data in the table below. Round the model with two-decimal-point accuracy.

x	1	2	3	4	5	6	7
y	2	4.08	5.3	6.16	6.83	7.38	7.84

11. **a.** Make a scatter plot of the data in the table below.
    b. Does it appear that a linear model or a logarithmic model is the best fit for the data?

x	1	3	5	7	9
y	−2	1	3	4	5

12. **a.** Find a logarithmic function that models the data in the table in Exercise 11. Round the model with three-decimal-point accuracy.
    b. Find a linear function that models the data.
    c. Visually determine which model is the better fit for the data.

3.4 EXERCISES

Report models accurate to three decimal places unless otherwise specified. Use the unrounded function to calculate and to graph the function.

Use the exponential form $y = a(1 + r)^x$ to model the information in Exercises 1–4.

1. *Inflation* Suppose the retail price of an automobile is $30,000 in 1995 and that it increases at 4% per year.
   a. Write the equation of the exponential function that models the retail price of the automobile t years after 1995.
   b. Use the model to predict the retail price of the automobile in 2010.

2. *Population* Suppose the population of a city is 190,000 in 2000 and that it grows at 3% per year.
   a. Write the equation of the exponential function that models the annual growth.
   b. Use the model to find the population of this city in 2010.

3. *Sales Decay* At the end of an advertising campaign, weekly sales amounted to $20,000 and then decreased by 2% each week after the end of the campaign.
   a. Write the equation of the exponential function that models the weekly sales.
   b. Find the sales 5 weeks after the end of the advertising campaign.

4. *Inflation* Suppose the average price of a house in a certain city is $220,000 in 2000, and that it increases at 3% per year.

   a. Write the equation of the exponential function that models the average price of a house t years after 2000.
   b. Use the model to predict the average price of a house in 2005.

5. *World Population* The table gives the world population for selected years from 1650 to 2001.
   a. Create an exponential function that models this data, with x representing the years after 1600 and y the population in millions. Round the model with four decimal place accuracy.
   b. Graph the data and the exponential function that models the data on the same axes with window [0, 402] by [0, 6000].

Years	Population (millions)	Years	Population (millions)
1650	503	1950	2406
1750	711	1996	5771
1800	913	1999	6000
1850	1131	2001	6200
1900	1590		

(Source: UN Population Reference Bureau)

6. *Electric Cooperatives* The figure gives the taxes paid by South Carolina's electric cooperatives for the years 1995 through 1999, in millions of dollars.

a. Find an exponential function that models the taxes paid in these years. Use the number of years after 1990 as your input.

b. Use the model to predict the taxes paid in 2005. Discuss the reliability of this prediction.

c. Graph the data and the function on the same axes.

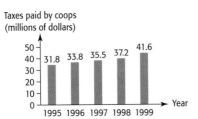

Taxes paid by coops
(millions of dollars)

(Source: "Cooperative Pride," *Living in South Carolina*, October 2000)

7. *Insurance Premiums* The table below gives the annual premiums required for a $250,000 20-year term life insurance policy on female nonsmokers of different ages.

a. Find an exponential function that models the monthly premium as a function of the age of the female nonsmoking policyholder.

b. Find the quadratic function that is the best fit for this data.

c. Graph each function on the same axes with the data points to determine visually which model is the better fit for the data. Use the window [30, 80] by [−10, 6800].

Age	Monthly Premium for a 20-Year Policy (dollars)	Age	Monthly Premium for a 20-Year Policy (dollars)
35	145	60	845
40	185	65	1593
45	253	70	2970
50	363	75	5820
55	550		

(Source: Quotesmith.com, December 1999)

8. *World Population* The United Nations has projected the world population from 1995 to 2150 to be as shown in the table.

a. Find the exponential function that models this data, using $x = 0$ in 1990.

b. Find the quadratic function that is the best fit for this data, using $x = 0$ in 1990.

c. Graph each function on the same axes with the data points to determine which model is the better fit for the data.

Year	Projected Population	Year	Projected Projection
1995	5666	2075	26,048
2000	6113	2100	52,508
2025	9069	2125	113,302
2050	14,421	2150	255,846

(Source: UN Department of Economic and Social Affairs)

9. *Smokers* The percent of all U.S. persons 18 years of age and older who smoked in selected years from 1965 to 1995 is given in the table below.

a. Graph the data, using $x = 0$ in 1960.

b. Can an exponential function be used to describe this data?

c. Find an exponential function that models the data, using an input of the number of years after 1960.

d. Find a power function that models the data, using an input of the number of years after 1960.

e. Which model is the better fit for the data?

Year	1965	1974	1979	1983	1985	1987
Percent	42.3	37.2	33.5	32.2	30.0	28.7

Year	1990	1991	1992	1993	1994	1995
Percent	25.4	25.4	26.4	25.0	25.5	24.7

(Source: Centers for Disease Control, www.cdc.gov/nchs/datawh/statabl/)

10. *Political Contributions* The figure below gives the total contributions, in millions of dollars, to federal political campaigns from the venture-capital industry for the years 1990 through 2000.

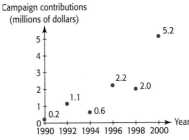

Campaign contributions
(millions of dollars)

(Source: *The Wall Street Journal*, October 31, 2000)

a. Find an exponential function that models the taxes paid for these years. Use the number of years after 1990 as the input.

b. Find the quadratic function that is the best fit for this data.

c. Graph each function on the same axes with the data points to determine which model is the best fit for the data.

11. *Life Span* The table below gives the life expectancy for people in the U.S. for the birth years 1910–1998.

a. Find the logarithmic function that models this data, with *x* equal to 0 in 1900.

b. Find the quadratic function that is the best fit for the data. Round the quadratic coefficient to five decimal places.

c. Graph each of these functions on the same axes with the data points to determine visually which of these functions is the better model for the data for the years 1910–1998.

d. Evaluate both models for the birth year 2010. Which model is better for prediction of life span after 2010?

Birth Year	Life Span (years)	Birth Year	Life Span (years)	Birth Year	Life Span (years)
1910	50.0	1981	74.2	1990	75.4
1920	54.1	1982	74.5	1991	75.5
1930	59.7	1983	74.6	1992	75.5
1940	62.9	1984	74.7	1993	75.5
1950	68.2	1985	74.7	1994	75.7
1960	69.7	1986	74.8	1995	75.8
1970	70.8	1987	75.0	1996	76.1
1975	72.6	1988	74.9	1997	76.5
1980	73.7	1989	75.2	1998	76.7

(Source: National Center for Health Statistics)

12. *Smokers* The percent of persons over age 18 who are smokers for selected years between 1965 and 1995 is given in the table.

a. Find a logarithmic function that models the data, using an input equal to 0 in 1960.

b. Use the model to estimate the percent of smokers in 1989.

c. A power model for this set of data was found to be $y = 74.575x^{-.298}$ in Exercise 9. Is this model as good a fit as the logarithmic function, or not?

Year	1965	1974	1979	1983	1985	1987
Smokers (percent)	42.3	37.2	33.5	32.2	30.0	28.7

Year	1990	1991	1992	1993	1994	1995
Smokers (percent)	25.4	25.4	26.4	25.0	25.5	24.7

(Source: Centers for Disease Control, www.cdc.gov/nchs/datawh/statabl/)

13. *Sexually Active Girls* The percent of girls of age *x* or younger who have been sexually active is given in the table below.

a. Create a logarithmic function that models the data, using an input equal to the age of the girls.

b. Use the model to estimate the percent for girls age 17 or younger who have been sexually active.

c. Find the quadratic function that is the best fit for the data.

d. Graph each of these functions on the same axes with the data points to determine which function is the better model for the data.

Age	Cumulative Percent Sexually Active Girls	Cumulative Percent Sexually Active Boys
15	5.4	16.6
16	12.6	28.7
17	27.1	47.9
18	44.0	64.0
19	62.9	77.6
20	73.6	83.0

(Source: "National Longitudinal Survey of Youth," *Risking the Future.* Washington D.C.: National Academy Press, 1987)

14. *Sexually Active Boys* The percent of boys age *x* or younger who have been sexually active is given in the table in Exercise 13.

a. Create a logarithmic function that models the data, using an input equal to the age of the boys.

b. Use the model to estimate the percent for boys age 17 or younger who have been sexually active.

c. Compare the percent that are sexually active for the two genders (see Exercise 13). What do you conclude?

3.5 Exponential Functions and Investing

Key Concepts

- Future value of investments
- Interest compounded annually
- Interest compounded k times per year
- Interest compounded continuously
  The number e
- Present value of an investment
- Investment models

If $1000 is invested in an account that earns 6% interest, with the interest added to the account at the end of each year, the interest is said to be **compounded annually**. The amount to which an investment grows over a period of time is called its **future value**. In this section, we use general formulas to find the future value of money invested when the interest is compounded at regular intervals of time and when it is compounded continuously. We will also find the lump sum that must be invested to have it grow to a specified amount in the future. This is called the **present value** of the investment.

Compound Interest

In Section 3.4, we found that if a set of data has initial value a and a constant percent change r for uniform inputs x, the data can be modeled by the exponential function

$$y = a(1 + r)^x.$$

Thus, if an investment of $1000 earns 6% interest compounded annually, the future value at the end of t years will be

$$S = \$1000(1.06)^t \text{ dollars.}$$

EXAMPLE 1   Future Value of an Account

To numerically and graphically view how $1000 grows at 6% interest compounded annually, do the following:

a. Construct a table that gives the future values $S = 1000(1 + .06)^t$ for $t = 0, 1, 2, 3, 4,$ and 5 years after the $1000 was invested.

b. Graph $S = 1000(1 + .06)^t$ for the values of t given in part (a).

c. Graph the equation $S = 1000(1 + .06)^t$ as a continuous function of t for $0 \le t \le 5$.

d. Is there another graph that more accurately represents the amount that would be in the account at any time during the first 5 years that the money is invested?

TABLE 3.16

t (years)	Future Value S (dollars)
0	1000.00
1	1060.00
2	1123.60
3	1191.02
4	1262.48
5	1338.23

Solution

a. Substitute the values of t into the equation $S = 1000(1 + .06)^t$. Because the output units for S are dollars, we round off to the nearest cent to obtain the values in Table 3.16.

b. Figure 3.34 shows the scatter plot of the data in Table 3.16.

c. Figure 3.35(a) shows a continuous graph of the function S.

d. The scatter plot in Figure 3.34 shows the amount in the account at the end of each year, but not at any other time during the 5-year period. The continuous graph in Figure 3.35(a) shows the amount continually increasing, but interest is paid only at the end of each year and the amount in the account is constant at the previous level until more interest is added. So the graph in Figure 3.35(b) is a most accurate graph

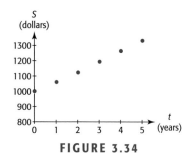

FIGURE 3.34

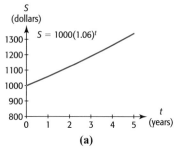

(a)

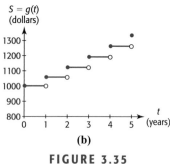

(b)

FIGURE 3.35

of the amount that would be in this account, because it shows the amount remaining constant until the end of each year, when interest is added.   ∎

The function g graphed in Figure 3.35(b) is a **step function**.

$$g(t) = \begin{cases} 1000.00 & \text{if } 0 \le t < 1 \\ 1060.00 & \text{if } 1 \le t < 2 \\ 1123.60 & \text{if } 2 \le t < 3 \\ 1191.02 & \text{if } 3 \le t < 4 \\ 1262.48 & \text{if } 4 \le t < 5 \\ 1338.23 & \text{if } t = 5 \end{cases} \text{dollars in } t \text{ years}$$

This step function is rather cumbersome to work with, so we will find the future value by using the graph in Figure 3.35(a) and the function $S = 1000(1 + .06)^t$ *interpreted discretely*. That is, we draw the graph of this function as it appears in Figure 3.35(a) and evaluate the formula for S with the understanding that the only inputs that make sense in this investment example are nonnegative integers. In general, we have the following.

> **FUTURE VALUE OF AN INVESTMENT WITH ANNUAL COMPOUNDING**
>
> If \$$P$ is invested at an interest rate r per year, compounded annually, the future value S at the end of t years is
>
> $$S = P(1 + r)^t.$$

The **annual interest rate** r is also called the **nominal interest rate**, or simply the **rate**. Remember that the interest rate is usually stated as a percent and that it is converted to a decimal when computing future value of investments.

If the interest is compounded more than once per year, then the additional compounding will result in a larger future value. For example, if the investment is compounded twice per year (semiannually), there would be twice as many compounding periods with the interest rate in each period equal to

$$r\left(\frac{1}{2}\right) = \frac{r}{2}$$

Thus the future value is found by doubling the number of periods and halving the interest rate if the compounding is done twice per year. In general, we use the following model, interpreted discretely, to find the amount of money that results when P dollars are invested and earn compound interest.

> **FUTURE VALUE OF AN INVESTMENT WITH PERIODIC COMPOUNDING**
>
> If \$$P$ is invested for t years at the annual interest rate r, where the interest is compounded k times per year, then the interest rate per period is $\dfrac{r}{k}$, the number of compounding periods is kt, and the future value that results is given by
>
> $$S = P\left(1 + \frac{r}{k}\right)^{kt} \text{dollars.}$$

For example, if $1000 is placed in an account that earns 6% interest per year, with the interest added to the account at the end of every month (compounded monthly), the future value of this investment in 5 years is given by

$$S = 1000\left(1 + \frac{0.06}{12}\right)^{12(5)} = \$1348.85$$

Recall from Table 3.16 that compounding annually gives $1338.23, so compounding monthly rather than yearly results in an additional $1348.85 − $1338.23 = $10.62 from the 5-year investment.

| **EXAMPLE 2** | Daily Versus Annual Compounding of Interest |

a. Write the equation that gives the future value of $1000 invested for t years at 8% compounded annually.

b. Write the equation that gives the future value of $1000 invested for t years at 8%, compounded daily.

c. Graph the equations from parts (a) and (b) on the same axes with t between 0 and 30.

d. What is the additional amount earned in 30 years from compounding daily rather than annually?

Solution

a. Substituting $P = 1000$ and $r = 0.08$ in $S = P(1 + r)^t$ gives

$$S = 1000(1 + 0.08)^t \text{ or } S = 1000(1.08)^t.$$

b. Substituting $P = 1000$, $r = 0.08$, and $k = 365$ in $S = P\left(1 + \frac{r}{k}\right)^{kt}$ gives

$$S = 1000\left(1 + \frac{0.08}{365}\right)^{365t}.$$

c. The graphs of both functions are shown in Figure 3.36.

d. Evaluating the two functions in parts (a) and (b) at $t = 30$, we calculate the future value of $1000 compounded annually at 8% to be $10,062.66 and the future value of $1000 at 8%, compounded daily, to be $11,020.28. The additional amount earned in 30 years from compounding daily rather than annually is the difference between these two amounts:

$$\$11,020.28 - \$10,062.66, \text{ or } \$957.62. \qquad \blacksquare$$

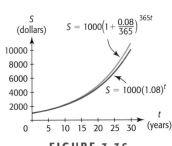

FIGURE 3.36

Continuous Compounding and the Number *e*

More frequent compounding results in interest being paid more often and, therefore, results in larger future values. The question becomes "how frequently can we compound the interest, and how much can we earn as a result of this compounding?" The answer to the first question is that we can let the number of compounding periods per year increase without bound, which is called **compounding continuously**. To answer the second question, we consider the future value of $1 invested at an annual rate of

100%, compounded for 1 year with different compounding periods. If we denote the number of periods per year by k, the model that gives the future value is

$$S = \left(1 + \frac{1}{k}\right)^k \text{ dollars.}$$

Table 3.17 shows the future value of this investment for different compounding periods.

TABLE 3.17

Type of Compounding	Number of Compounding Periods per Year	Future Value (in dollars)
Annually	1	$\left(1 + \frac{1}{1}\right)^1 = 2$
Quarterly	4	$\left(1 + \frac{1}{4}\right)^4 = 2.44140625$
Monthly	12	$\left(1 + \frac{1}{12}\right)^{12} \approx 2.61303529022$
Daily	365	$\left(1 + \frac{1}{365}\right)^{365} \approx 2.71456748202$
Hourly	8760	$\left(1 + \frac{1}{8760}\right)^{8760} \approx 2.71812669063$
Each minute	525,600	$\left(1 + \frac{1}{525,600}\right)^{525,600} \approx 2.7182792154$
x times per year	x	$\left(1 + \frac{1}{x}\right)^x$

As Table 3.17 indicates, the future value increases (but not very rapidly) as the number of compounding periods during the year increases. As x gets very large, the future value approaches the number e. In general, the outputs that result from larger and larger inputs in the function

$$f(x) = \left(1 + \frac{1}{x}\right)^x$$

approach the number e. (In calculus, e is defined as the *limit* of $\left(1 + \frac{1}{x}\right)^x$ as x approaches ∞.) The approximation (to nine decimal places) of the irrational number e is 2.718281828.

This definition of e permits us to create a new model for the future value of an investment when interest is compounded continuously. We can let $m = \dfrac{k}{r}$ to rewrite the formula

$$S = P\left(1 + \frac{r}{k}\right)^{kt} \text{ as } S = P\left(1 + \frac{1}{m}\right)^{mrt} \text{ or } P\left[\left(1 + \frac{1}{m}\right)^m\right]^{rt}.$$

Now as the compounding periods k increase without bound, $m = k/r$ increases without bound and

$$\left[\left(1 + \frac{1}{m}\right)^{m}\right]$$

approaches e, so $S = P\left[\left(1 + \frac{1}{m}\right)^{m}\right]^{rt}$ approaches Pe^{rt}.

FUTURE VALUE OF AN INVESTMENT WITH CONTINUOUS COMPOUNDING

If $\$P$ is invested for t years at an annual interest rate r, compounded continuously, then the future value S is given by

$$S = Pe^{rt} \text{ dollars.}$$

For a given principal and interest rate, the function $S = Pe^{rt}$ is a continuous function whose domain consists of real numbers greater than or equal to 0. Unlike the other compound interest functions, this one is not discretely interpreted because of the continuous compounding.

EXAMPLE 3   Future Value and Continuous Compounding

a. What is the future value of $2650 invested for 8 years at 12% compounded continuously?

b. How much will be earned on this investment?

Solution

a. The future value of this investment is $S = 2650e^{0.12(8)} = 6921.00$

b. The interest earned on this investment is the future value minus the original investment:

$$\$6921 - \$2650 = \$4271. \qquad \blacksquare$$

EXAMPLE 4   Continuous Versus Annual Compounding of Interest

a. For each of 9 years, compare the future value of an investment of $1000 at 8%, compounded annually, and of $1000 at 8%, compounded continuously.

b. Graph the functions for annual compounding and for continuously compounding on the same axes.

c. What conclusion can be made regarding compounding annually and compounding continuously?

Solution

a. The future value of $1000, compounded annually, is given by the function $S = 1000(1 + 0.08)^{t}$ and the future value of $1000, compounded continuously, is given by $S = 1000e^{0.08t}$. By entering these formulas in an Excel spreadsheet and evaluating them for each of 9 years (see Table 3.18), we can compare the future value for each of these 9 years. At the end of the 9 years, we see that compounding continuously results in $2054.43 − $1999.00 = $55.43 more than compounding annually yields.

TABLE 3.18

	A	B	C
1	YEAR	$S = 1000*(1.08)^t$	$S = 1000*e^{(.08t)}$
2	1	1080	1083.287068
3	2	1166.4	1173.510871
4	3	1259.712	1271.249150
5	4	1360.48896	1377.127764
6	5	1469.328077	1491.824698
7	6	1586.874323	1616.07440
8	7	1713.824269	1750.672500
9	8	1850.93021	1896.480879
10	9	1999.004627	2054.433211
11			

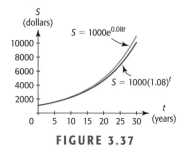

FIGURE 3.37

b. The graphs are shown in Figure 3.37.

c. The value of the investment increases more under continuous compounding than under annual compounding. ∎

Present Value of an Investment

We can write the formula for the future value S when a lump sum P is invested for $n = kt$ periods at interest rate $i = \dfrac{r}{k}$ per compounding period as

$$S = P(1 + i)^n.$$

We sometimes need to know the lump sum investment P that will give an amount S in the future. This is called the **present value** P, and it can be found by solving $S = P(1 + i)^n$ for P. Solving for P gives the present value as follows.

PRESENT VALUE

The lump sum that will give future value S in n compounding periods at rate i per period is the present value

$$P = \frac{S}{(1 + i)^n} \quad \text{or} \quad P = S(1 + i)^{-n}$$

EXAMPLE 5   Present Value

What lump sum must be invested at 10%, compounded semiannually, for the investment to grow to $15,000 in 7 years?

Solution

We have the future value $S = 15,000$, $i = \dfrac{0.10}{2} = 0.05$, and $n = 2 \cdot 7 = 14$, so the present value of this investment is

$$P = \frac{15,000}{(1 + 0.05)^{14}} = 7576.02 \text{ dollars.} \qquad \blacksquare$$

Investment Models

In addition to developing investment formulas, we can use graphing utilities to model actual investment data.

EXAMPLE 6   Mutual Fund Growth

TABLE 3.19

Time (years)	Average Annual Total Return (dollars)
1	1.23
3	2.24
5	3.11
10	6.82

(Source: *FundStation.*)

The data in Table 3.19 gives the annual return on an investment of $1.00 made on March 31, 1990, in the AIM Value Fund, Class A Shares. The values reflect reinvestment of all distributions and changes in net asset value but exclude sales charges.

a. Use an exponential function to model this data.

b. Use the model to find what the fund will amount to on March 31, 2001 if $10,000 is invested March 31, 1990 and the fund continues to follow this model after 2000.

c. Is it likely that this fund continued to grow at this rate in 2002?

Solution

a. Using technology to create an exponential model for the data gives

$$A(x) = 1.1609(1.2004)^x.$$

Thus $1.00 invested in this fund grows to $A(x) = 1.1609(1.2004)^x$ dollars x years after March 31, 1990.

b. March 31, 2001 is 11 years after March 31, 1990, so using the equation from part (a) with $x = 11$ gives $A(11) = 1.1609(1.2004)^{11}$ as the future value an investment of $1.00.

Thus the future value of an investment of $10,000 in 11 years is

$$\$10,000[1.1609(1.2004)^{11}] = \$86,572.64.$$

c. No. Most stocks and mutual funds decreased in value in 2002. $\blacksquare$

3.5 SKILLS CHECK

Evaluate the expressions in Exercises 1–10. Write approximate answers rounded two decimal places.

1. $15,000e^{0.06(20)}$ **2.** $8000e^{0.05(10)}$

3. $3000(1.06^x)$ for $x = 30$

4. $20,000(1.07^x)$ for $x = 20$

5. $12,000\left(1 + \dfrac{0.10}{k}\right)^{kn}$ for $k = 4$ and $n = 8$

6. $23,000\left(1 + \dfrac{0.08}{k}\right)^{kn}$ for $k = 12$ and $n = 20$

7. $P\left(1 + \dfrac{r}{k}\right)^{kn}$ for $P = 3000$, $r = 8\%$, $k = 2$, $n = 18$

8. $P\left(1 + \dfrac{r}{k}\right)^{kn}$ for $P = 8000$, $r = 12\%$, $k = 12$, $n = 8$

9. $300\left[\dfrac{1.02^n - 1}{0.02}\right]$ for $n = 240$

10. $2000\left[\dfrac{1.10^n - 1}{0.10}\right]$ for $n = 12$

11. Find $g(2.5)$, $g(3)$, and $g(3.5)$ if

$$g(x) = \begin{cases} 1000.00 & \text{if } 0 \leq x < 1 \\ 1060.00 & \text{if } 1 \leq x < 2 \\ 1123.60 & \text{if } 2 \leq x < 3 \\ 1191.00 & \text{if } 3 \leq x < 4 \\ 1262.50 & \text{if } 4 \leq x < 5 \\ 1338.20 & \text{if } x = 5 \end{cases}$$

12. Solve $S = P + Prt$ for P.

3.5 EXERCISES

1. *Future Value* If $8800 is invested for x years at 8% interest compounded annually, find the future value that results in
  a. 8 years.
  b. 30 years.

2. *Investments* Suppose $6400 is invested for x years at 7% interest compounded annually. Find the future value of this investment at the end of
  a. 10 years.
  b. 30 years.

3. *Future Value* If $3300 is invested for x years at 10% interest compounded annually, the future value that results is $S = 3300(1.10)^x$ dollars.
  a. Graph the function for $x = 0$ to $x = 8$.
  b. Use the graph to estimate when the money in the account will double.

4. *Investments* If $5500 is invested for x years at 12% interest compounded quarterly, the future value that results is $S = 5500(1.03)^{4x}$ dollars.
  a. Graph this function for $0 \leq x \leq 8$.

b. Use the graph to estimate when the money in the account will double.

5. *Future Value* If $10,000 is invested at 12% interest compounded quarterly, find the future value in 10 years.

6. *Future Value* If $8800 is invested at 6% interest compounded semiannually, find the future value earned in 10 years.

7. *Future Value* An amount of $10,000 is invested at 12% interest compounded daily.
  a. Find the future value in 10 years.
  b. How does this future value compare with the future value in Exercise 5? Why are they different?

8. *Future Value* A total of $8800 is invested at 6% interest compounded daily.
  a. Find the future value in 10 years.
  b. How does this future value compare with the future value in Exercise 6? Why are they different?

9. *Compound Interest* If $10,000 is invested at 12% interest compounded monthly, find the interest earned in 15 years.

10. *Compound Interest* If $20,000 is invested at 8% interest compounded quarterly, find the interest earned in 25 years.

11. *Continuous Compounding* Suppose $10,000 is invested for t years at 6% interest compounded continuously. Give the future value at the end of
    a. 12 years.
    b. 18 years.

12. *Continuous Compounding* If $42,000 is invested for t years at 7% interest compounded continuously, find the future value in
    a. 10 years.
    b. 20 years.

13. *Continuous Compounding* If $8000 is invested for t years at 8% interest compounded continuously, the future value is given by $S = 8000e^{0.08t}$ dollars.
    a. Graph this function for $0 \leq t \leq 15$.
    b. Use the graph to estimate when the future value will be $20,000.

14. *Continuous Compounding* If $35,000 is invested for t years at 9% interest compounded continuously, the future value is given by $S = 35,000e^{0.09t}$ dollars.
    a. Graph this function for $0 \leq t \leq 8$.
    b. Use the graph to estimate when the future value will be $60,000.

15. *Doubling Time* Use a spreadsheet, a table, or a graph to estimate how long it takes for an investment to double if it is invested at 10% interest
    a. Compounded annually.
    b. Compounded continuously.

16. *Doubling Time* Use a spreadsheet, a table, or a graph to estimate how long it takes for an investment to double if it is invested at 6% interest
    a. Compounded annually.
    b. Compounded continuously.

17. *Future Value* Suppose $2000 is invested in an account paying 5% interest compounded annually. What is the future value of this investment
    a. after 8 years?
    b. after 18 years?

18. *Future Value* If $12,000 is invested in an account that pays 8% interest compounded quarterly, find the future value of this investment
    a. after 2 quarters.
    b. after 10 years.

19. *Investments* Suppose $3000 is invested in an account that pays 6% interest compounded monthly. What is the future value of this investment after 12 years?

20. *Investments* If $9000 is invested in an account that pays 8% interest compounded quarterly, find the future value of this annuity
    a. after 0.5 year.
    b. after 15 years.

21. *Doubling Time* If money is invested at 10% interest compounded quarterly, the future value of the investment doubles approximately every 7 years.
    a. Use this information to complete the table below for an investment of $1000 at 10% interest compounded quarterly.
    b. Create an exponential function, rounded to three decimal places, that models the discrete function defined by the table.
    c. Because the interest is compounded quarterly, this model must be interpreted discretely. Use the rounded function to find the value of the investment in 5 years and in $10\frac{1}{2}$ years after the money was invested.

Years	0	7	14	21	28
Future Value (in dollars)	1000				

22. *Doubling Time* If money is invested at 11.6% interest compounded monthly, the future value of the investment doubles approximately every 6 years.
    a. Use this information to complete the table below for an investment of $1000 at 11.6% interest compounded monthly.
    b. Create an exponential function, rounded to three decimal places, that models the discrete function defined by the table.
    c. Because the interest is compounded monthly, this model must be interpreted discretely. Use the rounded model to find the value of the investment in 2 months, in 4 years, and in $12\frac{1}{2}$ years.

Years	0	6	12	18	24
Future Value (in dollars)	1000				

3.6 Annuities; Loan Repayment

Key Concepts

- Annuity
  Ordinary annuity

- Present value of an
  investment
  *Present value of a lump
  sum investment*
  *Present value of an
  ordinary annuity*

- Loan repayment
  Amortization

To prepare for future retirement, people frequently invest a fixed amount of money at the end of each month into an account that pays interest that is compounded monthly. Such an investment plan, or any other characterized by regular payments, is called an **annuity**. In this section, we will develop a model to find **future values** of annuities.

A person planning retirement may also want to know how much money is needed to provide regular payments to him or her during retirement, and a couple may want to know what lump sum to put in an investment to pay future college expenses for their child. The amount of money they need to invest to receive a series of payments in the future is the **present value** of an annuity.

We can also use the present value concept to determine the payments that are necessary to repay a loan with equal payments made on a regular schedule, which is called **amortization**. The formulas used to determine the future value and the present value of annuities and the payments necessary to repay loans are applications of exponential functions.

Future Value of an Annuity

An **annuity** is a financial plan characterized by regular payments. We can view an annuity as a savings plan where regular payments are made to an account, and we can use an exponential function to determine what the future value of the account will be. One type of annuity, called an **ordinary annuity**, is a financial plan where equal payments are contributed at the end of each period to an account that pays a fixed rate of interest compounded at the same time as payments are made. For example, suppose that $1000 is invested at the end of each year for 5 years in an account that pays interest at 10%, compounded annually. To find the future value of this annuity, we can think of it as the sum of 5 investments of $1000, one that draws interest for 4 years, (from the end of the first year to the end of the fifth year), one that draws interest for 3 years, etc. (See Table 3.20.)

TABLE 3.20

Investment	Future Value
Invested at end of 1st year	$1000(1.10)^4$
Invested at end of 2nd year	$1000(1.10)^3$
Invested at end of 3rd year	$1000(1.10)^2$
Invested at end of 4th year	$1000(1.10)^1$
Invested at end of 5th year	1000

Note that the last investment is at the end of the fifth year, so it earned no interest. The future value of the annuity is the sum of the separate future values. It is

$$S = 1000 + 1000(1.10)^1 + 1000(1.10)^2 + 1000(1.10)^3 + 1000(1.10)^4 = $$

$$6105.10 \text{ dollars.}$$

In Section 6.4 we will develop the following formula to find the future value of an annuity.

FUTURE VALUE OF AN ORDINARY ANNUITY

If R dollars are contributed at the end of each period for n periods into an annuity that pays interest at rate i at the end of the period, the future value of the annuity is

$$S = R\left[\frac{(1 + i)^n - 1}{i}\right] \text{ dollars.}$$

Note that the interest rate used in the above formula is the rate per period, not the rate for a year. We can use this formula to find the future value of an annuity in which $1000 is invested at the end of each year for 5 years into an account that pays interest at 10%, compounded annually. This future value, which we found earlier without the formula, is

$$S = R\left[\frac{(1 + i)^n - 1}{i}\right] = 1000\left[\frac{(1 + 0.10)^5 - 1}{0.10}\right] = 6105.10 \text{ dollars.}$$

EXAMPLE 1   Future Value of an Ordinary Annuity

Find the 5-year future value of an ordinary annuity with a contribution of $500 per quarter into an account that pays 12% per year, compounded quarterly.

Solution

The payments and interest compounding occur quarterly, so the interest rate per period is $\frac{0.12}{4} = 0.03$ and the number of compounding periods is $4(5) = 20$. Substituting the information into the formula for the future value of an annuity gives

$$S = 500\left[\frac{(1 + 0.03)^{20} - 1}{0.03}\right] = 13{,}435.19.$$

Thus the future value of this investment is $13,435.19.   ■

EXAMPLE 2   Future Value of an Annuity

Suppose $200 is deposited at the end of each month into an account that pays interest 12% per year, compounded monthly. Find the future value for every 4-month period, for up to 36 months.

Solution

To find the future value of this annuity, we note that the interest rate per period (month) is 12%/12 = .01, so the future value at the end of n months is given by the function

$$S = 200\left(\frac{1.01^n - 1}{0.01}\right) \text{dollars.}$$

This model for the future value of an ordinary annuity is a continuous function with discrete interpretation because the future value changes only at the end of each period. A graph of the function and the table of future values for every 4 months up to 36 months are shown on page 304 in Figure 3.38 and Table 3.21, respectively. Keep in mind that the graph should be interpreted discretely, that is, only positive integer inputs and corresponding outputs make sense.

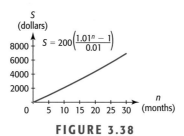

$$S = 200\left(\frac{1.01^n - 1}{0.01}\right)$$

FIGURE 3.38

TABLE 3.21

Number of Months	Value of Annuity (dollars)
4	812.08
8	1657.13
12	2536.50
16	3451.57
20	4403.80
24	5394.69
28	6425.82
32	7498.81
36	8615.38

Present Value of an Annuity

Many retirees purchase annuities that give them regular payments through a period of years. Suppose that a lump sum of money is invested to return payments of R at the end of each of n periods, with the investment earning interest at rate i per period, after which \$0 will remain in the account. This lump sum is the **present value** of this ordinary annuity, and it is equal to the sum of the present values of each of the future payments. Table 3.22 shows the present value of each payment of this annuity.

TABLE 3.22

Payment	Present Value of This Payment
Paid at end of 1st period	$R(1 + i)^{-1}$
Paid at end of 2nd period	$R(1 + i)^{-2}$
Paid at end of 3rd period	$R(1 + i)^{-3}$
.	.
.	.
Paid at end of $(n - 1)$st period	$R(1 + i)^{-(n-1)}$
Paid at end of nth period	$R(1 + i)^{-n}$

We must have enough money in the account to provide the present value of each of these payments, so the present value of the annuity is

$$A = R(1 + i)^{-1} + R(1 + i)^{-2} + R(1 + i)^{-3} + \ldots + R(1 + i)^{-(n-1)} + R(1 + i)^{-n}$$

We will show in Section 6.4 that this sum can be written in the form

$$A = R\left[\frac{1 - (1 + i)^{-n}}{i}\right]$$

PRESENT VALUE OF AN ORDINARY ANNUITY

If a payment of R is to be made at the end of each period for n periods from an account that earns interest at a rate of i per period, then the account is an **ordinary annuity**, and the **present value** is

$$A = R\left[\frac{1 - (1 + i)^{-n}}{i}\right].$$

EXAMPLE 3 Present Value of an Annuity

Suppose a retiring couple wants to establish an annuity that will provide $2000 at the end of each month for 20 years. If the annuity earns 6%, compounded monthly, how much must the couple put in the account to establish the annuity?

Solution

We seek the present value of an annuity that pays $2000 at the end of each month for $12(20) = 240$ months, with interest at $\frac{6\%}{12} = \frac{0.06}{12} = 0.005$ per month. The present value is

$$A = 2000\left[\frac{1 - (1 + 0.005)^{-240}}{0.005}\right] = 279{,}161.54 \text{ dollars.}$$

Thus the couple can receive a payment of $2000 at the end of each month for 20 years if they put a lump sum of $279,161.54 in an annuity. (Note that the sum of the money they will receive over the 20 years is $2000 \cdot 20 \cdot 12 = \$480{,}000$.)   ∎

EXAMPLE 4 Home Mortgage

A couple that wants to purchase a home has $30,000 for a down payment and wants to make monthly payments of $2200. If the interest rate for a 25-year mortgage is 6% per year on the unpaid balance, how much can they spend on a house?

Solution

The amount of money that they can pay for the house is the sum of the down payment and the present value of the payments that they can afford to pay. The present value of the $12(25) = 300$ monthly payments of $2200 with interest at $\frac{6\%}{12} = \frac{0.06}{12} = 0.005$ per month is

$$A = 2200\left[\frac{1 - (1 + .005)^{-300}}{.005}\right] \approx 341{,}455$$

The total amount that they can spend on a house is $341{,}455 + 30{,}000 = \$371{,}455$.   ∎

Loan Repayment

When money is borrowed, the borrower must repay the total amount that was borrowed (the debt) plus interest on that debt. Prior to the passage of a "truth in lending" law, there were several types of repayment plans, some of which were misleading to consumers. Most loans (including those for houses, for cars, and for other consumer goods) now require regular payments on the debt plus payment of interest on the unpaid balance of the loan. That is, loans are now paid off by a series of partial payments with interest charged on the unpaid balance at the end of each period. The stated interest rate (the **nominal** rate) is the *annual* rate. The rate per period is the nominal rate divided by the number of payment periods per year.

There are two popular repayment plans for these loans. One plan applies an equal amount to the debt each payment period plus the interest for the period, which is the interest on the unpaid balance. For example, a loan of $240,000 for 10 years at 12% could be repaid with 120 monthly payments of $2000 plus 1% (1/12 of 12%) of the unpaid balance each month. When this payment method is used, the payments will decrease as the unpaid balance decreases. A few sample payments are shown in Table 3.23.

TABLE 3.23

Payment Number (month)	Unpaid Balance	Interest on the Unpaid Balance	Balance Reduction	Monthly Payment	New Balance
1	$240,000	$2400	$2000	$4400	$238,000
2	238,000	2380	2000	4380	236,000
.					
51	140,000	1400	2000	3400	138,000
.					
101	40,000	400	2000	2400	38,000
.					
118	6000	60	2000	2060	4000

A loan of this type can also be repaid by making all payments (including the payment on the balance and the interest) of equal size. The process of repaying the loan in this way is called **amortization**. If a bank makes a loan of this type, it uses an amortization formula that gives the size of the equal payments. This formula is developed by considering the loan as an annuity purchased from the borrower by the bank, with the annuity paying a fixed return to the bank each payment period. The lump sum that the bank gives to the borrower (the principal of the loan) is the present value of this annuity, and each payment that the bank receives from the borrower is a payment from this annuity. To find the size of these equal payments, we solve the formula for the present value of an annuity,

$$A = R\left[\frac{1 - (1 + i)^{-n}}{i}\right],$$ for R. This gives the following formula.

AMORTIZATION FORMULA

If a debt of $\$A$, with interest at a rate of i per period, is amortized by n equal periodic payments made at the end of each period, then the size of each payment is

$$R = A\left[\frac{i}{1 - (1 + i)^{-n}}\right].$$

We can use this formula to find the *equal* monthly payments that will amortize the loan of $240,000 for 10 years at 12%. The interest rate per month is .01, and there are 120 periods.

$$R = A\left[\frac{i}{1 - (1 + i)^{-n}}\right] = 240,000\left[\frac{0.01}{1 - (1 + 0.01)^{-120}}\right] \approx 3443.303$$

Thus repaying this loan requires 120 equal payments of $3443.31. If we compare this payment with the selected payments in Table 3.23, we see that the first payment plan in Table 3.23 is nearly $1000 more than the equal amortization payment and that every payment after the 101st is $1000 less than the amortization payment. The advantage of the amortization method of repaying the loan is that the borrower can budget better by knowing what the payment will be each month. Of course, the balance of the loan will not be reduced as fast with this payment method, because most of the monthly payment will be needed to pay interest in the first few months. The partial spreadsheet in Table 3.24 shows how the outstanding balance is reduced.

TABLE 3.24

	A	B	C	D	E	F
1	Payment	Unpaid		Interest	Balance	New
2	Number	Balance	Interest	Payment	Reduction	Balance
3	1	240,000	2400	3443.31	1043.31	238,956.69
4	2	238,956.69	2389.57	3443.31	1053.74	237,902.95
.	.					
53	51	172,745.38	1727.45	3443.31	1715.86	171,029.52
.	.					
103	101	63,136.30	631.36	3443.31	2811.95	60,324.35

EXAMPLE 5 | Home Mortgage

A couple that wants to purchase a home with a price of $230,000 has $50,000 for a down payment. If they can get a 25-year mortgage at 9% per year on the unpaid balance:

a. What will their monthly payments be?
b. What is the total amount they will pay before they own the house outright?
c. How much interest will they pay?

Solution

a. The amount of money that they must borrow is $230,000 − $50,000 = $180,000. The number of monthly payments is 12(25) = 300, and the interest rate is $\frac{9\%}{12} = \frac{0.09}{12} = 0.0075$ per month. The monthly payment is

$$R = 180,000\left[\frac{.0075}{1 - (1.0075)^{-300}}\right] \approx 1510.553,$$

so the required payment would be $1510.56.

b. The amount they must pay before owning the house is the down payment plus the total of the 300 payments:

$$\$50{,}000 + 300(\$1510.56) = \$503{,}168.$$

c. The total interest is the total amount paid minus the price paid for the house. The interest is

$$\$503{,}168 - \$230{,}000 = \$273{,}168. \qquad \blacksquare$$

When the interest paid is the periodic interest rate on the unpaid balance of the loan, the nominal rate is also called the **annual percentage rate** (APR). You will occasionally see an advertisement for a loan with a stated interest rate and a different, larger APR. This can happen if a lending institution charges fees (sometimes called points) in addition to the stated interest rate. Federal law states that all these charges must be included in computing the APR, which is the true interest rate that is being charged on the loan.

3.6 SKILLS CHECK

Give answers to two decimal places.

1. Solve $S = P(1 + i)^n$ for P with positive exponent n.

2. Solve $S = P(1 + i)^n$ for P, with no denominator.

3. Solve $Ai = R[1 - (1 + i)^{-n}]$ for A.

4. Evaluate $2000\left[\dfrac{1 - (1 + .01)^{-240}}{.01}\right]$.

5. Solve $A = R\left[\dfrac{1 - (1 + i)^{-n}}{i}\right]$ for R.

6. Evaluate $240{,}000\left[\dfrac{.01}{1 - (1 + .10)^{-120}}\right]$.

3.6 EXERCISES

----

1. *Present Value* What lump sum investment will grow to $10,000 in 10 years, if it is invested at 6%, compounded annually?

2. *Present Value* What lump sum investment will grow to $30,000 if it is invested for 15 years at 7%, compounded annually?

3. *College Tuition* New parents want to put a lump sum into a money market fund to provide $30,000 in 18 years, to help pay for college tuition for their child. If the fund averages 10% per year, compounded monthly, how much should they invest?

4. *Trust Fund* Grandparents decide to put a lump sum of money into a trust fund on their granddaughter's tenth birthday so that she will have $1,000,000 on her 60th birthday. If the fund pays 11%, compounded monthly, how much money must they put in the account?

5. *Annuities* Find the present value of an annuity that will pay $1000 at the end of each year for 10 years if the interest rate is 7%, compounded annually.

6. *Annuities* Find the present value of an annuity that will pay $500 at the end of each year for 20 years if the interest rate is 9%, compounded annually.

7. *Lottery Winnings* The winner of a "million dollar" lottery is to receive $50,000 plus $50,000 per year for 19 years, or the present value of this annuity in cash. How much cash would she receive, if money is worth 8%, compounded annually?

8. *College Tuition* A couple wants to establish a fund that will provide $3000 for tuition at the end of each 6-month period for 4 years. If a lump sum can be placed in an account that pays 8%, compounded semiannually, what lump sum is required?

9. *Insurance Payment* A man is disabled in an accident and wants to receive an insurance payment that will provide him with $3000 at the end of each month for 30 years. If the payment can be placed in an account that pays 9%, compounded monthly, what size payment should he seek?

10. *Auto Leasing* A woman wants to lease rather than buy a car but does not want to make monthly payments. A dealer has the car she wants, which leases for $400 at the end of each month for 48 months. If money is worth 8%, compounded monthly, what lump sum should she offer the dealer to keep the car for 48 months?

11. *Business Sale* A man can sell his Thrifty Electronics business for $800,000 cash or for $100,000 plus $122,000 at the end of each year for 9 years.
    a. Find the present value of the annuity that is offered if money is worth 10%, compounded annually.
    b. If he takes the $800,000, spends $100,000 of it, and invests the rest in a 9-year annuity at 10%, compounded annually, what size annuity payment will he receive at the end of each year?
    c. Which is better, taking the $100,000 and the annuity or taking the cash settlement? Discuss the advantages of your choice.

12. *Sale of a Practice* A physician can sell her practice for $1,200,000 cash or for $200,000 plus $250,000 at the end of each year for 5 years.
    a. Find the present value of the annuity that is offered if money is worth 7%, compounded annually.
    b. If she takes the $1,200,000, spends $200,000 of it, and invests the rest in a 5-year annuity at 7%, compounded annually, what size annuity payment will she receive at the end of each year?
    c. Which is better, taking the $200,000 and the annuity or taking the cash settlement? Discuss the advantages of your choice.

13. *Home Mortgage* A couple wants to buy a house and can afford to pay $1600 per month.
    a. If they can get a loan for 30 years with interest at 9% per year on the unpaid balance, how much can they pay for a house?
    b. What is the total amount paid over the life of the loan?
    c. What is the total interest paid on the loan?

14. *Auto Loan* A man wants to buy a car and can afford to pay $400 per month.
    a. If he can get a loan for 48 months with interest at 12% per year on the unpaid balance, how much can he pay for a car?
    b. What is the total amount paid over the life of the loan?
    c. What is the total interest paid on the loan?

15. *Loan Repayment* A loan of $10,000 is to be amortized with quarterly payments over 4 years. If the interest on the loan is 8% per year, paid on the unpaid balance:
    a. What is the interest rate charged each quarter on the unpaid balance?
    b. How many payments are made to repay the loan?
    c. What payment is required quarterly to amortize the loan?

16. *Loan Repayment* A loan of $36,000 is to be amortized with monthly payments over 6 years. If the interest on the loan is 6% per year, paid on the unpaid balance:
    a. What is the interest rate charged each month on the unpaid balance?
    b. How many payments are made to repay the loan?
    c. What payment is required each month to amortize the loan?

17. *Home Mortgage* A couple who wants to purchase a home with a price of $350,000 has $100,000 for a down payment. If they can get a 30-year mortgage at 6% per year on the unpaid balance:
    a. What will their monthly payments be?
    b. What is the total amount they will pay before they own the house outright?
    c. How much interest will they pay over the life of the loan?

18. *Business Loan* Business partners want to purchase a restaurant that costs $750,000. They have $300,000 for a down payment and they can get a 25-year business loan for the remaining funds at 8% per year on the unpaid balance, with quarterly payments.
    a. What will the payments be?
    b. What is the total amount that they will pay over the 25-year period?
    c. How much interest will they pay over the life of the loan?

3.7 Logistic and Gompertz Functions

DVD players entered U.S. households faster than any other piece of home-electronics equipment in history. The past and projected sales from 1999 to 2004 are shown in Figure 3.39. Notice that the graph is an S-shaped curve with a relatively slow start, then a steep climb followed by a leveling off. It is reasonable that sales would eventually level off, even if everyone in the United States bought one. Very seldom will exponential growth continue indefinitely, so functions of this type better represent many types of sales and organizational growth. Two functions that are characterized by rapid growth followed by leveling off are **logistic functions** and **Gompertz functions**. We consider applications of these functions in this section.

Logistic Functions

When growth begins slowly, then increases at a rapid rate and finally slows over time to a rate that is almost zero, the amount (or number) present at any given time frequently fits on an S-shaped curve. Growth of business organizations and the spread of a virus or disease sometimes occur according to this pattern. For example, the total number y of people on a college campus infected by a virus can be modeled by

$$y = \frac{10,000}{1 + 9999e^{-0.99t}},$$

where t is the number of days after an infected student arrives on campus. Table 3.25 shows the number infected on selected days. Note that the number infected grows rapidly, and then levels off near 10,000. The graph of this function is shown in Figure 3.40.

Boffo Sales!

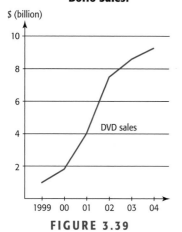

FIGURE 3.39

Source: *Newsweek* August 20, 2001

TABLE 3.25

Days	1	3	5	7	9	11	13	15	17
Number Infected	3	19	139	928	4255	8429	9749	9965	9995

A function that can be used to model growth of this type is called a **logistic function**; its graph is similar to the one shown in Figure 3.40.

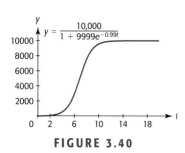

FIGURE 3.40

> ## LOGISTIC FUNCTION
>
> For real numbers a, b, and c, the function
>
> $$f(x) = \frac{c}{1 + ae^{-bx}}$$
>
> is a logistic function.
>
> If $a > 0$, a logistic function increases when $b > 0$ and decreases when $b < 0$.

The parameter c is often called the **limiting value*** or **upper limit** because the line $y = c$ is a horizontal asymptote for the logistic function. As seen in Figure 3.41, the line $y = 0$

---

*The logistic function fit by most technologies is a best-fit (least squares) logistic equation. There may therefore be data values that are larger than the value of c. To avoid confusion, we talk about the limiting value of the logistic function or the upper limit in the problem context.

is also a horizontal asymptote. Logistic growth begins at a rate that is near zero, rapidly increases in an exponential pattern, and then slows to a rate that is again near zero.

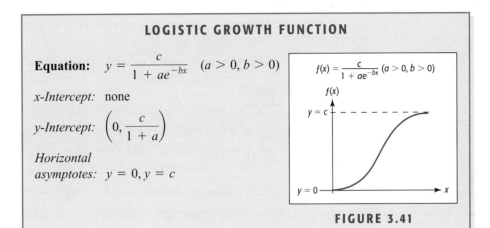

LOGISTIC GROWTH FUNCTION

Equation: $y = \dfrac{c}{1 + ae^{-bx}}$ $\quad (a > 0, b > 0)$

x-Intercept: none

y-Intercept: $\left(0, \dfrac{c}{1 + a}\right)$

Horizontal
asymptotes: $y = 0, y = c$

$f(x) = \dfrac{c}{1 + ae^{-bx}}$ $(a > 0, b > 0)$

FIGURE 3.41

Even though the graph of an increasing logistic function has the elongated S-shape indicated in Figure 3.41, it may be the case that we have data showing only a portion of the S. Consider the following example, which illustrates this.

EXAMPLE 1   Life Span

The data in Table 3.26 shows the expected life span of people for certain birth years in the United States. A technology-determined logistic function* for the life span data is

$$y = \frac{79.514}{1 + 0.835e^{-0.0298x}} \text{ years}$$

TABLE 3.26

Birth Year	Life Span (in years)	Year	Life Span (in years)
1920	54.1	1988	74.9
1930	59.7	1989	75.1
1940	62.9	1990	75.4
1950	68.2	1991	75.5
1960	69.7	1992	75.5
1970	70.8	1993	75.5
1975	72.6	1994	75.7
1980	73.7	1995	75.8
1987	75.0	1996	76.1

---

*Note that because of the exponential term in a logistic model, you must align input data to reasonably small values before using technology to fit a logistic model to data.

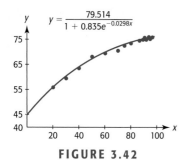

FIGURE 3.42

where x is the number of birth years after 1900. The graph of this logistic function is shown in Figure 3.42.

Use this model to:

a. Estimate the expected life of a person born in the United States in 1955 and a person born in 1980.

b. Find an upper limit for a person's expected life span according to this model.

Solution

a. To find the expected life span for people born in 1955, we evaluate the function for $x = 55$. This gives an expected life span of $y = \dfrac{79.514}{1 + 0.835e^{-0.0298(55)}} = 68.4$, or approximately 68 years. The expected life span for people born in 1980 is found by evaluating y when $x = 80$. This evaluation gives a life span of $y = \dfrac{79.514}{1 + 0.835e^{-0.0298(80)}} = 73.8$, or approximately 74 years.

b. By substituting larger values for x in the equation $y = \dfrac{79.514}{1 + 0.835e^{-0.0298x}}$, we see that the outputs increase very slowly and approach the value 79.514. (See Figure 3.43). Thus the upper limit of life span, according to this model, is about 79.5 years.

	A	B
1	100	76.279
2	150	78.761
3	200	79.343
4	250	79.475
5	300	79.505
6	350	79.512
7	400	79.514

X	Y₁
100	76.279
150	78.761
200	79.343
250	79.475
300	79.505
350	79.512
400	79.514

X=400

FIGURE 3.43 ■

A logistic function can also decrease; this behavior occurs when $b < 0$. Demand functions for products often follow such a pattern. Decreasing logistic functions can also be used to represent decay over time. Figure 3.44 shows a decreasing logistic function.

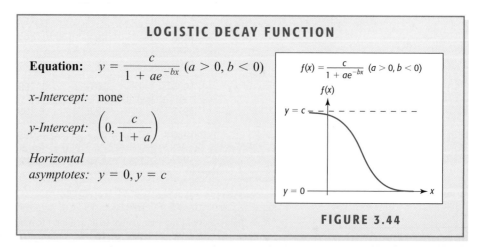

LOGISTIC DECAY FUNCTION

Equation: $y = \dfrac{c}{1 + ae^{-bx}}$ $(a > 0, b < 0)$

x-Intercept: none

y-Intercept: $\left(0, \dfrac{c}{1 + a}\right)$

Horizontal asymptotes: $y = 0, y = c$

$f(x) = \dfrac{c}{1 + ae^{-bx}}$ $(a > 0, b < 0)$

FIGURE 3.44

The properties of logarithms can also be used to solve for a variable in an exponent in a logistic function.

EXAMPLE 2   Expected Life Span

The expected life span at birth of people born in the United States can be modeled by the equation

$$y = \frac{79.514}{1 + 0.835e^{-0.0298x}}$$

where x is the number of years from 1900.

Use this model to estimate the birth year after 1900 that gives an expected life span of 65 years with

a. graphical methods.

b. analytical methods.

Solution

a. To solve the equation

$$65 = \frac{79.514}{1 + 0.835e^{-0.0298x}}$$

with the x-intercept method, we solve

$$0 = \frac{79.514}{1 + 0.835e^{-0.0298x}} - 65.$$

Graphing the function

$$y = \frac{79.514}{1 + 0.835e^{-0.0298x}} - 65$$

on the window $[0, 55]$ by $[-2, 3]$ shows a point where the graph crosses the x-axis (see Figure 3.45(a)). Finding the x-intercept gives the approximate solution

$$x = 44.26.$$

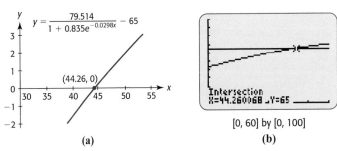

(a)   (b)

FIGURE 3.45

Because 44 years after the end of 1900 would be the end of 1944, we conclude that people born in 1945 have an expected life span of 65 years.

We can also use the intersection method to solve the equation graphically. To do this, we graph the equations $y_1 = 65$ and $y_2 = \dfrac{79.514}{1 + 0.835e^{-0.0298x}}$ and find the point of intersection of the graphs. (See Figure 3.45(b).)

b. To show that there is only one solution, we solve the equation

$$65 = \frac{79.514}{1 + 0.835e^{-0.0298x}}$$

analytically. To solve this equation analytically, we first multiply both sides by the denominator and then solve for $e^{-0.0298x}$:

$$65(1 + 0.835e^{-0.0298x}) = 79.514$$

$$65 + 54.275e^{-0.0298x} = 79.514$$

$$54.275e^{-0.0298x} = 14.514$$

$$e^{-0.0298x} = \frac{14.514}{54.275}$$

Taking the natural logarithm of both sides of the equation and using Logarithmic Property 3 are the next steps in the solution:

$$\ln e^{-0.0298x} = \ln \frac{14.514}{54.275}$$

$$-0.0298x = \ln \frac{14.514}{54.275}$$

$$x = \frac{\ln \dfrac{14.514}{54.275}}{-0.0298} \approx 44.260$$

This solution is the same solution that was found (more easily) by using the graphical method in Figure 3.45(a); 1945 is the birth year for people with an expected life span of 65 years. ■

Gompertz Functions

Another type of function that models rapid growth that eventually levels off is called a **Gompertz function**. This type of function can also be used to describe human growth and development, and the growth of organizations in a limited environment. These functions have equations of the form

$$N = Ca^{R^t},$$

where t represents the time, R $(0 < R < 1)$ is the expected rate of growth of the population, a represents the proportion of initial growth, and C is the maximum possible number of individuals.

For example, the equation

$$N = 3000(0.2)^{0.6^t}$$

could be used to predict the number of employees t years after the opening of a new facility. Here the maximum number of employees C would be 3000, the proportion of the ini-

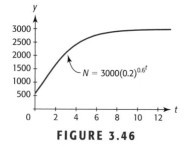

FIGURE 3.46

tial growth a is 0.2, and R is 0.6. The graph of this function is shown in Figure 3.46. Observe that the initial number of employees, at $t = 0$, is

$$N = 3000(0.2)^{0.6^0} = 3000(0.2)^1 = 3000(0.2) = 600.$$

Because $0.6 < 1$, higher powers of t make 0.6^t smaller, with 0.6^t approaching 0 as t approaches ∞. Thus $N \to 3000(0.2)^0 = 3000(1) = 3000$, so the maximum number possible of employees is 3000.

EXAMPLE 3   **Deer Population**

The Gompertz equation

$$N = 1000(0.06)^{0.2^t}$$

predicts the size of a deer herd on a small island t decades from now.

a. What is the number of deer on the island now ($t = 0$)?

b. How many deer are predicted to be on the island 1 decade from now ($t = 1$)?

c. Graph the function.

d. What is the maximum number of deer predicted by this model?

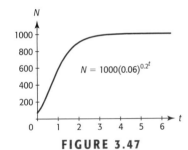

FIGURE 3.47

Solution

a. If $t = 0$, $N = 1000(0.06)^{0.2^0} = 1000(0.06)^1 = 60$.

b. If $t = 1$, $N = 1000(0.06)^{0.2^1} = 1000(0.06)^{0.2} = 570$ (approximately).

c. The graph is shown in Figure 3.47.

d. According to the model, the maximum number of deer is predicted to be 1000. Evaluating N as t gets large, we see the values of N approach, but never reach, 1000. We say that $N = 100$ is an asymptote for this graph. ∎

EXAMPLE 4   **Company Growth**

A new Dotcom company starts with 3 owners and 5 employees, but tells investors that it will grow rapidly, with the number of people in the company given by the model

$$N = 2000(.004)^{0.5^t}$$

where t is the number of years from the present. Use graphical methods to determine the year in which they predict that the number of employees will be 1000.

Solution

If the company has 1000 employees plus the 3 owners, the total number in the company will be 1003; so we seek to solve the equation

$$1003 = 2000(.004)^{0.5^t}.$$

Entering the equations $y_1 = 1003$ and $y_2 = 2000(0.004)^{0.5^x}$ in a graphing utility* and using the intersection method gives $x = 3$. (See Figure 3.48.) Because x represents t, the number of years, the owners predict that they will have 1000 employees in 3 years.

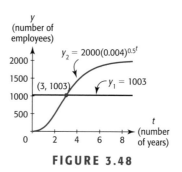

FIGURE 3.48 ∎

3.7 SKILLS CHECK

Give approximate answers to two decimal places.

1. Evaluate $\dfrac{79.514}{1 + 0.835e^{-0.0298(80)}}$.

2. If $y = \dfrac{79.514}{1 + 0.835e^{-0.0298x}}$ find
  a. y when $x = 10$.
  b. y when $x = 50$.

3. Evaluate $1000(0.06)^{0.2^t}$ for $t = 4$ and for $t = 6$.

4. Evaluate $2000(.004)^{0.5^t}$ for $t = 5$ and for $t = 10$.

5. a. Graph the function $f(x) = \dfrac{100}{1 + 3e^{-x}}$ for $x = 0$ to $x = 15$.

  b. Find $f(0)$ and $f(10)$.
  c. Is this function increasing or decreasing?

d. What is the limiting value of this function?

6. a. Graph $f(t) = \dfrac{1000}{1 + 9e^{-0.9t}}$ for $t = 0$ to $t = 15$.
  b. Find $f(2)$ and $f(5)$.
  c. What is the limiting value of this function?

7. a. Graph $y = 100(0.05)^{0.3^x}$ for $x = 0$ to $x = 10$.
  b. What is the initial value of this function (the y-value when $x = 0$)?
  c. What is the limiting value of this function?

8. a. Graph $N = 2000(.004)^{0.5^t}$ for $t = 0$ to $t = 10$.
  b. What is the initial value of this function (the y-value when $x = 0$)?
  c. What is the limiting value of this function?

3.7 EXERCISES

1. *Spread of a Disease* The spread of a highly contagious virus in a high school can be described by the logistic function $y = \dfrac{5000}{1 + 1000e^{-0.8x}}$, where x is the number of days after the virus is identified in the school and y is the total number of people who are infected by the virus.
  a. Graph the function for $0 \le x \le 15$.
  b. How many students had the virus when it was first discovered?

  c. What is the upper limit of the number infected by the virus during this period?

2. *Population Growth* Suppose that the size of a population of an island is given by $p(t) = \dfrac{98}{1 + 4e^{-0.1t}}$ thousand people, where t is the number of years after 1988.
  a. Graph this function for $0 \le x \le 30$.
  b. Find and interpret $p(10)$.
  c. Find and interpret $p(100)$.

*Enter $y_2 = 2000(0.004)$^(0.5^x) in a graphing utility.

d. What appears to be an upper limit for the size of this population?

3. *Sexually Active Boys* The percent of boys between ages 15 and 20 that have been sexually active at some time (the cumulative percent) can be modeled by the logistic function $y = \dfrac{89.786}{1 + 4.6531e^{-0.8256x}}$, where x is the number of years past age 15.

a. Graph the function for $0 \le x \le 5$.

b. What does the model estimate the cumulative percent to be for boys whose age is 16?

c. What cumulative percent does the model estimate for boys of age 21?

d. What is the limiting value implied by this model?

(Source: "National Longitudinal Survey of Youth," *Risking the Future*. Washington D.C.: National Academy Press, 1987)

4. *Sexually Active Girls* The percent of girls between ages 15 and 20 that have been sexually active at some time (the cumulative percent) can be modeled by the logistic function $y = \dfrac{83.84}{1 + 13.9233e^{-0.9248x}}$, where x is the number of years past age 15.

a. Graph the function for $0 \le x \le 5$.

b. What does the model estimate the cumulative percent to be for girls of age 16?

c. What cumulative percent does the model estimate for girls of age 20?

d. What is the upper limit implied by the given logistic model?

(Source: "National Longitudinal Survey of Youth," *Risking the Future*. Washington D.C.: National Academy Press, 1987)

5. *Spread of a Rumor* The number of people in a small town who are reached by a rumor about the mayor and an intern is given by $N = \dfrac{10,000}{1 + 100e^{-0.8t}}$, where t is the number of days after the rumor begins.

a. How many people will have heard the rumor by the end of the first day?

b. How many will have heard the rumor by the end of the fourth day?

c. Use graphical or numerical methods to find the day in which 7300 people in town have heard the rumor.

6. *Sexually Active Girls* The cumulative percent of sexually active girls ages 15 to 20 is given in the table below.

a. Find the logistic function that models the data, with x equal to the number of years past age 15.

b. Does this model agree with the model used in Exercise 4?

c. Find the linear function that is the best fit for this data.

d. Graph each function on the same axes with the data points to determine which model appears to be the better fit for the data.

Age (years)	Cumulative Percent Sexually Active
15	5.4
16	12.6
17	27.1
18	44.0
19	62.9
20	73.6

(Source: "National Longitudinal Survey of Youth," *Risking the Future*. Washington D.C.: National Academy Press, 1987)

7. *Sexually Active Boys* The cumulative percent of sexually active boys ages 15 to 20 is given in the table below.

a. Create the logistic function that models the data, with x equal to the number of years past age 15.

b. Does this agree with the model used in Exercise 3?

c. Find the linear function that is the best fit for this data.

d. Graph each function on the same axes with the data points to determine which model appears to be the better fit for the data.

Age (years)	Cumulative Percent Sexually Active
15	16.6
16	28.7
17	47.9
18	64.0
19	77.6
20	83.0

(Source: "National Longitudinal Survey of Youth," *Risking the Future*. Washington D.C.: National Academy Press, 1987)

8. *Japan's Population* The table on page 318 gives the population of Japan for the years 1984–1999.

a. Find the logistic function that models the population N, using an input x equal to the number of years from 1980.

b. Comment on the goodness of fit of the model to the data.

Year	Population (in millions)	Year	Population (in millions)
1984	120.235	1992	124.452
1985	121.049	1993	124.764
1986	121.672	1994	125.034
1987	122.264	1995	125.570
1988	122.783	1996	125.864
1989	123.255	1997	126.166
1990	123.611	1998	126.486
1991	124.043	1999	126.686

(Source: www.jinjapan.org/stat/)

9. *Organizational Growth* A new community college predicts that its student body will grow rapidly at first and then begin to level off according to the Gompertz curve with equation $N = 10{,}000(0.4)^{0.2^t}$ students, where t is the number of years after the college opens.
  a. What does this model predict the number of students to be when the college opens?
  b. How many students are predicted to attend the college after 4 years?
  c. Graph the equation for $0 \le t \le 10$ and estimate an upper limit on the number of students at the college.

10. *Organizational Growth* A new technology company started with 6 employees and predicted that its number of employees would grow rapidly at first and then begin to level off according to the Gompertz function

$$N = 150(0.04)^{0.5^t},$$

where t is the number of years after the company started.
  a. What does this model predict the number of employees to be in 8 years?
  b. Graph the function for $0 \le t \le 10$ and estimate the maximum predicted number of employees.

11. *Sales Growth* The president of a company predicts that sales will increase rapidly after a new product is brought to the market and that the number of units sold monthly can be modeled by $N = 40{,}000(0.2)^{0.4^t}$, where t represents the number of months after the product is introduced.
  a. How many units will be sold by the end of the first month?

b. Graph the function for $0 \le t \le 10$.
c. What is the predicted upper limit on sales?

12. *Company Growth* Because of a new research grant, the number of employees in a firm is expected to grow, with the number of employees modeled by $N = 1600(0.6)^{0.2^t}$, where t is the number of years after the grant was received.
  a. How many employees did the company have when the grant was received?
  b. How many employees did the company have at the end of 3 years after the grant was received?
  c. What is the expected upper limit on the number of employees?
  d. Graph the function.

13. *Company Growth* Suppose that the number of employees in a new company is expected to grow, with the number of employees modeled by $N = 1000(0.01)^{0.5^t}$, where t is the number of years after the company was formed.
  a. How many employees did the company have when it started?
  b. How many employees did the company have at the end of 1 year?
  c. What is the expected upper limit on the number of employees?
  d. Use graphical or numerical methods to find the year in which 930 people are employed.

14. *Sales Growth* The president of a company predicts that sales will increase rapidly after a new advertising campaign, with the number of units sold weekly modeled by $N = 8000(0.1)^{0.3^t}$, where t represents the number of weeks after the advertising campaign begins.
  a. How many units per week were sold at the beginning of the campaign?
  b. How many units were sold at the end of the first week?
  c. What is the expected upper limit on the number of units sold per week?
  d. Use graphical or numerical methods to find the first week in which 6500 units were sold.

15. *Spread of Disease* An employee brings a contagious disease to an office with 150 employees. The number of employees infected by the disease t days after the employees are first exposed to it is given by

$$N = \frac{100}{1 + 79e^{-0.9t}}.$$

Use graphical or numerical methods to find the number of days until 99 employees have been infected.

16. *Advertisement* The number of people in a community of 15,360 who are reached by a particular advertisement t weeks after it begins is given by

$$N(t) = \frac{14,000}{1 + 100e^{-0.6t}}.$$

Use graphical or numerical methods to find the number of days until at least half of the community is reached by this advertisement.

17. *Population Growth* A pair of deer are introduced on a small island, and the population grows until the food supply and natural enemies of the deer on the island limit the population. If the number of deer is

$$N = \frac{180}{1 + 89(2^{-0.8t})},$$

where t is the number of years after the deer are introduced, how long does it take for the deer population to reach 150?

18. *Spread of Disease* A student brings a contagious disease to an elementary school of 1200 students. If the number of students infected by the disease t days after the students are first exposed to it is given by

$$N = \frac{800}{1 + 799e^{-0.9t}},$$

use numerical or graphical methods to find in how many days at least 500 students will be infected.

CHAPTER 3 *Summary*

In this chapter we discussed exponential and logarithmic functions and their applications. We explored the fact that logarithmic and exponential functions are inverses of each other. Many applications in today's world involve exponential and logarithmic functions, including the pH of substances, stellar magnitude, intensity of earthquakes, the growth of bacteria, the decay of radioactive isotopes, and compound interest. Logarithmic and exponential equations are solved by using both analytical and graphical solution methods.

Key Concepts and Formulas

Section 3.1 Exponential Functions

Exponential function   If a is a positive real number, $a \neq 1$, then the function $f(x) = a^x$ is an exponential function.

Exponential growth   Whenever the base b of $f(x) = a(b^x)$ is greater than 1 and $a > 0$, the exponential function is increasing and can be used to model exponential growth. The graph of an exponential growth model resembles the graph at the right. The x-axis (on the left) is a horizontal asymptote for the graph.

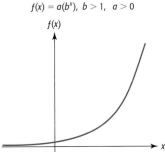

$f(x) = a(b^x),\ b > 1,\ a > 0$

$f(x)$

Exponential decay

Whenever the base b of $f(x) = a(b^{-x})$ is greater than 1 and $a > 0$, the exponential function is decreasing and can be used to model exponential decay. The graph of an exponential decay model resembles the graph at the right. The x-axis (on the right) is a horizontal asymptote for the graph.

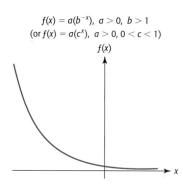

$f(x) = a(b^{-x})$, $a > 0$, $b > 1$
(or $f(x) = a(c^x)$, $a > 0$, $0 < c < 1$)

$f(x)$

Section 3.2 Logarithmic Functions

Logarithmic function

For $x > 0$, $a > 0$, and $a \neq 1$, the logarithmic function to the base a is $y = \log_a x$ which is defined by $x = a^y$.

The graph of $y = \log_a x$ for the base $a > 0$ is shown at the right. The y-axis is a vertical asymptote for the graph.

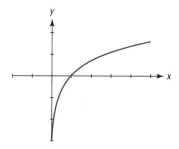

Logarithmic and exponential forms

These forms are equivalent:

$$y = \log_a x \text{ and } x = a^y$$

Common logarithms

Logarithms with a base of 10; $\log_{10} x$ is denoted $\log x$.

Natural logarithms

Logarithms with a base of e; $\log_e x$ is denoted $\ln x$.

Change of base

If $a > 0$, $a \neq 1$, $b > 0$, $b \neq 1$, and $x > 0$, then $\log_b x = \dfrac{\log_a x}{\log_a b}$.

Section 3.3 Solving Exponential Equations; Properties of Logarithms

Solving equations

Many exponential and logarithmic equations can be solved by converting between exponential and logarithmic form, that is, $y = \log_a x$ and $x = a^y$.

Properties of logarithms

For $a > 0$, $a \neq 1$, k a real number, and M and N positive real numbers,

Property 1. $\log_a a = 1$

Property 2. $\log_a 1 = 0$

Property 3. $\log_a a^x = x$

Property 4. $a^{\log_a x} = x$

Property 5. $\log_a(MN) = \log_a M + \log_a N$

Property 6. $\log_a\left(\dfrac{M}{N}\right) = \log_a M - \log_a N$

Property 7. $\log_a M^k = k \log_a M$

Exponential inequalities	Solving inequalities involving exponential functions involves rewriting the related equation in logarithmic form and solving the related logarithmic equation. Then graphical methods are used to find the values of the variable that satisfy the inequality.

Section 3.4 *Exponential and Logarithmic Models*

Exponential models	Many sources provide data that can be modeled by exponential growth and decay functions according to the model $y = ab^x$ for real numbers a and b.
Logarithmic models	Models can be created involving $\ln x$. Data that exhibits an initial rapid increase and then has a slow rate of growth can often be described by the function $f(x) = a + b \ln x$ for $b > 0$.

3.5 *Exponential Functions and Investing*

Future value with annual compounding	If P dollars are invested for n years at an interest rate r compounded annually, then the future value of P is given by $S = P(1 + r)^n$ dollars.
Future value of an investment	If P dollars are invested for t years at the annual interest rate r, where the interest is compounded k times per year, then the interest rate per period is $\frac{r}{k}$, the number of compounding periods is kt, and the future value that results is given by $S = P\left(1 + \frac{r}{k}\right)^{kt}$ dollars.
The number e	The number e is an irrational number that is approximated by 2.718281828, to nine decimal places.
Future value with continuous compounding	If P dollars are invested for t years at an annual interest rate r compounded continuously, then the future value S is given by $S = Pe^{rt}$ dollars.
Present value of an investment	The lump sum P that is necessary to have it grow to the future value S if invested for n periods at interest rate i per compounding period is

$$P = \frac{S}{(1 + i)^n} \quad \text{or} \quad P = S(1 + i)^{-n}$$

Section 3.6 *Annuities; Loan Repayment*

Future value of an ordinary annuity	If R dollars are contributed at the end of each period for n periods to an annuity that pays at rate i at the end of the period, the future value of the annuity is

$$S = R\left[\frac{(1 + i)^n - 1}{i}\right].$$

Present value of an ordinary annuity	If a payment of \$$R$ is to be made at the end of each period for n periods from an account that earns interest at a rate of i per period, then the account is an ordinary annuity, and the present value is

$$A = R\left[\frac{1 - (1 + i)^{-n}}{i}\right].$$

Amortization formula	If a debt of \$$A$, with interest at a rate of i per period, is amortized by n equal periodic payments made at the end of each period, then the size of each payment is

$$R = A\left[\frac{i}{1 - (1 + i)^{-n}}\right].$$

Section 3.7 *Logistic and Gompertz Functions*

Logistic function

For real numbers a, b, and c, the function

$$f(x) = \frac{c}{1 + ae^{-bx}}$$

is a logistic function.

A logistic function increases when $a > 0$ and $b > 0$. The graph of a logistic function resembles the graph at right.

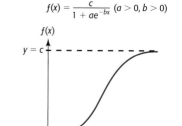

Logistic modeling

If data points indicate that growth begins slowly, then increases very rapidly and finally slows over time to a rate that is almost zero, the data may fit the logistic model

$$f(x) = \frac{c}{1 + ae^{-bx}}$$

Gompertz function

For real numbers C, a, and R $(0 < R < 1)$, the function $N = Ca^{R^t}$ is a Gompertz function. A Gompertz function increases rapidly then levels off as it approaches a limiting value. The graph of a Gompertz function resembles the graph at right.

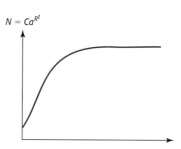

Chapter 3 Skills Check

1. a. Graph the function $f(x) = 4e^{-0.3x}$.
    b. Find $f(-10)$ and $f(10)$.

2. Is $f(x) = 4e^{-0.3x}$ an increasing or a decreasing function?

3. Graph $f(x) = 3^x$ on $[-10, 10]$ by $[-10, 20]$.

4. Graph $y = 3^{(x-1)} + 4$ on $[-10, 10]$ by $[-10, 20]$.

5. How does the graph of Exercise 4 compare with the graph of Exercise 3?

6. Is $y = 3^{(x-1)} + 4$ an increasing or a decreasing function?

7. a. If $y = 1000(2^{-0.1x})$, find the value of y when $x = 10$.
    b. Use graphical or numerical methods to determine the value of x that gives $y = 250$ if $y = 1000(2^{-0.1x})$.

In Exercises 8 and 9, write the following exponential equations in logarithmic form.

8. $x = 6^y$ **9.** $y = 7^{3x}$

In Exercises 10–12, write the following logarithmic equations in exponential form.

10. $y = \log_4 x$ **11.** $y = \log(x)$

12. $y = \ln x$

13. Write the inverse of $y = 4^x$ in logarithmic form.

In Exercises 14–16, evaluate the following logarithms, if possible. Give approximate solutions to four decimal places.

14. $\log 22$ **15.** $\ln 56$

16. $\log 10$

In Exercises 17–19, find the value of the following logarithms without using a calculator.

17. $\log_2 16$ **18.** $\ln(e^4)$

19. $\log 0.001$

In Exercises 20 and 21, use a change of base formula to evaluate each of the following logarithms. Give approximate solutions to four decimal places.

20. $\log_3(54)$ **21.** $\log_8(56)$

In Exercises 22 and 23, graph the following functions with technology.

22. $y = \ln(x - 3)$ **23.** $y = \log_3 x$

In Exercises 24–27, solve the following exponential equations. Give approximate solutions to four decimal places.

24. $340 = e^x$ **25.** $1500 = 300e^{8x}$

26. $9200 = 23(2^{3x})$ **27.** $4(3^x) = 36$

28. Rewrite $\ln\dfrac{(2x - 5)^3}{x - 3}$ as the sum, difference, or product of logarithms, and simplify if possible.

29. Rewrite $6 \log_4 x - 2 \log_4 y$ as a single logarithm.

30. Determine whether a linear or an exponential function is the better fit for the data in the table. Then find the equation of the function with three-decimal-point accuracy.

x	1	2	3	4	5
y	2	5	12	30	75

31. Evaluate $P\left(1 + \dfrac{r}{k}\right)^{kn}$ for $P = 1000$, $r = 8\%$, $k = 12, n = 20$, to two decimal places.

32. Evaluate $1000\left[\dfrac{1 - 1.03^{-240+n}}{0.03}\right]$ for $n = 120$, to two decimal places.

33. a. Graph $f(t) = \dfrac{2000}{1 + 8e^{-0.8t}}$ for $t = 0$ to $t = 10$.
   b. Find $f(0)$ and $f(8)$, to two decimal places.
   c. What is the limiting value of this function?

34. a. Graph $y = 500(0.1^{0.2^x})$ for $x = 0$ to $x = 10$.
   b. What is the initial value (the value of y when $x = 0$) of this function?
   c. What is the limiting value of this function?

Chapter 3 Review

1. *Prisoners* The total number of prisoners (in thousands) in state and federal prisons from 1970 to 1997 can be modeled by the function $y = 84.7518(1.0746^x)$, where x is the number of years after 1960. According to this model, how many prisoners were incarcerated in 1990? (Source: Index of Leading Cultural Indicators)

2. *Sales Decay* At the end of an advertising campaign, the daily sales (in dollars) declined, with daily sales given by the equation $y = 2000(2^{-0.1x})$, where x is the number of weeks after the end of the campaign. According to this model, what will sales be 4 weeks after the end of the ad campaign?

3. *Corporate Revenue* Prior to the November 1994 $1.7 billion takeover proposal by Quaker Oats, Snapple Beverage Corporation's annual revenues were given by the function $B(t) = 1.337e^{0.718t}$ million dollars, where t is the number of years after 1985. When does this model indicate that the annual revenue is more than $10 million? (Source: *U.S. News and World Report*, November 14, 1994)

4. *Earthquakes*
   a. If an earthquake has an intensity of 1000 times I_0, what is the magnitude of the earthquake?
   b. An earthquake that measured 6.5 on the Richter scale occurred in Pakistan in February 1991. Express this intensity in terms of I_0.

5. *Earthquakes* On January 23, 2001, an earthquake registering 7.9 hit western India, killing more than 15,000 people. On the same day, an earthquake registering 4.8 hit the United States, killing no one. How much more intense was the Indian earthquake than the American earthquake?

6. *Investments* The number of years needed for an investment of $10,000 to grow to $30,000, when money is invested at 12% compounded annually, is given by

$$t = \log_{1.12}\frac{30,000}{10,000} = \log_{1.12}3.$$

In how many years will this occur?

7. *Investments* If $1000 is invested at 10% compounded quarterly, the future value of the investment is given by $S = 1000(2^{x/7})$, where x is the number of years after the investment is made.
   a. Use the logarithmic form of this function to solve the equation for x.
   b. Find when the future value of the investment is $19,504.

8. *Sales Decay* At the end of an advertising campaign, the daily sales (in dollars) declined, with daily sales given by the equation $y = 2000(2^{-0.1x})$, where x is the number of weeks after the end of the campaign. In how many weeks will sales be half of the sales at the end of the ad campaign?

9. *Prisoners* The rate (number per 100,000 people in the U.S. population) of prisoners in state and federal prisons from 1970 to 1997 can be modeled by the function $y = 45.6786(1.06386^x)$, where x is the number of years from 1960. In what year does this model indicate a rate of 275 per 100,000?

10. *Mobile Home Sales* A company that buys and sells used mobile homes estimates its cost to be given by $C(x) = 2x + 50$ thousand dollars when x mobile homes are purchased. The same company estimates that its revenue from the sale of x mobile homes is given by $R(x) = 10(1.26)^x$ thousand dollars.
   a. Combine C and R into a single function that gives the profit for the company when x used mobile homes are bought and sold.
   b. How many mobile homes must the company sell so that revenue is at least $30,000 more than cost?

11. *Sales Decay* At the end of an advertising campaign, the sales (in dollars) declined, with weekly sales given by the equation $y = 40,000(3^{-0.1x})$, where x is the number of weeks after the end of the campaign. In how many weeks will sales be less than half of the sales at the end of the ad campaign?

12. *Carbon-14 Dating* An exponential decay function can be used to model the number of atoms of a radioactive material that remain after a period of time. Carbon-14 decays over time, with the amount remaining after t years given by

    $$y = y_0 e^{-0.00012378t}$$

    where y_0 is the original amount.
   a. If a sample of carbon-14 weighs 100 grams originally, how many grams will be present in 5000 years?

   b. If a sample of wood at an archeological site contains 36% as much carbon-14 as living wood, determine when the wood was cut.

13. *Purchasing Power* If a retired couple has a fixed income of $60,000 per year, the purchasing power (adjusted value of the money) after t years of 5% inflation is given by the equation $P = 60,000e^{-0.05t}$. In how many years will the purchasing power of their income fall below half of their current income?

14. *Investments* If $2000 is invested at 8% compounded continuously, the future value of the investment after t years is given by $S = 2000e^{0.08t}$ dollars. What is the future value of this investment in 10 years?

15. *Investments* If $3300 is invested for x years at 10% compounded annually, the future value that will result is $S = 3300(1.10)^x$ dollars. In how many years will the investment result in $13,784.92?

16. *Smokers* The table below gives the percent of black females over 18 years of age who are smokers in each of the years 1965–1995.
   a. Find an exponential equation that models this percent, with x equal to 0 in 1960.
   b. Use the model to estimate the percent of black females who are smokers in 1972 and in 1996.

Years	Percent
1965	38.1
1974	34.1
1979	33.8
1983	35.5
1985	30.4
1987	26.1
1990	22.5
1991	22.4
1992	24.9
1993	22.9
1994	25.2
1995	21.8

(Source: Centers for Disease Control, www.cdc.gov/nchs/datawh/statabl/)

17. *Smokers* The percent of the U.S. population who are smokers for the years 1965–1995 is given in the table.
   a. Find an exponential model for this data, where x is the number of years from 1960.
   b. Use the unrounded model to estimate the percent in 2003.

Year	Percent	Year	Percent
1965	51.6	1990	28
1974	42.9	1991	27.5
1979	37.2	1992	28.2
1983	34.7	1993	27.5
1985	32.1	1994	27.8
1987	31	1995	26.7

(Source: Centers for Disease Control, www.cdc.gov/nchs/datawh/statabl/)

18. *Out-of-Wedlock Births* The percent of all teen mothers who were unmarried during the years 1960–1996 is shown in the table below.
   a. Find an exponential function to fit the data, using as input the number of years after 1950.
   b. Find a logistic function to fit this data, using as input the number of years after 1950.
   c. Graph both models with the data to determine which model is the better fit for the data.

Year	Teen Mothers Who Are Unwed (percent)	Year	Teen Mothers Who Are Unwed (percent)
1960	15	1991	69
1970	30	1992	71
1980	48	1993	72
1986	61	1994	76
1990	68	1996	76

19. *Pupil Expenditures in Schools* The per pupil expenditure on public elementary and secondary schools, in constant 1997–98 dollars, is given in the table.
   a. Find an exponential function to fit the data, using as input the number of years after 1950.
   b. Find a logistic function to fit this data, using as input the number of years after 1950.

c. Graph both models with the data to determine which model is the better fit for the data.

Years	Spending per Pupil	Years	Spending per Pupil
1960	2422	1992	6587
1965	3077	1993	6587
1970	3764	1994	6633
1975	4444	1995	6676
1980	4770	1996	6745
1985	5285	1997	6814
1990	6591	1998	6943
1991	6626		

(Source: U.S. Department of Education)

20. *Mothers in the Workforce* The data shown in the table below gives the percent of moms who returned to the workforce within 1 year after they had a child (for selected years 1976 to 1998). Find a logarithmic model for the data, with x equal to the number of years after 1975.

Year	Returning Mothers (percent)	Year	Returning Mothers (percent)
1976	31	1988	51
1978	36	1990	52
1980	39	1992	53
1982	43	1994	52
1984	48	1995	55
1986	50	1998	59

(Source: Associated Press)

21. *Japan's Population* The table on page 326 gives the population of Japan, in millions, for the years 1984–1999.
   a. Find a logarithmic function that models the data, with input equal to the number of years after 1980.
   b. What does the model predict as the population in 2004?

Year	Population (in millions)	Year	Population (in millions)
1984	120.235	1992	124.452
1985	121.049	1993	124.764
1986	121.672	1994	125.034
1987	122.264	1995	125.570
1988	122.783	1996	125.864
1989	123.255	1997	126.166
1990	123.611	1998	126.486
1991	124.043	1999	126.686

(Source: www.jinjapan.org/stat/)

22. *Hospital Discharges* The number of discharges per 1000 people in nonfederal short-stay hospitals is given in the table below.
(Source: National Hospital Discharge Survey)
   a. Align the data as the number of years after 1979 and find a logarithmic model to fit the data.
   b. Graph the model and the data on the same graph.
   c. Use the model in part (a) to predict the number of hospital discharges per 1000 people in 2005.

Year	Discharges per 1000 People
1980	158.5
1985	137.7
1988	117.6
1990	113.0
1992	110.5
1994	106.5
1996	102.3

23. *Investment* Suppose that $12,500 is invested in an account earning 5% annual interest compounded continuously. What is the future value in 10 years?

24. *Investments* Find the 7-year future value of an investment of $20,000 placed into an account that pays 6% compounded annually.

25. *Annuities* At the end of each quarter, $1000 is placed into an account that pays 12%, compounded quar-

terly. What is the future value of this annuity in 6 years?

26. *Annuities* Find the 10-year future value of an ordinary annuity with a contribution of $1500 at the end of each month, placed into an account that pays 8% compounded monthly.

27. *Present Value* Find the present value of an annuity that will pay $2000 at the end of each month for 15 years if the interest rate is 8%, compounded monthly.

28. *Present Value* Find the present value of an annuity that will pay $500 at the end of each 6-month period for 12 years if the interest rate is 10%, compounded semiannually.

29. *Loan Amortization* A debt of $2000 with interest at 12%, compounded monthly, is amortized by equal monthly payments for 36 months. What is the size of each payment?

30. *Loan Amortization* A debt of $120,000 with interest at 6%, compounded monthly, is amortized by equal monthly payments for 25 years. What is the size of each monthly payment?

31. *Out-of-Wedlock Births* The percent of all teen mothers who were unmarried during the years 1960–1996 can be modeled by the logistic function

$$y = \frac{96.3641}{1 + 12.3313e^{-0.0844x}}$$

where x is the number of years after 1950.
   a. Use this model to estimate the percent in 1990 and in 1996.
   b. What is the upper limit of the percent of teen mothers who were unmarried according to this model?

32. *Spread of Disease* A student brings a contagious disease to an elementary school of 2000 students. The number of students infected by the disease is given by

$$n = \frac{1400}{1 + 200e^{-0.5x}}$$

where x is the number of days after the student brings the disease.
   a. How many students will be infected in 14 days?
   b. How many days will it take for 1312 students to be infected?

33. *Organizational Growth* The president of a new campus of a university predicts that the student body will grow rapidly after the campus is open, with the number of students at the beginning of year *t* given by

$$N = 4000\left(.06^{0.4^{t-1}}\right).$$

a. How many students does this model predict for the beginning of the second year (*t* = 2)?

b. How many students are predicted for the beginning of the tenth year?

c. What is the limit on the number of students that can attend this campus, according to the model?

34. *Sales Growth* The number of units of a new product that were sold each month after the product was introduced is given by

$$N = 18,000\left(.03^{0.4^{t}}\right),$$

where *t* is the number of months.

a. How many units were sold 10 months after the product was introduced?

b. What is the limit on sales if this model is accurate?

35. *Cell Phones* The table shows the number of cellular telephone sites at the end of the indicated years and the number of people working in the cellular telephone industry.

Year	Cell Sites (in thousands)	Employees
1990	5.62	21,382
1992	10.31	34,348
1994	17.92	53,902
1996	30.05	84,191
1998	65.89	134,754

(Source: Cellular Telecommunications Industry Association)

a. Fit an exponential function *C*(*t*) with input *t* equal to the number of years after 1990 and output the cell sites (in thousands). Round to three decimal places.

b. Find a quadratic function *E*(*C*) with input *C* equal to the number of cell sites (in thousands) and output the number of employees. Round to three decimal places.

c. Find and interpret the function *E*(*C*(*t*)), using the functions reported in part (a) and part (b).

d. Use the result of part (c) and the table to find *E*(*C*(7)).

e. Estimate the number of cellular telephone employees in 1997.

GROUP ACTIVITY/EXTENDED APPLICATION *1*

Suppose that you receive a chain letter asking you to send $1 to the person at the top of the list of 6 names, then to add your name to the bottom of the list, and finally to send the revised letter to 6 people. The promise is that when the letters have been sent to each of the people above you on the list, you will be at the top of the list and you will receive a large amount of money. To investigate whether this is worth the dollar you are asked to spend, create the requested models and answer the following questions.

1. Suppose that each of the 6 persons on the original list sent 6 letters. How much money would the person on the top of the list receive?

2. Consider the person who is second on the original list. How much money would this person receive?

3. Complete the partial table of amounts that will be sent to the person on the top of the list during each "cycle" of the chain letter.

Cycle Number (after original 6 names)	Money Sent to Person on Top of List ($)
1	6 × 6 = 36
2	6 × 36 = 216
3	
4	
5	

4. If you were going to start such a chain letter (don't, it's illegal), would you put your name at the top of the first 6 names or as number 5?

5. Find the best quadratic, power, and exponential models for the data in the table. Which of these function types gives the best model for the data?

6. Use the exponential function found in part (5) to determine the amount of money the person who was at the bottom of the original list would receive if all people contacted send the $1 and mail the letter to 6 additional people.

7. How many people will have to respond to the chain letter for the sixth person on the original list to receive all the money that was promised?

8. If you receive the letter in its tenth cycle, how many other people have been contacted, along with you, assuming that everyone who receives the letter cooperates with its suggestions?

9. Who remains in the United States for you to send your letter to, if no one sends a letter to someone who has already received it?

10. Why do you think the federal government has made it illegal to send chain letters in the U.S. mail?

GROUP ACTIVITY/EXTENDED APPLICATION 2

Your mission is to find real-world data for company sales, a stock price, biological growth, or some sociological trend over a period of years that fits an exponential, logistic, or logarithmic function. To do this, you can look at a statistical graph called a histogram which describes the situation or at a graph or table describing it. Once you have found data consisting of numerical values, you are to find the equation that is the best exponential, logistic, or logarithmic fit for the data and then write the model that describes the relationship. Some helpful steps for this process are given in the instructions below.

1. Look in newspapers, periodicals, on the Internet, or in statistical abstracts for a graph or table of data containing at least six data points. The data must have numerical values for the independent and dependent variables.

2. If you decide to use a relation determined by a graph you have found, read the graph very carefully to determine the data points, or (for full credit) contact the source of the data to obtain the data from which the graph was drawn.

3. If you have a table of values for different years, create a graph of the data to determine the type of function that will be the best fit for the data. To do this, first align the independent variable by letting the input represent the number of years from some convenient year and then enter the data into your graphing technology and draw a scatter plot. Check to see if the points on the scatter plot lie near some curve that could be described by an exponential, logistic, or logarithmic function. If not, save the data for possible later use and renew your search. You can also test to see if the percent change of the outputs is nearly constant for equally spaced inputs. If so, the data can be modeled by an exponential function.

4. Use your calculator to create the equation of the function that is the best fit for the data. Graph this equation and the data points on the same axes to see if the equation is reasonable.

5. Write some statements about the data you have been working with. For example, describe how the quantity is increasing or decreasing over periods of years, why it is changing as it is, or when this model is no longer appropriate, and why.

6. Your completed project should include the following:
   a. A proper bibliographical reference for the source of your data.
   b. An original copy or photocopy of the data being used.
   c. A scatter plot of the data and reasons why you chose the model you did to fit it.
   d. The equation that you have created and labeled with appropriate units of measure and variable descriptions.
   e. A graph of the scatter plot and the model on the same axes.
   f. One or more statements about how you think the model you have created could actually be used. Include any restrictions that should be placed on the model.

Higher Degree Polynomial and Rational Functions

The amount of money spent by domestic tourists can be modeled by the cubic function

$$y = 0.0504x^3 - 1.4659x^2 + 28.2691x - 96.8781$$

where y represents the amount spent in billions of dollars and x represents the number of years after 1980. This is an example of an application whose data can be modeled by a **polynomial function** of degree higher than 2.

If the total cost of producing x units of a product is $C(x) = 100 + 30x + 0.1x^2$, the function that gives the average cost per unit for this product is the **rational function**

$$\overline{C}(x) = \frac{100 + 30x + 0.1x^2}{x}.$$

The main concepts and skills discussed in this chapter, along with some of their applications, follow.

	TOPICS	APPLICATIONS
4.1 Higher Degree Polynomial Functions	Polynomial functions; cubic functions, quartic functions; local extrema	United Nations debt, U.S. foreign-born population, alcohol-related traffic fatalities
4.2 Modeling Cubic and Quartic Functions	Modeling with cubic functions; spreadsheets; modeling with quartic functions; model comparisons; third and fourth differences	Sales of tobacco, tourism, teen pregnancy, foreign-born population
4.3 Solution of Polynomial Equations	Solving polynomial equations by factoring; the root method; estimating solutions with technology	Photosynthesis, maximizing volume, cost, future value of an investment, juvenile crime arrests, stock prices
4.4 Solution of Polynomial Equations Using Synthetic Division	Synthetic division; solving cubic equations; combining graphical and analytical methods; rational solutions test; solving quartic equations	Break-even revenue, births
4.5 Complex Solutions; Fundamental Theorem of Algebra	Imaginary and complex numbers; operations with complex numbers; quadratic equations with complex solutions; polynomial equations with complex solutions; Fundamental Theorem of Algebra	Fractals
4.6 Rational Functions and Rational Equations	Graphs of rational functions; analytical and graphical solution of rational equations	Average cost of production, cost-benefit, deer population, advertising and sales
4.7 Polynomial and Rational Inequalities	Polynomial inequalities; rational inequalities	Average cost, constructing a box

ALGEBRA *Toolbox*

Polynomials

Recall from Chapter 2 that an expression containing a finite number of additions, subtractions, and multiplications of constants and positive integer powers of variables is called a **polynomial**. The general form of a polynomial in x is

$$a_n x^n + a_{n-1} x^{n-1} + \cdots + a_1 x + a_0,$$

where a_0 and each coefficient of x are real numbers and each power of x is positive. If $a_n \neq 0$, n is the highest power of x, a_n is called the **leading coefficient**, and n is the **degree** of the polynomial. Thus $5x^4 + 3x^2 - 6$ is a fourth degree (quartic) polynomial with leading coefficient 5.

EXAMPLE 1

What are the degree and the leading coefficient of each of the following polynomials?

a. $4x^2 + 5x^3 - 17$ **b.** $30 - 6x^5 + 3x^2 - 7$

Solution

a. It is a third-degree (cubic) polynomial. The leading coefficient is 5, the coefficient of the highest degree term.

b. It is a fifth-degree polynomial. The leading coefficient is -6, the coefficient of the highest degree term.   ∎

Factoring Higher Degree Polynomials

Some higher degree polynomials can be factored by expressing them as the product of a monomial and another polynomial, which sometimes can be factored to complete the factorization.

EXAMPLE 2

Factor   **a.** $3x^3 - 21x^2 + 36x$ **b.** $-6x^4 - 10x^3 + 4x^2$

Solution

a. $3x^3 - 21x^2 + 36x = 3x(x^2 - 7x + 12) = 3x(x - 3)(x - 4)$

b. $-6x^4 - 10x^3 + 4x^2 = -2x^2(3x^2 + 5x - 2) = -2x^2(3x - 1)(x + 2)$ ∎

Some polynomials can be written in *quadratic form* by using substitution. Then quadratic factoring methods can be used to begin the factorization.

EXAMPLE 3

Factor   **a.** $x^4 - 6x^2 + 8$ **b.** $x^4 - 18x^2 + 81$

Solution

a. The expression $x^4 - 6x^2 + 8$ is not a quadratic polynomial, but substituting u for x^2 converts it into the quadratic expression $u^2 - 6u + 8$, so we say that the original expression is in *quadratic form*. We factor it using quadratic methods, as follows:

$$u^2 - 6u + 8 = (u - 2)(u - 4).$$

Replacing u with x^2 gives

$$(x^2 - 2)(x^2 - 4).$$

Completing the factorization by factoring $x^2 - 4$ gives

$$x^4 - 6x^2 + 8 = (x^2 - 2)(x - 2)(x + 2).$$

b. Substituting u for x^2 converts $x^4 - 18x^2 + 81$ into the quadratic expression $u^2 - 18u + 81$. Factoring this expression gives

$$u^2 - 18u + 81 = (u - 9)(u - 9).$$

Replacing u with x^2 and completing the factorization gives

$$x^4 - 18x^2 + 81 = (x^2 - 9)(x^2 - 9)$$
$$= (x - 3)(x + 3)(x - 3)(x + 3) = (x - 3)^2(x + 3)^2. \quad \blacksquare$$

Rational Expressions

An expression that is the quotient of two polynomials is called a **rational expression**. For example, $\dfrac{3x + 1}{x^2 - 2}$ is a rational expression. A rational expression is undefined when the denominator equals 0.

   We simplify a rational expression by factoring the numerator and denominator and dividing both the numerator and denominator by any common factors. We will assume that all rational expressions are defined for those values of the variables that do not make any denominator 0.

EXAMPLE 4

Simplify the rational expression:

a. $\dfrac{2x^2 - 8}{x + 2}$

b. $\dfrac{3x}{3x + 6}$

c. $\dfrac{3x^2 - 14x + 8}{x^2 - 16}$

Solution

a. $\dfrac{2x^2 - 8}{x + 2} = \dfrac{2(x^2 - 4)}{x + 2} = \dfrac{2\overset{1}{\cancel{(x + 2)}}(x - 2)}{\underset{1}{\cancel{(x + 2)}}} = 2(x - 2)$

b. Note that we cannot divide both numerator and denominator by $3x$, because $3x$ is not a factor of $3x + 6$. Instead, we must factor the denominator.

$$\frac{3x}{3x + 6} = \frac{\overset{1}{\cancel{3}x}}{\underset{1}{\cancel{3}(x + 2)}} = \frac{x}{x + 2}$$

c. $\dfrac{3x^2 - 14x + 8}{x^2 - 16} = \dfrac{(3x - 2)\overset{1}{\cancel{(x - 4)}}}{(x + 4)\underset{1}{\cancel{(x - 4)}}} = \dfrac{3x - 2}{x + 4}$ ∎

Multiplying and Dividing Rational Expressions

To multiply rational expressions, we write the product of the numerators divided by the product of the denominators, and then simplify. We may also simplify prior to multiplying; this is usually easier than multiplying first.

EXAMPLE 5

Multiply and simplify:

a. $\dfrac{4x^2}{5y^2} \cdot \dfrac{25y}{12x}$

b. $\dfrac{6 - 3x}{2x + 4} \cdot \dfrac{4x - 20}{2 - x}$

Solution

a. $\dfrac{4x^2}{5y^2} \cdot \dfrac{25y}{12x} = \dfrac{4x^2 \cdot 25y}{5y^2 \cdot 12x} = \left(\dfrac{4 \cdot 25}{5 \cdot 12}\right)\left(\dfrac{x^2}{x}\right)\left(\dfrac{y}{y^2}\right) = \left(\dfrac{5}{3}\right)\left(\dfrac{x}{1}\right)\left(\dfrac{1}{y}\right) = \dfrac{5x}{3y}$

b. $\dfrac{6 - 3x}{2x + 4} \cdot \dfrac{4x - 20}{2 - x} = \dfrac{(6 - 3x)(4x - 20)}{(2x + 4)(2 - x)} = \dfrac{3\overset{1}{\cancel{(2 - x)}} \cdot \overset{2}{\cancel{4}}(x - 5)}{2(x + 2) \cdot \underset{1}{\cancel{(2 - x)}}} = \dfrac{6x - 30}{x + 2}$ ∎

To divide rational expressions, we multiply by the reciprocal of the divisor.

EXAMPLE 6

Divide and simplify.

a. $\dfrac{a^2b}{c} \div \dfrac{ab^3}{c^2}$

b. $\dfrac{x^2 + 7x + 12}{2 - x} \div \dfrac{x^2 - 9}{x^2 - x - 2}$

Solution

a. $\dfrac{a^2b}{c} \div \dfrac{ab^3}{c^2} = \dfrac{a^2b}{c} \cdot \dfrac{c^2}{ab^3} = \left(\dfrac{a^2}{a}\right)\left(\dfrac{b}{b^3}\right)\left(\dfrac{c^2}{c}\right) = \dfrac{ac}{b^2}$

b. $\dfrac{x^2 + 7x + 12}{2 - x} \div \dfrac{x^2 - 9}{x^2 - x - 2} = \dfrac{x^2 + 7x + 12}{2 - x} \cdot \dfrac{x^2 - x - 2}{x^2 - 9}$

$= \dfrac{(x + 3)(x + 4)(x - 2)(x + 1)}{-(x - 2)(x - 3)(x + 3)} = \dfrac{(x + 4)(x + 1)}{-(x - 3)} = \dfrac{x^2 + 5x + 4}{3 - x}$ ■

Adding and Subtracting Rational Expressions

We can add or subtract two fractions with the same denominator by adding or subtracting the numerators over their common denominator. If the denominators are not the same, we can write each fraction as an equivalent fraction with the common denominator. Our work is easier if we use the **least common denominator**, or LCD. The least common denominator is the lowest degree polynomial that all denominators will divide into. For example, if several denominators are $3x$, $6x^2y$, and $9y^3$, the lowest degree polynomial that all three denominators will divide into is $18x^2y^3$. We can find the least common denominator as follows.

FINDING THE LEAST COMMON DENOMINATOR OF A SET OF FRACTIONS

1. Completely factor all the denominators.

2. The LCD is the product of each factor used the maximum number of times it occurs in any one denominator.

EXAMPLE 7

Find the LCD of the fractions $\dfrac{3x}{x^2 - 4x - 5}$ and $\dfrac{5x - 1}{x^2 - 2x - 3}$.

Solution

The factored denominators are $(x - 5)(x + 1)$ and $(x - 3)(x + 1)$. The factors $(x - 5)$, $(x + 1)$, and $(x - 3)$ each occur a maximum of one time in any one denominator. Thus the LCD is $(x - 5)(x + 1)(x - 3)$. ■

The procedure for adding or subtracting rational expressions follows.

ADDING OR SUBTRACTING RATIONAL EXPRESSIONS

1. Find the LCD of all the rational expressions.

2. Write the equivalent of each rational expression with the LCD as its denominator.

3. Combine the like terms in the numerators and write this expression in the numerator with the LCD in the denominator.

4. Simplify the rational expression, if possible.

EXAMPLE 8

Add $\dfrac{3}{x^2y} + \dfrac{5}{xy^2}$.

Solution

The factor x occurs a maximum of two times in any denominator and the factor y occurs a maximum of two times in any denominator, so the LCD is x^2y^2. We rewrite the fractions with the common denominator by multiplying the numerator and denominator of the first fraction by y and multiplying the numerator and denominator of the second fraction by x, as follows:

$$\frac{3}{x^2y} + \frac{5}{xy^2} = \frac{3 \cdot y}{x^2y \cdot y} + \frac{5 \cdot x}{xy^2 \cdot x} = \frac{3y}{x^2y^2} + \frac{5x}{x^2y^2}.$$

We can now add the fractions by adding the numerators over the common denominator:

$$\frac{3y}{x^2y^2} + \frac{5x}{x^2y^2} = \frac{3y + 5x}{x^2y^2}.$$

This result cannot be simplified because there is no common factor of the numerator and denominator. ∎

EXAMPLE 9

Subtract $\dfrac{x + 1}{x^2 + x - 6} - \dfrac{2x - 1}{x - 2}$.

Solution

We factor the first denominator to get the LCD.

$$x^2 + x - 6 = (x + 3)(x - 2), \text{ so the LCD is } (x + 3)(x - 2).$$

We rewrite the second fraction so that it has the common denominator and then combine the like terms of the numerators.

$$\frac{x + 1}{x^2 + x - 6} - \frac{2x - 1}{x - 2} = \frac{x + 1}{(x - 2)(x + 3)} - \frac{(2x - 1)(x + 3)}{(x - 2)(x + 3)}$$

$$= \frac{x + 1}{(x - 2)(x + 3)} - \frac{2x^2 + 5x - 3}{(x - 2)(x + 3)}$$

$$= \frac{x + 1 - (2x^2 + 5x - 3)}{(x - 2)(x + 3)}$$

$$= \frac{x + 1 - 2x^2 - 5x + 3}{(x - 2)(x + 3)}$$

$$= \frac{-2x^2 - 4x + 4}{(x - 2)(x + 3)} \qquad ∎$$

Division of Polynomials

When the divisor has a degree less than the dividend, the division of one polynomial by another is done in a manner similar to long division in arithmetic. The degree of the quotient polynomial will be less than the degree of the dividend polynomial, and the degree of the remainder will be less than the degree of the divisor. (If the remainder is 0, the divisor is a factor of the dividend). The procedure follows.

LONG DIVISION OF POLYNOMIALS

Divide $(4x^3 + 4x^2 + 5)$ by $(2x^2 + 1)$.

1. Write both the divisor and dividend in descending order of a variable. Include missing terms with 0 coefficient in the dividend.

$$2x^2 + 1 \overline{\smash{)}4x^3 + 4x^2 + 0x + 5}$$

2. Divide the highest power of the divisor into the highest power of the dividend, and write this partial quotient above the dividend. Multiply the partial quotient times the divisor, write the product under the dividend, subtract, and bring down the next term, getting a new dividend.

$$
\begin{array}{r}
2x \\
2x^2 + 1 \overline{\smash{)}4x^3 + 4x^2 + 0x + 5} \\
\underline{4x^3 + 2x} \\
4x^2 - 2x
\end{array}
$$

3. Repeat until the degree of the new dividend is less than the degree of the divisor. If a nonzero expression remains, it is the remainder from the division.

$$
\begin{array}{r}
2x + 2 \\
2x^2 + 1 \overline{\smash{)}4x^3 + 4x^2 + 0x + 5} \\
\underline{4x^3 + 2x} \\
4x^2 - 2x + 5 \\
\underline{4x^2 + 2} \\
-2x + 3
\end{array}
$$

Thus the quotient is $2x + 2$, with remainder $-2x + 3$.

EXAMPLE 10

Perform the indicated division: $\dfrac{x^4 + 6x^3 - 4x + 2}{x + 2}$.

Solution

$$
\begin{array}{r}
x^3 + 4x^2 - 8x + 12 \\
x + 2 \overline{\smash{)}x^4 + 6x^3 + 0x^2 - 4x + 2} \\
\underline{x^4 + 2x^3 } \\
4x^3 + 0x^2 \\
\underline{4x^3 + 8x^2 } \\
-8x^2 - 4x \\
\underline{-8x^2 - 16x } \\
12x + 2 \\
\underline{12x + 24} \\
-22
\end{array}
$$

Thus the quotient is $x^3 + 4x^2 - 8x + 12$, with remainder -22. We can write the result as

$$\frac{x^4 + 6x^3 - 4x + 2}{x + 2} = x^3 + 4x^2 - 8x + 12 - \frac{22}{x + 2}$$

∎

Toolbox EXERCISES

For Exercises 1–4, (a) give the degree of the polynomial,
(b) give the leading coefficient.

1. $3x^4 - 5x^2 + \dfrac{2}{3}$ **2.** $5x^3 - 4x + 7$

3. $7x^2 - 14x^5 + 16$ **4.** $2y^5 + 7y - 8y^6$

In Exercises 5–8, factor the polynomials.

5. $4x^3 - 8x^2 - 140x$ **6.** $4x^2 + 7x^3 - 2x^4$

7. $x^4 - 13x^2 + 36$ **8.** $2x^4 - 8x^2 + 8$

In Exercises 9–11, simplify each rational expression.

9. $\dfrac{x - 3y}{3x - 9y}$ **10.** $\dfrac{x^2 - 6x + 8}{x^2 - 16}$

11. $\dfrac{3x^2 - 7x - 6}{x^2 - 4x + 3}$

In Exercises 12–20, perform the indicated operations and
simplify.

12. $\dfrac{6x^3}{8y^3} \cdot \dfrac{15y^4}{x^2}$

13. $\dfrac{4x + 4}{x - 4} \div \dfrac{8x^2 + 8x}{x^2 - 6x + 8}$

14. $(x^2 - 4) \cdot \dfrac{2x - 3}{x + 2}$

15. $\dfrac{6x^2}{4x^2y - 12xy} \div \dfrac{3x^2 + 12x}{x^2 + x - 12}$

16. $\dfrac{2x}{x^2 - 4} + \dfrac{4}{x^2 - 4}$

17. $3 + \dfrac{1}{x^2} - \dfrac{2}{x^3}$

18. $\dfrac{a}{a - 2} - \dfrac{a - 2}{a}$

19. $\dfrac{x - 1}{x + 1} - \dfrac{2}{x^2 + x}$

20. $\dfrac{2x + 1}{4x - 2} + \dfrac{5}{2x} - \dfrac{x + 1}{2x^2 - x}$

In Exercises 21–24, perform the long division.

21. $(x^5 + x^3 - 1) \div (x + 1)$

22. $(a^4 + 3a^3 + 2a^2) \div (a + 2)$

23. $(3x^5 - x^4 + 5x - 1) \div (x^2 - 2)$

24. $\dfrac{3x^4 + 2x^2 + 1}{3x^2 - 1}$

4.1 Higher Degree Polynomial Functions

Although the United States is one of many countries that are members of the United Nations, the United States pays most of the operating expense of the United Nations. Table 4.1 shows the amount of U.S. debt and total debt (in $million) owed by all members to run the United Nations for the years 1990 to 2000, and the percent of the debt owed by the United States for each of these years.

TABLE 4.1

Year	U.S. Debt	Total Debts	Percent U.S. Share
1990	296.2	403	73.5
1991	266.4	439.4	60.6
1992	239.5	500.6	47.8
1993	260.4	478	54.5
1994	248	480	51.6
1995	414.4	564	73.5
1996	376.8	510.7	73.8
1997	373.2	473.6	78.8
1998	315.7	417	75.7
1999	167.9	244.2	68.8
2000	164.6	222.4	74.0

(Source: United Nations, Institute for Global Communication)

The data can be used to find a **polynomial function**

$$y = -1.832x^3 + 22.111x^2 - 55.367x + 290.659$$

that models U.S. debt to the United Nations, y, in millions of dollars, as a function of the number of years, x, after 1990.

In Chapter 1 and Chapter 2 we solved many real problems by modeling data with linear functions and with quadratic functions. (Recall that $ax + b$ is a first-degree polynomial and $ax^2 + bx + c$ is a second-degree polynomial.) Both of these functions are types of **polynomial functions**. **Higher degree polynomial functions** are functions with degree higher than 2.

Examples are

$$y = x^3 - 16x^2 - 2x, \quad f(x) = 3x^4 - x^3 + 2, \quad y = 2x^5 - x^3$$

Cubic Functions

Using data from 1990 to 2000, the function

$$y = -1.832x^3 + 22.111x^2 - 55.367x + 290.659$$

can be used to model U.S. debt to the United Nations, in millions of dollars, with x equal to the number of years after 1990. This function is a third-degree function, or **cubic function**, because the highest degree term in the function has degree 3. The graph of this function is shown in Figure 4.1.

$$y = -1.832x^3 + 22.111x^2 - 55.367x + 290.659$$

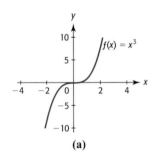

FIGURE 4.1

A **cubic function** in the variable x has an equation of the form

$$f(x) = ax^3 + bx^2 + cx + d, \text{ with } a \neq 0.$$

The basic cubic function $f(x) = x^3$ was discussed in Section 2.3. The graph of this function, shown in Figure 4.2(a), is one of the possible shapes of the graph of a cubic function. The graphs of two other cubic functions are shown in Figure 4.2(b) and Figure 4.2(c).

    The graph of a cubic function will have 0 or 2 turning points, as shown in Figures 4.1 and 4.2. Also, it is characteristic of a cubic polynomial function to have a graph that has one of the end behaviors shown in these figures. This end behavior can be described as "one end opening up and one end opening down." The specific end behavior is determined by the leading coefficient (the coefficient of the third-degree term); the curve opens up on the right if the leading coefficient is positive, and it opens down on the right if the leading coefficient is negative.

    Notice that the graph of the cubic function shown in Figure 4.2(a) has one x-intercept, the graph of the cubic function shown in Figure 4.2(b) has three x-intercepts, and in Figure 4.2(c) the graph of the cubic function has two x-intercepts.

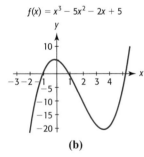

(a)

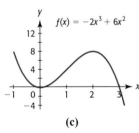

(b)

> In general, the graph of a polynomial function of degree n has at most n x-intercepts.

Recall that the graph of a quadratic (second-degree) function is a parabola with one "turning point," called a vertex, and that the vertex is a maximum point or a minimum point. Notice that in each of the graphs of the cubic (third-degree) functions shown in Figure 4.2(b) and Figure 4.2(c), there are two points where the function changes from increasing to decreasing, or from decreasing to increasing. (Recall that a curve is increasing on an interval if the curve rises as it moves from left to right on that interval.) These *turning points* are called the **local extrema points** of the graph of the function. In particular, a point where the curve changes from decreasing to increasing is called a **local minimum point**, and a point where the curve changes from increasing to decreasing is called a **local maximum point**. For example, in Figure 4.2(c), the point

(c)

FIGURE 4.2

(0, 0) is a local minimum point and the point (2, 8) is a local maximum point. Notice that the graph in Figure 4.2(a) has no local extrema points. In general, a cubic equation will have 0 or 2 local extrema points.

The highest point on the graph over an interval is called the **absolute maximum point** on the interval, and the lowest point on the graph over an interval is called the **absolute minimum point** on that interval. The graph in Figure 4.2(c) is shown on the x-interval $[-1.2, 3.2]$; the absolute maximum point on this interval is $(-1.2, 12.096)$, and the absolute minimum point is $(3.2, -4.096)$ on this interval. Note also that at the point $(0, 0)$, the function changes from decreasing to increasing, and at the point $(2, 8)$, the function changes from increasing to decreasing.

EXAMPLE 1   Cubic Graph

a. Using an appropriate window, graph $y = x^3 - 27x$.

b. Find the local maximum and local minimum, if possible.

Solution

a. This function is a third-degree (cubic) function, so it has 0 or 2 local extrema. We want to graph the function in a window that allows us to see any turning points that exist and recognize the end behavior of the function. Using the window $[-40, 40]$ by $[-100, 100]$ we get the graph shown in Figure 4.3(a). This graph has a shape like Figure 4.2(b), it has three x-intercepts, and it has two local extrema, so it is a complete graph. A smaller viewing window can be used to provide a more detailed view of the turning points of the graph (see Figure 4.3(b)).

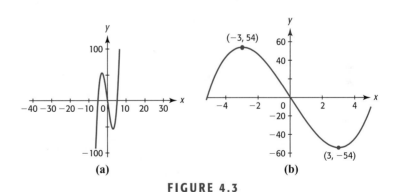

(a)   (b)

FIGURE 4.3

b. Using the ⎡MAXIMUM⎤ and ⎡MINIMUM⎤ features of a graphing utility shows that there is a local maximum at $(-3, 54)$ and a local minimum at $(3, -54)$.   ∎

EXAMPLE 2   U.S. Foreign-Born Population

Table 4.2 gives the percents of U.S. population that were foreign born for the years 1900 to 1999.

TABLE 4.2

Years	Percent of Foreign Born	Years	Percent of Foreign Born
1900	13.6	1960	5.4
1910	14.7	1970	4.7
1920	13.2	1980	6.2
1930	11.6	1990	8.0
1940	8.8	1999	9.7
1950	6.9		

(Source: U.S. Bureau of the Census)

This data can be modeled by the cubic function

$$f(x) = 0.0000584x^3 - 0.00667x^2 + 0.0504x + 14.153,$$

where y is the percent and x is the number of years after 1900.

a. Graph the data points and this function on the same axes, using the window [0, 100] by [0, 16].

b. Use the model and technology to find the approximate percent of the U.S. population that was foreign born in 1935.

c. Interpret the y-intercept of the graph of this function.

d. Find the local maximum and the local minimum. What do these two points on the graph indicate?

e. Over what interval is the function (and thus the percent) decreasing? What does this mean?

Solution

a. The graph is shown in Figure 4.4(a).

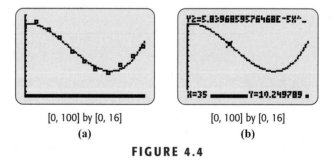

[0, 100] by [0, 16]
(a)

[0, 100] by [0, 16]
(b)

FIGURE 4.4

b. The approximate percent of the U.S. population that was foreign born in 1935 is $f(35) = 10.25$ percent (see Figure 4.4(b)).

c. The y-intercept, 14.153, gives the percent of the U.S. population that was foreign born in 1900, according to the model.

d. Using ⟦ MINIMUM ⟧ and ⟦ MAXIMUM ⟧ commands on a graphing utility gives the local minimum (72.16, 5.00) and the local maximum (3.99, 14.25). (See Figure 4.5.) Because the model is interpreted at the end of each year, we test $x = 72$ and $x = 73$ to find the minimum. The model indicates that the minimum percent occurs in 1972 and is approximately 5%. The x-value of the local maximum, $3.99 \approx 4$, corresponds to the year 1904, so the maximum percent occurs in 1904 and is approximately 14.25%, according to the model.

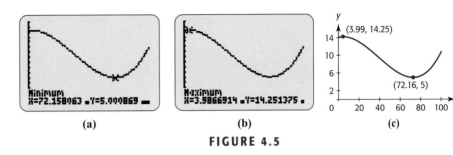

(a)   **(b)**   **(c)**

FIGURE 4.5

e. As x increases from 3.99 to 72.16, y decreases from the maximum at 14.25 to the minimum at 5. (See Figure 4.5(c).) This means that the percent decreases from 14.25% to 5% in the years from 1904 to 1972. ∎

Quartic Functions

A polynomial function of the form $f(x) = ax^4 + bx^3 + cx^2 + dx + e$, $a \neq 0$, is a fourth-degree, or **quartic**, function. The basic quartic function $f(x) = x^4$ has the graph shown in Figure 4.6(a). Notice that this graph resembles a parabola, although it is not actually a parabola (recall the shape of a parabola, which is discussed in Section 2.1). The graph in Figure 4.6(a) is one of the possible shapes of the graph of a quartic function. The graphs of two additional quartic functions are shown in Figures 4.6(b) and 4.6(c).

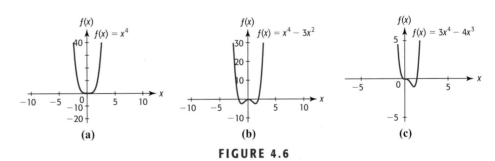

(a)   **(b)**   **(c)**

FIGURE 4.6

Notice that in the graph of the quartic function shown in Figure 4.6(b), there are 3 turning points and that in the graphs shown in Figures 4.6(a) and 4.6(c), there is 1 turning point each. The end behavior of a quartic function can be described as both ends opening up (if the leading coefficient is positive) or both ends opening down (if the leading coefficient is negative). Variations in the curvature may occur for different quartic functions, but these are the only possible end behaviors.

In Table 4.3, we summarize the information about turning points and end behavior for cubic and quartic functions as well as for linear and quadratic functions.

TABLE 4.3

Function	Possible Graphs	Degree	Turning Points	End Behavior positive leading coefficient	negative leading coefficient	Degree Even or Odd
Linear	 Positive slope Negative slope	1	0			odd
Quadratic	Positive leading coefficient Negative leading coefficient	2	1			even
Cubic Positive leading coefficient	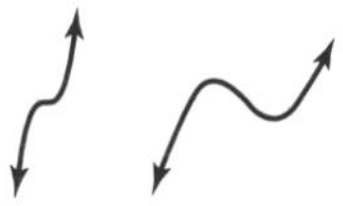	3	2 or 0			odd
Negative leading coefficient	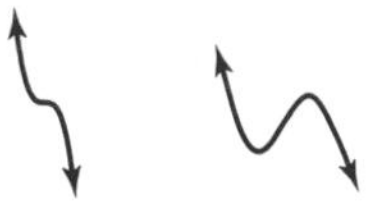					
Quartic Positive leading coefficient		4	3 or 1			even
Negative leading coefficient	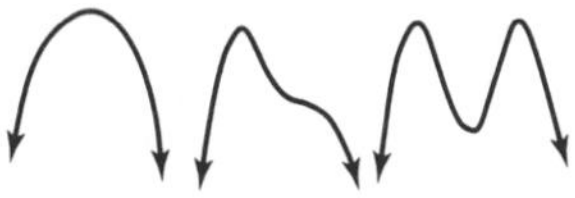					

Table 4.3 helps us observe the following.

POLYNOMIAL GRAPHS

1. The graph of a polynomial function of degree n has at most $n - 1$ turning points.
2. The end behavior of a polynomial function with odd degree can be described as "one end opening up and one end opening down."
3. The end behavior of a polynomial function with even degree can be described as "both ends opening up" or "both ends opening down."

EXAMPLE 3   Alcohol-Related Traffic Fatalities

Using data from 1982 to 1999, the total number of people killed in alcohol-related crashes can be modeled by the function

$$y = -0.3718x^4 + 21.9933x^3 - 395.6523x^2 + 1880.2042x + 21,524.7647,$$

where x is the number of years from 1982.

a. Graph the function over the x-values $0 \le x \le 18$.

b. Use the graph and technology to approximate the year in which the number of alcohol-related traffic deaths is a maximum and in which year the number is a minimum during this period.

c. The graph in part (a) shows 2 turning points, and the graph of a fourth-degree function must have 1 or 3 turning points. Graph the function in a window with a maximum x-value of 40 to see if a third turning point exists for positive x-values.

d. Is it likely that this model can be used to estimate the total number of people killed in alcohol-related crashes for long after 1999?

Solution

a. The graph is shown in Figure 4.7.

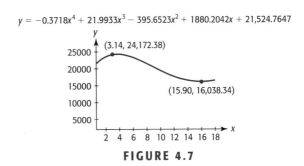

$y = -0.3718x^4 + 21.9933x^3 - 395.6523x^2 + 1880.2042x + 21,524.7647$

FIGURE 4.7

b. The maximum occurs at (3.14, 24,172.38). (See Figure 4.7.) Testing end of year values indicates that the number of traffic deaths is a maximum in the year 1985. The minimum occurs at (15.90, 16,038.34). Testing end of year values indicates that the number of traffic deaths is a minimum in the year 1998.

c. There are 2 turning points shown, and because the function has degree four, there can only be 1 or 3, so a third turning point must exist. The leading coefficient is negative, so the complete graph of this function will have both ends opening down. Because the graph appears to bend upward toward the right of the graph shown in Figure 4.7, it would seem likely that the graph has its third turning point in that direction. Graphing the function in a window with a maximum x-value of 40 allows the third turning point to be seen. (See Figure 4.8.)

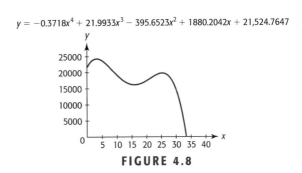

$y = -0.3718x^4 + 21.9933x^3 - 395.6523x^2 + 1880.2042x + 21,524.7647$

FIGURE 4.8

  d. Observing the graph, we see that it drops very rapidly after $x = 25$ (which corresponds to 2007), with outputs becoming negative before $x = 34$ (which corresponds to the year 2116). Thus the model ceases to be valid eventually, certainly before 2116. ■

4.1 SKILLS CHECK

1. Graph the function $h(x) = 3x^3 + 5x^2 - x - 10$ on the windows given in parts (a) and (b). Which window gives a complete graph?
   a. $[-5, 5]$ by $[-5, 5]$
   b. $[-5, 5]$ by $[-20, 20]$

2. Graph the function $f(x) = 2x^3 - 3x^2 - 6x$ on the windows given in parts (a) and (b). Which window gives a complete graph?
   a. $[-5, 5]$ by $[-5, 5]$
   b. $[-10, 10]$ by $[-10, 10]$

3. Graph the function $g(x) = 3x^4 - 12x^2$ on the windows given in parts (a), (b), and (c). Which window gives a complete graph?
   a. $[-5, 5]$ by $[-5, 5]$
   b. $[-10, 10]$ by $[-10, 10]$
   c. $[-5, 5]$ by $[-20, 20]$

4. Graph the function $g(x) = 3x^4 - 4x^2 + 10$ on the windows given in parts (a) and (b). Which window gives a complete graph?
   a. $[-10, 10]$ by $[-10, 10]$
   b. $[-5, 5]$ by $[-20, 20]$

For Exercises 5–8, use the given graph of the polynomial function to
a. *estimate the x-intercept(s).*
b. *state whether the leading coefficient is positive or negative.*
c. *determine the minimum degree of the polynomial function.*

5.

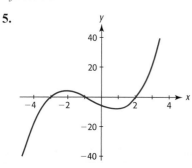

6.

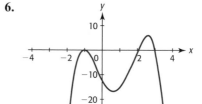

7.

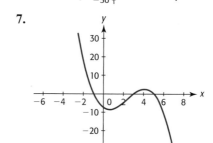

8.

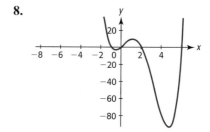

For Exercises 9–12, use the equation of the polynomial function to
a. *state the degree and the leading coefficient.*
b. *predict the end behavior of the graph of the function.*
c. *support your answer by graphing the function.*

9. $f(x) = 2x^3 - x$

10. $g(x) = 0.3x^4 - 6x^2 + 17x$

11. $f(x) = -2(x - 1)(x^2 - 4)$

12. $g(x) = -3(x - 3)^2(x - 1)^2$

13. **a.** Graph the function $y = x^3 - 3x^2 - x + 3$ using the window $[-10, 10]$ by $[-10, 10]$.
    b. Is the graph complete?

14. Graph the function $y = x^3 - 16x$
    a. using the window $[-10, 10]$ by $[-10, 10]$.
    b. using a window that shows two turning points.

15. Graph the function $y = x^4 - 4x^3 + 4x^2$
    a. using the window $[-10, 10]$ by $[-10, 10]$.
    b. using the window $[-4, 4]$ by $[-4, 4]$.
    c. Which window gives a more detailed view of the graph near the turning points?

16. Sketch a graph of any polynomial function that has degree 3, a positive leading coefficient, and three x-intercepts.

17. Sketch a graph of any cubic polynomial function that has a negative leading coefficient and one x-intercept.

18. Sketch a graph of any polynomial function that has degree 4, a positive leading coefficient, and two x-intercepts.

19. Sketch a graph of any cubic polynomial function that has a positive leading coefficient and three x-intercepts.

20. Use technology to find a local maximum and two local minima (minimums) of the graph of the function $y = x^4 - 4x^3 + 4x^2$.

21. a. Graph $y = x^3 + 4x^2 + 5$ on a window that shows a local maximum and a local minimum.
    b. A local maximum occurs at what point?
    c. A local minimum occurs at what point?

22. a. Graph the function $y = -x^3 - x^2 + 9x$ using the window $[-5, 5]$ by $[-15, 10]$.
    b. Graph the function on an interval with $x \geq 0$ and with $y \geq 0$.
    c. The graph of what other type of function resembles the *piece* of the graph of $y = -x^3 - x^2 + 9x$ shown in part (b)?

4.1 EXERCISES

-----

1. *Daily Revenue* The daily revenue from the sale of a product is given by $R = -0.1x^3 + 11x^2 - 100x$ dollars, where x is the number of units sold.
   a. Graph this function on the window $[-100, 100]$ by $[-5000, 25,000]$. How many turning points do you see?
   b. Because x represents the number of units sold in the context of the problem, what restriction should be placed on x? What restriction should be placed on y?
   c. Using your result from part (b), graph the function on a new window that makes sense for the problem.
   d. What does this function give as the revenue if 50 units are produced?

2. *Weekly Revenue* A firm has total weekly revenue for its product given by $R(x) = 2000x + 30x^2 - 0.3x^3$, where x is the number of units sold.
   a. Graph this function on the window $[-100, 150]$ by $[-30,000, 220,000]$.
   b. Because x represents the number of units sold, what restrictions should be placed on x in the context of the problem? What restrictions should be placed on y?
   c. Using your result from part (b), graph the function on a new window that makes sense for the problem.
   d. What does this function give as the revenue if 60 units are produced?

3. *Daily Revenue* The daily revenue from the sale of a product is given by $R = 600x - 0.1x^3 + 4x^2$ dollars, where x is the number of units sold.
   a. Graph this function on the window $[0, 100]$ by $[0, 30,000]$.
   b. Use the graph of this function to estimate the number of units that will give the maximum daily revenue and the maximum possible daily revenue.
   c. Find a window that will show a complete graph, that is, will show all of the turning points and intercepts. Graph the function using this window.
   d. Does the graph in part (a) or the graph in part (c) better represent the revenue from the sale of x units of a product?
   e. Over what interval of x-values is the revenue increasing, if $x \geq 0$?

4. *Weekly Revenue* A firm has total weekly revenue for its product given by $R(x) = 2800x - 8x^2 - x^3$, where x is the number of units sold.
   a. Graph this function on the window $[0, 50]$ by $[0, 51,000]$.
   b. Use technology to find the maximum possible revenue and the number of units that gives the maximum revenue.
   c. Find a window that will show a complete graph, that is, will show all of the turning points and intercepts. Graph the function using this window.

d. Does the graph in part (a) or the graph in part (c) better represent the revenue from the sale of x units of a product?

e. Over what interval of x-values is the revenue increasing, if $x \geq 0$?

5. *Investment* If $2000 is invested for 3 years at rate r, compounded annually, the future value of this investment is given by $S = 2000(1 + r)^3$, where r is the rate written as a decimal.

a. Complete the following table to see how increasing the interest rate affects the future value of this investment.

r (rate)	S (future value)
0%	
5%	
10%	
15%	
20%	

b. Graph this function for $0 \leq r \leq 0.24$.

c. Use the table and/or graph to compare the future value if $r = 10\%$ and if $r = 20\%$. How much more money is earned at 20%?

d. Which interest rate, 10% or 20%, is more realistic for an investment?

6. *Investment* The future value of $10,000 invested for 5 years at rate r, compounded annually, is given by $S = 10,000(1 + r)^5$, where r is the rate written as a decimal.

a. Complete the following table to see how increasing the interest rate affects the future value of this investment.

r (rate)	S (future value)
0%	
5%	
7%	
12%	
18%	

b. Graph this function for $0 \leq r \leq 0.24$.

c. Use the graph to compare the future value if $r = 10\%$ and if $r = 24\%$. How much more money is earned at 24%?

d. Is it more likely that you can get an investment paying 10% or 24%?

7. *Homicide Rate* The number of homicides per 100,000 people is given by the equation $y = -0.000280x^3 + 0.0152x^2 - 0.0360x + 4.144$, with x equal to the number of years after 1950.

a. Graph this function on a window with values for x representing the years 1950–2000.

b. What does this model give as the number of homicides per 100,000 people in 1990?

c. Use this model and technology to estimate the year after 1950 in which the number of homicides is at a maximum.

d. What was the average rate of change in the number of homicides per 100,000 people for 1970–1990?

(Source: U.S. Department of Justice, Bureau of Justice Statistics)

8. *Drunk Driving Fatalities* Suppose the total number of fatalities in drunk driving crashes in South Carolina can be modeled by the function $y = 0.4566x^3 - 14.3085x^2 + 117.2978x + 107.8456$, using data for 1982–1998, where x is the number of years after 1980.

a. Graph this function on a window with values for x representing the years 1980–1998.

b. How many such fatalities occurred in 1992, according to this model?

c. Use this model and technology to estimate the year in which the number of fatalities from drunk driving crashes was at a maximum.

d. Use this model and technology to estimate the year after 1981 in which the number of fatalities from drunk driving crashes was at a minimum.

(Source: Anheuser-Busch, www.beeresponsible.com)

9. *Population* Using data for 1950–1992, the population of young adults, from 22 to 24 years of age, can be modeled by the equation $y = -0.6182x^3 + 54.956x^2 - 1270.1312x + 15,025.612$, where x is the number of years after 1950 and y is in thousands.

a. Graph the function on an interval for the years 1950–1992.

b. What does the model predict the population of young adults aged 22 to 24 to be in 1980?

c. During what year does the model estimate this population to be at a minimum?

(Source: U.S. Census Bureau)

10. *Criminal Executions* Using data for 1982–1998, the number of criminal executions can be modeled by the function $y = 0.0402x^3 - 0.9270x^2 + 8.3279x - 8.2059$, with x equal to the number of years after 1980.

a. Graph the function on an interval for the years 1982–1998.

b. Does this model indicate that the number of executions increased or decreased during the given period?

c. What does the model predict as the number of executions in 1999?

(Source: www.editonnine.deathrowbook.com)

11. *United Nations Debt* The cubic function

$$y = -1.832x^3 + 22.111x^2 - 55.367x + 290.659$$

models the U.S. debt to the United Nations, in millions of dollars, as a function of the number of years after 1990.

a. Graph this function using the window [0, 10] by [0, 400].

b. According to the model, what was the approximate U.S. debt in 1998?

c. Use technology to find the local maximum and the local minimum. What does this indicate about the maximum and minimum U.S. debt over this period?

4.2 Modeling Cubic and Quartic Functions

Key Concepts
- Cubic models
- Quartic models
- Third and fourth differences

The Canadian government has used numerous ways of informing citizens of the dangers of smoking, and consequently sales of tobacco have decreased. Table 4.4 gives the domestic sales, in kilograms, of fine-cut tobacco in Canada from 1987 to 1997. The scatter plot of this data, shown in Figure 4.9(a) shows that neither a line nor a parabola would fit these data points well, and it appears that a cubic function would be a good fit.

TABLE 4.4

Year	Sales of Fine Tobacco (kilograms)	Year	Sales of Fine Tobacco (kilograms)
1987	7.413	1992	6.360
1988	7.863	1993	5.080
1989	8.027	1994	3.829
1990	7.749	1996	3.964
1991	6.658	1997	3.830

FIGURE 4.9(a)

(Source: Canadian Council on Smoking and Health)

The use of graphing utilities and spreadsheets permits us to find cubic and quartic functions that model nonlinear data.

Modeling with Cubic Functions

The scatter plot of the data in Figure 4.9(a) indicates that the data may fit close to one of the possible shapes of the graph of a cubic function, so it is possible to find a cubic function that models the data.

EXAMPLE 1 Sales of Tobacco

Use the data in Table 4.4 to do the following:

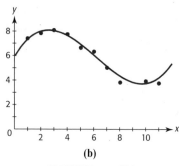

(b)

FIGURE 4.9(b)

a. Find a function that models the data in the table, with x equal to the number of years after 1986 and y equal to the number of kilograms.

b. Graph the function reported in part (a) and the data on the same axes to see how well the model fits the data.

c. Use the model reported in part (a) to predict sales in 2000.

Solution

a. A cubic function that models the data is $y = 0.024x^3 - 0.442x^2 + 1.797x + 5.974$ where y is in kilograms and x is the number of years after 1986.

b. Figure 4.9(b) shows the graph of the function and the data on the same axes. The model is a good visual fit for the data.

c. The sales in 2000 are predicted from the model with $f(14) = 10.356$ kilograms. ∎

Sometimes a set of data points can be modeled reasonably well by both a linear function and a cubic function. Consider the following example.

EXAMPLE 2 Tourism

The importance of the tourist industry in the United States is illustrated in Table 4.5, which shows the money spent by domestic travelers in the United States for each of the years 1987 through 1996.

TABLE 4.5

Year	Domestic Traveler Spending (billions of dollars)	Year	Domestic Traveler Spending (billions of dollars)
1987	235	1992	308
1988	258	1993	322
1989	273	1994	339
1990	291	1995	360
1991	296	1996	383

(Source: *1998 World Almanac*)

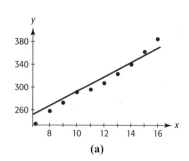

(a)

(b)

FIGURE 4.10

a. Find the linear function that is the best fit for the data with x in years after 1980 and y in $billion. Graph the function and the data points on the same axes.

b. Find the cubic function that is the best fit for the data, and graph the function and the data points on the same axes.

c. Use the graphs to determine which function appears to be the better model for the data.

Solution

a. The linear function that is the best fit for the data is

$$y = 15.036x + 133.582.$$

The graphs of this function and the data points are shown in Figure 4.10(a).

b. The cubic function that is the best fit for the data is

$$y = 0.267x^3 - 8.910x^2 + 110.236x - 190.944,$$

with x equal to the number of years after 1980. The graphs of this function and the data points are shown in Figure 4.10(b).

c. The graphs indicate that the cubic function is a better fit than the linear function.

■

Spreadsheet Solution

We can use graphing calculators, software programs, and spreadsheets to find the polynomial function that is the best fit for data. Table 4.6 shows the Excel spread-sheet for the data of Example 2. Selecting the cells containing the data, using Chart Wizard to get the scatter plot of the data, selecting $\boxed{\text{ADD TRENDLINE}}$ and picking $\boxed{\text{POLYNOMIAL}}$ with the order (degree) of the polynomial gives the equation of the polynomial function that is the best fit for the data, along with the scatter plot and the graph of the best-fitting curve. Figure 4.11 shows an Excel worksheet with the data and the graph of the cubic function that fits the data.

TABLE 4.6

	A	B
	Year after 1980 x	Spending ($billion) y
1	Year after 1980 x	Spending ($billion) y
2	7	235
3	8	258
4	9	273
5	10	291
6	11	296
7	12	308
8	13	322
9	14	339
10	15	360
11	16	383

$y = 0.2667x^3 - 8.9097x^2 + 110.24x - 190.94$

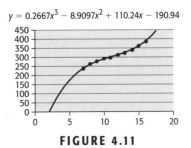

FIGURE 4.11

Modeling with Quartic Functions

If a scatter plot of data indicates that the data may fit close to one of the possible shapes that are graphs of quartic functions, it is possible to find a quartic function that models the data. Consider the following example.

EXAMPLE 3   Teen Pregnancy

Table 4.7 gives the number of pregnancies per 1000 U.S. females from 15 to 19 years of age for the years 1972 to 1996.

TABLE 4.7

Year	Pregnancies/1000 Teens	Year	Pregnancies/1000 Teens	Year	Pregnancies/1000 Teens
1972	95	1981	110	1990	117
1973	96	1982	110	1991	116
1974	99	1983	109	1992	112
1975	101	1984	108	1993	109
1976	101	1985	109	1994	106
1977	105	1986	107	1995	101
1978	105	1987	107	1996	97
1979	109	1988	111		
1980	111	1989	115		

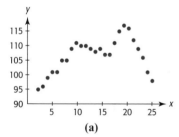

(a)

(b)

FIGURE 4.12

a. Make a scatter plot of the data with $x = 0$ in 1970 to determine if the data can be modeled by a quartic function.

b. Find the quartic function that models the data, with $x = 0$ in 1970.

c. Graph the model and the scatter plot of the data on the same axes.

d. Use the model to estimate the number of pregnancies per thousand in 1997.

Solution

a. The scatter plot of the data is shown in Figure 4.12(a).

b. The quartic function that models this data is

$$f(x) = -0.001x^4 + 0.052x^3 - 0.960x^2 + 8.173x + 79.664.$$

c. The graph of the model and the scatter plot of the data are shown in Figure 4.12(b).

d. Because $f(27) \approx 88.46$, the estimated rate of pregnancies in 1997 is about 88 per 1000.   ■

Model Comparisons

Even when a scatter plot does not show all the characteristics of the graph of a quartic function, a quartic function may be a good fit for the data. Consider the following example, which shows a comparison of a cubic and a quartic model for a set of data. It also shows that many decimal places are sometimes required in reporting a model, so that the leading coefficient is not written as zero.

EXAMPLE 4   Foreign-Born Population

The percent of U.S. population that was foreign born for selected years during 1900 to 1999 is shown in Figure 4.13.

a. Create a scatter plot for the data, using the number of years after 1900 as the input x and the percent as the output y.

b. Find a cubic function to model the data and graph the function on the same axes with the data points. Round coefficients to seven decimal places.

c. Find a quartic function to model the data and graph the function on the same axes with the data points. Round coefficients to seven decimal places.

d. Use the graphs of the equations reported in parts (b) and (c) to determine which function is the better fit for the data.

Year	Percent
1900	13.6
1910	14.7
1920	13.2
1930	11.6
1940	8.8
1950	6.9
1960	5.4
1970	4.8
1980	6.2
1990	8
1999	9.7

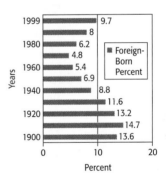

(Source: U.S. Bureau of the Census)

FIGURE 4.13

Solution

a. The scatter plot of the data is shown in Figure 4.14.

b. A cubic function that models the data is

$$y = 0.0000579x^3 - 0.0066066x^2 + 0.0486239x + 14.1587916,$$

with $x = 0$ in 1900. Figure 4.15(a) shows the graph of the unrounded model on the same axes with the data points.

FIGURE 4.14

c. A quartic function that models the data is

$$y = -0.0000009x^4 + 0.0002271x^3 - 0.0170674x^2 + 0.2554588x + 13.5733822,$$

with $x = 0$ representing 1900. Figure 4.15(b) shows the graph of the unrounded model on the same axes with the data points. Note that the graph of this quartic function is not complete on this window because only two turning points are shown. Increasing the x-interval on the window would show another turning point.

d. Although both models are relatively good fits for the data, the graph of the quartic model appears to be a slightly better fit for the data than the cubic model.

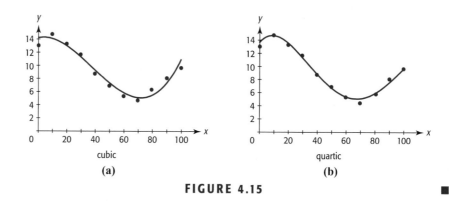

(a)

(b)

FIGURE 4.15 ∎

Third and Fourth Differences

Recall that for equally spaced inputs, the first differences of outputs are constant for data fitting linear functions, and the second differences are constant for data fitting quadratic functions. In a like manner, the third differences of outputs are constant for data fitting cubic functions, and the fourth differences are constant for data fitting quartic functions, if the inputs are equally spaced.

EXAMPLE 5 ## Model Comparisons

Table 4.8 below has an equally spaced set of inputs x in column 1 and three sets of outputs in columns 2, 3, and 4. Use third and/or fourth differences to determine if each set of data can be modeled by a cubic or quartic function; find the function that is the best fit for each set of outputs and the input, using:

a. Output Set I   **b.** Output Set II   **c.** Output Set III.

TABLE 4.8

(Inputs) x	Output Set I	Output Set II	Output Set III
1	3	4	−16
2	14	4	−13
3	47	18	1
4	114	88	33
5	227	280	87
6	398	684	173

Solution

a. The inputs (*x*-values in column 1) are uniform. The third differences for Output Set I data in column 2 follow.

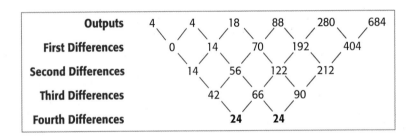

Outputs		3		14		47		114		227		398
First Differences			11		33		67		113		171	
Second Differences				22		34		46		58		
Third Differences					12		12		12			

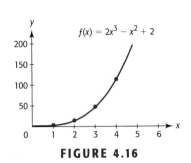

$$f(x) = 2x^3 - x^2 + 2$$

FIGURE 4.16

The third differences are constant, so the data points fit exactly on the graph of a cubic function. The function that models the data is $f(x) = 2x^3 - x^2 + 2$. Figure 4.16 shows the graph of this function for $0 \le x \le 6$.

b. The third and fourth differences for Output Set II data in column 3 follow.

Outputs		4		4		18		88		280		684
First Differences			0		14		70		192		404	
Second Differences				14		56		122		212		
Third Differences					42		66		90			
Fourth Differences						24		24				

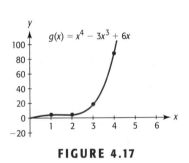

$$g(x) = x^4 - 3x^3 + 6x$$

FIGURE 4.17

The fourth differences are constant, so the data points fit exactly on the graph of a quartic function. The function that models the data is $g(x) = x^4 - 3x^3 + 6x$. (See Figure 4.17.)

c. The third differences for Output Set III data in column 4 follow.

Outputs		-16		-13		1		33		87		173
First Differences			3		14		32		54		86	
Second Differences				11		18		22		32		
Third Differences					7		4		10			

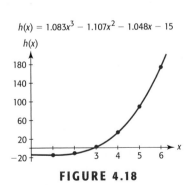

$$h(x) = 1.083x^3 - 1.107x^2 - 1.048x - 15$$

FIGURE 4.18

The third differences are not constant, but are relatively close, so the data points can be approximately fitted by a cubic function. The cubic function that is the best fit for the data is $h(x) = 1.083x^3 - 1.107x^2 - 1.048x - 15$. Figure 4.18 shows that the graph is a good fit for the data points. ■

Note that the third (or fourth) differences will not be constant unless the data points fit the cubic (or quartic) function exactly, and real data points will rarely fit a function exactly. As with other models, statistical measures exist to measure the "goodness of fit" of the model to the data.

4.2 SKILLS CHECK

1. Find the cubic function that models the data in the table below.

x	−2	−1	0	1	2	3	4
y	−16	−3	0	−1	0	9	32

2. Find the quartic function that models the data in the table below.

x	−2	−1	0	1	2	3	4
y	0	−3	0	−3	0	45	192

3. **a.** Make a scatter plot of the data in the table below.
   b. Does it appear that a linear model or a cubic model is the better fit for the data?

x	1	2	3	4	5	6
y	−2	−1	0	4	8	16

4. **a.** Find a cubic function that models the data in the table in Exercise 3.
   b. Find a linear function that models the data.
   c. Visually determine which model is the better fit for the data.

5. **a.** Find the cubic function that is the best fit for the data.
   b. Find the quartic function that is the best fit for the data.

x	1	2	3	4	5	6	7
y	−2	0	4	16	54	192	1500

6. **a.** Graph each of the functions found in Exercise 5 on the same axis with the data in the table.
   b. Does it appear that a cubic model or a quartic model is the better fit for the data?

7. Find the quartic function that is the best fit for the data in the table below.

x	−3	−2	−1	0	1	2	3
y	55	7	1	1	−5	−5	37

8. The following table has the inputs, x, and the outputs for two functions, f and g. Use third differences to determine whether each function is exactly cubic.

x	0	1	2	3	4	5
$f(x)$	0	1	5	24	60	110
$g(x)$	0	.5	4	13.5	32	62.5

9. Find the cubic function that is the best fit for $f(x)$ defined by the table in Exercise 8.

10. Find the cubic function that is the best fit for $g(x)$ defined by the table in Exercise 8.

4.2 EXERCISES

Use unrounded models for graphing and calculations unless otherwise stated.

1. *College Enrollment* The following table gives the number of thousands of college enrollments by recent high school graduates, where x is the number of years after 1960.

   a. Use the data to find a cubic function that models the number of these enrollments as a function of the number of years after 1960.
   b. According to the model, how many college enrollments by recent high school graduates in the U.S. are predicted for 2000?

Year	Enrollments (thousands)	Year	Enrollments (thousands)
1960	1679	1988	2673
1965	2659	1989	2454
1970	2757	1990	2355
1975	3186	1991	2276
1980	3089	1992	2398
1982	3100	1993	2338
1983	2964	1994	2517
1984	3012	1995	2599
1985	2666	1996	2660
1986	2786	1997	2769
1987	2647		

(Source: U.S. Center for Education Statistics)

2. *Japanese Economy* Japan experienced an economic setback and some economists were fearful in 2002 that the Japanese economy was slipping into recession. The percent change in gross domestic product for each year since 1997, with forecasts for 2001 and 2002, is given in the table below.
   a. Find a quartic function that models the data, where x equals the number of years after 1990 and y is the percent change.
   b. Use the model to estimate the percent change in gross domestic product for 2002.

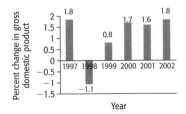

3. *Hotel Revenue* The revenues of Amerihost Properties for the years 1995–1999 are given in the table below.
   a. Find the quartic model that is the best fit for this data, with y equal to the revenue in thousands of dollars and x equal to the number of years after 1995.
   b. Use the model to estimate the revenue in 1998, and compare it with the data in the table.

Year	Revenue (thousands of dollars)
1995	51,962
1996	68,342
1997	62,666
1998	68,618
1999	76,058

(Source: "Financial Highlights," Amerihost Properties, Inc. Annual Report)

4. *Homicide Rate* The number of homicides per 100,000 people for selected years from 1950–1997 is given in the table below.

Year	Homicides per 100,000	Year	Homicides per 100,000
1950	4.6	1980	10.2
1955	4.1	1985	7.9
1960	5.1	1990	9.4
1965	5.1	1995	8.2
1970	7.9	1997	6.8
1975	9.6		

(Source: FBI, Uniform Crime Statistics)

The scatter plot of this data is shown in the figure below, with x equal to the number of years after 1950.
   a. Does the scatter plot of the data indicate that a quadratic, cubic, or quartic function is the best fit for the data?
   b. Use technology to find the equation that models the data and graph the data and the model on the same axes. Round to four decimal places.
   c. What is the number of homicides per 100,000 in 1990, according to the unrounded model? How does this compare with the given data?

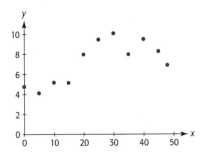

5. *Single Parent Families* The table gives the single parent families as a percent of all families for the years 1960 through 1998.
   a. Graph the data points from the table, with x equal to the number of years after 1950 and y equal to the percent of all families.
   b. Find the cubic model that is the best fit for this data with x equal to the number of years after 1950.
   c. Find the power function that is the best fit for this data with x equal to the number of years after 1950.
   d. Does the cubic model appear to be a better fit for the data, or do the models appear to fit equally well?

Year	Single Parent Families (percent)	Year	Single Parent Families (percent)
1960	9.1	1992	29.7
1965	11.0	1993	30.2
1970	12.9	1994	30.8
1980	21.5	1995	31.0
1985	26.3	1996	31.6
1990	28.1	1997	32.0
1991	28.9	1998	31.7

(Source: U.S. Census Bureau)

6. *Congressional Grants* Congress earmarks millions of dollars each year for academic projects. These awards are often the result of congressional connections. The table below gives the congressional money earmarked for academic projects, in millions of dollars, for the years 1990–2000.
   a. Use a scatter plot, with x equal to the number of years after 1990, to determine the degree of the function that is the best fit for the data.
   b. Use technology to find a function that models the millions of dollars in earmarked grants as a function of the number of years after 1990.
   c. Graph the unrounded model on the same axes with the data. Does the model appear to be an excellent fit, a good fit, or a poor fit for the data?

Year	$ million	Year	$ million
1990	230	1992	640
1991	450	1993	680

Year	$ million	Year	$ million
1994	610	1998	470
1995	600	1999	800
1996	640	2000	1040
1997	410		

(Source: Associated Press)

7. *Births* The numbers of births to females in the U.S. between 15 and 17 years of age for certain years are shown in the table below.
   a. Find the cubic function that models the data, where y is the number of thousands of births and x is the number of years from 1960. Round the coefficients to four decimal places.
   b. Use the rounded model reported in part (a) to estimate the number of births that occurred in 1994.

Year	Births (thousands)	Year	Births (thousands)
1960	182.408	1991	188.226
1970	223.590	1992	187.549
1980	198.222	1993	190.535
1986	168.572	1994	195.169
1990	183.327		

(Source: Child Trends, Inc. MSNBC, 1998)

8. *Births* The numbers of births to females in the U.S. under 15 years of age for certain years are shown in the table below.
   a. Find the cubic function that models the data, where y is the number of thousands of births and x is the number of years from 1960. Round the coefficients to four decimal places.
   b. Use the rounded model reported in part (a) to estimate the number of such births that occurred in 1996.

Year	Births (thousands)	Year	Births (thousands)
1960	6.780	1991	12.014
1970	11.752	1992	12.220
1980	10.169	1993	12.554
1986	10.176	1994	12.901
1990	11.657	1995	12.318

(Source: Child Trends, Inc. MSNBC, 1998)

9. *Alcohol-Related Traffic Fatalities* The table below shows the total number of people killed in alcohol-related traffic accidents for the years 1983–1999.
   a. Find a cubic function that models this data, where x is the number of years after 1982 and y is the number of fatalities.
   b. Find a quartic model that models this data, where x is the number of years after 1982 and y is the number of fatalities.
   c. Graph both models on $0 \le x \le 18$ to determine which of these models is the best fit for the data.

Alcohol-Related Traffic Fatalities

Year	Total Killed	Year	Total Killed
1983	23,646	1992	17,858
1984	23,758	1993	17,473
1985	22,716	1994	16,580
1986	24,045	1995	17,247
1987	23,641	1996	17,218
1988	23,646	1997	16,485
1989	22,404	1998	16,020
1990	22,084	1999	15,786
1991	19,887		

(Source: www.madd.org/stats/fatalities)

10. *Property crimes* The table below gives the rate of total property crimes as the number per 1000 households for each of the years 1973–1999.
    a. Find a cubic model that is the best fit for this data, with x equal to the number of years after 1970.
    b. Find a quartic model that is the best fit for this data, with x equal to the number of years from 1970.
    c. Which model appears to be a better fit for the data?
    d. What does the cubic model estimate the total property crime rate to be for 2000?

Year	Total Property Crime (number per 1000 households)	Year	Total Property Crime (number per 1000 households)
1973	519.9	1976	544.2
1974	551.5	1977	544.1
1975	553.6	1978	532.6

Year	Total Property Crime (number per 1000 households)	Year	Total Property Crime (number per 1000 households)
1979	531.8	1990	348.9
1980	496.1	1991	353.7
1981	497.2	1992	325.3
1982	468.3	1993	318.9
1983	428.4	1994	310.2
1984	399.2	1995	290.5
1985	385.4	1996	266.3
1986	372.7	1997	248.3
1987	379.6	1998	217.4
1988	378.4	1999	198.0
1989	373.4		

(Source: U.S. Dept. of Justice, www.ojp.usdoj.gov)

11. *Movie Tickets* The average price of a U.S. movie ticket is given in the table below.
    a. Write the cubic function that models the data, with x equal to the number of years after 1990.
    b. Use graphical or numerical methods to find when the model estimates that the price was $5.70.

Year	Ticket Price ($)	Year	Ticket Price ($)
1991	4.21	1996	4.42
1992	4.15	1997	4.59
1993	4.14	1998	4.69
1994	4.18	1999	5.08
1995	4.35		

(Source: www.hollywood.com/nato/s stats3.html)

12. *Box Office Gross* The total U.S. box office grosses (in millions of dollars) is given in the table on page 358.
    a. Write the cubic function that models the data, with x equal to the number of years after 1980.
    b. Use graphical or numerical methods with the model to estimate when the box office grossed $4,653,000,000.

Year	Total U.S. Box Office Grosses ($ million)	Year	Total U.S. Box Office Grosses ($ million)
1987	4252	1994	5396
1988	4458	1995	5493
1990	5021	1996	5911
1991	4803	1997	6365
1992	4871	1998	6949
1993	5154	1999	7448

(Source: www.hollywood.com/nato/s stats3.html)

13. *Juvenile Crime* The table below gives the annual number of arrests for crime per 100,000 juveniles from 10 to 17 years of age (with some juveniles having multiple arrests).
    a. Use a scatter plot to determine the degree of the function that is the best fit for the data, with x equal to the number of years after 1980.
    b. Create a cubic function that gives the annual number of arrests for crime per 100,000 juveniles 10 to 17 years of age, with $x = 0$ in 1980.

c. Graph the data and the model reported in part (b) on the same axes.
d. Using the model reported in part (b), what is the ratio of juvenile crimes in 1999 to those in 1990?

Year	Number of Arrests per 100,000	Year	Number of Arrests per 100,000
1980	334.09	1990	428.48
1981	322.64	1991	461.06
1982	314.48	1992	482.14
1983	295.98	1993	504.48
1984	297.46	1994	526.64
1985	302.97	1995	517.74
1986	316.70	1996	460.27
1987	310.55	1997	442.46
1988	326.48	1998	369.64
1989	381.56	1999	339.14

(Source: U.S. Department of Justice, Bureau of Justice Statistics, 2000)

4.3 Solution of Polynomial Equations

Key Concepts

■ Solving polynomial equations using
  Factoring
  Factoring by grouping
  The root method

■ Estimating solutions with technology using
  The intersection method
  The x-intercept method

The closing price of a share of Worldcom, Inc. stock for the years 1994 to 2000 can be modeled by the equation $y = -2.56x^3 + 53.73x^2 - 356.47x + 772.74$, where x represents the number of years after 1990.

  The graph of this model, shown in Figure 4.19, shows that in 1995, the Worldcom stock value dropped to nearly $10, then increased in value, and then experienced a sharp decline in its value after 1999. If this trend were to continue, in what year will the closing price again be worth $15?

(Source: S&P Comstock)

FIGURE 4.19

To answer this question, we would solve the equation

$$-2.56x^3 + 53.73x^2 - 356.47x + 772.74 = 15.$$

The technology of graphing utilities allows us to find approximate solutions of equations such as this one.

While many equations derived from real data frequently require technology to find or to approximate solutions, some higher degree equations can be solved by factoring, and some can be solved by the root method. In this section we discuss these methods for solving polynomial equations, and in the next section we will discuss additional methods of solving higher degree polynomial equations.

Solving Polynomial Equations by Factoring

Some polynomial equations can be solved by factoring and using the zero-product property, much like the method used to solve quadratic equations described in Chapter 2. From our discussion in Chapter 2, we know that the following statements regarding a function f are equivalent:

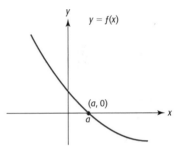

- $(x - a)$ is a factor of $f(x)$.
- a is a zero of the function f.
- a is a solution to the equation $f(x) = 0$.
- a is an x-intercept of the graph of $y = f(x)$.
- The graph crosses the x-axis at the point $(a, 0)$.

EXAMPLE 1   **Factors, Zeros, Intercepts, and Solutions**

If $P(x) = (x - 1)(x + 2)(x + 5)$, find

a. the zeros of $P(x)$.

b. the solutions of $P(x) = 0$.

c. the x-intercepts of the graph of $y = P(x)$.

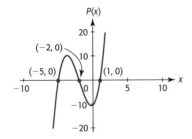

FIGURE 4.20

Solution

a. The factors of $P(x)$ are $(x - 1)$, $(x + 2)$, and $(x + 5)$; thus the zeros of $P(x)$ are $x = 1$, $x = -2$, and $x = -5$.

b. The solutions to $P(x) = 0$ are the zeros of $P(x)$: $x = 1$, $x = -2$, and $x = -5$.

c. The x-intercepts of the graph of $y = P(x)$ are $1, -2$, and -5. (See Figure 4.20.) ■

EXAMPLE 2   **Finding the Maximum Volume**

A box is to be formed by cutting squares x inches on each side from each corner of a square piece of cardboard that is 24 inches on each side and folding up the sides. This will give a box whose height is x inches, with the length of each side of the bottom $2x$

less than the original length of the cardboard. (See Figure 4.21.) Using $V = lwh$, its volume is given by

$$V = (24 - 2x)(24 - 2x)x,$$

where x is length of the side of the square that is cut out.

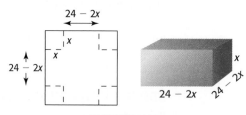

FIGURE 4.21

a. Use the factors of V to find the values of x that give volume 0.

b. Graph the function that gives the volume as a function of the side of the square that is cut out over an x-interval that includes the x-values found in part (a).

c. What input values make sense for this problem (that is, actually result in a box)?

d. Graph the function using the input values that make sense for the problem (found in part (c)).

e. Use technology to determine the size of the squares that should be cut out to give the maximum volume.

Solution

a. We use factoring to solve the equation $0 = (24 - 2x)(24 - 2x)x$.

$$0 = (24 - 2x)(24 - 2x)x$$

$$\underline{x = 0 \text{ or } 24 - 2x = 0}$$

$$x = 0 \quad \begin{vmatrix} -2x = -24 \\ x = 12 \end{vmatrix}$$

b. The graph of $V(x)$ including the values $x = 0$ and $x = 12$ is shown in Figure 4.22(a).

$V = x(24 - 2x)(24 - 2x)$

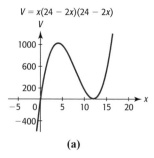

(a)

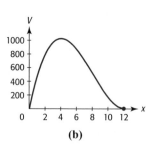

(b)

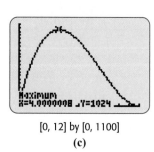

[0, 12] by [0, 1100]

(c)

FIGURE 4.22

c. The box can exist only if each of the dimensions is positive, resulting in a volume that is positive. Thus, $24 - 2x > 0$, or $x < 12$, so the box can only exist if the value of x is greater than 0 and less than 12. Note that if squares 12 inches on a side are removed, no material remains to make a box, and no squares larger than 12 inches can be removed from all four corners.

d. The graph of $V(x)$ over the x-interval $0 < x < 12$ is shown in Figure 4.22(b).

e. By observing the graph, we see that the function has a maximum value, which gives the maximum volume over the interval $0 < x < 12$. Using the $\boxed{\text{MAXIMUM}}$ command on a graphing utility, we see that the maximum occurs at or near $x = 4$ (see Figure 4.22(c)). Cutting the squares of side 4 inches from each corner and folding up the sides gives a box that is 16 inches by 16 inches by 4 inches. The volume of this box is $16 \cdot 16 \cdot 4 = 1024$ cubic inches, and testing shows that this is the maximum possible volume. ∎

EXAMPLE 3 Photosynthesis

The amount y of photosynthesis that takes place in a certain plant depends on the intensity x of the light (in lumens) present, according to the function $y = 120x^2 - 20x^3$. The model is valid only for nonnegative x-values that produce a positive amount of photosynthesis.

a. Graph this function with a viewing window large enough to see two turning points.

b. Use factoring to find the x-intercepts of the graph.

c. What nonnegative values of x will give positive photosynthesis?

d. Graph the function with a viewing window that contains only the values of x found in part (c) and nonnegative values of y.

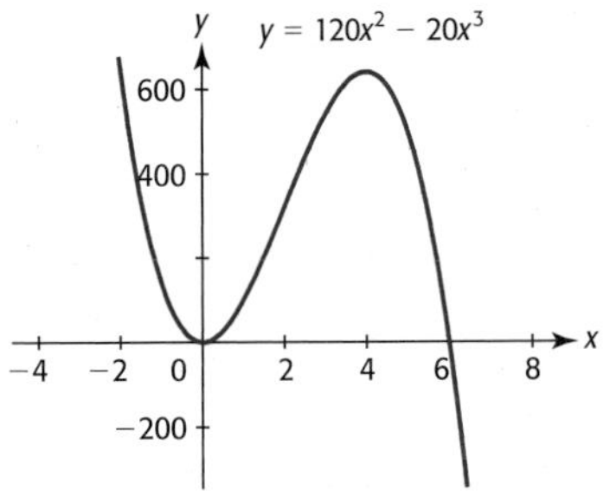

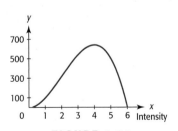

FIGURE 4.23

Solution

a. The graph in Figure 4.23 shows the function with two turning points.

b. We can find the x-intercepts of the graph by solving $120x^2 - 20x^3 = 0$ for x. Factoring gives

$$20x^2(6 - x) = 0$$
$$20x^2 = 0 \quad \text{or} \quad 6 - x = 0$$
$$x = 0 \quad \text{or} \quad x = 6$$

Thus, the x-intercepts are 0 and 6.

c. From the graph of $y = 120x^2 - 20x^3$, in Figure 4.23, we see that the positive output values result when the input values are between 0 and 6. Thus, the nonnegative x-values that produce positive amounts of photosynthesis are $0 < x < 6$.

d. Graphing the function in a viewing window that is [0, 6] by [0, 700] will show the region that is appropriate for the model. (See Figure 4.24.) ∎

FIGURE 4.24

Solution Using Factoring by Grouping

Some higher degree equations can be solved by using the method of factoring by grouping.* Consider the following examples.

---

*Factoring by grouping was discussed in the Chapter 2 Algebra Toolbox.

EXAMPLE 4   Cost

The cost of producing a product is given by the function

$$C(x) = x^3 - 12x^2 + 3x + 9 \text{ thousand dollars,}$$

where x is the number of hundreds of units produced. How many units must be produced to give a total cost of $45,000?

Solution

Because $C(x)$ is the cost in thousands of dollars, a cost of $45,000 is represented by $C(x) = 45$. So we seek to solve

$$45 = x^3 - 12x^2 + 3x + 9 \quad \text{or} \quad 0 = x^3 - 12x^2 + 3x - 36.$$

This equation appears to have a form that permits factoring by grouping. The solution steps follow.

$0 = (x^3 - 12x^2) + (3x - 36)$	Separate the terms into two groups, each having a common factor.
$0 = x^2(x - 12) + 3(x - 12)$	Factor x^2 from the first group, then 3 from the second group.
$0 = (x - 12)(x^2 + 3)$	Factor the common factor $(x - 12)$ from each of the two terms.
$x - 12 = 0$ or $x^2 + 3 = 0$	Set each factor $= 0$ and solve for x.
$x = 12$ or $x^2 = -3$	
$x = 12$	

Because no real number x can satisfy $x^2 = -3$, the only solution is 12 hundred units. Thus producing 1200 units gives a total cost of $45,000. ∎

The Root Method

We solved quadratic equations of the form $x^2 = C$ in Chapter 2 by taking the square root of both sides and using a $\pm$ symbol to indicate that we get both the positive and the negative root: $x = \pm\sqrt{C}$. In the same manner, we can solve $x^3 = C$ by taking the cube root of both sides. There will not be a $\pm$ sign because there is only one real cube root of a number. In general, there is only one real nth root if n is odd, and there are zero or two real roots if n is even.

ROOT METHOD

The real solutions of the equation $x^n = C$ are found by taking the nth root of both sides:

$$x = \sqrt[n]{C} \text{ if } n \text{ is odd, and}$$

$$x = \pm\sqrt[n]{C} \text{ if } n \text{ is even and } C \geq 0.$$

EXAMPLE 5   Root Method

Solve the following equations.

a. $x^3 = 125$ **b.** $5x^4 = 80$ **c.** $4x^2 = 18$

Solution

a. Taking the cube root of both sides gives

$$x^3 = 125$$
$$x = \sqrt[3]{125}$$
$$x = 5$$

No $\pm$ sign is needed because the root is odd.

b. We divide both sides by 5 to isolate the x^4 term

$$5x^4 = 80$$
$$x^4 = 16$$

and take the fourth root of both sides, using a $\pm$ sign (because the root is even).

$$x = \pm\sqrt[4]{16} = \pm 2$$

The solutions are 2 and -2.

c. We divide both sides by 4 to isolate the x^2 term:

$$4x^2 = 18$$
$$x^2 = \frac{9}{2}$$

Then we take the square root of both sides and use a $\pm$ sign (because the root is even).

$$x = \pm\sqrt{\frac{9}{2}} = \pm\frac{3}{\sqrt{2}}$$

The *exact* solutions are $\pm\dfrac{3}{\sqrt{2}}$ and these solutions are *approximately* 2.121 and -2.121. ∎

EXAMPLE 6   **Future Value of an Investment**

- -

The future value of \$10,000 invested for 4 years at interest rate r, compounded annually, is given by $S = 10{,}000(1 + r)^4$. Find the rate r, as a percent, for which the future value is \$14,641.

Solution

We seek to solve $14{,}641 = 10{,}000(1 + r)^4$ for r. We can rewrite this equation with $(1 + r)^4$ on one side by dividing both sides by 10,000.

$$14{,}641 = 10{,}000(1 + r)^4$$
$$1.4641 = (1 + r)^4$$

We can now remove the exponent by taking the fourth root of both sides of the equation, and using a $\pm$ sign.

$$\pm\sqrt[4]{1.4641} = 1 + r$$
$$\pm 1.1 = 1 + r$$

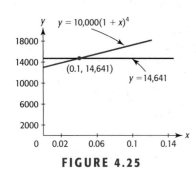

FIGURE 4.25

Solving for r gives

$$
\begin{array}{c|c}
1.1 = 1 + r & -1.1 = 1 + r \\
0.1 = r & -2.1 = r
\end{array}
$$

The negative value cannot represent an interest rate, so the interest rate that gives this future value is

$$r = 0.1 = 10\%.$$

The graphs of $y_1 = 10,000(1 + x)^4$ and $y_2 = 14,641$ shown in Figure 4.25 support the conclusion that $r = .10$ gives the future value 14,641.* ∎

Estimating Solutions with Technology

While we can solve some higher degree equations by using factoring, many real problems do not have "nice" solutions, and factoring is not an appropriate solution method. Thus many equations resulting from real data applications require graphical or numerical methods of solution.

EXAMPLE 7   Juvenile Crime Arrests

The annual number of arrests for crime per 100,000 juveniles from 10 to 17 years of age can be modeled by the function

$$f(x) = -0.357x^3 + 9.417x^2 - 51.852x + 361.208,$$

where x is the number of years after 1980.[†] The number of arrests per 100,000 peaked in 1994 and then decreased. Use this model to graphically estimate the year after 1980 in which the number of arrests fell to 234 per 100,000 juveniles.

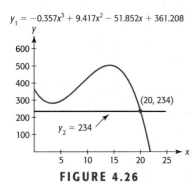

FIGURE 4.26

Solution

To find the year when the number of arrests fell to 234 per 100,000 juveniles, we solve

$$234 = -0.357x^3 + 9.417x^2 - 51.852x + 361.208.$$

To solve this equation graphically, we can use either the intersection method or the x-intercept method. To use the intersection method, we graph

$$y_1 = -0.357x^3 + 9.417x^2 - 51.852x + 361.208 \text{ and } y_2 = 234$$

for $x \geq 0$, and find the x-coordinate of the point of intersection. (See Figure 4.26.)

The graphs intersect at the point where $x \approx 20$. Thus, according to the model, the number of arrests for crime fell to 234 per 100,000 juveniles during the year 2000. ∎

EXAMPLE 8   Stock Prices

The closing price of a share of Worldcom, Inc. stock for the years 1994 to 2000 can be modeled by the equation $y = -2.56x^3 + 53.73x^2 - 356.47x + 772.74$, where x represents the number of years after 1990.

---

*The graph of the quartic function $y = 10,000(1 + x)^4$ is actually curved, but within this small window it appears to be a line.

[†]Note that some juveniles have multiple arrests.

If this trend were to continue, graphically determine the years after 1990 in which the closing price was \$15 (roughly the cost of the stock in 1997).
(Source: S&P Comstock)

Solution

To answer this question, we would solve the equation

$$-2.56x^3 + 53.73x^2 - 356.47x + 772.74 = 15, \quad \text{or}$$

$$-2.56x^3 + 53.73x^2 - 356.47x + 757.74 = 0$$

To solve this equation graphically with the x-intercept method, we graph

$$y = -2.56x^3 + 53.73x^2 - 356.47x + 757.74$$

and find the x-intercepts of the graph. See Figure 4.27(a).

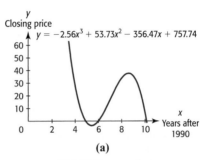

(a)

FIGURE 4.27(a)

It appears that this graph crosses the x-axis three times, so there are three solutions to the equation $-2.56x^3 + 53.73x^2 - 356.47x + 757.74 = 0$. The approximate solutions are shown in Figures 4.27(b), 4.27(c), and 4.27(d).

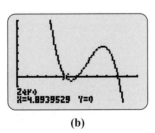
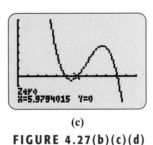
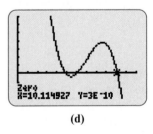

(b) **(c)** **(d)**

FIGURE 4.27(b)(c)(d)

We see that $y = 15$ when $x \approx 5$, $x \approx 6$, or $x \approx 10$. Thus, we might expect the closing price to be approximately \$15 in 1995, in 1996, and in 2000. Worldcom declared bankruptcy in 2002, after accounting irregularities were discovered. ∎

4.3 SKILLS CHECK

In Exercises 1–8, solve the polynomial equations by factoring and check the solutions graphically.

1. $x^3 - 16x = 0$

2. $2x^3 - 8x = 0$

3. $x^4 - 4x^3 + 4x^2 = 0$ **4.** $x^4 - 6x^3 + 9x^2 = 0$

5. $x^3 - 4x^2 - 9x + 36 = 0$

6. $x^3 + 5x^2 - 4x - 20 = 0$

7. $4x^3 - 4x = 0$

8. $x^4 - 3x^3 + 2x^2 = 0$

In Exercises 9–12, solve the polynomial equations by using the root method and check the solutions graphically.

9. $2x^3 - 16 = 0$ **10.** $3x^3 - 81 = 0$

11. $\frac{1}{2}x^4 - 8 = 0$ **12.** $2x^4 - 162 = 0$

In Exercises 13–18, use factoring and the root method to solve the polynomial equations.

13. $4x^4 - 8x^2 = 0$ **14.** $3x^4 - 24x^2 = 0$

15. $0.5x^3 - 12.5x = 0$ **16.** $0.2x^3 - 24x = 0$

17. $x^4 - 6x^2 + 9 = 0$ **18.** $x^4 - 10x^2 + 25 = 0$

In Exercises 19 and 20, use the graph of the polynomial function $f(x)$ to solve $f(x) = 0$.

19. $f(x) = x^3 - 2x^2 - 11x + 12$

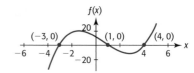
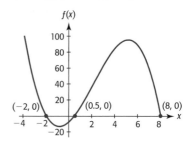

20. $f(x) = -x^3 + 6.5x^2 + 13x - 8$

4.3 EXERCISES

1. *Revenue* The revenue from the sale of a product is given by the function $R = 400x - x^3$.
   a. Use factoring to find the number of units that must be sold to give zero revenue.
   b. Does the graph of the revenue function verify this solution?

2. *Revenue* The revenue from the sale of a product is given by the function $R = 12,000x - 0.003x^3$.
   a. Use factoring and the root method to find the number of units that must be sold to give zero revenue.
   b. Does the graph of the revenue function verify this solution?

3. *Revenue* The price for a product is given by $p = 100,000 - 0.1x^2$, where x is the number of units sold, so the revenue function for the product is $R = px = (100,000 - 0.1x^2)x$.
   a. Find the number of units that must be sold to give zero revenue.
   b. Does the graph of the revenue function verify this solution?

4. *Revenue* If the price from the sale of x units of a product is given by the function $p = 100x - x^2$, the revenue function for this product is given by $R = px = (100x - x^2)x$.
   a. Find the number of units that must be sold to give zero revenue.
   b. Does the graph of the revenue function verify this solution?

5. *Future Value* The future value of $2000 invested for 3 years at rate r, compounded annually, is given by $S = 2000(1 + r)^3$.
   a. Graph this function on the window [0, 0.24] by [0, 5000].
   b. Use the root method to find the rate r, as a percent, for which the future value is $2662.
   c. Use the root method to find the rate r, as a percent, for which the future value is $3456.

6. *Future Value* The future value of $5000 invested for 4 years at rate r, compounded annually, is given by $S = 5000(1 + r)^4$.
   a. Graph this function on the window [0, 0.24] by [0, 12,000].
   b. Use the root method to find the rate r, as a percent, for which the future value is $10,368.
   c. What rate, as a percent, gives $2320.50 in interest on this investment?

7. *Constructing a Box* A box can be formed by cutting squares out of each corner of a piece of tin and folding the "tabs" up. Suppose the piece of tin is 18 inches by 18 inches and each side of the square that is cut out has length x.

a. Write an expression for the height of the box that is constructed.

b. Write an expression for the dimensions of the base of the box that is constructed.

c. Use the formula $V = lwh$ to verify that an equation that represents the volume of the box is $V = 324x - 72x^2 + 4x^3$.

d. Use the equation that you constructed to find the values of x that make $V = 0$.

e. For which of these values of x does a box exist if squares of length x are cut out and the tabs are folded up?

8. *Constructing a Box* A box can be formed by cutting squares out of each corner of a piece of cardboard and folding the "tabs" up. If the piece of cardboard is 12 inches by 12 inches and each side of the square that is cut out has length x, the function that gives the volume of the box is $V = 144x - 48x^2 + 4x^3$.
   a. Find the values of x that make $V = 0$.
   b. For each of the values of x in part (a), discuss what happens to the box if squares of length x are cut out.
   c. For what values of x does a box exist?
   d. Graph the function that gives the volume as a function of the side of the square that is cut out using an x-interval that makes sense for the problem (see part (c)).

9. *Profit* The profit function for a product is given by $P(x) = -x^3 + 2x^2 + 400x - 400$, where x is the number of units produced and sold and P is in hundreds of dollars. Use factoring by grouping to find the number of units that will give a profit of $40,000.

10. *Cost* The total cost function for a product is given by $C(x) = 3x^3 - 6x^2 - 300x + 1800$, where x is the number of units produced and C is the cost in hundreds of dollars. Use factoring by grouping to find the number of units that will give a total cost of $120,000.

11. *Ballistics* Ballistics experts are able to identify the weapon that fired a certain bullet by studying the markings on the bullet after it is fired. They test the rifling (the grooves in the bullet as it travels down the barrel of the gun) by comparing it to that of a second bullet fired into a bale of paper. The speed, s, in centimeters per second, that the bullet travels through the paper is given by $s = 30(3 - 10t)^3$, where $t \le 0.3$ second.
   a. Complete the table to find the speed for given values of t.

t	0	0.1	0.2	0.3
s (cm/sec)				

   b. Use the root method to find the number of seconds to give $s = 0$. Does this agree with the data in the table?

12. *Homicide Rate* The number of homicides per 100,000 people is given by the function $y = -0.000280x^3 + 0.0152x^2 - 0.0360x + 4.144$, with x equal to the number of years from 1950. Use technology to find the year or years in which the number of homicides per 100,000 people was 8.1.
   (Source: U.S. Dept. of Justice, Bureau of Justice Statistics)

13. *Arrests* The annual number of arrests for driving infractions by the New York State Police is given by $y = 1.646x^3 - 18.494x^2 + 39.403x + 838.950$ where y is in thousands and x is the number of years from 1990.
   a. What does the model estimate the number of arrests to be in 1995?
   b. Use technology to find the year after 1995 when there will be 813,310 arrests, according to this model.

14. *Stock Price* The closing stock price of Priceline.com is given by $y = -0.104x^3 + 3.369x^2 - 36.268x + 184.703$, where x is the number of months after March 31, 1999 and y is in dollars.
   a. Use graphical or numerical methods to find the value of x that gives a closing price of $54.98.
   b. At the end of what month does this model say the Priceline.com stock is $54.98?
   (Source: Priceline.com Inc, PCLN Ticker)

15. *Japanese Economy* The percent change y in the Japanese gross domestic product for years since 1997 is given by $y = 0.148x^4 - 5.829x^3 + 85.165x^2 - 545.851x + 1293.932$, where x is the number of years after 1990.
   a. Use graphical or numerical methods with this model to find the year before 1999 when the change in gross domestic product was approximately 1.3%.
   b. According to this model, will the change in gross domestic product ever be 1.3% again?
   (Source: *USA Today*, April 2, 2001)

4.4

Solution of Polynomial Equations Using Synthetic Division

Suppose the profit function for a product is given by

$$P(x) = -x^3 + 98x^2 - 700x - 1800 \text{ dollars,}$$

where x is the number of units produced and sold. To find the number of units that gives break-even for the product, we must solve $P(x) = 0$. Solving this equation by factoring would be difficult. Finding all the zeros of $P(x)$ by using a graphical method may also be difficult without knowing an appropriate viewing window. In general, if we seek to find the exact solutions to a polynomial equation, a number of steps are involved if the polynomial is not easily factored. In this section we investigate methods of finding additional solutions to polynomial equations after one or more solutions have been found. Specifically, if we know that a is one of the solutions to a polynomial equation, we know that $(x - a)$ is a factor of $P(x)$, and we can divide $P(x)$ by $(x - a)$ to find a second factor of $P(x)$ that may give additional solutions.

Division of Polynomials; Synthetic Division

Suppose $f(x)$ is a cubic function and we know that a is a solution of $f(x) = 0$. We can write $(x - a)$ as a factor of $f(x)$ and, if we divide this factor into the cubic function, the quotient will be a quadratic factor of $f(x)$. If there are additional real solutions to $f(x) = 0$, this quadratic factor can be used to find the remaining solutions. Division of a polynomial by a binomial was discussed in the Algebra Toolbox for this chapter.

When we divide a polynomial by a binomial of the form

$$x - a,$$

the division process can be simplified by using a technique called **synthetic division**. Note in the division of $x^4 + 6x^3 + 5x^2 - 4x + 2$ by $x + 2$ on the left below, many terms are recopied. Note also that the divisor is linear, so the quotient has degree one less than the degree of the dividend and that the degree decreases by one in each remaining term. This is true in general, so we can simplify the division process by omitting all variables and writing only the coefficients. If we rewrite the work without the terms that were recopied in the division and without the x's, the process looks like the one below on the right.

$$
\begin{array}{r}
x^3 + 4x^2 - 3x + 2 \\
x + 2{\overline{\smash{\big)}\,x^4 + 6x^3 + 5x^2 - 4x + 2}} \\
\underline{x^4 + 2x^3} \\
4x^3 + 5x^2 \\
\underline{4x^3 + 8x^2} \\
-3x^2 - 4x \\
\underline{-3x^2 - 6x} \\
2x + 2 \\
\underline{2x + 4} \\
-2
\end{array}
$$

$$
\begin{array}{r}
1 + 4 - 3 + 2 \\
+2{\overline{\smash{\big)}\,1 + 6 + 5 - 4 + 2}} \\
\underline{2} \\
4 \\
\underline{8} \\
-3 \\
\underline{-6} \\
2 \\
\underline{4} \\
-2
\end{array}
$$

Look at the process on page 368 on the right and observe that the coefficient of the first term of the dividend and the differences found in each step are identical to the coefficients of the quotient; the last difference, -2, is the remainder. Because of this, we can compress the division process as follows:

$$+2\overline{)1 + 6 + 5 - 4 + 2}$$
$$\underline{2 + 8 - 6 + 4}$$
$$1 + 4 - 3 + 2 - 2$$

Note that the last line in this process gives the coefficients of the quotient, with the last number equal to the remainder. Changing the sign in the divisor allows us to use addition instead of subtraction in the process, as follows:

$$-2\overline{)1 + 6 + 5 - 4 + 2}$$
$$\underline{-2 - 8 + 6 - 4}$$
$$1 + 4 - 3 + 2 - 2$$

The bottom line of this process gives the coefficients of the quotient, and the degree of the quotient is one less than the degree of the dividend, so the quotient is

$$x^3 + 4x^2 - 3x + 2, \text{ with remainder } -2,$$

just as it was in the original division.

The synthetic division process uses the following steps.

SYNTHETIC DIVISION STEPS TO DIVIDE A POLYNOMIAL BY $x - a$

Divide $2x^3 - 9x - 27$ by $x - 3$.

1. Arrange the coefficients in descending powers of x, with a 0 for any missing power. Place a from $x - a$ to the left of the coefficients.

$$3\overline{)2 + 0 - 9 - 27}$$

2. Bring down the first coefficient to the third line. Multiply the last number in the third line by a and write the product in the second line under the next term.

$$3\overline{)2 + 0 - 9 - 27}$$
$$\underline{6}$$
$$2$$

3. Add the last number in the second line to the number above in the first line. Continue this process until all numbers in the first line are used.

$$3\overline{)2 + 0 - 9 - 27}$$
$$\underline{6 + 18 + 27}$$
$$2 + 6 + 9 + 0$$

4. The third line represents the coefficients of the quotient, with the last number the remainder. The quotient is a polynomial of degree one less than the dividend.

The quotient is
$$2x^2 + 6x + 9.$$

If the remainder is 0, $x - a$ is a factor of the polynomial, and the polynomial can be written as the product of the divisor $x - a$ and the quotient.

The remainder is 0, so $x - 3$ is a factor of $2x^3 - 9x - 27$.
$$2x^3 - 9x - 27 =$$
$$(x - 3)(2x^2 + 6x + 9).$$

Using Synthetic Division to Solve Cubic Equations

If we know one solution of a cubic equation $P(x) = 0$, we can divide by the corresponding factor using synthetic division to obtain a second quadratic factor that can be used to find any additional solutions of the polynomial.

EXAMPLE 1   Solving an Equation with Synthetic Division

If $x = -2$ is a solution of $x^3 + 8x^2 + 21x + 18 = 0$, find the remaining solutions.

Solution

If -2 is a solution of the equation $x^3 + 8x^2 + 21x + 18 = 0$, then $x + 2$ is a factor of

$$f(x) = x^3 + 8x^2 + 21x + 18.$$

Synthetically dividing $f(x)$ by $x + 2$ gives

$$
\begin{array}{r}
-2\,\overline{)\,1 + 8 + 21 + 18} \\
\underline{-2 - 12 - 18} \\
1 + 6 + 9 + 0
\end{array}
$$

Thus the quotient is $x^2 + 6x + 9$, with remainder 0, so $x + 2$ is a factor of $x^3 + 8x^2 + 21x + 18$ and

$$x^3 + 8x^2 + 21x + 18 = (x + 2)(x^2 + 6x + 9).$$

Because $x^2 + 6x + 9 = (x + 3)(x + 3)$, we have

$$
\begin{aligned}
x^3 + 8x^2 + 21x + 18 &= 0 \\
(x + 2)(x^2 + 6x + 9) &= 0 \\
(x + 2)(x + 3)(x + 3) &= 0
\end{aligned}
$$

and the solutions to the equation are $x = -2$, $x = -3$, and $x = -3$.

Because one of the solutions of this equation, -3, occurs *twice*, we say that it is a double solution, or a solution of **multiplicity** 2.

To check the solutions, we can graph the function $f(x) = x^3 + 8x^2 + 21x + 18$ and observe x-intercepts at -2 and -3. (See Figure 4.28.) Observe that the graph touches but does not cross the x-axis at $x = -3$, where a *double* solution to the equation (and zero of the function) occurs. ∎

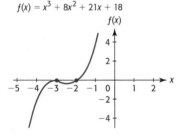

$f(x) = x^3 + 8x^2 + 21x + 18$

FIGURE 4.28

A graph touches but does not cross the x-axis at a zero of even multiplicity, and it crosses the x-axis at a zero of odd multiplicity.

Graphs and Solutions

Sometimes a combination of graphical methods and factoring can be used to solve a polynomial equation. If we are seeking exact solutions to a cubic equation and one solution can be found graphically, we can use that solution and division to find the remaining quadratic factor and to find the remaining two solutions, if they are real. Consider the following example.

EXAMPLE 2   Combining Graphical and Analytical Methods

Solve the equation $x^3 + 9x^2 - 610x + 600 = 0$.

Solution

If we graph $P(x) = x^3 + 9x^2 - 610x + 600$ in a standard window, shown in Figure 4.29, we see that the graph appears to cross the x-axis at $x = 1$, so one solution to $P(x) = 0$ may be $x = 1$. (Notice that this window does not give a complete graph of the cubic polynomial function.)

To verify that $x = 1$ is a solution to $P(x) = 0$, we use synthetic division to divide

$$x^3 + 9x^2 - 610x + 600 \text{ by } x - 1.$$

$$
\begin{array}{r}
1)\overline{1 + 9 - 610 + 600} \\
\underline{1 + 10 - 600} \\
1 + 10 - 600 + 0
\end{array}
$$

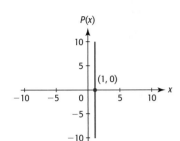

$P(x)$

(1, 0)

FIGURE 4.29

The fact that the division results in a 0 remainder verifies that 1 is a solution to the equation $P(x) = 0$. The remaining factor is $x^2 + 10x - 600$, so the remaining solution to $P(x) = 0$ can be found by factoring or by using the quadratic formula to solve $x^2 + 10x - 600 = 0$.

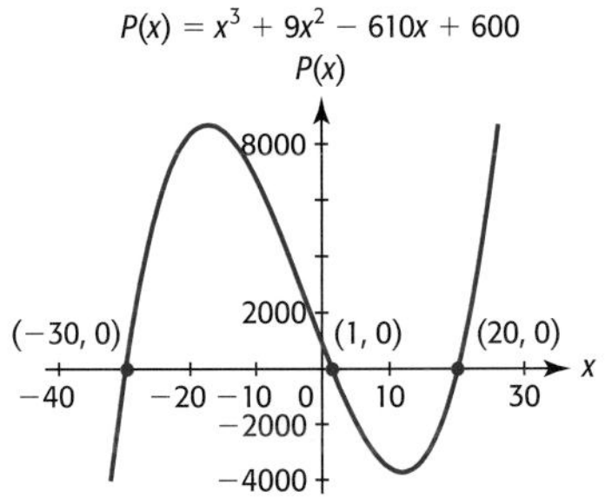

$P(x) = x^3 + 9x^2 - 610x + 600$

$P(x)$

8000

2000

(−30, 0)   (1, 0)   (20, 0)

−2000

−4000

FIGURE 4.30

$$x^3 + 9x^2 - 610x + 600 = 0$$

$$(x - 1)(x^2 + 10x - 600) = 0$$

$$x - 1 = 0 \quad \text{or} \quad x^2 + 10x - 600 = 0$$

$$x = 1 \quad\quad\quad (x + 30)(x - 20) = 0$$

$$x = 1 \quad\quad\quad x = -30 \text{ or } x = 20$$

Graphing the function $P(x) = x^3 + 9x^2 - 610x + 600$ using a viewing window that includes x-values of -30, 1, and 20 verifies that -30, 1, and 20 are zeros of the function and thus solutions to the equation $x^3 + 9x^2 - 610x + 600 = 0$. (See Figure 4.30.) ∎

EXAMPLE 3   Break-Even

The weekly profit for a product is $P(x) = -0.1x^3 + 11x^2 - 80x - 2000$ thousand dollars, where x is the number of thousands of units produced and sold. To find the number of units that gives break-even:

a. Graph the function using a window representing up to 50 thousand units and find one x-intercept of the graph.

b. Use synthetic division to find a quadratic factor of $P(x)$.

c. Find all of the zeros of $P(x)$.

d. Determine the levels of production that give break-even.

Solution

a. Break-even occurs where the profit is 0, so we seek the solution to

$$0 = -0.1x^3 + 11x^2 - 80x - 2000$$

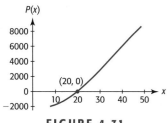

FIGURE 4.31

for $x \geq 0$ (because x represents the number of units). Because $x = 50$ represents 50 thousand units and we seek an x-value where $P(x) = 0$, we use the viewing window $[0, 50]$ by $[-2000, 8000]$. The graph of

$$P(x) = -0.1x^3 + 11x^2 - 80x - 2000$$

is shown in Figure 4.31. (Note that this is not a complete graph of the polynomial function.)

An x-intercept of this graph appears to be $x = 20$ (see Figure 4.31).

b. To verify that $P(x) = 0$ at $x = 20$ and to determine if any other x-values give break-even, we divide

$$P(x) = -0.1x^3 + 11x^2 - 80x - 2000$$

by the factor $x - 20$.

$$
\begin{array}{r}
20) \ \overline{-0.1 + 11 - 80 - 2000} \\
\underline{- 2 + 180 + 2000} \\
-0.1 + 9 + 100 + 0
\end{array}
$$

Thus $P(x) = 0$ at $x = 20$ and the quotient is the quadratic factor $-0.1x^2 + 9x + 100$, so

$$-0.1x^3 + 11x^2 - 80x - 2000 = (x - 20)(-0.1x^2 + 9x + 100).$$

c. The solution to $P(x) = 0$ follows.

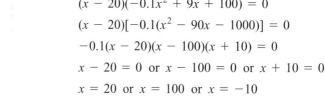

$$-0.1x^3 + 11x^2 - 80x - 2000 = 0$$
$$(x - 20)(-0.1x^2 + 9x + 100) = 0$$
$$(x - 20)[-0.1(x^2 - 90x - 1000)] = 0$$
$$-0.1(x - 20)(x - 100)(x + 10) = 0$$
$$x - 20 = 0 \text{ or } x - 100 = 0 \text{ or } x + 10 = 0$$
$$x = 20 \text{ or } x = 100 \text{ or } x = -10$$

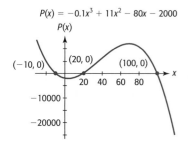

FIGURE 4.32

Thus the zeros are 20, 100 and -10. Figure 4.32 shows the graph of $P(x) = -0.01x^3 + 11x^2 - 80x - 2000$ on a window that shows all three x-intercepts.

d. The solutions indicate that break-even occurs at $x = 20$ and at $x = 100$. The value $x = -10$ also makes $P(x) = 0$, but a negative number of units does not make sense in the context of this problem.   ■

Combining graphical methods and synthetic division, as we did in Example 3, is especially useful in finding exact solutions to polynomial equations. If some of the solutions are irrational or nonreal solutions, the quadratic formula can be used rather than factoring. Remember that the quadratic formula can always be used to solve a quadratic equation, and factoring cannot always be used.

Rational Solutions Test

We can gain more information about the rational solutions of a polynomial equation by using the following:

RATIONAL SOLUTIONS TEST

The rational solutions of the polynomial equation

$$a_n x^n + a_{n-1} x^{n-1} + \cdots + a_1 x + a_0 = 0$$

with integer coefficients, must be of the form $\dfrac{p}{q}$, where p is a factor of the constant term a_0 and q is a factor of a_n, the leading coefficient.

For example, the rational solutions of $x^3 + 8x^2 + 21x + 18 = 0$ must be numbers that are factors of 18 divided by factors of 1. Thus the possible rational solutions are $\pm 1, \pm 18, \pm 2, \pm 9, \pm 3,$ and ± 6. A viewing window containing x-values from -18 to 18 will show all possible rational solutions. Note that the solutions to this equation, found in Example 1 to be -2 and -3, are contained in this list. Many cubic and quartic equations can be solved by using the following steps.

SOLVING CUBIC AND QUARTIC EQUATIONS OF THE FORM $f(x) = 0$

1. Determine the possible rational solutions of $f(x) = 0$.

2. Graph $y = f(x)$ to see if any of the values from Step 1 are x-intercepts. The x-intercepts are also solutions to $f(x) = 0$.

3. Find the factors associated with the x-intercepts from Step 2.

4. Use synthetic division to divide $f(x)$ by the factors from Step 3 to confirm the graphical solutions and find additional factors. Continue until a quadratic factor remains.

5. Use factoring or the quadratic formula to find the solutions associated with the quadratic factor. These solutions are also solutions to $f(x) = 0$.

Example 4 will show how the rational solutions test is used when the leading coefficient of the polynomial is not 1.

EXAMPLE 4 Solving a Quartic Equation

Solve the equation $2x^4 + 10x^3 + 13x^2 - x - 6 = 0$.

Solution

The rational solutions of this equation must be factors of the constant -6 divided by factors of 2, the leading coefficient. Thus the possible rational solutions are*

$$\pm 1, \pm 2, \pm 3, \pm 6, \text{ and } \pm \frac{1}{2} \pm \frac{2}{2}, \pm \frac{3}{2}, \pm \frac{6}{2}.$$

---

*Some of these possible solutions are duplicates.

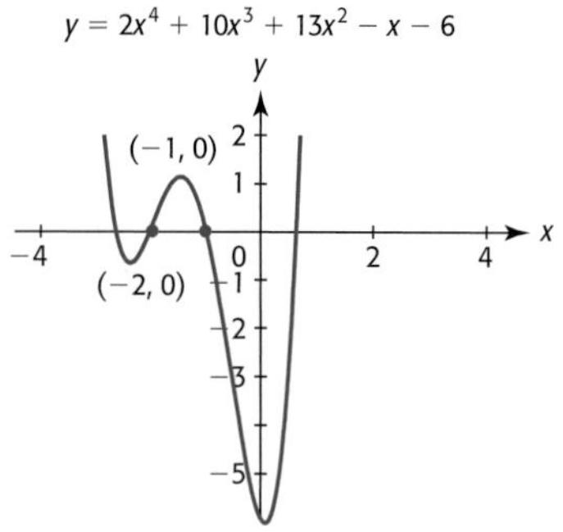

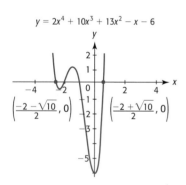

$y = 2x^4 + 10x^3 + 13x^2 - x - 6$

FIGURE 4.33

All of these possible rational solutions are between -6 and 6, so we graph

$$P(x) = 2x^4 + 10x^3 + 13x^2 - x - 6$$

in a window with x-values from -6 to 6. Figure 4.33 shows that the graph appears to cross the x-axis at $x = -1$ and at $x = -2$, so two of the solutions to $P(x) = 0$ appear to be $x = -1$ and $x = -2$. We will use synthetic division to confirm that these values are solutions to the equation and to find the two remaining solutions. (We know there are two additional solutions, which may be irrational, because the graph crosses the x-axis at two other points.)

We can confirm that $x = -1$ is a solution and find the remaining factor of $P(x)$ by using synthetic division to divide $2x^4 + 10x^3 + 13x^2 - x - 6$ by $x + 1$.

$$
\begin{array}{r}
-1\overline{)\,2 + 10 + 13 - 1 - 6} \\
\underline{-\ 2 - 8 - 5 + 6} \\
2 + 8 + 5 - 6 + 0
\end{array}
$$

Thus we have $2x^4 + 10x^3 + 13x^2 - x - 6 = (x + 1)(2x^3 + 8x^2 + 5x - 6)$.

We can also confirm that $x = -2$ is a solution of $P(x)$ by showing that $x + 2$ is a factor of $2x^3 + 8x^2 + 5x - 6$ and thus a factor of $P(x)$. Dividing $x + 2$ into $2x^3 + 8x^2 + 5x - 6$ using synthetic division gives a quadratic factor.

$$
\begin{array}{r}
-2\overline{)\,2 + 8 + 5 - 6} \\
\underline{-\ 4 - 8 + 6} \\
2 + 4 - 3 + 0
\end{array}
$$

So the quotient is the quadratic factor $2x^2 + 4x - 3$ and $P(x)$ has the following factorization.

$$2x^4 + 10x^3 + 13x^2 - x - 6 = (x + 1)(x + 2)(2x^2 + 4x - 3)$$

Thus $P(x) = 0$ is equivalent to $(x + 1)(x + 2)(2x^2 + 4x - 3) = 0$, so the solutions can be found by solving $x + 1 = 0$, $x + 2 = 0$, and $2x^2 + 4x - 3 = 0$.

The two remaining x-intercepts of the graph of $y = 2x^2 + 4x - 3$ do not appear to cross the x-axis at integer values, and $2x^2 + 4x - 3$ does not appear to be factorable. Using the quadratic formula is the obvious method to solve $2x^2 + 4x - 3 = 0$ to find the two remaining solutions.

$y = 2x^4 + 10x^3 + 13x^2 - x - 6$

$\left(\dfrac{-2 - \sqrt{10}}{2}, 0\right)$ $\left(\dfrac{-2 + \sqrt{10}}{2}, 0\right)$

FIGURE 4.34

$$x = \frac{-4 \pm \sqrt{16 - 4(2)(-3)}}{2(2)} = \frac{-4 \pm \sqrt{40}}{4} = \frac{-2 \pm \sqrt{10}}{2}$$

So the solutions are $-1, -2, \dfrac{-2 + \sqrt{10}}{2}$, and $\dfrac{-2 - \sqrt{10}}{2}$. These two solutions can also be verified graphically. (See Figure 4.34.) Note that two of these solutions are rational and two are irrational, approximated by 0.58114 and -2.58114. ∎

Note that some graphing utilities and spreadsheets have commands that can be used to find or to approximate solutions to polynomial equations.

4.4 SKILLS CHECK

In Exercises 1–4, use synthetic division to find the quotient and the remainder.

1. $(x^4 - 4x^3 + 3x + 10) \div (x - 3)$

2. $(x^4 + 2x^3 - 3x^2 + 1) \div (x + 4)$

3. $(2x^4 - 3x^3 + x - 7) \div (x - 1)$

4. $(x^4 - 1) \div (x + 1)$

In Exercises 5 and 6, determine whether the second polynomial is a factor of the first polynomial.

5. $(-x^4 - 9x^2 + 3x); (x + 3)$

6. $(2x^4 + 5x^3 - 6x - 4); (x + 2)$

In Exercises 7 and 8, determine whether the given constant is a solution to the given polynomial equation.

7. $2x^4 - 4x^3 + 3x + 18 = 0; 3$

8. $x^4 + 3x^3 - 10x^2 + 8x + 40 = 0; -5$

In Exercises 9 and 10, one solution of a polynomial equation is given. Use synthetic division to find any remaining solutions.

9. $-x^3 + x^2 + x - 1 = 0; -1$

10. $x^3 + 4x^2 - x - 4 = 0; 1$

11. $x^4 + 2x^3 - 21x^2 - 22x = -40; -5$

12. $2x^4 - 17x^3 + 51x^2 - 63x + 27; 3$

In Exercises 13–15, find one solution graphically and then find the remaining solutions using synthetic division.

13. $x^3 + 3x^2 - 18x - 40 = 0$

14. $x^3 - 3x^2 - 9x - 5 = 0$

15. $3x^3 + 2x^2 - 7x + 2 = 0$

In Exercises 16–18, determine all possible rational solutions of the polynomial equation.

16. $x^3 - 6x^2 + 5x + 12 = 0$

17. $4x^3 + 3x^2 - 9x + 2 = 0$

18. $9x^3 + 18x^2 + 5x - 4 = 0$

In Exercises 19–21, find all rational zeros of the polynomial function.

19. $x^3 - 6x^2 + 5x + 12 = 0$

20. $4x^3 + 3x^2 - 9x + 2 = 0$

21. $9x^3 + 18x^2 + 5x - 4 = 0$

4.4 EXERCISES

In Exercises 1–6, use synthetic division and factoring to solve the problems.

1. *Break-Even* The profit function for a product is given by $P(x) = -0.2x^3 + 66x^2 - 1600x - 60,000$ dollars, where x is the number of units produced and sold. If break-even occurs when 50 units are produced and sold:
   a. Use synthetic division to find a quadratic factor of $P(x)$.
   b. Use factoring to find a number of units other than 50 that gives break-even for the product, and verify your answer graphically.

2. *Break-Even* The profit function for a product is given by $P(x) = -x^3 + 98x^2 - 700x - 1800$ dollars, where x is the number of units produced and sold. If break-even occurs when 10 units are produced and sold:

   a. Use synthetic division to find a quadratic factor of $P(x)$.
   b. Find a number of units other than 10 that gives break-even for the product, and verify your answer graphically.

3. *Break-Even* The weekly profit for a product is $P(x) = -0.1x^3 + 50.7x^2 - 349.2x - 400$ thousand dollars, where x is the number of thousands of units produced and sold. To find the number of units that gives break-even:
   a. Graph the function on the window $[-10, 20]$ by $[-100, 100]$.
   b. Graphically find one x-intercept of the graph.
   c. Use synthetic division to find a quadratic factor of $P(x)$.
   d. Find all of the zeros of $P(x)$.
   e. Determine the levels of production and sale that give break-even.

4. *Break-Even* The weekly profit for a product is $P(x) = -0.1x^3 + 10.9x^2 - 97.9x - 108.9$ thousand dollars, where x is the number of thousands of units produced and sold. To find the number of units that gives break-even:
   a. Graph the function on the window $[-10, 20]$ by $[-100, 100]$.
   b. Graphically find one x-intercept of the graph.
   c. Use synthetic division to find a quadratic factor of $P(x)$.
   d. Find all of the zeros of $P(x)$.
   e. Determine the levels of production and sale that give break-even.

5. *Revenue* The revenue from the sale of a product is given by $R = 1810x - 81x^2 - x^3$. If the sale of 9 units gives a total revenue of $9000, use synthetic division to find another number of units that will give $9000 in revenue.

6. *Revenue* The revenue from the sale of a product is given by $R = 250x - 5x^2 - x^3$. If the sale of 5 units gives a total revenue of $1000, use synthetic division to find another number of units that will give $1000 in revenue.

7. *Drunk Driving Crashes* Suppose the total number of fatalities in drunk driving crashes in South Carolina is modeled, using data from 1982–1998, by the function $y = 0.4566x^3 - 14.3085x^2 + 117.2978x + 107.8456$, where x is the number of years from 1980. This model indicates that the number of fatalities was 244 in 1992. To find any other years that the number was 244:
   a. Set the function equal to 244, and rewrite the resulting equation with 0 on one side.
   b. Graph the function from part (a) on the window $[-10, 25]$ by $[-75, 200]$ and use the graph to find an integer solution to this equation.
   c. Use synthetic division to find a quadratic factor.
   d. Use the quadratic formula to find any other solutions to this equation.

   e. Use this information to determine the years in which the number of fatalities in drunk driving crashes was 244.
   (Source: Anheuser-Busch, www.beeresponsible.com)

8. *College Enrollment* Using data from 1960–1997, the number of thousands of college enrollments by recent high school graduates is given by $f(x) = 0.20x^3 - 13.71x^2 + 265.06x + 1612.56$, where x is the number of years from 1960. To find the year or years after 1960 when the number of enrollments was 2862 thousand:
   a. Set the function equal to 2862, and rewrite the resulting equation with 0 on one side.
   b. Graph this function on the window $[25, 50]$ by $[-50, 50]$ and use the graph to find an integer solution to this equation.
   c. Use synthetic division to find a quadratic factor.
   d. Use the quadratic formula to find any other solutions to this equation.
   e. Use this information to determine the years in which the number of enrollments was 2,862,000.
   (Source: U.S. Census Bureau)

9. *Births* The number of births to females in the U.S. under 15 years of age can be modeled by the function $y = 0.000876x^3 - 0.0492x^2 + 0.831x + 6.922$, where y is the number of thousands of births and x is the number of years from 1960. If the model indicates that 11,224 births occurred in 1990, find two other years when the model indicates that 11,224 births occurred.
   (Source: Child Trends, Inc. MSNBC, 1998)

10. *Births* The number of births to females in the U.S. between 15 and 17 years of age can be modeled by the cubic function $y = 0.0146x^3 - 0.799x^2 + 10.717x + 182.353$, where y is the number of thousands of births and x is the number of years from 1960. If the model indicates that 178,963 births occurred in 1990, find another year after 1960 when the model indicates that 178,963 births occurred.
    (Source: Child Trends, Inc. MSNBC, 1998)

4.5 Complex Solutions; Fundamental Theorem of Algebra

FIGURE 4.35

As we have seen, not all quadratic equations have solutions in the real number system. But if we extend our number system to include **complex numbers**, quadratic equations will always have solutions. In fact, the **Fundamental Theorem of Algebra** tells us that every polynomial equation has at least one solution in the complex numbers.

The complex number system is an extension of the real numbers that includes **imaginary numbers**. The term "imaginary number" seems to imply that the numbers do not exist, but in fact they do have important theoretical and technical applications. Complex numbers are used in the design of electrical circuits and airplanes, and they were used in the development of quantum physics. For example, one way of accounting for the amount as well as the phase of the current or voltage in an alternating current electrical circuit involves complex numbers. Special sets of complex numbers can be graphed on the **complex coordinate system** to create pictures called **fractal images**. (Figure 4.35 shows a fractal image called the Mandelbrot set, which can be generated with the complex number i.)

In this section we will find complex solutions to some quadratic and higher degree polynomial equations.

Imaginary and Complex Numbers

The numbers discussed up to this point are real numbers (either rational numbers such as 2, -3, $\frac{5}{8}$ and $-\frac{2}{3}$ or irrational numbers such as $\sqrt{3}$, $\sqrt[3]{6}$, and π.). We have seen that some equations do not have real solutions. For example, if

$$x^2 + 1 = 0 \text{ then } x^2 = -1,$$

but there is no real number which, when squared, will equal -1. However, we can denote one solution to this equation as the **imaginary unit i**, defined by

$$i = \sqrt{-1}.$$

The set of **complex numbers** is formed by adding real numbers and multiples of i.

COMPLEX NUMBER

The number $a + bi$, in which a and b are real numbers, is said to be a **complex number** in **standard form**. The a is the real part of the number and b is the imaginary part. If $b = 0$, the number $a + bi = a$ is a real number, and if $b \neq 0$, the number $a + bi$ is an **imaginary number**. If $a = 0$, bi is a **pure imaginary number**.

The complex number system includes the real numbers as well as the imaginary numbers. (See Figure 4.36.) Examples of imaginary numbers are $3 + 2i$, $5 - 4i$, and $\sqrt{3} - \frac{1}{2}i$; examples of pure imaginary numbers are $-i$, $2i$, $12i$, $-4i$, $i\sqrt{3}$, and πi.

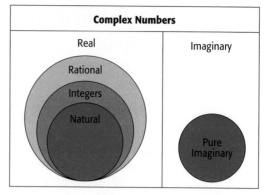

FIGURE 4.36

Operations with Complex Numbers

Complex numbers can be added and subtracted by combining their real and imaginary parts.

> ### ADDITION AND SUBTRACTION OF COMPLEX NUMBERS
>
> The sum and difference of the complex numbers $a + bi$ and $c + di$ are:
>
> $$(a + bi) + (c + di) = (a + c) + (b + d)i$$
> $$(a + bi) - (c + di) = (a - c) + (b - d)i$$

EXAMPLE 1 Sums and Differences of Complex Numbers

Find the sums and differences.

a. $(3 - 5i) + (-12 - 3i)$

b. $(2 + 6i) - (7 - 9i)$

Solution

a. $(3 - 5i) + (-12 - 3i) = (3 + (-12)) + (-5i) + (-3i)$

$= -9 + (-5 - 3)i = -9 - 8i$

b. $(2 + 6i) - (7 - 9i) = (2 - 7) + (6i + 9i) = -5 + (6 + 9)i = -5 + 15i$ ∎

Because of the definition of i, $i \cdot i = i^2 = (\sqrt{-1})^2 = -1$. The product of two complex numbers of the form $a + bi$ can be found by multiplying them as if they were binomials (using the FOIL method) and substituting -1 for i^2. The general procedure follows:

> $$(a + bi)(c + di) = ac + adi + bci + bdi^2$$
> $$= ac + (ad + bc)i + bd(-1)$$
> $$= (ac - bd) + (ad + bc)i$$

EXAMPLE 2   Products of Complex Numbers

Find the following products.

a. $(4 + 3i)(2 - 5i)$

b. $(3 + 2i)(3 - 2i)$

Solution

a. $(4 + 3i)(2 - 5i) = 8 - 20i + 6i - 15i^2 = 8 - 14i - 15(-1) = 23 - 14i$

b. $(3 + 2i)(3 - 2i) = 9 - 6i + 6i - 4i^2 = 9 - 4(-1) = 13$ ∎

Notice that the two factors multiplied in Example 2, part (b) have identical real parts and that their imaginary parts are negatives; such a pair of complex numbers are called **conjugates**. As we saw in Example 2, part (b), the product of two conjugates is always a real number.

> **COMPLEX CONJUGATES**
>
> Two complex numbers with the same real parts and imaginary parts with opposite signs are called **conjugates**.
>
> The product of $a + bi$ and $a - bi$ is the real number $a^2 + b^2$, that is, $(a + bi)(a - bi) = a^2 + b^2$.

The conjugate of the divisor can be used to find the quotient of two complex numbers.

EXAMPLE 3   Quotient of Two Complex Numbers

Find the quotient $\dfrac{5 - 3i}{1 - 2i}$ by writing the fraction as a complex number in standard form.

Solution

The quotient is found by converting the denominator to a real number. This is done by multiplying the numerator and denominator by the conjugate of the denominator, $1 + 2i$, and then simplifying.

$$\frac{5 - 3i}{1 - 2i} = \frac{5 - 3i}{1 - 2i} \cdot \frac{1 + 2i}{1 + 2i} = \frac{5 + 10i - 3i - 6i^2}{1 + 2i - 2i - 4i^2} = \frac{5 + 7i + 6}{1 + 4}$$

$$= \frac{11 + 7i}{5} = \frac{11}{5} + \frac{7}{5}i$$ ∎

Some graphing calculators have a complex number mode that can be used to perform operations with complex numbers. Figure 4.37 shows some of the computations from Examples 1 to 3 done with technology.

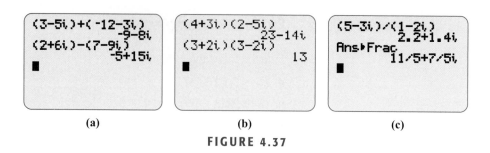

(a) (b) (c)

FIGURE 4.37

Fundamental Theorem of Algebra

As we learned in Section 2.2, not all quadratic equations have real solutions. However, every quadratic equation has complex solutions. If the complex number a is a solution to $P(x) = 0$, then a is a complex zero of P. The **Fundamental Theorem of Algebra** states that every polynomial function has a complex solution.

> **FUNDAMENTAL THEOREM OF ALGEBRA**
>
> If $f(x)$ is a polynomial function of degree $n \geq 1$, then f has at least one complex zero.

As a result of this theorem, it is possible to prove the following.

> **COMPLEX SOLUTIONS**
>
> Every polynomial function $f(x)$ of degree $n \geq 1$ has exactly n complex zeros. Some of these solutions may be imaginary, and some may be repeated.

Equations with Complex Solutions

Recall that the solutions of the quadratic equation $x^2 = C$ are $x = \pm\sqrt{C}$, so that the solutions of the equation $x^2 = -a$ for $a > 0$ are

$$x = \pm\sqrt{-a} = \pm\sqrt{-1}\sqrt{a} = \pm i\sqrt{a}$$

EXAMPLE 4 Solution Using the Root Method

Solve the equations

a. $x^2 = -9$ **b.** $3x^2 + 24 = 0$

Solution

a. Taking the square root of both sides of the equation gives the solution of $x^2 = -9$

$$x = \pm\sqrt{-9} = \pm\sqrt{-1}\sqrt{9} = \pm 3i$$

b. We solve $3x^2 + 24 = 0$ using the root method, as follows:

$$3x^2 = -24$$

$$x^2 = -8$$

$$x = \pm\sqrt{-8} = \pm\sqrt{-1}\sqrt{4 \cdot 2} = \pm 2i\sqrt{2} \qquad \blacksquare$$

We used the root method to solve the equations in Example 4 because neither equation contained a first-degree term (that is, a term containing x to the first power). We can also find complex solutions by using the quadratic formula.* Recall that the solutions of the quadratic equation $ax^2 + bx + c = 0$ with $a \neq 0$, are given by the formula

$$x = \frac{-b \pm \sqrt{b^2 - 4ac}}{2a}.$$

EXAMPLE 5   Solution with the Quadratic Formula

Solve the equations

a. $x^2 - 3x + 5 = 0$

b. $3x^2 + 4x = -3$

Solution

a. Using the quadratic formula, with $a = 1$, $b = -3$, and $c = 5$, gives

$$x = \frac{-(-3) \pm \sqrt{(-3)^2 - 4(1)(5)}}{2(1)} = \frac{3 \pm \sqrt{-11}}{2} = \frac{3 \pm i\sqrt{11}}{2}.$$

Note that the solutions can also be written in the form $\dfrac{3}{2} \pm \dfrac{\sqrt{11}}{2}i$.

Thus the solutions are the complex numbers $\dfrac{3}{2} + \dfrac{\sqrt{11}}{2}i$ and $\dfrac{3}{2} - \dfrac{\sqrt{11}}{2}i$.

b. Writing $3x^2 + 4x = -3$ in the form $3x^2 + 4x + 3 = 0$ gives $a = 3$, $b = 4$, and $c = 3$, so the solution is

$$x = \frac{-4 \pm \sqrt{(4)^2 - 4(3)(3)}}{2(3)} = \frac{-4 \pm \sqrt{-20}}{6} = \frac{-4 \pm \sqrt{-1}\sqrt{4}\sqrt{5}}{6}$$

$$= \frac{-4 \pm 2i\sqrt{5}}{6} = \frac{2(-2 \pm i\sqrt{5})}{2 \cdot 3} = \frac{-2 \pm i\sqrt{5}}{3}$$

Thus,

$$x = -\frac{2}{3} + \frac{\sqrt{5}}{3}i \text{ and } x = -\frac{2}{3} - \frac{\sqrt{5}}{3}i. \qquad \blacksquare$$

Note that the complex solutions in Example 5 occurred in conjugate pairs. We can formalize this as follows.

CONJUGATE PAIRS

Suppose $f(x)$ is a polynomial with real coefficients. If $a + bi$ is a zero of f, then the complex conjugate $a - bi$ is also a zero of f.

---

*The solutions could also be found by completing the square. Recall that the quadratic formula was developed by completing the square on the quadratic equation $ax^2 + bx + c = 0$.

EXAMPLE 6 — Solution in the Complex Number System

Find the complex solutions to $x^3 + 2x^2 - 3 = 0$.

Solution

Because the function is cubic, there are three solutions, and because the nonreal solutions occur in conjugate pairs, at least one of the solutions must be real. To estimate that solution, we graph

$$y = x^3 + 2x^2 - 3$$

on a window that includes the possible rational solutions, which are ± 1 and ± 3. (See Figure 4.38.)

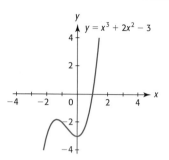

FIGURE 4.38

The graph appears to cross the x-axis at 1, so we (synthetically) divide $x^3 + 2x^2 - 3$ by $x - 1$.

$$
\begin{array}{r}
1\overline{)1 + 2 + 0 - 3} \\
\underline{1 + 3 + 3} \\
1 + 3 + 3 + 0
\end{array}
$$

The division confirms that 1 is a solution to $x^3 + 2x^2 - 3 = 0$ and shows that

$$x^2 + 3x + 3$$

is a factor of $x^3 + 2x^2 - 3$. Thus any remaining solutions to the equation satisfy

$$x^2 + 3x + 3 = 0.$$

Using the quadratic formula gives the remaining solutions.

$$x = \frac{-3 \pm \sqrt{3^2 - 4(1)(3)}}{2(1)} = \frac{-3 \pm \sqrt{-3}}{2} = \frac{-3 \pm i\sqrt{3}}{2}$$

Thus the solutions are $1, \dfrac{3}{2} + \dfrac{\sqrt{3}}{2}i$, and $-\dfrac{3}{2} - \dfrac{\sqrt{3}}{2}i$.

Because there can only be three solutions to a cubic equation, we have all of the solutions. ∎

4.5 SKILLS CHECK

In Exercises 1–4, identify each number as one or more of real, imaginary, or pure imaginary.

1. $2 - i\sqrt{2}$ **2.** $5i$

3. $4 + 0i$ **4.** $2 - 5i^2$

In Exercises 5–8, find values for a and b that make the statement true.

5. $a + bi = 4$ **6.** $a + bi = 2 + 4i$

7. $a + 3i = 15 - bi$ **8.** $a + 2i = bi$

In Exercises 9–20, perform the indicated operations and write the answer in the form a + bi.

9. $(3 + 7i) + (2i - 4)$ **10.** $(4 - 2i) - (2 - 3i)$

11. $(6 - 4i) + (-6 + 4i)$

12. $(2 + i\sqrt{3}) - (3 - i\sqrt{3})$

13. $2(3 - i) - (7 - 2i)$ **14.** $2(2i + 3) - (4i + 5)$

15. $(2 - i)(2 + i)$ **16.** $(6 - 3i)(4 + 5i)$

17. $(2i - 1)^2$ **18.** $-3i(2 - i)(3 + i)$

19. $\dfrac{4 - 3i}{3 - 3i}$

20. $\dfrac{\sqrt{3} - 2i}{\sqrt{5} + i}$

In Exercises 21–30, simplify the radicals by writing them in the form a, bi, or a + bi.

21. $\sqrt{-36}$

22. $\sqrt{-98}$

23. $(\sqrt{-3})^2$

24. $(\sqrt{-4})^2$

25. $\sqrt{-6}\sqrt{-15}$

26. $(3 + \sqrt{-4}) + (2 - \sqrt{-9})$

27. $(2 + \sqrt{-2}) - (3 - \sqrt{-8})$

28. $(1 + \sqrt{-4})(2 - \sqrt{-9})$

29. $(2 - 3\sqrt{-1})^2$

30. $\dfrac{1 + \sqrt{-1}}{\sqrt{-1}}$

In Exercises 31–40, (a) find the exact solutions to $f(x) = 0$ in the complex numbers, and (b) confirm that

the solutions are not real by showing that the graph of $y = f(x)$ does not cross the x-axis.

31. $x^2 + 25 = 0$

32. $2x^2 + 40 = 0$

33. $(x - 1)^2 = -4$

34. $(2x + 1)^2 + 7 = 0$

35. $x^2 + 4x + 8 = 0$

36. $x^2 - 5x + 7 = 0$

37. $2x^2 - 8x + 9 = 0$

38. $x^2 - x + 1 = 0$

39. $4x^2 - 5x + 3 = 0$

40. $4x^2 + 2x + 1 = 0$

Solve each of the equations in Exercises 41–46 exactly in the complex number system. (Find one solution graphically and then use the quadratic formula.)

41. $x^3 = 10x - 7x^2$

42. $t^3 - 2t^2 + 3t = 0$

43. $w^3 - 5w^2 + 6w - 2 = 0$

44. $2w^3 + 3w^2 + 3w + 2 = 0$

45. $z^3 - 8 = 0$

46. $x^3 + 1 = 0$

4.6 Rational Functions and Rational Equations

Key Concepts

- Graphs of rational functions
  Vertical asymptotes
  Horizontal asymptotes
  Slant asymptotes
  Holes

- Solving rational equations
  Analytically
  Graphically

The function that gives the daily average cost per light for the production of Sta Glo lights is formed by dividing the total cost function for the production of the lights,

$$C(x) = 25 + 13x + x^2,$$

by x, the number of units produced. Thus the average cost per light is given by

$$\overline{C}(x) = \frac{25 + 13x + x^2}{x}, \quad x > 0,$$

where x is the number of units produced. The graph of this function with x restricted to positive values is shown in Figure 4.39(a). The graph of the function

$$\overline{C}(x) = \frac{25 + 13x + x^2}{x}$$

with x not restricted is shown in Figure 4.39(b).

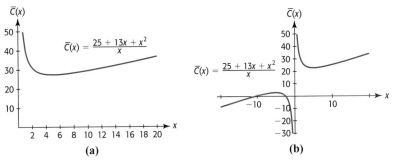

(a)   **(b)**

FIGURE 4.39

This function, and any other function that is formed by taking the quotient of two polynomials, is called a **rational function**.

Graphs of Rational Functions

A rational function is defined as follows.

RATIONAL FUNCTION

The function f is a rational function if

$$f(x) = \frac{P(x)}{Q(x)}$$

where $P(x)$ and $Q(x)$ are polynomials and $Q(x) \neq 0$.

In Section 2.5, we discussed a simple rational function, $y = \dfrac{1}{x}$, whose graph is shown in Figure 4.40.

Because division by 0 is not possible, those real values of x for which $Q(x) = 0$ are not in the domain of the rational function $f(x) = \dfrac{P(x)}{Q(x)}$. Observe that the graph approaches, but does not touch, the y-axis. Thus the y-axis is a **vertical asymptote**. The graph of a rational function frequently has vertical asymptotes.

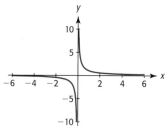

FIGURE 4.40

VERTICAL ASYMPTOTE

A vertical asymptote occurs in the graph of $f(x) = \dfrac{P(x)}{Q(x)}$ at those values of x where $Q(x) = 0$ and $P(x) \neq 0$; that is, at values of x where the denominator equals 0 but the numerator does not equal 0.

To find where the vertical asymptote(s) occur, set the denominator equal to 0 and solve for x. If any of these values of x do not also make the numerator equal to 0, a vertical asymptote occurs at those values.

EXAMPLE 1 Rational Function

Find the vertical asymptote and sketch the graph of $y = \dfrac{x + 1}{(x + 2)^2}$.

Solution

Setting the denominator equal to 0, taking the square root of both sides of the equation, and solving gives

$$(x + 2)^2 = 0$$
$$(x + 2) = 0$$
$$x = -2$$

The value $x = -2$ makes the denominator of the function equal to 0 and does not make the numerator equal to 0, so a vertical asymptote occurs at $x = -2$. The graph of the function $y = \dfrac{x + 1}{(x + 2)^2}$ is shown in Figure 4.41.

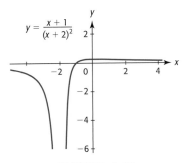

$$y = \frac{x + 1}{(x + 2)^2}$$

FIGURE 4.41

EXAMPLE 2   Cost-Benefit

Suppose that for a specified interval of the variable p, the function

$$C(p) = \frac{800p}{100 - p}$$

can be used to model the cost of removing p percent of the particulate pollution from the exhaust gases at an industrial site.

a. Graph this function on the window $[-100, 200]$ by $[-5000, 5000]$.

b. Does the graph of this function have a vertical asymptote on this window? Where?

c. Over what p-interval does this function serve as a model for the cost of removing particulate pollution?

d. Graph the model on the p-interval determined in part (c).

e. What does the part of the graph near the vertical asymptote tell us about the cost of removing particulate pollution?

Solution

a. The graph is shown in Figure 4.42(a).

b. The denominator is equal to 0 at $p = 100$, and the numerator does not equal 0 at $p = 100$, so a vertical asymptote occurs at $p = 100$.

c. Because p represents the percent of pollution, p must be limited to the interval from 0 to 100. The domain cannot include 100, so the p-interval is $[0, 100]$.

d. The graph of the function on the window $[0, 100]$ by $[0, 5000]$ is shown in Figure 4.42(b).

e. From Figure 4.42(b) we see that the graph is increasing rapidly as it approaches the vertical asymptote at $p = 100$, which tells us that the cost of removing pollution gets extremely high as the amount of pollution removed approaches 100%. It is impossible to remove 100% of the particulate pollution.

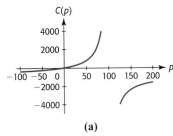

(a)

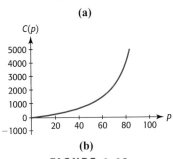

(b)

FIGURE 4.42

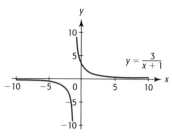

FIGURE 4.43

We see in Figure 4.43 that the graph of

$$y = \frac{3}{x + 1}$$

approaches the x-axis asymptotically on the left and on the right. Because the x-axis is a horizontal line, we say that the x-axis is a **horizontal asymptote** for this graph.

We can study the graph of a rational function to determine if the curve approaches a horizontal asymptote. If the graph approaches the horizontal line $y = a$ as $|x|$ gets very large, the graph has a horizontal asymptote at $y = a$. We can denote this as follows.

HORIZONTAL ASYMPTOTE

If y approaches a as x approaches $+\infty$ or as x approaches $-\infty$, the graph of $y = f(x)$ has a horizontal asymptote at $y = a$.

We can also compare the degree of the numerator and denominator of the rational function to determine if the curve approaches a horizontal asymptote, and to find the horizontal asymptote if it exists.

DETERMINING THE HORIZONTAL ASYMPTOTES OF A RATIONAL FUNCTION

Consider the rational function

$$f(x) = \frac{P(x)}{Q(x)} = \frac{a_n x^n + \cdots + a_1 x + a_0}{b_m x^m + \cdots + b_1 x + b_0}$$

1. If $n < m$ (that is, if the degree of the numerator is less than the degree of the denominator), a horizontal asymptote occurs at $y = 0$ (the x-axis).

2. If $n = m$ (that is, if the degree of the numerator is equal to the degree of the denominator), a horizontal asymptote occurs at $y = \dfrac{a_n}{b_m}$. (This is the ratio of the leading coefficients.)

3. If $n > m$ (that is, if the degree of the numerator is greater than the degree of the denominator), there is no horizontal asymptote.

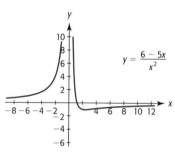

FIGURE 4.44

In the function $y = \dfrac{3}{x + 1}$, the degree of the numerator is less than the degree of the denominator, so a horizontal asymptote occurs at $y = 0$ (the x-axis). (See Figure 4.43.)

In the function $C(p) = \dfrac{800p}{100 - p}$ of Example 2, the degrees of the numerator and the denominator are equal, so the graph has a horizontal asymptote at $y = \dfrac{800}{-1} = -800$. Observing the graph of the function, shown in Figure 4.42(a), we see that this horizontal asymptote is reasonable for the graph of $C(p) = \dfrac{800p}{100 - p}$.

While it is impossible for a graph to cross a vertical asymptote (because the function is undefined at that value of x), the curve *may* cross a horizontal asymptote, because the horizontal asymptote describes the end behavior (as x approaches ∞ or x

approaches $-\infty$) of a graph. See Figure 4.44, which shows the graph of $y = \dfrac{6 - 5x}{x^2}$; it crosses the x-axis but approaches the line $y = 0$ (the x-axis) as x approaches ∞ and as x approaches $-\infty$.

EXAMPLE 3   Average Cost

Graph the function $\overline{C}(x) = \dfrac{25 + 13x + x^2}{x}$, which represents the daily average cost per light for the production of Sta Glo lights, with x equal to the number of lights produced:

a. Using the window $[-20, 20]$ by $[-30, 50]$.

b. Using the window $[0, 20]$ by $[0, 50]$.

c. Does the graph of the function using the window in part (a) or part (b) better model the average cost function? Why?

d. Use technology to find the minimum average cost, and the number of units that gives the minimum average cost.

Solution

a. The graph using the window $[-20, 20]$ by $[-30, 50]$ is shown in Figure 4.45(a). Note that the graph has a vertical asymptote at $x = 0$ (the y-axis).

b. The graph using the window $[0, 20]$ by $[0, 50]$ is shown in Figure 4.45(b).

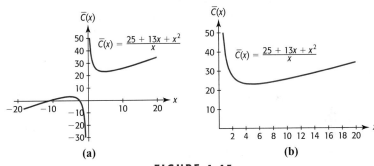

(a)   (b)

FIGURE 4.45

c. Because the number of units produced cannot be negative, the window used in part (b) gives a better representation of the graph of the average cost function.

d. It appears that the graph reaches a low point somewhere between $x = 4$ and $x = 6$. By using technology we can find the minimum point. Figure 4.46 shows that the minimum value of y is 23 at $x = 5$. Thus the minimum average cost is \$23 per unit when 5 units are produced. ∎

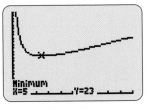

FIGURE 4.46

The average cost function from Example 3,

$$\overline{C}(x) = \frac{25 + 13x + x^2}{x},$$

does not have a horizontal asymptote, because the degree of the numerator is greater than the degree of the denominator. However, dividing each term of the numerator by x gives

$$\overline{C}(x) = \frac{25}{x} + 13 + x.$$

Because $\dfrac{25}{x}$ approaches 0 as x approaches ∞ or x approaches $-\infty$, the graph of $\overline{C}(x) = \dfrac{25}{x} + 13 + x$ approaches the graph of

$$y = 13 + x,$$

which we call the **slant asymptote** for the graph of $\overline{C}(x) = \dfrac{25 + 13x + x^2}{x}$. (See Figure 4.47.)

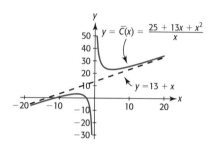

FIGURE 4.47

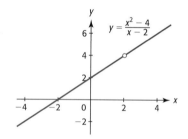

FIGURE 4.48

There is another type of rational function that is undefined at one or more values of x. Note that if there is a value a such that $Q(a) = 0$ and $P(a) = 0$, then the function

$$f(x) = \frac{P(x)}{Q(x)}$$

is undefined at $x = a$, but its graph may have a "hole" in it at $x = a$ rather than an asymptote. Figure 4.48 shows the graph of such a function,

$$y = \frac{x^2 - 4}{x - 2}.$$

Observe that the function is undefined at $x = 2$, but the graph does not have an asymptote at this value of x. Notice that for all values of x except 2,

$$y = \frac{x^2 - 4}{x - 2} = \frac{(x + 2)(x - 2)}{x - 2} = x + 2.$$

Thus the graph of this function looks like the graph of $y = x + 2$, except that it has a missing point (hole) at $x = 2$. (See Figure 4.48.)

Analytical and Graphical Solution of Rational Equations

To solve an equation involving rational functions analytically, we multiply both sides of the equation by the least common denominator (LCD) of the fractions in the equation. This will result in a linear or polynomial equation which can be solved by using the methods discussed earlier in the text. For these types of equations it is essential that all solutions be checked in the original equation, because some solutions to the resulting polynomial equation may not be solutions to the original equation. In particular, some solutions to the polynomial equation may result in a zero in the denominator of the original equation and thus cannot be solutions to the equation. Such solutions are called **extraneous solutions**.

The steps used to solve a rational equation follow.

SOLVING A RATIONAL EQUATION

To solve a rational equation:

1. Multiply both sides of the equation by the LCD of the fractions in the equation.

2. Solve the resulting polynomial equation for the variable analytically and/or graphically.

3. Check each solution in the original equation. Some solutions to the polynomial equation may not be solutions to the original rational equation (these are called extraneous solutions).

EXAMPLE 4 Solving a Rational Equation

Solve the equation $x^2 + \dfrac{x}{x-1} = x + \dfrac{x^3}{x-1}$ for x.

Solution

We first multiply both sides of the equation by $x - 1$, the LCD of the fractions in the equation, and then we solve the resulting polynomial equation.

$$x^2 + \frac{x}{x-1} = x + \frac{x^3}{x-1}$$

$$(x-1)\left(x^2 + \frac{x}{x-1}\right) = (x-1)\left(x + \frac{x^3}{x-1}\right)$$

$$(x-1)x^2 + (x-1)\frac{x}{x-1} = (x-1)x + (x-1)\frac{x^3}{x-1}$$

$$x^3 - x^2 + x = x^2 - x + x^3$$

$$0 = 2x^2 - 2x$$

$$0 = 2x(x-1)$$

$$x = 0, \, x = 1$$

We must check these answers in the original equation.

Check at $x = 0$:

$$0^2 + \frac{0}{0-1} = 0 + \frac{0^3}{0-1} \quad \text{checks}$$

Check at $x = 1$:

$$1^2 + \frac{1}{1-1} = 1 + \frac{1^3}{1-1}, \quad \text{or} \quad 1 + \frac{1}{0} = 1 + \frac{1}{0}$$

But $1 + \dfrac{1}{0}$ cannot be evaluated, so $x = 1$ does not check in the original equation. This means that 1 is an extraneous solution of the equation. Thus the only solution is $x = 0$. ∎

As with polynomial equations derived from real data, some equations involving rational functions may require technology to find or to approximate their solutions. The following example shows how applied problems involving rational functions can be solved analytically and graphically.

EXAMPLE 5   Advertising and Sales

Monthly sales y (in thousands of dollars) for Yang products are related to monthly advertising expenses x (in thousands of dollars) according to the function

$$y = \frac{300x}{15 + x}.$$

Determine the amount of money that must be spent on advertising to generate $100,000 in sales

a. Analytically.

b. Graphically.

Solution

a. To solve the equation $100 = \dfrac{300x}{15 + x}$ analytically, we multiply both sides of the equation by $15 + x$, getting

$$100(15 + x) = 300x.$$

Solving this equation gives

$$1500 + 100x = 300x$$
$$1500 = 200x$$
$$x = 7.5$$

Checking this solution in the original equation gives

$$100 = \frac{300(7.5)}{15 + 7.5} \quad \text{or} \quad 100 = \frac{2250}{22.5},$$

so the solution checks.

Thus spending 7.5 thousand dollars ($7500) per month results in monthly sales of $100,000.

b. To solve the equation graphically, we graph the rational function $y = \dfrac{300x}{15 + x}$ and $y = 100$ on the same axes on an interval with $x \geq 0$ (see Figure 4.49). We then find the point of intersection $(7.5, 100)$, which gives the same solution.   ■

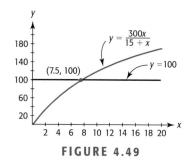

FIGURE 4.49

4.6 SKILLS CHECK

Give the equations of any (a) vertical and (b) horizontal asymptotes for the graphs of the rational functions in Exercises 1–6.

1. $f(x) = \dfrac{3}{x - 5}$

2. $f(x) = \dfrac{7}{x - 4}$

3. $f(x) = \dfrac{x - 4}{5 - 2x}$

4. $f(x) = \dfrac{2x - 5}{3 - x}$

5. $f(x) = \dfrac{x^3 + 4}{x^2 - 1}$

6. $f(x) = \dfrac{x^2 + 6}{x^2 + 3}$

7. Which of the following has a graph that does not have a vertical asymptote? Why?

a. $f(x) = \dfrac{x - 3}{x^2 - 4}$

b. $f(x) = \dfrac{3}{x^2 - 4x}$

c. $f(x) = \dfrac{3x^2 - 6}{x^2 + 6}$

d. $f(x) = \dfrac{x + 4}{(x - 6)(x + 2)}$

8. Which of the following has a graph that does not have a horizontal asymptote? Why?

a. $f(x) = \dfrac{2x + 3}{x - 4}$ **b.** $f(x) = \dfrac{5x}{x^2 - 16}$

c. $f(x) = \dfrac{x^3}{3x^2 + 2}$ **d.** $f(x) = \dfrac{5}{(x + 2)(x - 3)}$

In Exercises 9–12, find (a) the horizontal asymptotes, (b) the vertical asymptotes, and (c) sketch a graph of the function.

9. $f(x) = \dfrac{x + 1}{x - 2}$ **10.** $f(x) = \dfrac{5x}{x - 3}$

11. $f(x) = \dfrac{5 + 2x}{1 - x^2}$ **12.** $f(x) = \dfrac{2x^2 + 1}{2 - x}$

13. Graph the function $y = \dfrac{x^2 - 9}{x - 3}$. What happens at $x = 3$?

14. Graph the function $y = \dfrac{x^2 - 16}{x + 4}$. What happens at $x = -4$?

In Exercises 15 and 16, use graphical methods to find any turning points of the graph of the function.

15. $f(x) = \dfrac{x^2}{x - 1}$ **16.** $f(x) = \dfrac{(x - 1)^2}{x}$

17. a. Graph $y = \dfrac{1 - x^2}{x - 2}$ on the window $[-10, 10]$ by $[-10, 10]$.
b. Use the graph to find y when $x = 1$ and when $x = 3$.
c. Use the graph to find the value(s) of x that give $y = -7.5$.
d. Use analytical methods to solve $-7.5 = \dfrac{1 - x^2}{x - 2}$.

18. a. Graph $y = \dfrac{2 + 4x}{x^2 + 1}$ on the window $[-10, 10]$ by $[-5, 5]$.
b. Use the graph to find y when $x = -3$ and when $x = 3$.
c. Use the graph to find the value(s) of x that give $y = 2$.
d. Use analytical methods to solve $2 = \dfrac{2 + 4x}{x^2 + 1}$.

19. a. Graph $y = \dfrac{3 - 2x}{x}$ on the window $[-10, 10]$ by $[-10, 10]$.
b. Use the graph to find y when $x = -3$ and when $x = 3$.
c. Use the graph to find the value(s) of x that give $y = -5$.
d. Use analytical methods to solve $-5 = \dfrac{3 - 2x}{x}$.

20. a. Graph $y = \dfrac{x^2}{(x + 1)^2}$ on the window $[-10, 10]$ by $[-6, 6]$.
b. Use the graph to find y when $x = 0$ and when $x = -2$.
c. Use the graph to find the value(s) of x that give $y = \dfrac{1}{4}$.
d. Use analytical methods to solve $\dfrac{1}{4} = \dfrac{x^2}{(x + 1)^2}$.

21. Use analytical methods to solve
$$\dfrac{x^2 + 1}{x - 1} + x = 2 + \dfrac{2}{x - 1}.$$

22. Use analytical methods to solve
$$\dfrac{x}{x - 2} - x = 1 + \dfrac{2}{x - 2}.$$

4.6 EXERCISES

1. *Average Cost* The average cost per unit for the production of a certain brand of DVD players is given by $\overline{C} = \dfrac{400 + 50x + 0.01x^2}{x}$, where x is the number of units produced.
  a. What is the average cost per unit when 500 units are produced?
  b. What is the average cost per unit when 60 units are produced?
  c. What is the average cost per unit when 100 units are produced?

  d. Is it reasonable to say that the average cost continues to fall as the number of units produced rises?

2. *Average Cost* The average cost per set for the production of a certain brand of television sets is given by $\overline{C}(x) = \dfrac{1000 + 30x + 0.1x^2}{x}$, where x is the number of units produced.
  a. What is the average cost per set when 30 sets are produced?

b. What is the average cost per set when 300 sets are produced?

c. What happens to the function when $x = 0$? What does this tell you about the average cost when 0 units are produced?

3. *Advertising and Sales* The monthly sales volume y (in thousands of dollars) of a product is related to monthly advertising expenditures x (in thousands of dollars) according to the equation $y = \dfrac{400x}{x + 20}$.

   a. What monthly sales will result if \$5000 is spent monthly on advertising?

   b. What value of x makes the denominator of this function 0? Will this ever happen in the context of this problem?

4. *Productivity* As an 8-hour shift progresses, the rate at which workers produce picture frames, in units per hour, changes during the day according to the equation

   $$f(t) = \frac{100(t^2 + 3t)}{(t^2 + 3t + 12)^2}, \; 0 \le t \le 8$$

   where t is the number of hours after the beginning of the shift.

   a. Is the rate of productivity higher near lunch ($t = 4$) or near quitting time ($t = 8$)?

   b. Graph this function on the window $[-10, 10]$ by $[-5, 5]$.

   c. Graph this function on the window $[0, 8]$ by $[0, 5]$.

   d. Does the graph in part (b) or on part (c) better represent the rate of change of productivity function? Why?

5. *Average Cost* The average cost per set for the production of television sets is given by $\overline{C}(x) = \dfrac{1000 + 30x + 0.1x^2}{x}$, where x is the number of hundreds of units produced.

   a. Graph this function using the window $[-10, 10]$ by $[-200, 300]$.

   b. Graph this function using the window $[0, 50]$ by $[0, 300]$.

   c. Which window makes sense for this application?

   d. Use the graph with the window $[0, 200]$ by $[0, 300]$ to find the minimum average cost, and the number of units that gives the minimum average cost.

6. *Average Cost* The average cost per unit for the production of a certain brand of DVD players is given by $\overline{C} = \dfrac{400 + 50x + 0.01x^2}{x}$, where x is the number of hundreds of units produced.

   a. Graph this function using the window $[-100, 100]$ by $[-100, 400]$.

b. Graph this function using the window $[0, 300]$ by $[0, 400]$.

c. Which window is more appropriate for this problem?

d. Use the graph to find the minimum average cost and the number of units that gives the minimum average cost.

7. *Population* Suppose that the number of employees of a start-up company is given by

   $$f(t) = \frac{30 + 40t}{5 + 2t}$$

   where t is the number of months after the company is organized.

   a. Graph this function on the window $[0, 20]$ by $[0, 20]$.

   b. Use the graph to find $f(0)$. What does this represent?

   c. Use the graph to find $f(12)$. What does this represent? If the population is a number of individuals, how should you report your answer?

8. *Drug Concentration* Suppose the concentration of a drug (as percent) in a patient's bloodstream t hours after injection is given by

   $$C(t) = \frac{200t}{2t^2 + 32}.$$

   a. Graph the function on the window $[0, 20]$ by $[0, 20]$.

   b. What is the drug concentration 1 hour after injection? 5 hours?

   c. What is the highest percent concentration? In how many hours will it occur?

   d. Describe how the end behavior of the graph of this function relates to the drug concentration.

9. *Cost-Benefit* Suppose that the cost C of removing p percent of the impurities from the waste water in a manufacturing process is given by

   $$C(p) = \frac{3600p}{100 - p}.$$

   a. Where does the graph of this function have a vertical asymptote?

   b. What does this tell us about removing the impurities from this process?

10. *Cost-Benefit* The percent p of particulate pollution that can be removed from the smokestacks of an industrial plant by spending C dollars is given by

    $$p = \frac{100C}{8300 + C}.$$

a. Find the percent of pollution that could be removed if spending were allowed to increase without bound.
b. Can 100% of the pollution be removed? Explain.

11. *Sales Volume* Suppose that the weekly sales volume (in thousands of units) for a product is given by

$$V = \frac{640}{(p + 2)^2}.$$

where p is the price in dollars per unit.
a. Graph this function on the p-interval $[-10, 10]$. Does the graph of this function have a vertical asymptote on this interval? Where?
b. Complete the table below.
c. What values of p give a weekly sales volume that makes sense? Does the graph of this function have a vertical asymptote on this interval?
d. What is the horizontal asymptote of the graph of the weekly sales volume? Explain what this means.

Price per Unit	5	20	50	100	200	500
Weekly Sales Volume						

12. *Population* Suppose that the number of employees of a start-up company is given by

$$N = \frac{30 + 400t}{5 + 2t}$$

where t is the number of months after the company is organized.
a. Does the graph of this function have a vertical asymptote on $[-10, 10]$? Where?
b. Does the graph of this function have a vertical asymptote for values of $t > 0$?
c. Does the graph of this function have a horizontal asymptote? Where?
d. What is the maximum number of employees predicted by this model?

13. *Demand* The quantity of a product demanded by consumers is defined by the function

$$p = \frac{100,000}{(q + 1)^2},$$

where p is the price and q is the quantity demanded.
a. Graph this function using the window $[0, 100]$ by $[0, 1000]$.
b. What is the horizontal asymptote that this graph approaches?
c. Use the graph to explain what happens to quantity demanded as price becomes lower.

14. *Advertising and Sales* Weekly sales y (in hundreds of dollars) are related to weekly advertising expenses x (in hundreds of dollars) according to the equation

$$y = \frac{800x}{20 + 5x}.$$

a. Graph this function on the window $[0, 100]$ by $[0, 300]$.
b. Find the horizontal asymptote for this function.
c. Complete the table below.
d. Use your results from part (a) and part (b) to determine the maximum weekly sales even if an unlimited amount of money is spent on advertising.

Weekly Expenses	0	50	100	200	300	500
Weekly Sales						

15. *Sales and Training* During the first 3 months of employment, the monthly sales S (in thousands of dollars) for an average new salesperson depends on the number of hours of training x, according to

$$S = \frac{40}{x} + \frac{x}{4} + 10, x \geq 4.$$

a. Combine the terms of this function over a common denominator to create a rational function.
b. How many hours of training should result in monthly sales of $21,000?

16. *Sales and Training* The average monthly sales volume (in thousands of dollars) for a company depends on the number of hours of training x of its sales staff, according to $S(x) = \frac{20}{x} + 40 + \frac{x}{2}$, with $4 \leq x \leq 120$.
a. Graph this function.
b. How many hours of training will give average monthly sales of $51,000?

17. *Worker Productivity* Suppose that the average time (in hours) that a new production team takes to assemble one unit of a product is given by

$$H = \frac{5 + 3t}{2t + 1},$$

where t is the number of days of training for the team.
a. Graph this function on the window $[0, 20]$ by $[0, 8]$.
b. What is the horizontal asymptote? What does this mean?
c. Use a graphical or numerical solution method to find the number of days of training necessary to reduce the production time to 1.6 hours.

18. *Advertising and Sales* Weekly sales y (in hundreds of dollars) are related to weekly advertising expenses x (in hundreds of dollars) according to the equation

$$y = \frac{800x}{20 + 5x}.$$

Use graphical or numerical methods to find the amount of weekly advertising expenses that will result in weekly sales of $14,000, according to this model.

19. *Farm workers* The percent of U.S. workers in farm occupations during certain years from 1820 is shown in the table below.

Year	Percent of Workers Who Are Farm Workers	Year	Percent of Workers Who Are Farm Workers
1820	71.8	1950	11.6
1850	63.7	1960	6.1
1870	53	1970	3.6
1900	37.5	1980	2.7
1920	27	1985	2.8
1930	21.2	1990	2.4
1940	17.4	1994	2.5

Source: *The World Almanac, 2001*

Assume that the percent can be modeled with the function

$$r(t) = \frac{-8091.2t + 1{,}558{,}900}{1.09816t^2 - 122.183t + 21{,}472.6},$$

where t is the number of years past 1800.
   a. Graph this function and the data points from the table on the same axes, using the window [0, 250] by [0, 80].
   b. Compare the percent of farm workers in 1930 from the table of values and the functional model.

20. *Fences* Suppose a rectangular field is to have an area of 51,200 square feet, and that it needs to be enclosed by fence on three of its four sides (see figure below). What is the minimum length of fence needed? What dimensions should this field have to minimize the amount of fence needed? To solve this problem:
   a. Write an equation that describes the area of the proposed field, with x representing the length and y representing the width of the field.
   b. Write an equation that gives the length of fence L as a function of the dimensions of the field, remembering that only three sides are fenced in.
   c. Use the equation from part (a) to write the length L as a function of one of the dimensions.
   d. Graph the model on the window that describes the problem: [0, 400] by [0, 1000].
   e. Use graphical or numerical methods to find the minimum length of fence needed and the dimensions required to give this length.

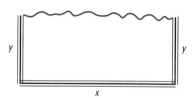

4.7 Polynomial and Rational Inequalities

Key Concepts

- Polynomial inequalities
  Graphical solutions
  Analytical solutions

- Rational inequalities
  Graphical solutions
  Analytical solutions

The average cost per set for the production of one brand of television sets is given by

$$\overline{C}(x) = \frac{5000 + 80x + x^2}{x},$$

where x is hundreds of units produced. To find the number of television sets that must be produced to keep the average cost to at most $590 per set, we solve the **rational inequality**

$$\frac{5000 + 80x + x^2}{x} \le 590.$$

In this section, we use analytical and graphical methods similar to those used in Section 2.7 to solve inequalities involving polynomial functions and rational functions.

Polynomial Inequalities

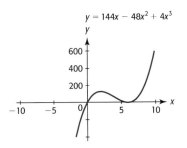

$y = 144x - 48x^2 + 4x^3$

FIGURE 4.50

The quadratic inequalities studied in Section 2.7 are examples of polynomial inequalities. Other examples of polynomial inequalities are

$$3x^3 + 3x^2 - 4x \geq 2, \quad x - 3 < 8x^4, \quad \text{and } x \geq (x - 2)^3.$$

We can solve the inequality $144x - 48x^2 + 4x^3 > 0$ with a graphing utility. The graph of the function $y = 144x - 48x^2 + 4x^3$ is shown in Figure 4.50.

We can determine that the x-intercepts of the graph are $x = 0$ and $x = 6$. The graph touches but does not cross the x-axis at $x = 6$, and the graph of the function is above the x-axis for values of $x > 0$ and $x \neq 6$. The solution to $144x - 48x^2 + 4x^3 > 0$ is the set of all x that give positive values for this function: $0 < x < 6$ and $6 < x < \infty$, that is, for x in the intervals $(0, 6)$ and $(6, \infty)$.

The steps used to solve polynomial inequalities analytically are similar to those used to solve quadratic inequalities.

SOLVING POLYNOMIAL INEQUALITIES

To solve a polynomial inequality:

1. Write an equivalent inequality with 0 on one side and with the function $f(x)$ on the other side.

2. Solve $f(x) = 0$.

3. Create a sign diagram that uses the solutions from Step 2 to divide the number line into intervals. Pick a test value in each interval to determine whether $f(x)$ is positive or negative in that interval.*

4. Identify the intervals that satisfy the inequality in Step 1. The values of x that define these intervals are the solutions to the original inequality.

We illustrate the use of this method in the following example.

EXAMPLE 1   Constructing a Box

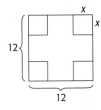

A box can be formed by cutting squares out of each corner of a piece of tin and folding the "tabs" up. If the piece of tin is 12 inches by 12 inches and each side of the square that is cut out has length x, the function that gives the volume of the box is $V(x) = 144x - 48x^2 + 4x^3$.

a. Use factoring to find the values of x that give positive values for $V(x)$.

b. Which of the values of x that give positive values for $V(x)$ result in a box with a positive volume?

---

*The numerical feature of your graphing utility can be used to test the x-values.

Solution

a. We seek those values of x that give positive values for $V(x)$; that is, for which

$$144x - 48x^2 + 4x^3 > 0.$$

Writing the related equation $f(x) = 0$ and solving for x gives

$$144x - 48x^2 + 4x^3 = 0$$

$$4x(x^2 - 12x + 36) = 0$$

$$4x(x - 6)(x - 6) = 0$$

$$x = 0 \text{ or } x = 6$$

These two values of x, 0 and 6, divide the number line into three intervals. Testing a value in each of the intervals (see Figure 4.51) determines the intervals where $f(x) > 0$, and thus determines the solutions to the inequality. The values of x that give positive values for V satisfy $0 < x < 6$ and $6 < x < \infty$. We found the same solutions to this inequality using graphical methods. (See Figure 4.50.)

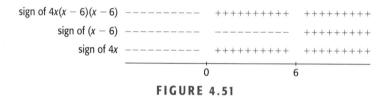

FIGURE 4.51

b. The box is formed by cutting 4 squares of length x inches from a piece of tin that is 12 inches by 12 inches, so it is impossible to cut squares of length is 6 inches or longer. Thus the solution that involves values of x greater than 6 does not apply to physical building of the box, and so the values of x that result in a box with a positive volume satisfy $0 < x < 6$.   ■

Rational Inequalities

To solve a **rational inequality** using an analytical method, we first get zero on the right side of the inequality; then, if necessary, we get a common denominator and combine the rational expressions on the left side. We find the numbers that make the numerator of the rational expression equal to zero and those that make the denominator equal to zero. These numbers are used to create a sign diagram much like we used when solving polynomial inequalities. Note that we should avoid multiplying both sides of an inequality by any term containing a variable, because we cannot easily determine when the expression is positive or negative.

EXAMPLE 2   **Average Cost**

The average cost per set for the production of one brand of television sets is given by $\overline{C}(x) = \dfrac{5000 + 80x + x^2}{x}$, where x is the number of hundreds of units produced. Find the number of television sets that must be produced to keep the average cost to at most $590 per set.

Solution

To find the number of television sets that must be produced to keep the average cost to at most $590 per set, we solve the rational inequality

$$\frac{5000 + 80x + x^2}{x} \leq 590.$$

To solve this inequality with analytical methods, we rewrite the inequality with 0 on the right side of the inequality, giving

$$\frac{5000 + 80x + x^2}{x} - 590 \leq 0.$$

Combining the terms over a common denominator gives

$$\frac{5000 + 80x + x^2 - 590x}{x} \leq 0$$

$$\frac{x^2 - 510x + 5000}{x} \leq 0$$

$$\frac{(x - 10)(x - 500)}{x} \leq 0$$

The values of x that make the numerator equal to 0 are 10 and 500. The value $x = 0$ makes the denominator equal to 0 and we cannot have a negative number of units, so the average cost is defined only for values of $x > 0$. The values 0, 10, and 500 divide the number line into three intervals. Next we find the sign of $f(x)$ in each interval. Testing a value in each interval determines which interval(s) satisfy the inequality. We do this by using the following sign diagram. (See Figure 4.52.)

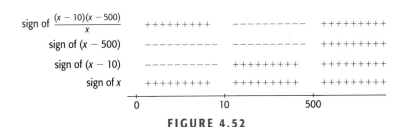

FIGURE 4.52

The function $f(x)$ is negative on $10 \leq x \leq 500$, so the solution to $\frac{5000 + 80x + x^2}{x} - 590 \leq 0$ and the original inequality is $10 \leq x \leq 500$.

Because x represents the number of hundreds, between 1000 and 50,000 television sets must be produced to keep the average cost to at most $590 per set. ∎

We can also use graphical methods to solve rational inequalities. For example, we can graphically solve the inequality

$$\frac{5000 + 80x + x^2}{x} \leq 590, \text{ for } x > 0$$

by graphing

$$y_1 = \frac{5000 + 80x + x^2}{x} \text{ and } y_2 = 590$$

on the same axes. We are only interested in the graph for positive values of x, so we restrict our graph to the first quadrant. The points of intersection can be found by the intersection method. (See Figure 4.53(a).) The x-interval that satisfies the inequality can be seen by shading above $y_1 = \dfrac{5000 + 80x + x^2}{x}$ and below $y_2 = 590$. (See Figure 4.53(b).)

As the shaded region shows, the inequality has solution $10 \le x \le 500$. Recall that this is the solution that we found using analytical methods in Example 2.

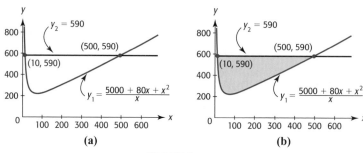

(a)

(b)

FIGURE 4.53

4.7 SKILLS CHECK

In Exercises 1–16, use graphical methods to solve the following inequalities.

1. $16x^2 - x^4 \ge 0$ **2.** $x^4 - 4x^2 \le 0$

3. $2x^3 - x^4 < 0$ **4.** $3x^3 \ge x^4$

5. $(x - 1)(x - 3)(x + 1) \ge 0$

6. $(x - 3)^2(x + 1) < 0$

7. $\dfrac{4 - 2x}{x} > 2$ **8.** $\dfrac{x - 3}{x + 1} \ge 3$

9. $\dfrac{x}{2} + \dfrac{x - 2}{x + 1} \le 1$ **10.** $\dfrac{x}{x - 1} \le 2x + \dfrac{1}{x - 1}$

11. $(x - 1)^3 > 27$ **12.** $(2x + 3)^3 \le 8$

13. $(x - 1)^3 < 64$ **14.** $x^3 + 10x^2 + 25x < 0$

15. $-x^3 - 10x^2 - 25x \le 0$

16. $(x + 4)^3 - 125 \ge 0$

For Exercises 17–19, use the graph of $y = f(x)$ to solve the requested inequality.

17.

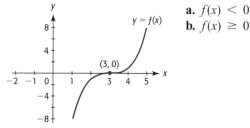

a. $f(x) < 0$
b. $f(x) \ge 0$

18.

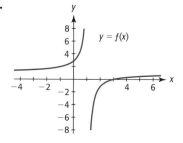

a. $f(x) < 0$
b. $f(x) \ge 0$

19.

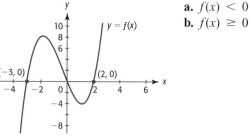

a. $f(x) < 0$
b. $f(x) \ge 0$

4.7 EXERCISES

Combine factoring with graphical and/or numerical methods to solve Exercises 1–6.

1. *Revenue* The revenue from the sale of a product is given by the function $R = 400x - x^3$. Selling how many units will give positive revenue?

2. *Revenue* The price for a product is given by $p = 1000 - 0.1x^2$, where x is the number of units sold.
   a. Form the revenue function for the product.
   b. Selling how many units gives positive revenue?

3. *Constructing a Box* A box can be formed by cutting squares out of each corner of a piece of cardboard and folding the "tabs" up. If the piece of cardboard is 36 cm by 36 cm and each side of the square that is cut out has length x cm, the function that gives the volume of the box is $V = 1296x - 144x^2 + 4x^3$.

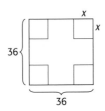

   a. Find the values of x that make $V > 0$.
   b. What size squares can be cut out to construct a box with positive volume?

4. *Constructing a Box* A box can be formed by cutting squares out of each corner of a piece of cardboard and folding the "tabs" up. If the piece of cardboard is 12 inches by 16 inches and each side of the square that is cut out has length x, the function that gives the volume of the box is $V = 192x - 56x^2 + 4x^3$. What size squares can be cut out to construct a box with positive volume?

5. *Cost* The total cost function for a product is given by $C(x) = 3x^3 - 6x^2 - 300x + 1800$, where x is the number of units produced and C is the cost in hundreds of dollars. Use factoring by grouping to find the number of units that will give a total cost of at least $120,000. Verify your conclusion with a graphing utility.

6. *Profit* The profit function for a product is given by $P(x) = -x^3 + 2x^2 + 400x - 400$, where x is the number of units produced and sold and P is in hundreds of dollars. Use factoring by grouping to find

the number of units that will give a profit of at least $40,000. Verify your conclusion with a graphing utility.

7. *Advertising and Sales* The monthly sales volume y (in thousands of dollars) is related to monthly advertising expenditures x (in thousands of dollars) according to the equation

$$y = \frac{400x}{x + 20}.$$

   Spending how much money on advertising will result in sales of at least $200,000 per month?

8. *Average Cost* The average cost per set for the production of a portable stereo system is given by $\overline{C} = \dfrac{100 + 30x + 0.1x^2}{x}$, where x is the number of hundreds of units produced. What number of units can be produced while keeping the average cost to at most $41?

9. *Future Value* The future value of $2000 invested for 3 years at rate r, compounded annually, is given by $S = 2000(1 + r)^3$. Find the rate r that gives a future value from $2662 to $3456, inclusive.

10. *Future Value* The future value of $5000 invested for 4 years at rate r, compounded annually, is given by $S = 5000(1 + r)^4$. Find the rate r that gives a future value of at least $6553.98.

11. *Revenue* The revenue from the sale of a product is given by the function $R = 4000x - 0.1x^3$ dollars. Use graphical or numerical methods to determine how many units should be sold to give a revenue of at least $39,990.

12. *Revenue* The price for a product is given by $p = 1000 - 0.1x^2$, where x is the number of units sold.
    a. Form the revenue function for the product.
    b. Use graphical or numerical methods to determine how many units should be sold to give a revenue of at most $37,500.

13. *Supply and Demand* Suppose the supply function for a product is given by $6p - q = 180$ and the demand is given by $(p + 20)q = 30,000$. Over what meaningful price interval does supply exceed demand?

14. *Drug Concentration* A pharmaceutical company claims that the concentration of a drug in a patient's bloodstream will be at least 10% for 8 hours. Suppose

that clinical tests show that the concentration of a drug (as percent) t hours after injection is given by

$$C(t) = \frac{200t}{2t^2 + 32}.$$

During what time period is the concentration at least 10%? Is its claim supported by the evidence?

15. *Population* Suppose that the number of employees of a start-up company is given by

$$f(t) = \frac{30 + 40t}{5 + 2t},$$

where t is the number of months after the company is organized.

a. For what values of t is $f(t) < 18$?

b. During what months is the number of employees below 18?

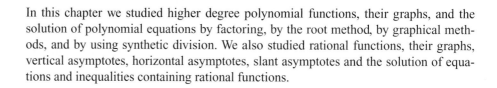

CHAPTER 4 *Summary*

In this chapter we studied higher degree polynomial functions, their graphs, and the solution of polynomial equations by factoring, by the root method, by graphical methods, and by using synthetic division. We also studied rational functions, their graphs, vertical asymptotes, horizontal asymptotes, slant asymptotes and the solution of equations and inequalities containing rational functions.

Key Concepts and Formulas

4.1 Higher Degree Polynomial Functions

Polynomial function

A polynomial function of degree n with independent variable x is a function in the form $P(x) = a_n x^n + a_{n-1} x^{n-1} + \cdots + a_1 x + a_0$. Each a_i represents a real number and each n is a positive integer.

Cubic function

A polynomial function of the form $f(x) = ax^3 + bx^2 + cx + d$, with $a \neq 0$.

Quartic function

A polynomial function of the form $f(x) = ax^4 + bx^3 + cx^2 + dx + e$, with $a \neq 0$.

Extrema points

"Turning points" on the graph of a polynomial function: local minimum, local maximum, absolute minimum, absolute maximum.

Graphs of cubic functions

End behavior: one end opening up and one end opening down.

Turning points: 0 or 2

Graphs of quartic functions

End behavior: Both ends opening up or both ends opening down.

Turning points: 1 or 3

4.2 Modeling Cubic and Quartic Functions

Cubic models	If a scatter plot of data indicates that the data may fit close to one of the possible shapes of the graph of a cubic function, it is possible to find a cubic function that is an approximate model for the data.
Quartic models	If a scatter plot of data indicates that the data may fit close to one of the possible shapes that are graphs of quartic functions, it is possible to find a quartic function that is an approximate model for the data.

4.3 Solution of Polynomial Equations

Solving polynomial equations

By factoring	Some polynomial equations can be solved by factoring and using the zero-product property.
By graphical methods	If a is an x-intercept of the graph of $y = f(x)$, then a is a solution to the equation $f(x) = 0$.
By the root method	The real solutions of the equation $x^n = C$ are $x = \pm\sqrt[n]{C}$ if n is even or $x = \sqrt[n]{C}$ if n is odd.
Approximating solutions	Many equations derived from real data require graphical or numerical methods for solution, using technology.

4.4 Solution of Polynomial Equations Using Synthetic Division

Division of polynomials	If $f(x)$ is a function and we know that $x = a$ is a solution of $f(x) = 0$, then we can write $x - a$ as a factor of $f(x)$. If we divide this factor into the function, the quotient will be another factor of $f(x)$.
Synthetic division	To divide a polynomial by a binomial of the form $x - a$ (or $x + a$), we can use synthetic division to simplify the division process by omitting the variables and writing only the coefficients.
Solving cubic equations	If we know one exact solution of a cubic equation $P(x) = 0$, we can divide by the corresponding factor using synthetic division, obtaining a second quadratic factor that can be used to find any additional solutions of the polynomial.
Combining graphical and analytical methods	To find exact solutions to a polynomial equation when one solution can be found graphically, we can use that solution and synthetic division to find the remaining factor.
Rational solutions test	The rational solutions of the polynomial equation $a_n x^n + a_{n-1}x^{n-1} + \cdots + a_1 x + a_0 = 0$ with integer coefficients must be of the form $\frac{p}{q}$, where p is a factor of the constant term a_0 and q is a factor of the leading coefficient a_n.

4.5 Complex Solutions; Fundamental Theorem of Algebra

Complex numbers	The standard form of a complex number is $a + bi$.
	Real numbers have the form: $a + bi$ with $b = 0$.
	Imaginary numbers have the form: $a + bi$ with $b \neq 0$.
	Pure imaginary numbers have the form: $a + bi$ with $a = 0$.

Operations with complex numbers	Complex numbers can be added or subtracted by combining their real and imaginary parts.
	The product of two complex numbers can be found by multiplying them as if they were binomials, and substituting -1 for i^2.
	The quotient of two complex numbers can be found by multiplying the numerator and denominator by the conjugate of the denominator and simplifying.
Fundamental Theorem of Algebra	If $f(x)$ is a polynomial function of degree $n \geq 1$, then f has at least one complex zero.
Conjugate pairs	Suppose $f(x)$ is a polynomial with real number coefficients. If $a + bi$ is a zero of f, then the complex conjugate $a - bi$ is also a zero of f.
Complex solutions of quadratic equations	The solutions of the equation $x^2 = -a$ for $a > 0$ can be found by the root method: $x = \pm\sqrt{-a} = \pm i\sqrt{a}$.
	The solutions of the equation $ax^2 + bx + c = 0$ can be solved by using the quadratic formula: $x = \dfrac{-b \pm \sqrt{b^2 - 4ac}}{2a}$.
Complex solutions of polynomial equations	Some polynomial equations can be solved by using synthetic division to find a quadratic factor and using the quadratic formula to complete the solution.

4.6 Rational Functions and Rational Equations

Rational function	The function f is a rational function if $f(x) = \dfrac{P(x)}{Q(x)}$ where $P(x)$ and $Q(x)$ are polynomials and $Q(x) \neq 0$.
Graphs of rational functions	The graph of a rational function may have vertical asymptotes and/or a horizontal asymptote. The graph may have a "hole" in it at $x = a$ rather than a vertical asymptote.
Vertical asymptotes	A vertical asymptote occurs in the graph of $f(x) = \dfrac{P(x)}{Q(x)}$ at those values of x where $Q(x) = 0$ and $P(x) \neq 0$; that is, where the denominator equals 0 but the numerator does not equal 0.
Horizontal asymptotes	Consider the rational function

$$f(x) = \frac{P(x)}{Q(x)} = \frac{a_n x^n + \cdots + a_1 x + a_0}{b_m x^m + \cdots + b_1 x + b_0}.$$

1. If $n < m$, a horizontal asymptote occurs at $y = 0$ (the x-axis).

2. If $n = m$, a horizontal asymptote occurs at $y = \dfrac{a_n}{b_m}$.

3. If $n > m$, there is no horizontal asymptote.

Missing point	If there is a value a such that $Q(a) = 0$ and $P(a) = 0$, then the graph of the function is undefined at $x = a$, but may have a "hole" in it at $x = a$ rather than an asymptote.
Solution of rational equations	
• **Analytically**	Multiplying both sides of the equation by the LCD of any fractions in the equation converts the equation to a polynomial equation that can be solved. All solutions must be checked in the original equation, as some may not satisfy it.
• **Graphically**	Solutions to rational equations can be found or approximated by graphical methods.

4.7 *Polynomial and Rational Inequalities*

Polynomial inequalities

The steps used to solve polynomial inequalities graphically are similar to those used to solve quadratic inequalities.

To solve a polynomial inequality analytically, solve the corresponding polynomial equation and use a sign diagram to complete the solution.

Rational inequalities

To solve a rational inequality, get 0 on the right side of the inequality; then, if necessary, get a common denominator and combine the rational expressions on the left side. Find the numbers that make the numerator of the rational expression equal to 0 and those that make the denominator equal to 0. Use these numbers to create a sign diagram. Graphical methods can be used to find or to verify the solutions.

Chapter 4 Skills Check

1. What is the degree of the polynomial $5x^4 - 2x^3 + 4x^2 + 5$?

2. What is the type of the polynomial function $y = 5x^4 - 2x^3 + 4x^2 + 5$?

3. Graph $y = -4x^3 + 4x^2 + 1$ on the window $[-10, 10]$ by $[-10, 10]$. Is the graph complete on this window?

4. Graph the function $y = x^4 - 4x^2 - 20$
   a. On the window $[-10, 10]$ by $[-10, 10]$.
   b. On a window that gives a complete graph.

5. a. Graph $y = x^3 - 3x^2 - 4$ using a window that shows a local maximum and local minimum.
   b. A local maximum occurs at what point?
   c. A local minimum occurs at what point?

6. Solve $y = x^3 - 15x^2 + 56x$ by factoring.

7. Use factoring by grouping to solve $4x^3 - 20x^2 - 4x + 20 = 0$.

8. Use the root method to solve the equation $(x - 4)^3 = 8$.

9. Solve $x^4 - 13x^2 + 36 = 0$.

10. Use technology to find the solutions to $x^4 - 3x^3 - 3x^2 + 7x + 6 = 0$.

11. Use synthetic division to divide $4x^4 - 3x^3 + 2x - 8$ by $x - 2$.

12. Use synthetic division to solve $2x^3 + 5x^2 - 11x + 4 = 0$, given that $x = 1$ is one solution.

13. Find one solution of $3x^3 - x^2 - 12x + 4 = 0$ graphically, and use synthetic division to find any additional solutions.

14. For the function $y = \dfrac{1 - x^2}{x + 2}$
    a. Find any x-intercepts and y-intercepts if they exist.
    b. Find any horizontal asymptotes and vertical asymptotes that exist.
    c. Sketch the graph of the function.

15. Use a graph and technology to find any local maxima and/or minima of the function $y = \dfrac{x^2}{x - 4}$.

16. a. Graph $y = \dfrac{1 + 2x^2}{x + 2}$, on the window $[-10, 10]$ by $[-30, 10]$.
    b. Use the graph to find y when $x = 1$ and when $x = 3$.
    c. Use the graph to find the value(s) of x that give $y = 9/4$.
    d. Use analytical methods to solve $\dfrac{9}{4} = \dfrac{1 + 2x^2}{x + 2}$.

In Exercises 17–20, perform the operations and write the answer in the form a + bi.

17. $(3 + 4i) + (2 - 3i)$

18. $(-4 + 2i) - (2 - 6i)$

19. $(2 + 5i)(-1 + 2i)$

20. $\dfrac{3 + 2i}{1 - 3i}$

In Exercises 21–24, find the exact solutions to $f(x) = 0$ in the complex numbers.

21. $z^2 - 4z + 6 = 0$

22. $w^2 - 4w + 5 = 0$

23. $x^3 + x^2 + 2x - 4 = 0$

24. $4x^3 + 10x^2 + 5x + 2 = 0$

25. Solve $x^3 - 5x^2 \geq 0$.

26. Use a graphing utility to solve
$x^3 - 5x^2 + 2x + 8 \geq 0$.

27. Solve $2 < \dfrac{4x - 6}{x}$.

Chapter 4 Review

1. *Revenue* The monthly revenue for a product is given by $R = -0.1x^3 + 15x^2 - 25x$, where x is the number of thousands of units sold.
  a. Graph this function on the window $[-100, 200]$ by $[-2000, 60{,}000]$.
  b. Graph the function on a window that makes sense for the problem, that is, with nonnegative x and R.
  c. What is the revenue when 50,000 units are produced?

2. *Revenue* The monthly revenue for a product is given by $R = -0.1x^3 + 13.5x^2 - 150x$, where x is the number of thousands of units sold. Use technology to find the number of units that gives the maximum possible revenue.

3. *Investment* If \$5000 is invested for 6 years at interest rate r, compounded annually, the future value of the investment is given by $S = 5000(1 + r)^6$ dollars.
  a. Find the future value of this investment for selected interest rates by completing the following table.
  b. Graph this function for $0 \leq r \leq 0.20$.

r (rate)	S (future value)
1%	
5%	
10%	
15%	

  c. Compute the future value if $r = 10\%$ and $r = 20\%$. How much more money is earned at 20%?

4. *Immigration* The percent of the U.S. population that is foreign born is given by the graph below. The data can be modeled by the function $P = 0.00006465x^3 - 0.0074x^2 + 0.0734x + 14.053$, where x is the number of years from 1900.

  a. Graph this function on the window $[0, 100]$ by $[0, 20]$.
  b. What does the model give as the percent in 1960? How does this compare with the data?
  c. Use graphical or numerical methods to find the years after 1900 during which the percent of the U.S. population that is foreign born is 7%.

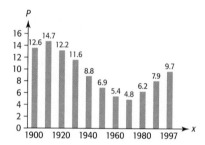

5. *United Nations Debt* The cubic function $y = -0.532x^3 - 1.003x^2 + 41.964x + 401.708$ models the total debt to the United Nations, in millions of dollars, as a function of the number of years from 1990.
  a. Graph this function using the window $[0, 10]$ by $[0, 600]$.
  b. According to the model, what was the approximate debt to the U.N. in 1998?
  c. Use graphical methods to estimate the maximum debt over the period from 1990–2000.
  (Source: United Nations, Institute for Global Communication)

6. *United Nations Debt* The table shows the amount of U.S. debt and total debt owed by all members to run the United Nations for the years 1990–2000, and the percent of the debt owed by the U.S. for each of these years.

Year	U.S. Debt ($ millions)	Total Debts ($ millions)	% U.S. Share
1990	296.2	403	73.5
1991	266.4	439.4	60.6
1992	239.5	500.6	47.8
1993	260.4	478	54.5
1994	248	480	51.6
1995	414.4	564	73.5
1996	376.8	510.7	73.8
1997	373.2	473.6	78.8
1998	315.7	417	75.7
1999	167.9	244.2	68.8
2000	164.6	222.4	74.0

(Source: United Nations, Institute for Global Communication)

a. Create a scatter plot from the table that gives U.S. debt to the United Nations, in millions of dollars, as a function of the number of years from 1990.

b. Find the cubic function that models U.S. debt to the United Nations, in millions of dollars, as a function of the number of years from 1990.

7. *Photosynthesis* The amount y of photosynthesis that takes place in a certain plant depends on the intensity x of the light present, according to the function $y = 120x^2 - 20x^3$. The model is only valid for x-values that are nonnegative and that produce a positive amount of photosynthesis. Use graphical or numerical methods to find what intensity allows the maximum amount of photosynthesis.

8. *Investment* The future value of an investment of $8000 at interest rate r compounded annually for 3 years is given by $S = 8000(1 + r)^3$. Use the root method to find the rate r that gives a future value of $9261.

9. *Constructing a Box* A box can be formed by cutting squares out of each corner of a piece of tin and folding the "tabs" up. If the piece of tin is 18 inches by 18 inches and each side of the square that is cut out has length x, the function that gives the volume of the box is $V = 324x - 72x^2 + 4x^3$.

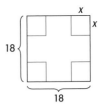

a. Use factoring to find the values of x that make $V = 0$.

b. For these values of x, discuss what happens to the box if squares of length x are cut out.

c. What values of x are reasonable for the squares that can be cut out to make a box?

10. *Break-Even* The daily profit for a product is given by $P(x) = -0.2x^3 + 20.5x^2 - 48.8x - 120$ dollars, where x is the number of hundreds of units produced. To find the number of units that gives break-even:

a. Graph the function on the window $[0, 50]$ by $[-20, 20]$, and graphically find an x-intercept of the graph.

b. Use synthetic division to find a quadratic factor of $P(x)$.

c. Find all of the zeros of $P(x)$.

d. Determine the levels of production that give break-even.

11. *Drugs in the Bloodstream* The concentration of a drug in a patient's bloodstream is given by $C = \dfrac{0.3t}{t^2 + 1}$, where t is the number of hours after the drug was ingested.

a. What is the horizontal asymptote for the graph of this function, if t satisfies $t \geq 0$?

b. What does this say about the concentration of the drug?

c. Use technology to find the maximum concentration of the drug, and when it occurs, for $0 \leq t \leq 4$.

12. *Average Cost* The average cost for production of a product is given by $\overline{C}(x) = \dfrac{50x + 5600}{x}$, where x is the number of units produced.

a. Find $\overline{C}(0)$, if it exists. What does this tell us about average cost?

b. Find the horizontal asymptote for the graph of this function. What does this tell us about the average cost of this product?

c. Does this function increase or decrease for $x > 0$?

13. *Average Cost* The average cost for production of a product is given by $\overline{C}(x) = \dfrac{30x^2 + 12{,}000}{x}$, where x is the number of units produced.

a. Graph the function for $x > 0$.

b. Use technology to find the number of units that gives the minimum average cost. What is the minimum average cost?

14. *Demand* The quantity of a certain product that is demanded is given by the equation $q = \dfrac{30,000}{p + 20}$, where p is the price per unit. Graph the function on a window that fits the context of the problem, with $p \le 300$.

15. *Printing* A printer has a contract to run 20,000 posters for a state fair. He can use any number of plates from 1 to 8 to run the posters, so that each stamp of the press produces a number of posters equivalent to the number of plates that have been created. The cost of printing all the posters is given by $C(x) = 200 + 20x + 180/x$, where x is the number of plates.

a. Combine the terms of this function over a common denominator to create a rational function.

b. Use numerical or graphical methods to find the number of plates that will make the cost $336.

c. How many plates should the printer create to produce the posters at the minimum cost?

16. *Computer Usage* The number of students per computer in U.S. public schools from the 1983-84 through 1998-99 school years is given in the table below. A function that can be used to model the data is $y = \dfrac{375.5 - 15x}{x + .03}$, where x is the number of years from the 1980-81 school year.

a. Graph the data from the table on the same axes as the model.

b. What is the number of students per computer in the 1993-94 school year, according to the model? How does this compare with the data in the table?

School Year	Students per Computer	School Year	Students per Computer
1983–84	125	1993–94	14
1985–86	50	1995–96	10
1987–88	32	1997–98	6.1
1989–90	22	1998–99	5.7
1991–92	18		

17. *Sales and Training* During the first month of employment, the monthly sales S (in thousands of dollars) for an average new salesperson depends on the number of hours of training x, according to

$$S = \frac{40}{x} + \frac{x}{4} + 10, \, x \ge 4.$$

How many hours of training should result in monthly sales greater than $23,300?

18. *Homicide Rate* The homicide rate per 100,000 people is given by the function $y = -0.00028x^3 + 0.0152x^2 - 0.0360x + 4.144$, with x equal to the number of years from 1950. Use technology to find the year or years after 1950 in which the homicide rate per 100,000 people is at least 8.13.
(Source: U.S. Department of Justice, Bureau of Justice Statistics)

19. *Revenue* The revenue from the sale of x units of a product is given by the function $R = 1200x - 0.003x^3$. Selling how many units will give a revenue of at least $59,625?

20. *Average Cost* The average cost per set for the production of television sets is given by

$$\overline{C} = \frac{100 + 30x + .1x^2}{x},$$

where x is the number of hundreds of units produced. For how many units is the average cost at most $37?

21. *Cost-Benefit* The percent p of particulate pollution that can be removed from the smokestacks of an industrial plant by spending C dollars per week is given by $p = \dfrac{100C}{9600 + C}$.
How much would it cost to remove at least 30.1% of the particulate pollution?

GROUP ACTIVITY/EXTENDED APPLICATION *1*

Unemployment Rates

The unemployment rate (the percent of the civilian labor force in a group) by race according to the Department of Labor, Bureau of Labor Statistics, December 2000, is given in the following table.

Unemployment Rates (Percents)

Period	White	Black	Nonwhite
1990	4.8	11.4	10.1
1991	6.1	12.5	11.1
1992	6.6	14.2	12.7
1993	6.1	13.0	11.7
1994	5.3	11.5	10.5
1995	4.9	10.4	9.6
1996	4.7	10.5	9.3
1997	4.2	10.0	8.8
1998	3.9	8.9	7.8
1999	3.7	8.0	8.0

1. Create a scatter plot using the number of years from 1990 as the input and unemployment rate for whites as the output.
2. Find the cubic function that models the unemployment rate for whites as a function of the number of years since 1990.
3. Graph the model of the unemployment rates for whites and the scatter plot on the same axes with the window [0, 9] by [0, 15].
4. Create a scatter plot using years from 1990 as input and unemployment rate for blacks as output.
5. Find the cubic function that models the unemployment rate for blacks as a function of the number of years since 1990.
6. Graph the model of the unemployment rates for blacks and the scatter plot on the same axes with the window [0, 9] by [0, 15].
7. Graph the models of unemployment rates for whites and blacks on the same axes with the window [0, 9] by [0, 15]. Is there any time in the 1990's when the black unemployment rate is not at least twice the white unemployment rate?
8. What conclusions can you make about unemployment rates of whites and blacks in the 1990's?
9. Graph the function formed by subtracting the white unemployment model from the black unemployment model. Use this graph to determine if the difference of the white unemployment rate and the black unemployment rate is increasing or decreasing in the 1990's.
10. Create the scatter plot using unemployment rate for whites as input and unemployment rate for blacks as output.
11. Find the linear function that models the black unemployment rate as a function of the white unemployment rate.
12. Graph the linear function found in part 11 on the same axes as the scatter plot in part 10.
13. Do your answers to the questions above indicate that the black unemployment rate will eventually fall to the same level as the white unemployment rate? Discuss "affirmative action" hiring in light of your answers.

GROUP ACTIVITY/EXTENDED APPLICATION 2

Printing

A printer has a contract to print 100,000 invitations for a political candidate. He can run the invitations by using any number of metal printing plates from 1 to 20 on his press. For example, if he prepares 10 plates, each impression of his press makes 10 invitations. Preparing each plate costs $8 and he can make 1000 impressions per plate per hour. If it costs $128 per hour to run the press, how many metal printing plates should the printer use to minimize the cost of printing the 100,000 invitations?

A. To understand the problem:
1. Determine how many hours it would take to print the 100,000 invitations with 10 plates.
2. How much would it cost to print the invitations with 10 plates?
3. Determine how much it would cost to prepare 10 printing plates.
4. What is the total cost to make the 10 plates and print the invitations?

B. To find the number of plates that will minimize the total cost of printing the invitations:
1. Find the cost of preparing x plates.
2. Find the number of invitations that can be made with one impression made by the x plates.
3. Find the number of invitations that can be made per hour with x plates, and the number of hours it would take to print 100,000 invitations.
4. Write a function of x that gives the total cost of producing the invitations.
5. Use technology to graph the cost function and to establish an estimate of the number of plates that will give the minimum cost of producing the plates.
6. Use numerical methods to find the number of plates that will give the minimum cost and the minimum possible cost of producing the invitations.

Systems of Equations and Matrices

Many applied problems can be solved easily by writing the relationships as a system of equations and finding the solution to the system by using graphical methods, substitution, or elimination. We can also represent systems with matrices (rectangular arrays of numbers) and use matrix reduction to solve the systems. Solving systems of equations is useful in finding break-even, finding market equilibrium, and solving investment problems, transportation problems, and traffic flow problems.

We can also use matrices to store data in rectangular arrays and to make comparisons of different types of data. We analyze data and make decisions using the operations of addition, subtraction, and multiplication for matrices. Operations with matrices can be used to solve pricing problems, advertising problems, and encoding and decoding messages.

	TOPICS	**APPLICATIONS**
5.1 Systems of Linear Equations in Two Variables	Graphical solution of systems; solution by substitution; solution by elimination; modeling systems of linear equations; inconsistent and dependent systems	Concerta market share, break-even, market equilibrium, investing
5.2 Systems of Linear Equations in Three Variables; Matrix Solution	Left-to-right elimination method; matrix representation of systems of equations; echelon forms of matrices; solving systems with matrices; Gauss-Jordan elimination; solution with technology	Product transportation, investments
5.3 Systems of Linear Equations with Nonunique Solutions	Nonunique solution; dependent systems; inconsistent systems	Purchasing, transportation, traffic flow
5.4 Matrices; Basic Operations	Addition and subtraction of matrices; multiplication of a matrix by a number; matrix multiplication; multiplication with technology	Life expectancy, pricing, purchasing, advertising, manufacturing
5.5 Inverse Matrices; Matrix Equations	Inverse matrices; inverses and technology; solving matrix equations; matrix equations and technology	Credit card security, encoding messages, decoding messages, manufacturing

ALGEBRA *Toolbox*

Proportion

Often in mathematics we need to express relationships between quantities. One relationship that is frequently used in applied mathematics occurs when two quantities are proportional. Two variables x and y are proportional to each other (or vary directly) if their quotient is constant. That is, y is **directly proportional** (or simply **proportional**) to x if y and x are related by the equation

$$\frac{y}{x} = k \quad \text{or equivalently, by} \quad y = kx,$$

where k is called the constant of proportionality.

EXAMPLE 1

Is the circumference of a circle proportional to the radius of a circle?

Solution

The circumference of a circle is proportional to the radius of a circle, because $C = 2\pi r$. In this case, C and r are the variables, and 2π is the constant of proportionality. We could write an equivalent equation,

$$\frac{C}{r} = 2\pi,$$

which says that the quotient of the circumference divided by the radius is constant, and that constant is 2π. ■

PROPORTIONALITY

One pair of numbers is said to be **proportional** to another pair of numbers if the ratio of one pair is equal to the ratio of the other pair. In general, the pair a, b is proportional to the pair c, d, if and only if

$$\frac{a}{b} = \frac{c}{d}, \quad b \neq 0, d \neq 0.$$

This proportion can be written $a{:}b$ as $c{:}d$.

If $b \neq 0, d \neq 0$, we can verify that $\dfrac{a}{b} = \dfrac{c}{d}$ by showing that the cross-products are equal, that is,

$$\frac{a}{b} = \frac{c}{d} \quad \text{if} \quad ad = bc.$$

EXAMPLE 2

Is the pair 4, 3 proportional to the pair 12, 9?

Solution

It is true that $\dfrac{4}{3} = \dfrac{12}{9}$, because we can reduce $\dfrac{12}{9}$ to $\dfrac{4}{3}$; or we can show that $\dfrac{4}{3} = \dfrac{12}{9}$ is true by showing that the cross-products are equal: $4 \cdot 9 = 3 \cdot 12$. Thus, the pair 4, 3 is proportional to 12, 9. ∎

Linear Equations in Two Variables

In Chapter 1 we graphed linear equations in two variables. Recall that an equation of the form $ax + by = c$ is called a *linear equation* because the graph of the ordered pairs (x, y) that satisfy the equation form a line. An equation of this form is also called a first-degree equation, because the power of each variable is the first power. Any pair of values for x and y that satisfies the equation is a *solution* to the equation. Each solution can be written as an ordered pair (x_1, y_1), where the first number x_1 is the replacement for x and the second number y_1 is the replacement for y.

EXAMPLE 3

a. Is $4x + 3y = 15$ a linear equation in two variables?

b. If so, give three solutions to the equation. How can these solutions be represented?

c. Can we list all the solutions to this equation?

Solution

a. The equation $4x + 3y = 15$ is a linear equation in the two variables x and y because it has the form $ax + by = c$, with $a = 4$, $b = 3$, and $c = 15$.

b. Because $4(3) + 3(1) = 15$, one solution to the equation $4x + 3y = 15$ is $x = 3$ and $y = 1$. Two other solutions include $x = 0, y = 5$ and $x = 3/2, y = 3$. We can represent the solutions as we have above, or as the ordered pairs $(3, 1)$, $(0, 5)$, $(3/2, 3)$. We can list solutions in a table or plot the ordered pairs as points on a graph.

c. It is impossible to list all the solutions because there are infinitely many of them. ∎

As with linear equations in one variable, equations in two variables are called equivalent if they have the same solutions. We can change an equation in two variables to an equivalent equation by multiplying both sides of the equation by the same nonzero number.

EXAMPLE 4

Write an equation that is equivalent to $4x + 3y = 15$, and that has 12 as its x-coefficient.

Solution

Multiplying both sides of the equation $4x + 3y = 15$ by 3 gives the equivalent equation $12x + 9y = 45$.

Note that the graphs of the two equations in Example 4 are the same line. (See Figure 5.1(a).)

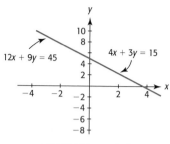

FIGURE 5.1(a) ■

In general, if two linear equations are equivalent, their graphs are the same line.

EXAMPLE 5

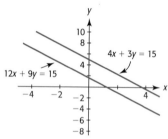

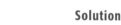

FIGURE 5.1(b)

What is the relationship of the graphs of $4x + 3y = 15$ and $12x + 9y = 15$?

Solution

The graphs of these equations appear to be parallel lines. (See Figure 5.1(b).) Solving both equations for y gives

$$y = -\frac{4}{3}x + 5 \quad \text{and} \quad y = -\frac{4}{3}x + \frac{5}{3}.$$

This shows that both lines have slope $-4/3$, but they have different y-intercepts, so the graphs are parallel lines. ■

Note that the respective coefficients of x and y in the equations of Example 5 are proportional, with 4:12 as 3:9, but that the constants are not in the same proportion. In general, if the coefficients of the variables of two linear equations are proportional but the constants are not in the same proportion, their graphs are parallel lines. If the respective coefficients of x and y of two linear equations are not proportional, the graphs of the equations will intersect at a point. (See Figure 5.2.)

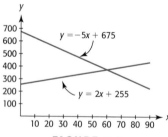

FIGURE 5.2

Linear Equations in Three Variables

If a, b, c, and d represent constants, then

$$ax + by + cz = d$$

is a first-degree equation in three variables. We also call these equations linear equations in three variables, but unlike linear equations in two variables, their graphs are not lines. When equations of this form are graphed in a three-dimensional coordinate system, their graphs are planes. As with the graphs of linear equations in two variables, the graphs of two linear equations in three variables will be parallel planes if the respective coefficients of x, y, and z are proportional and the constants are not in the same proportion. If the respective coefficients of x, y, and z in two equations are not proportional, then the two planes will intersect along a line. And if the two equations are equivalent, the equations represent the same plane.

EXAMPLE 6

Determine if the planes represented by the following pairs of equations represent the same plane, parallel planes, or neither.

a. $3x + 4y + 5z = 1$ and $6x + 8y + 10z = 2$

b. $2x - 5y + 3z = 4$ and $-6x + 10y - 6z = -12$

c. $3x - 6y + 9z = 12$ and $x - 2y + 3z = 36$

Solution

a. If we multiply both sides of the first equation by 2, the second equation results, so the equations are equivalent and thus the graphs of both equations are the same plane.

b. The respective coefficients of the variables are not in proportion, so the graphs are not the same plane and are not parallel planes, so the planes representing these equations must intersect.

c. The respective coefficients of the variables are in proportion and the constants are not in the same proportion, so the graphs are parallel planes. ■

Toolbox EXERCISES

1. If y is proportional to x, and $y = 12x$, what is the constant of proportionality?

2. If x is proportional to y, and $y = 10x$, what is the constant of proportionality?

3. Is the ratio 5:6 proportional to 10:18?

4. Is the ratio 8:3 proportional to 24:9?

5. Is the pair of numbers 4, 6 proportional to the pair 2, 3?

6. Is the pair of numbers 5.1, 2.3 proportional to the pair 51, 23?

7. Suppose y is proportional to x, and when $x = 2$, $y = 18$. Find y when $x = 11$.

8. Suppose x is proportional to y, and when $x = 2$, $y = 18$. Find y when $x = 11$.

9. Find three solutions to $10x - 8y = 24$.

10. Find three solutions to $5x + 3y = 20$.

11. a. Are the equations $5x - 4y = 14$ and $10x - 8y = 24$ equivalent equations?
   b. Are the graphs of the equations $5x - 4y = 14$ and $10x - 8y = 24$ the same line, parallel lines, or neither?

12. a. Are the equations $3x - 5y = 10$ and $3x - 10y = 20$ equivalent equations?
   b. Are the graphs of the equations $3x - 5y = 10$ and $3x - 10y = 20$ the same line, parallel lines, or neither?

13. Determine if the planes represented by the following pair of equations represent the same plane, parallel planes, or neither: $x + 3y - 7z = 12$ and $2x + 6y - 14z = 18$.

14. Determine if the planes represented by the following pair of equations represent the same plane, parallel planes, or neither: $x - 3y - 2z = 15$ and $2x - 6y - 4z = 30$.

15. Determine if the planes represented by the following pair of equations represent the same plane, parallel planes, or neither: $2x - y - z = 2$ and $2x + y - 4z = 8$.

5.1 Systems of Linear Equations in Two Variables

The new drug Concerta contains a time-release version of methylphenidate, the main ingredient in Ritalin. Primarily because its effects last 12 to 14 hours as opposed to Ritalin's 4 to 5 hours, sales of this drug increased rapidly while Ritalin sales decreased slightly. Looking at the graphs of the market shares of the two drugs in Figure 5.3(a), we can see that they intersect at a point, and we interpret this to mean that Concerta market share equaled that of Ritalin in the first few months after it was released. The graphs appear to intersect near the point representing 7 weeks after Aug. 25 (2000) and a 7% market share. (Other drugs, mostly generic, account for the remaining market share.) If we wrote the equations that represented these graphs with x representing weeks and y representing market share percent, the point of intersection of the two graphs would represent the *simultaneous* solution of the two equations because both equations would be satisfied by the coordinates of the point.

In this section we solve systems of linear equations in two variables graphically, by substitution, and by the elimination method.

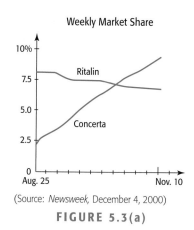

Weekly Market Share

(Source: *Newsweek,* December 4, 2000)

FIGURE 5.3(a)

Graphical Solution of Systems

In Chapter 1 we used the intersect method to solve a linear equation by first graphing functions representing the expressions on each side of the equation and then finding the intersection of these graphs. For example, to solve

$$3000x - 7200 = 5800x - 8600,$$

we can graph

$$y_1 = 3000x - 7200 \text{ and } y_2 = 5800x - 8600,$$

and find the point of intersection to be $(0.5, -5700)$. The x-coordinate of the point of intersection of the lines is the value of x that satisfies the original equation, $3000x - 7200 = 5800x - 8600$. Thus the solution to this equation is $x = 0.5$. (See Figure 5.3(b).) In this example, we were actually using a graphical method to solve a **system of two equations in two variables** denoted by

$$\begin{cases} y = 3000x - 7200 \\ y = 5800x - 8600 \end{cases}$$

The coordinates of the point of intersection of the two graphs give the x and y values that satisfy both equations **simultaneously**, and these values are called the **solution** to

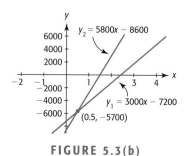

FIGURE 5.3(b)

the system. The following example uses the graphical method to solve a system of equations in two variables.

EXAMPLE 1   Break-Even

Suppose a company has its total revenue for a product given by $R = 5585x$ and its total cost given by $C = 61,740 + 440x$, where x is the number of thousands of tons of the product that is produced and sold per year. The company is said to break even when the total revenue equals the total cost; that is, when $R = C$. Find the number of thousands of tons of the product that gives break-even and how much the revenue and cost are at that level of production.

Solution

We graph the revenue function as $y_1 = 5585x$ and the cost function as $y_2 = 61,740 + 440x$ with a graphing utility on a window that contains the point of intersection. (See Figure 5.4(a).) We find the point of intersection to be $(12, 67,020)$. (See Figure 5.4(b).)

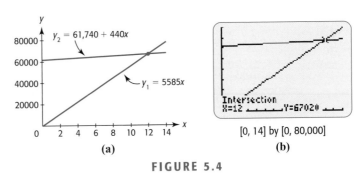

(a)   (b)

FIGURE 5.4

Thus the company will break even on this product if 12 thousand tons of the product are sold, when both the cost and revenue equal \$67,020.   ■

It is frequently necessary to solve each equation for a variable so that the equation can be graphed with a graphing utility. It is also necessary to find a viewing window that contains the point of intersection. Consider the following example.

EXAMPLE 2   Solving a System of Linear Equations

Solve the system

$$\begin{cases} 3x - 4y = 21 \\ 2x + 5y = -9 \end{cases}$$

Solution

The coefficients of the variables are not proportional, that is, $\dfrac{3}{2} \neq \dfrac{-4}{5}$, so there is a point of intersection. To solve this system with a graphing utility, we first solve both equations for y. The solution follows:

$$3x - 4y = 21$$
$$-4y = 21 - 3x$$
$$y = \frac{21 - 3x}{-4} = \frac{3x - 21}{4}$$

$$2x + 5y = -9$$
$$5y = -9 - 2x$$
$$y = \frac{-9 - 2x}{5}$$

Graphing these equations with a window that contains the point of intersection (see Figure 5.5(a)) and finding the point of intersection (see Figure 5.5(b)) gives $x = 3$, $y = -3$, so the solution is $(3, -3)$.

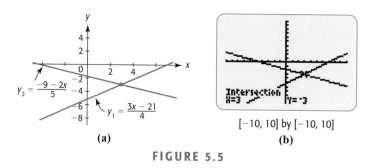

(a)   (b)

FIGURE 5.5

The system of linear equations discussed in Example 2 has a unique solution, shown as the point of intersection of the graphs. It is possible that two equations in a system describe the same line. When this happens, the equations are equivalent, and the values that satisfy one equation are also solutions to the other equation, and to the system. Such a system is called a **dependent system**. If a system contains two equations whose graphs are parallel lines, they have no point in common and thus the system has no solution. Such a system of equations is called **inconsistent**. Figure 5.6(a), (b), and (c) represents these 3 situations: systems that have a unique solution, many solutions (dependent system), and no solutions (inconsistent system), respectively. Note that the coefficients of the variables are proportional and slopes of the lines are equal in Figure 5.6(b) and in Figure 5.6(c).

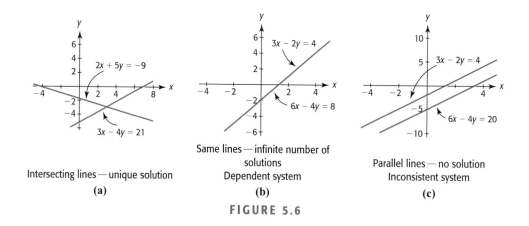

Intersecting lines — unique solution

(a)

Same lines — infinite number of solutions
Dependent system

(b)

Parallel lines — no solution
Inconsistent system

(c)

FIGURE 5.6

Solution by Substitution

Solving systems of equations by graphing with a graphing utility is not always the easiest method to use, because the equations must be solved for y to be entered in the utility, and an appropriate window must be used. A second solution method for a system of linear equations is called the **substitution method**, where one equation is solved for a variable and that variable is replaced by the equivalent expression in the other equation.

EXAMPLE 3   Concerta

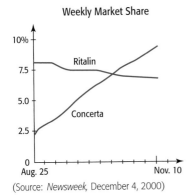

Weekly Market Share

(Source: *Newsweek,* December 4, 2000)

FIGURE 5.7

The graphs that compare the weekly market share of the two drugs Concerta and Ritalin are shown in Figure 5.7. If the linear equation $y = 0.690x + 2.4$ models the percent of market share of Concerta and $y = -0.072x + 7.7$ models the percent of market share of Ritalin, where x is the number of weeks since Concerta was made available, find the number of weeks past the release date that the weekly market share of Concerta reached that of Ritalin.

Solution

To find the number of weeks (x) until the Concerta market share equals the Ritalin share, we solve the system

$$\begin{cases} y = 0.690x + 2.4 \\ y = -0.072x + 7.7 \end{cases}$$

We can solve this system by substitution. From the first equation we see that $0.690x + 2.4$ is equal to y, so we substitute this expression for y in the second equation. This substitution gives the equation

$$0.690x + 2.4 = -0.072x + 7.7.$$

Solving this equation for x gives

$$0.762x = 5.3$$
$$x = 6.96 \approx 7 \text{ (weeks)}$$

Thus the weekly market share of Concerta reached that of Ritalin 7 weeks after it was released. ∎

The substitution method is also illustrated by the following example, which finds market equilibrium. The quantity of a product that is demanded by consumers is called the demand for the product and the quantity that is supplied is called the supply. In a free economy, both demand and supply are related to the price, and the price where the number of units demanded equals the number of units supplied is called the equilibrium price.

EXAMPLE 4   Market Equilibrium

Suppose that the daily demand for a product is given by $p = 200 - 2q$, where q is the number of units demanded and p is the price per unit in dollars, and that the daily supply is given by $p = 60 + 5q$, where q is the number of units supplied and p is the price in dollars. If a price results in more units being supplied than demanded, we say there is a *surplus,* and if the price results in fewer units being supplied than demanded, we say there is a *shortfall.* **Market equilibrium** occurs when the supply quantity equals the demand quantity (and when the prices are equal); that is, when q and p both satisfy the system

$$\begin{cases} p = 200 - 2q \\ p = 60 + 5q \end{cases}$$

a. If the price is $140, how many units are supplied and how many are demanded?

b. Does this price give a surplus or a shortfall of the product?

c. What price gives market equilibrium?

Solution

a. If the price is \$140, the number of units demanded satisfies $140 = 200 - 2q$, or $q = 30$, and the number of units supplied satisfies $140 = 60 + 5q$, or $q = 16$.

b. At this price, the quantity supplied is less than the quantity demanded, so a shortfall occurs.

c. Because market equilibrium occurs where q and p both satisfy the system

$$\begin{cases} p = 200 - 2q \\ p = 60 + 5q \end{cases}$$

we seek the solution to this system.

We can solve this system by substitution. Substituting $60 + 5q$ for p in the first equation gives the equation

$$60 + 5q = 200 - 2q.$$

Solving this equation gives

$$60 + 5q = 200 - 2q$$
$$7q = 140$$
$$q = 20$$

Thus market equilibrium occurs when the number of units is 20, and the equilibrium price is

$$p = 60 + 5(20) = 160 \text{ dollars per unit.} \qquad \blacksquare$$

The substitution in Example 4 was not difficult because both equations were solved for p. When this is not the case, we use the following steps to solve systems of two equations in two variables by substitution.

SOLUTION OF SYSTEMS OF EQUATIONS BY SUBSTITUTION

1. Solve one of the equations for one of the variables in terms of the other variable.

2. Substitute the expression from Step 1 into the other equation to give an equation in one variable.

3. Solve the linear equation for the variable.

4. Substitute this solution into the equation from Step 1 or into one of the original equations and solve this equation for the second variable.

5. Check the solution in both original equations or check graphically.

EXAMPLE 5   Solution by Substitution

Solve the system $\begin{cases} 3x + 4y = 10 \\ 4x - 2y = 6 \end{cases}$ by substitution.

Solution

To solve this system by substitution, we can solve either equation for either variable and substitute the resulting expression into the other equation. Solving the second equation for y gives

$$4x - 2y = 6$$
$$-2y = -4x + 6$$
$$y = 2x - 3$$

Substituting this expression for y in the first equation gives

$$3x + 4(2x - 3) = 10.$$

Solving this equation gives

$$3x + 8x - 12 = 10$$
$$11x = 22$$
$$x = 2$$

Substituting $x = 2$ into $y = 2x - 3$ gives $y = 2(2) - 3 = 1$, so the solution to the system is $x = 2, y = 1$, or $(2, 1)$.

Checking shows that this solution satisfies both original equations. ∎

Solution by Elimination

A second analytical method, called the **elimination method**, is frequently an easier method to use to solve a system of linear equations. The elimination method is based on rewriting one or both of the equations in an equivalent form that allows us to eliminate one of the variables by adding or subtracting the equations.

SOLVING A SYSTEM OF TWO EQUATIONS IN TWO VARIABLES BY ELIMINATION

1. If necessary, multiply one or both equations by a nonzero number that will make the coefficients of one of the variables in the equations equal, except perhaps for sign.
2. Add or subtract the equations to eliminate one of the variables.
3. Solve for the variable in the resulting equation.
4. Substitute the solution from Step 3 into one of the original equations and solve for the second variable.
5. Check the solutions in the remaining original equation, or check graphically.

EXAMPLE 6 Solution by Elimination

Use the elimination method to solve the system

$$\begin{cases} 3x + 4y = 10 \\ 4x - 2y = 6 \end{cases}$$

and check the solution graphically.

Solution

The goal is to convert one of the equations into an equivalent equation of a form so that addition of the two equations will eliminate one of the variables. Notice that the coefficient of y in the second equation, -2, is a factor of the coefficient of y in the first equation, 4. If we multiply both sides of the second equation by 2 and add the two equations, this will eliminate the y-variable.

$$\begin{cases} 3x + 4y = 10 & \text{(Equation 1)} \\ 4x - 2y = 6 & \text{(Equation 2)} \end{cases}$$

Multiply 2 times equation (2), getting equivalent equation (3).

$$\begin{cases} 3x + 4y = 10 & \text{(Equation 1)} \\ 8x - 4y = 12 & \text{(Equation 3)} \end{cases}$$

Add equations (1) and (3) to eliminate y. $11x \qquad = 22$
Solve the new equation for x. $x \qquad = 2$

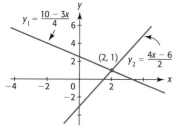

$y_1 = \dfrac{10 - 3x}{4}$

$(2, 1)$ $y_2 = \dfrac{4x - 6}{2}$

FIGURE 5.8

Substituting $x = 2$ in the first equation gives $3(2) + 4y = 10$, or $y = 1$. Thus the solution to the system is $x = 2, y = 1$, or $(2, 1)$.

To check graphically, we solve both equations for y, getting $y_1 = \dfrac{10 - 3x}{4}$ and $y_2 = \dfrac{4x - 6}{2}$, then graph the equations and find the point of intersection to be $(2, 1)$. (See Figure 5.8.)   ■

Modeling Systems of Linear Equations

Solution of some real problems requires us to create two or more equations whose simultaneous solution is the solution to the problem. Consider the following examples.

EXAMPLE 7   Investments

An investor has $300,000 to invest, part at 12% and the remainder in a less risky investment at 7%. If her investment goal is to have an annual income of $27,000, how much should she put in each investment?

Solution

If we denote the amount invested at 12% as x and the amount invested at 7% as y, the sum of the investments is $x + y$, so we have the equation

$$x + y = 300,000.$$

The annual income from the 12% investment is $0.12x$ and the annual income from the 7% investment is $0.07y$. Thus the desired annual income from the two investments is

$$0.12x + 0.07y = 27,000.$$

Thus we can write the given information as a system of equations.

$$\begin{cases} x + y = 300,000 \\ 0.12x + 0.07y = 27,000 \end{cases}$$

To solve this system, we multiply the first equation by -0.12 and add the two equations. This results in an equation with one variable.

$$\begin{cases} -0.12x - 0.12y = -36{,}000 \\ 0.12x + 0.07y = 27{,}000 \end{cases}$$

$$-0.05y = -9000$$
$$y = 180{,}000$$

Substituting 180,000 for y into the first original equation and solving for x gives $x = 120{,}000$. Thus \$120,000 should be invested at 12% and \$180,000 should be invested at 7%.

To check this solution, we see that the total investment is \$120,000 + \$180,000, which equals \$300,000. The interest earned at 12% is \$120,000(.12) = \$14,400, and the interest earned at 7% is \$180,000(.07) = \$12,600. The total interest is \$14,400 + \$12,600, which equals \$27,000. This agrees with the given information. ∎

EXAMPLE 8 Market Equilibrium

A group of wholesalers will buy 75 dryers per month if the price is \$300 and 45 per month if the price is \$450. The manufacturer is willing to supply 30 dryers per month if the price is \$315 and 45 if the price is \$345. Assuming that the resulting supply and demand functions are linear, find the quantity and price that give market equilibrium for these dryers.

Solution

Representing the price by p and the quantity by q, we can use the point-slope form to write the equation of the line through the points (75, 300) and (45, 450) that represents the demand function. The slope is

$$m_1 = \frac{450 - 300}{45 - 75} = -5.$$

Using this slope and either one of the points gives the demand equation we seek.

$$p - 300 = -5(q - 75)$$
$$p = -5q + 375 + 300$$
$$p = -5q + 675$$

Similarly, we can write the equation of the line through the points (30, 315) and (45, 345) that represents the supply function:

$$m_2 = \frac{345 - 315}{45 - 30} = 2$$
$$p - 345 = 2(q - 45)$$
$$p = 2q - 90 + 345$$
$$p = 2q + 255$$

Because the price for demand and supply are equal at market equilibrium, we solve

$$\begin{cases} p = -5q + 675 \\ p = 2q + 255 \end{cases}$$

by substitution

p

$p = -5q + 675$

(60, 375)

$p = 2q + 255$

q

FIGURE 5.9

$$-5q + 675 = 2q + 255$$
$$-7q = -420$$
$$q = 60$$
$$p = 2(60) + 255 = 375$$

Thus the equilibrium quantity is 60 units, when the price is $375. We say the equilibrium point is (60, 375). The graphs of these functions are shown in Figure 5.9. ■

Inconsistent and Dependent Systems

EXAMPLE 9   Systems with Nonunique Solutions

Use the elimination method to solve each of the following systems, if possible. Verify the solution graphically.

a. $\begin{cases} 2x - 3y = 4 \\ 6x - 9y = 12 \end{cases}$ **b.** $\begin{cases} 2x - 3y = 4 \\ 6x - 9y = 36 \end{cases}$

Solution

a. To solve $\begin{cases} 2x - 3y = 4 \\ 6x - 9y = 12 \end{cases}$, we multiply the first equation by -3 and add the equations, getting the following:

$$\begin{cases} -6x + 9y = -12 \\ 6x - 9y = 12 \end{cases}$$
$$0 = 0$$

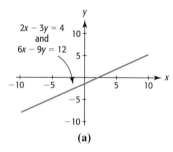

This indicates that the graphs of the equations intersect when $0 = 0$, *which is always true*. Thus any value that satisfies one of these equations also satisfies the other, and there are *infinitely many* solutions. Figure 5.10(a) shows that the graphs of the equations lie on the same line. Notice that the second equation is a multiple of the first, so the equations are equivalent. This system is called *dependent*.

b. To solve $\begin{cases} 2x - 3y = 4 \\ 6x - 9y = 36 \end{cases}$, we multiply the first equation by -3 and add the equations, getting the following:

$$\begin{cases} -6x + 9y = -12 \\ 6x - 9y = 36 \end{cases}$$
$$\overline{0 = 24}$$

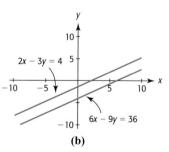

This indicates that the equations intersect when $0 = 24$, *which is never true*. Thus no values of x and y satisfy both of the equations. Figure 5.10(b) shows that the graphs of the equations are parallel. Notice that the coefficients of the variables are proportional, but the constants are not in the same proportion. This system is called *inconsistent*. ■

FIGURE 5.10

5.1 SKILLS CHECK

1. What are the coordinates of the point of intersection of $y = 3x - 2$ and $y = 3 - 2x$?

2. Give the coordinates of the point of intersection of $3x + 2y = 5$ and $5x - 3y = 21$.

3. Does the system $\begin{cases} 2x + 5y = 6 \\ x + 2.5y = 3 \end{cases}$ have a unique solution, no solution, or many solutions? What does this mean graphically?

4. Does the system $\begin{cases} 6x + 4y = 3 \\ 3x + 2y = 3 \end{cases}$ have a unique solution, no solution, or many solutions? What does this mean graphically?

In Exercises 5–7, solve the following systems of equations graphically.

5. $\begin{cases} y = 3x - 12 \\ y = 4x + 2 \end{cases}$ 6. $\begin{cases} 2x - 4y = 6 \\ 3x + 5y = 20 \end{cases}$

7. $\begin{cases} 4x - 3y = -4 \\ 2x - 5y = -4 \end{cases}$

In Exercises 8–10, solve the following systems of equations by substitution.

8. $\begin{cases} 2x - 3y = 2 \\ 5x - y = 18 \end{cases}$ 9. $\begin{cases} x - 5y = 12 \\ 3x + 4y = -2 \end{cases}$

10. $\begin{cases} 2x - 3y = 5 \\ 5x + 4y = 1 \end{cases}$

In Exercises 11–14, solve the following systems of equations by elimination, if a solution exists.

11. $\begin{cases} x + 3y = 5 \\ 2x + 4y = 8 \end{cases}$ 12. $\begin{cases} 3x + 6y = 12 \\ 2x + 4y = 8 \end{cases}$

13. $\begin{cases} 3x + 3y = 5 \\ 2x + 4y = 8 \end{cases}$ 14. $\begin{cases} 5x + 3y = 8 \\ 2x + 4y = 8 \end{cases}$

In Exercises 15–18, solve the following systems of equations by any convenient method.

15. $\begin{cases} y = 3x - 2 \\ y = 5x - 6 \end{cases}$ 16. $\begin{cases} 4x + 6y = 4 \\ x = 4y + 8 \end{cases}$

17. $\begin{cases} 3x - 7y = -1 \\ 4x + 3y = 11 \end{cases}$ 18. $\begin{cases} 5x - 3y = 12 \\ 3x - 5y = 8 \end{cases}$

5.1 EXERCISES

1. *Break-Even* A manufacturer of kitchen sinks has total revenue given by the function $R = 76.50x$ and has total cost given by $C = 2970 + 27x$, where x is the number of sinks produced and sold. To find the number of units that gives break-even for the product, solve the equation $R = C$.

2. *Break-Even* A jewelry maker has total revenue for her bracelets given by $R = 89.75x$ and incurs a total cost of $C = 23.50x + 1192.50$, where x is the number of bracelets produced and sold. To find the number of units that gives break-even for the product, solve the equation $R = C$.

3. *Supply and Demand* A certain product has supply and demand functions given by $p = 5q + 20$ and $p = 128 - 4q$, respectively.
   a. If the price p is $60, how many units q are supplied and how many are demanded?
   b. Does this give a surplus or a shortfall of product?

c. What price gives market equilibrium and how many units are demanded and supplied at this price?

4. *Market Equilibrium* The demand for a brand of clock radio is given by $p + 2q = 320$ and the supply for these radios is given by $p - 8q = 20$, where p is the price and q is the quantity demanded at price p. Solve the system containing these two equations to find the price at which the quantity demanded equals the quantity supplied, and the equilibrium quantity.

5. *Military* The number of active-duty U.S. Navy personnel is given by $y = -5.686x + 676.173$ thousand, and the number of active-duty U.S. Air Force personnel is given by $y = -11.997x + 847.529$ thousand, where x is the number of years from 1960.
   a. Use graphical methods to find the year in which the number of Navy personnel reaches the number of Air Force personnel.

b. How many will be in each service when the numbers of personnel are equal?
(Source: *The World Almanac*)

6. *College Enrollment* Suppose the percent of males who enrolled in college within 12 months of high school graduation is given by $y = -0.126x + 55.72$ and the percent of females enrolled in college within 12 months of high school graduation is given by $y = 0.73x + 39.7$, where x is the number of years after 1960. Use graphical methods to find the year these models indicate that the percent of females equaled the percent of males.
(Source: *Statistical Abstract of the U.S., 2000*)

7. *Snack Sales* The graphs below make it appear that mint sales have equaled gum sales in 1999. Using this data, the best-fitting line for mint sales has equation $y = 24.5x + 93.5$ (million dollars) and the best-fitting line for gum sales is $y = -0.2x + 1007$ (million dollars), where x is the number of years after 1990.

a. For what value of x does mint sales equal gum sales according to these models?

b. In what year do these models predict that mint sales will reach gum sales?

c. Are these graphs misleading even if they are correct? How?

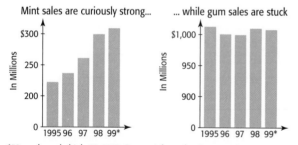

Mint sales are curiously strong... ... while gum sales are stuck

*52 weeks ended July 18, 1999 Source: Information Resources Inc.

(Source: *Newsweek*, November 1, 1999)

8. *Nursing Homes* The number of nursing homes in Massachusetts between 1993 and 1997 can be described by $m(x) = 3.37x - 6212.34$, and the number of nursing homes during that same time period in Illinois is given by $n(x) = 12.34x - 23,802.09$. In each equation, x represents the year. Assuming the pattern indicated by these equations is valid prior to 1993, during what year did Massachusetts and Illinois have the same number of nursing homes? How many homes were there in each state when the number of nursing homes was equal?
(Source: Health Care Financing Administration)

9. *Stock Prices* The sum of the high and low prices of a share of stock for The Gap, Inc. in 2000 is $83.5 and the difference of these two prices in 2000 is $21.88. Find the high and low prices.
(Source: The Gap 2000 Annual Report)

10. *Revenue* The sum of the 1999 revenue and twice the 1998 revenue for Papa John's International, Inc. is $2144.9 million. The difference in the 1999 and 1998 revenues is $135.5 million. If Papa John's revenue between 1990 and 1999 is an increasing function, find the 1998 and 1999 revenues.
(Source: Papa John's 1999 Annual Report)

11. *Pricing* A concert promoter needs to make $84,000 from the sale of 2400 tickets. The promoter charges $30 for some tickets and $45 for the others.

a. If there are x of the $30 tickets sold and y of the $45 tickets sold, write an equation that states that the sum of the tickets sold is 2400.

b. How much money is received from the sale of x tickets for 30 dollars each?

c. How much money is received from the sale of y tickets for 45 dollars each?

d. Write an equation that states that the total amount received from the sale is 84,000 dollars.

e. Solve the equations simultaneously to find how many tickets of each type must be sold to yield the $84,000.

12. *Rental Income* A woman has $250,000 invested in two rental properties. One yields an annual return of 10% of her investment and the other returns 12% per year on her investment. Her total annual return from the two investments is $26,500. Let x represent the amount of the 10% investment and y represent the amount of the 12% investment.

a. Write an equation that states that the sum of the investments is 250,000 dollars.

b. What is the annual return on the 10% investment?

c. What is the annual return on the 12% investment?

d. Write an equation that states that the sum of the annual returns is 26,500 dollars.

e. Solve these two equations simultaneously to find how much is invested in each property.

13. *Investment* One safe investment pays 8% per year, and a more risky investment pays 12% per year.

a. How much must be invested in each account if the investor of $100,000 would like a return of $9000 per year?

b. Why might the investor use two accounts rather than put all money in the 12% investment?

14. *Investment* A woman invests $52,000 in two different mutual funds, one that averages 10% per year and another that averages 14% per year. If her average annual return on the two mutual funds is $5720, how much did she invest in each fund?

15. *Nutrition* A glass of skim milk supplies 0.1 mg of iron and 8.5 g of protein. A quarter pound of lean meat provides 3.4 mg of iron and 22 g of protein. If a person on a special diet is to have 7.1 mg of iron and 69.5 g of protein, how many glasses of skim milk and how many quarter-pound servings of meat will provide this?

16. *Nutrition* Each ounce of substance A supplies 6% of the nutrient a patient needs, and each ounce of substance B supplies 10% of the required nutrient. If the total number of ounces given to the patient is 14, and 100% of the nutrient is supplied, how many ounces of each substance was given?

17. *Medication* A nurse has two solutions that contain different concentrations of a certain medication. One is a 10% concentration and the other is a 5% concentration. How many cubic centimeters of each should he mix to obtain 20 cubic centimeters of an 8% solution?

18. *Medication* A nurse has two solutions that contain different concentrations of a certain medication. One is a 30% concentration and the other is a 15% concentration. How many cubic centimeters (cc) of each should she mix to obtain 45 cc of a 20% solution?

19. *Supply and Demand* The table below gives the quantity of graphing calculators demanded and the quantity supplied for selected prices.
   a. Find the linear equation that gives the price as a function of the quantity demanded.
   b. Find the linear equation that gives the price as a function of the quantity supplied.
   c. Use these equations to find the market equilibrium price.

Price (dollars)	Quantity Demanded (thousands)	Quantity Supplied (thousands)
50	210	0
60	190	40
70	170	80
80	150	120
100	110	200

20. *Market Analysis* The supply function and the demand function for a product are linear and are determined by the tables that follow. Create the supply and demand functions and find the price that gives market equilibrium.

Supply Price	Function Quantity	Demand Price	Function Quantity
200	400	400	400
400	800	200	800
600	1200	0	1200

21. *Alcohol use* The percent of the U.S. population who admitted to using alcohol at least once during the month prior to being asked about alcohol use is as shown in the table below.

Year	1994	1995	1996	1997
18- to 25-year-olds (%)	63	61	60	58
26- to 34-year-olds (%)	65	63	62	60

(Source: National Household Survey on Drug Abuse)

Letting x represent the number of years after 1990, the models are:

18- to 25-year-olds: $y = -1.6x + 69.3$

26- to 34-year-olds: $y = -1.6x + 71.3$

Solve this system of equations, if possible, to find when the percent of 18- to 25-year-olds who admitted to using alcohol equals the percent of 26- to 34-year-olds.

22. *Tax Rate* The federal tax rate is equivalent to a flat rate of 34% for taxable federal income between $35,000 and $1,000,000 and the Alabama tax rate is 5% of the taxable state income. Assume an income of $1,000,000 before either tax is paid, and use the fact that Alabama's income tax can be deducted from federal taxable income and that the federal tax can be deducted from Alabama's taxable income.
   a. Create an equation that gives the federal tax x after the Alabama tax y is deducted.
   b. Create an equation that gives the Alabama tax y after the federal tax x is deducted.
   c. Write the equations created in part (a) and part (b) as a system of equations.
   d. Solve the system of equations to find the amount paid in taxes.

5.2 Systems of Linear Equations in Three Variables; Matrix Solution

A manufacturer of furniture has three models of chairs, Anderson, Blake, and Colonial. The numbers of hours required for framing, upholstery, and finishing for each type of chair are given in the table below. The company has 1500 hours per week for framing, 2100 hours for upholstery, and 850 hours for finishing.

TABLE 5.1

	Anderson	Blake	Colonial
Framing	2	3	1
Upholstery	1	2	3
Finishing	1	2	1/2

Finding how many of each type of chair can be produced under these conditions involves solving a system with three linear equations in three variables. It is not possible to solve this equation graphically, and new analytical methods are needed. In this section, we discuss how to solve a system of equations by using the left-to-right elimination method and then discuss how matrices can be used to solve such systems. We also discuss other applications of matrices.

Left-to-Right Elimination Method

We used the elimination method to reduce a system of two equations in two variables to one equation in one variable. The **left-to-right elimination** method is an extension of this method and can be used to solve systems with three or more variables. The steps in this solution method follow.

LEFT-TO-RIGHT ELIMINATION METHOD OF SOLVING SYSTEMS OF LINEAR EQUATIONS IN THREE VARIABLES

1. If necessary, interchange two equations or use multiplication to make the coefficient of the first variable in the first equation a 1.

2. Add a multiple of the first equation to each of the following equations so that the coefficients of the first variable in the new second and third equations become 0.

3. Multiply (or divide) both sides of the second equation by a number that makes the coefficient of the second variable in the second equation equal to 1. Add a multiple of the (new) second equation to the (new) third equation so that the coefficient of the second variable in the newest third equation becomes 0.

4. Multiply (or divide) both sides of the third equation by a number that makes the coefficient of the third variable in the third equation equal to 1. This gives the solution for the third variable of the system of equations.

5. Use the solution for the third variable to solve for the second variable in the second equation. Then substitute values for the second and third variables to solve for the first variable in the first equation. (This is called *back substitution*.)

EXAMPLE 1 | Solution by Elimination

Solve the system

$$\begin{cases} 2x - 3y + z = -1 & \text{(Equation 1)} \\ x - y + 2z = -3 & \text{(Equation 2)} \\ 3x + y - z = 9 & \text{(Equation 3)} \end{cases}$$

Solution

1. To write the system with x-coefficient 1 in the first equation, interchange equations (1) and (2).

$$\begin{cases} x - y + 2z = -3 \\ 2x - 3y + z = -1 \\ 3x + y - z = 9 \end{cases}$$

2. To eliminate x in the second equation, multiply the (new) first equation by -2 and add it to the second equation. To eliminate x in the third equation, multiply the (new) first equation by -3 and add it to the third equation. This gives the equivalent system

$$\begin{cases} x - y + 2z = -3 \\ -y - 3z = 5 \\ 4y - 7z = 18 \end{cases}$$

3. To get 1 as the coefficient of y in the second equation, multiply the second equation by -1. To eliminate y from the third equation, multiply the (new) second equation by -4 and add it to the third equation.

$$\begin{cases} x - y + 2z = -3 \\ y + 3z = -5 \\ -19z = 38 \end{cases}$$

4. To solve for z, divide both sides of the third equation by -19.

$$\begin{cases} x - y + 2z = -3 \\ y + 3z = -5 \\ z = -2 \end{cases}$$

This gives the solution $z = -2$.

5. Substituting -2 for z in the second equation and solving for y gives $y + 3(-2) = -5$ or $y = 1$. Substituting -2 for z and 1 for y in the first equation and solving for x gives $x - 1 + 2(-2) = -3$, so $x = 2$, and the solution is $(2, 1, -2)$.

The process used in Step 5 is called *back substitution* because we use the solution for the third variable to work back to the solutions of the first and second variables. Checking in the original system verifies that the solution satisfies all the equations.

$$\begin{cases} 2(2) - 3(1) + (-2) = -1 \\ 2 - 1 + 2(-2) = -3 \\ 3(2) + 1 - (-2) = 9 \end{cases}$$

∎

Matrix Representation of Systems of Equations

We can represent the coefficients and constants from a system of linear equations in a rectangular array, with each row of the array representing an equation and each column of the array representing a variable. Such an array is called a **matrix** (the plural name is matrices). The matrix A below has 3 rows (horizontal) and 4 columns (vertical), so its **dimension** is 3×4 (three by four). Matrix B is a 2×3 matrix and Matrix C is a 3×3 matrix. Matrices like C, with the same number of rows and columns, are called **square matrices**. A square matrix that has 1's down its diagonal and 0's everywhere else, as in Matrix I below, is called an **identity matrix**. Matrix I is a 3×3 identity matrix.

$$A = \begin{bmatrix} 1 & 4 & 1 & -2 \\ 3 & -7 & 0 & 3 \\ 4 & 2 & -1 & 5 \end{bmatrix} \qquad B = \begin{bmatrix} 1 & 6 & -9 \\ 3 & 2 & 5 \end{bmatrix}$$

$$C = \begin{bmatrix} 1 & 4 & 3 \\ 6 & -5 & 3 \\ -1 & 4 & 2 \end{bmatrix} \qquad I = \begin{bmatrix} 1 & 0 & 0 \\ 0 & 1 & 0 \\ 0 & 0 & 1 \end{bmatrix}$$

The system of equations that was solved in Example 1,

$$\begin{cases} 2x - 3y + z = -1 \\ x - y + 2z = -3 \\ 3x + y - z = 9 \end{cases}$$

can be written in a **matrix** as follows:

$$\left[\begin{array}{ccc|c} 2 & -3 & 1 & -1 \\ 1 & -1 & 2 & -3 \\ 3 & 1 & -1 & 9 \end{array} \right]$$

Note that the first column (vertical) contains the coefficients of x for each of the equations, the second column contains the coefficients of y, and the third column contains the coefficients of z. The column to the right of the vertical line, containing the constants of the equations, is called the **augment** of the matrix, and a matrix containing an augment is called an **augmented matrix**. The matrix that contains only the coefficients of the variables is called a **coefficient matrix**. The coefficient matrix for this system is

$$\begin{bmatrix} 2 & -3 & 1 \\ 1 & -1 & 2 \\ 3 & 1 & -1 \end{bmatrix}$$

This coefficient matrix is called a 3×3 matrix because it has three rows and 3 columns. The augmented matrix above has 3 rows and 4 columns, so it is a 3×4 matrix; we say that it has **dimension** 3×4.

Echelon Forms of Matrices; Solving Systems with Matrices

The key to solving a system of linear equations with matrices is finding an equivalent matrix from which the solution to the system can be found easily. Notice that when we solved the system of equations in Example 1, the system was reduced to the equivalent system

$$\begin{cases} x - y + 2z = -3 \\ y + 3z = -5 \\ z = -2 \end{cases}$$

The matrix that represents this equivalent system is

$$\begin{bmatrix} 1 & -1 & 2 & -3 \\ 0 & 1 & 3 & -5 \\ 0 & 0 & 1 & -2 \end{bmatrix}$$

Note the relationship of the matrix with the reduced system of equations: we can see that the last row of the matrix represents $z = -2$, and that row 1 and row 2 represent the first equation and the second equation of the system, respectively. The coefficient part of this augmented matrix has all 1's on its diagonal and all 0's below its diagonal.

> Any augmented matrix that has 1's or 0's on the diagonal of its coefficient part and 0's below the diagonal is said to be in **row-echelon form**.

Matrix D below is another matrix in row-echelon form.

$$D = \begin{bmatrix} 1 & 3 & -2 & 3 \\ 0 & 1 & 3 & -1 \\ 0 & 0 & 1 & 2 \end{bmatrix}$$

If the rows of an augmented matrix represent the equations of a system of linear equations, we can perform **row operations** on the matrices that are equivalent to operations that we performed on equations to solve a system of equations. These operations are as follows.

MATRIX ROW OPERATIONS

Row Operation	**Corresponding Equation Operation**
1. Interchanging two rows of the matrix.	1. Interchanging two equations.
2. Multiplying a row by any nonzero constant.	2. Multiplying an equation by any nonzero constant.
3. Adding a multiple of one row to another row.	3. Adding a multiple of one equation to another.

When a new matrix results from one or more of these row operations performed on a matrix, the new matrix is **equivalent** to the original matrix.

EXAMPLE 2   Matrix Solution of Linear Systems

Use matrix row operations to solve the system of equations

$$\begin{cases} 2x - 3y + z = -1 & \text{(Equation 1)} \\ x - y + 2z = -3 & \text{(Equation 2)} \\ 3x + y - z = 9 & \text{(Equation 3)} \end{cases}$$

Solution

We begin by writing the augmented matrix that represents the system.

$$\left[\begin{array}{rrr|r} 2 & -3 & 1 & -1 \\ 1 & -1 & 2 & -3 \\ 3 & 1 & -1 & 9 \end{array}\right]$$

To get a first row that has 1 in the first column, we interchange row 1 and row 2. (Compare this to the equivalent system of equations in Step 1 of Example 1, which is to the right.)

$$\left[\begin{array}{rrr|r} 1 & -1 & 2 & -3 \\ 2 & -3 & 1 & -1 \\ 3 & 1 & -1 & 9 \end{array}\right] \qquad \begin{cases} x - y + 2z = -3 \\ 2x - 3y + z = -1 \\ 3x + y - z = 9 \end{cases}$$

To get a 0 as the first entry in the second row, we multiply the first row by -2 and add it to the second row. To get a 0 as the first entry in the third row, we multiply the first row by -3 and add it to the third row. (Compare this to the equivalent system of equations in Step 2 of Example 1, which is to the right.)

$$\left[\begin{array}{rrr|r} 1 & -1 & 2 & -3 \\ 0 & -1 & -3 & 5 \\ 0 & 4 & -7 & 18 \end{array}\right] \qquad \begin{cases} x - y + 2z = -3 \\ -y - 3z = 5 \\ 4y - 7z = 18 \end{cases}$$

To get 1 as the second entry in the second row, we multiply the second row by -1. To get a 0 as the second entry in the third row, we multiply the new second row by -4 and add it to the third row. (Compare this to the equivalent system of equations in Step 3 of Example 1, which is to the right.)

$$\left[\begin{array}{rrr|r} 1 & -1 & 2 & -3 \\ 0 & 1 & 3 & -5 \\ 0 & 0 & -19 & 38 \end{array}\right] \qquad \begin{cases} x - y + 2z = -3 \\ y + 3z = -5 \\ -19z = 38 \end{cases}$$

To get 1 as the third entry in the third row, we multiply both sides of the third row by $-\frac{1}{19}$. (Compare this to the equivalent system of equations in Step 4 of Example 1, which is to the right.)

$$\left[\begin{array}{rrr|r} 1 & -1 & 2 & -3 \\ 0 & 1 & 3 & -5 \\ 0 & 0 & 1 & -2 \end{array}\right] \qquad \begin{cases} x - y + 2z = -3 \\ y + 3z = -5 \\ z = -2 \end{cases}$$

The matrix is now in row-echelon form. We see that the equivalent system can be solved by back substitution, giving the solution $(2, 1, -2)$, the same solution as was found in Example 1. ∎

Gauss-Jordan Elimination

One method that can be used to solve a system of equations is called the **Gauss-Jordan elimination method**. To use this method, we attempt to reduce the coefficient part of an augmented matrix to one that contains 1's on its diagonal and 0's everywhere else.

> The augmented matrix is said to be in **reduced row-echelon form** if it has 1's or 0's on the diagonal of its coefficient part and 0's everywhere else.

When the original augmented matrix represents a system with a unique solution, this method will reduce the coefficient matrix of the augmented matrix to an identity matrix. When this happens, we can easily find the solution to the system, as we will see below.

Rather than use back substitution to solve the system of Example 2, we can apply several additional row operations to reduce the matrix to reduced row-echelon form. That is, we can reduce the matrix

$$\begin{bmatrix} 1 & -1 & 2 & | & -3 \\ 0 & 1 & 3 & | & -5 \\ 0 & 0 & 1 & | & -2 \end{bmatrix}$$

to reduced row-echelon form with the following steps:

1. Add row 2 to row 1:

$$\begin{bmatrix} 1 & 0 & 5 & | & -8 \\ 0 & 1 & 3 & | & -5 \\ 0 & 0 & 1 & | & -2 \end{bmatrix}$$

2. Add (-5) times row 3 to row 1 and (-3) times row 3 to row 2:

$$\begin{bmatrix} 1 & 0 & 0 & | & 2 \\ 0 & 1 & 0 & | & 1 \\ 0 & 0 & 1 & | & -2 \end{bmatrix}$$

3. The coefficient matrix part of this reduced row-echelon form is the identity matrix, and the solutions to the system can be easily "read" from the reduced augmented matrix. The rows of this reduced row-echelon matrix translate into the equations

$$\begin{cases} x + 0y + 0z = 2 \\ 0x + y + 0z = 1 \\ 0x + 0y + z = -2 \end{cases} \quad \text{or} \quad \begin{cases} x = 2 \\ y = 1 \\ z = -2 \end{cases}$$

This solution agrees with the solution to the system found in Example 2.

We can use a **special notation** to denote the steps in Gauss-Jordan elimination in a compact form. We show below how the steps used to solve the system of Example 2 are denoted with this notation (with R_1, R_2, and R_3 representing rows 1, row 2, and row 3).

$$\begin{bmatrix} 2 & -3 & 1 & | & -1 \\ 1 & -1 & 2 & | & -3 \\ 3 & 1 & -1 & | & 9 \end{bmatrix} \xrightarrow{R_1 \leftrightarrow R_2} \begin{bmatrix} 1 & -1 & 2 & | & -3 \\ 2 & -3 & 1 & | & -1 \\ 3 & 1 & -1 & | & 9 \end{bmatrix}$$

$$\xrightarrow[\substack{(-2)R_1 + R_2 \rightarrow R_2 \\ (-3)R_1 + R_3 \rightarrow R_3}]{} \begin{bmatrix} 1 & -1 & 2 & | & -3 \\ 0 & -1 & -3 & | & 5 \\ 0 & 4 & -7 & | & 18 \end{bmatrix}$$

$$\xrightarrow[(-1)R_2 \rightarrow R_2]{} \begin{bmatrix} 1 & -1 & 2 & | & -3 \\ 0 & 1 & 3 & | & -5 \\ 0 & 4 & -7 & | & 18 \end{bmatrix} \xrightarrow[\substack{R_2 + R_1 \rightarrow R_1 \\ (-4)R_2 + R_3 \rightarrow R_3}]{} \begin{bmatrix} 1 & 0 & 5 & | & -8 \\ 0 & 1 & 3 & | & -5 \\ 0 & 0 & -19 & | & 38 \end{bmatrix}$$

$$\xrightarrow[(-1/19)R_3 \rightarrow R_3]{} \begin{bmatrix} 1 & 0 & 5 & | & -8 \\ 0 & 1 & 3 & | & -5 \\ 0 & 0 & 1 & | & -2 \end{bmatrix} \xrightarrow[\substack{(-5)R_3 + R_1 \rightarrow R_1 \\ (-3)R_3 + R_2 \rightarrow R_2}]{} \begin{bmatrix} 1 & 0 & 0 & | & 2 \\ 0 & 1 & 0 & | & 1 \\ 0 & 0 & 1 & | & -2 \end{bmatrix}$$

EXAMPLE 3 Manufacturing

A manufacturer of furniture has three models of chairs, Anderson, Blake, and Colonial. The numbers of hours required for framing, upholstery, and finishing for each type of chair are given in Table 5.2. The company has 1500 hours per week for framing, 2100 hours for upholstery, and 850 hours for finishing. How many of each type of chair can be produced under these conditions?

TABLE 5.2

	Anderson	Blake	Colonial
Framing	2	3	1
Upholstery	1	2	3
Finishing	1	2	1/2

Solution

If we represent the number of units of the Anderson model by x, the number of units of the Blake model by y, and the number of units of Colonial by z, then we have the following equations.

$$\begin{aligned} \text{Framing:} \quad & 2x + 3y + z = 1500 \\ \text{Upholstery:} \quad & x + 2y + 3z = 2100 \\ \text{Finishing:} \quad & x + 2y + \frac{1}{2}z = 850 \end{aligned}$$

We seek the values of x, y, and z that satisfy all three equations above. That is, we seek the solution to the system

$$\begin{cases} 2x + 3y + z = 1500 \\ x + 2y + 3z = 2100 \\ x + 2y + \frac{1}{2}z = 850 \end{cases}$$

We can represent this system with the augmented matrix

$$\begin{bmatrix} 2 & 3 & 1 & 1500 \\ 1 & 2 & 3 & 2100 \\ 1 & 2 & 1/2 & 850 \end{bmatrix}.$$

We solve the system by finding the reduced row-echelon form of the matrix. We first interchange row 1 and row 2.

$$\begin{bmatrix} 2 & 3 & 1 & 1500 \\ 1 & 2 & 3 & 2100 \\ 1 & 2 & 1/2 & 850 \end{bmatrix} \xrightarrow{R_1 \leftrightarrow R_2} \begin{bmatrix} 1 & 2 & 3 & 2100 \\ 2 & 3 & 1 & 1500 \\ 1 & 2 & 1/2 & 850 \end{bmatrix}$$

$$\xrightarrow[\begin{subarray}{l} (-2)R_1 + R_2 \to R_2 \\ (-1)R_1 + R_3 \to R_3 \end{subarray}]{} \begin{bmatrix} 1 & 2 & 3 & 2100 \\ 0 & -1 & -5 & -2700 \\ 0 & 0 & -5/2 & -1250 \end{bmatrix}$$

$$\xrightarrow{(-1)R_2 \rightarrow R_2} \begin{bmatrix} 1 & 2 & 3 & | & 2100 \\ 0 & 1 & 5 & | & 2700 \\ 0 & 0 & -5/2 & | & -1250 \end{bmatrix}$$

$$\xrightarrow{(-2)R_2 + R_1 \rightarrow R_1} \begin{bmatrix} 1 & 0 & -7 & | & -3300 \\ 0 & 1 & 5 & | & 2700 \\ 0 & 0 & -5/2 & | & -1250 \end{bmatrix}$$

$$\xrightarrow{(-2/5)R_3 \rightarrow R_3} \begin{bmatrix} 1 & 0 & -7 & | & -3300 \\ 0 & 1 & 5 & | & 2700 \\ 0 & 0 & 1 & | & 500 \end{bmatrix}$$

$$\begin{matrix} 7R_3 + R_1 \rightarrow R_1 \\ (-5)R_3 + R_2 \rightarrow R_2 \end{matrix} \xrightarrow{} \begin{bmatrix} 1 & 0 & 0 & | & 200 \\ 0 & 1 & 0 & | & 200 \\ 0 & 0 & 1 & | & 500 \end{bmatrix}$$

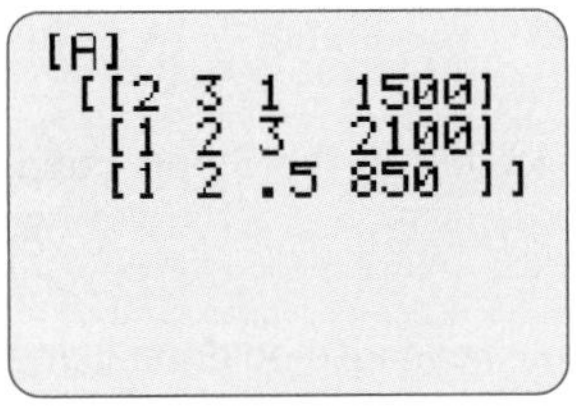

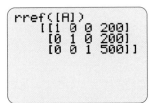

FIGURE 5.11

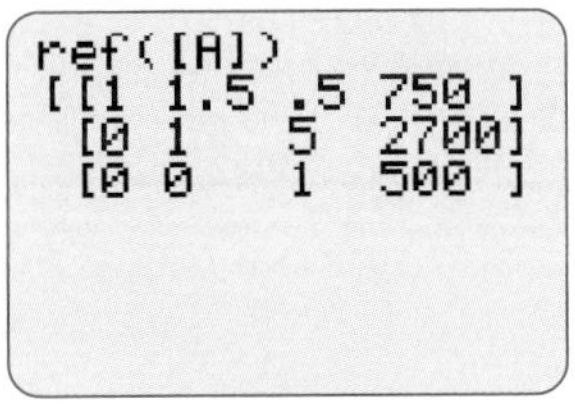

ref([A])
[[1 1.5 .5 750]
[0 1 5 2700]
[0 0 1 500]]

FIGURE 5.12

rref([A])
[[1 0 0 200]
[0 1 0 200]
[0 0 1 500]]

FIGURE 5.13

The coefficient matrix part of the reduced row-echelon form is the identity matrix, so the solutions to the systems can be easily "read" from the reduced matrix. The rows of the reduced matrix translate into the equations

$$x + 0y + 0z = 200 \Rightarrow x = 200$$
$$0x + y + 0z = 200 \Rightarrow y = 200.$$
$$0x + 0y + z = 500 \Rightarrow z = 500$$

Thus the manufacturer can produce 200 units of the Anderson model, 200 units of the Blake model, and 500 units of Colonial. ∎

Solution with Technology

Spreadsheets, computer programs, and graphing calculators are very useful in performing the matrix row operations used to reduce the augmented matrix and solve systems of linear equations. Some types of technology have a direct command that finds the equivalent row-echelon form of an augmented matrix. Figure 5.11 shows the augmented matrix for the system of equations in Example 3. We enter this matrix in a graphing utility as a 3 × 4 matrix, because the matrix has 3 rows and 4 columns.

The row-echelon form of this augmented matrix that results from the use of technology is shown in Figure 5.12. We can also use technology to find the **reduced row-echelon** form of an augmented matrix. Figure 5.13 shows the reduced row-echelon form found by using technology on the augmented matrix shown in Figure 5.11. This matrix is the same as the one found in Example 3.

EXAMPLE 4	Investment

The Trust Department of Century Bank divided a $150,000 investment among three mutual funds with different levels of risk and return. The Potus Fund returns 10% per year, the Stong Fund returns 8% per year, and the Franklin Fund returns 7%. If the annual return from the combined investments is $12,900 and if the investment in the Potus Fund has $20,000 less than the sum of the investments in the other two funds, how much is invested in each fund?

Solution

If we represent the amount invested in the Potus Fund by x, the amount invested in the Stong Fund by y, and the amount invested in the Franklin Fund by z, the equations that represent this situation are:

$$x + y + z = 150{,}000$$
$$0.10x + 0.08y + 0.07z = 12{,}900$$
$$x = y + z - 20{,}000$$

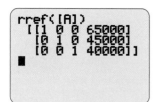

FIGURE 5.14

Writing these equations in a system and creating the augmented matrix gives

$$\begin{cases} x + \quad y + \quad z = \quad 150{,}000 \\ 0.10x + 0.08y + 0.07z = \quad 12{,}900 \\ x - \quad y - \quad z = -20{,}000 \end{cases} \qquad \begin{bmatrix} 1 & 1 & 1 & 150{,}000 \\ 0.10 & 0.08 & 0.07 & 12{,}900 \\ 1 & -1 & -1 & -20{,}000 \end{bmatrix}$$

Entering the matrix in technology and finding the reduced row-echelon form gives the matrix from which the solution to the system can be found. (See Figure 5.14.) This gives $x = 65{,}000$, $y = 45{,}000$, and $z = 40{,}000$.

Thus the investments are \$65,000 in the Potus Fund, \$45,000 in the Stong Fund, and \$40,000 in the Franklin Fund. ■

Solving systems of linear equations in more than three variables can also be done using Gauss-Jordan elimination. Because of the increased number of steps involved, technology is especially useful in solving larger systems.

EXAMPLE 5   Four Equations in Four Variables
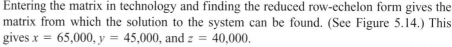

Solve the system

$$\begin{cases} x + \quad y + z + \quad w = \quad 3 \\ x - 2y + z - 4w = -5 \\ x \quad\quad - z + \quad w = \quad 0 \\ \quad\quad y + z + \quad w = \quad 2 \end{cases}$$

Solution

This system can be represented by the augmented matrix

$$\begin{bmatrix} 1 & 1 & 1 & 1 & 3 \\ 1 & -2 & 1 & -4 & -5 \\ 1 & 0 & -1 & 1 & 0 \\ 0 & 1 & 1 & 1 & 2 \end{bmatrix}$$

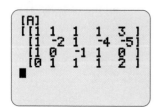

We can enter this augmented matrix into a graphing calculator and reduce the matrix to its reduced row-echelon form. (See Figure 5.15.) The solution to the system can be "read" from the reduced matrix:

$$x = 1, y = 11, z = -4, \text{ and } w = -5, \text{ or } (1, 11, -4, -5). \qquad ■$$

It should be noted that a system with fewer equations than variables cannot have a unique solution. We will consider systems with nonunique solutions in the next section.

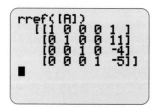

FIGURE 5.15

5.2 SKILLS CHECK

1. Write the augmented matrix associated with the system
$$\begin{cases} x + y - z = 4 \\ x - 2y - z = -2. \\ 2x + 2y + z = 11 \end{cases}$$

2. Write the augmented matrix associated with the system
$$\begin{cases} 5x - 3y + 2z = 12 \\ 3x + 6y - 9z = 4 \; . \\ 2x + 3y - 4z = 9 \end{cases}$$

3. The matrix $\begin{bmatrix} 1 & 0 & 0 & -1 \\ 0 & 1 & 0 & 4 \\ 0 & 0 & 1 & -2 \end{bmatrix}$ is an augmented matrix used in the solution of a system of linear equations in the variables x, y, and z. What is the solution to the system?

4. The matrix $\begin{bmatrix} 1 & 0 & 0 & 3 \\ 0 & 1 & 0 & 2 \\ 0 & 0 & 1 & 1 \end{bmatrix}$ is an augmented matrix used in the solution of a system of linear equations in the variables x, y, and z. What is the solution to the system?

5. The matrix $\begin{bmatrix} 1 & 1 & -1 & 4 \\ 1 & -2 & -1 & -2 \\ 2 & 2 & 1 & 11 \end{bmatrix}$ is the augmented matrix associated with a system of linear equations in x, y, and z. Find the solution to the system.

6. The matrix $\begin{bmatrix} 2 & -3 & 4 & 13 \\ 1 & -2 & 1 & 3 \\ 2 & -3 & 1 & 4 \end{bmatrix}$ is the augmented matrix associated with a system of linear equations in x, y, and z. Find the solution to the system.

Solve the systems in Exercises 7–14.

7. $\begin{cases} x + y - z = 0 \\ x - 2y - z = 6 \\ 2x + 2y + z = 3 \end{cases}$

8. $\begin{cases} x - 2y + z = -5 \\ 2x - y + 2z = 6 \\ 3x + 2y - z = 1 \end{cases}$

9. $\begin{cases} 3x - 2y + 5z = 15 \\ x - 2y - 2z = -1 \\ 2x - 2y = 0 \end{cases}$

10. $\begin{cases} x + 3y + 5z = 5 \\ 2x + 4y + 3z = 9 \\ 2x + 3y + z = 1 \end{cases}$

11. $\begin{cases} 2x + 3y + 4z = 5 \\ 6x + 7y + 8z = 9 \\ 2x + y + z = 1 \end{cases}$

12. $\begin{cases} 4x + 3y + 8z = 1 \\ 2x + 3y + 8z = 5 \\ 2x + 5y + 5z = 6 \end{cases}$

13. $\begin{cases} x - y + z - w = -2 \\ 2x + 4z + w = 5 \\ 2x - 3y + z = -5 \\ y + 2z + 20w = 4 \end{cases}$

14. $\begin{cases} x - 2y + z - 3w = 10 \\ 2x - 3y + 4z + w = 12 \\ 2x - 3y + z - 4w = 7 \\ x - y + z + w = 4 \end{cases}$

5.2 EXERCISES

1. *Rental Cars* A car rental agency rents compact, midsize, and luxury cars. Its goal is to purchase 60 cars with a total of $1,400,000 and to earn a daily rental of $2200 from all the cars. The compact cars cost $15,000 and earn $30 per day in rental, the midsize cars cost $25,000 and earn $40 per day, and the luxury cars cost $45,000 and earn $50 per day. To find the number of each type of car the agency should purchase, solve the system of equations

$$\begin{cases} x + y + z = 60 \\ 15{,}000x + 25{,}000y + 45{,}000z = 1{,}400{,}000 \\ 30x + 40y + 50z = 2200 \end{cases}$$

where x, y, and z are the numbers of compact cars, midsize cars, and luxury cars, respectively.

2. *Ticket Pricing* A theater owner wants to divide an 1800 seat theater into three sections, with tickets costing $20, $35, and $50, depending on the section. He wants to have twice as many $20 tickets as the sum of the other tickets, and he wants to earn $48,000 from a full house. To find how many seats he should have in each section, solve the system of equations

$$\begin{cases} x + y + z = 1800 \\ x = 2(y + z) \\ 20x + 35y + 50z = 48{,}000 \end{cases}$$

where x, y, and z are the numbers of $20 tickets, $35 tickets, and $50 tickets, respectively.

3. *Pricing* A concert promoter needs to make $120,000 from the sale of 2600 tickets. The promoter charges

$40 for some tickets and $60 for the others.
   a. If there are x of the $40 tickets and y of the $60 tickets, write an equation that states that the total number of the tickets sold is 2600.
   b. How much money is made from the sale of x tickets for $40 each?
   c. How much money is made from the sale of y tickets for $60 each?
   d. Write an equation that states that the total amount made from the sale is $120,000.
   e. Solve the equations simultaneously to find how many tickets of each type must be sold to yield the $120,000.

4. *Investment* A trust account manager has $500,000 to be invested in three different accounts. The accounts pay 8%, 10%, and 14%, respectively, and the goal is to earn $49,000 with the amount invested at 8% equal to the sum of the other two investments. To accomplish this, assume that x dollars are invested at 8%, y dollars are invested at 10%, and z dollars are invested at 14%.
   a. Write an equation that describes the sum of money in the three investments.
   b. Write an equation that describes the total amount of money earned by the three investments.
   c. Write an equation that describes the relationship among the three investments.
   d. Solve the system of equations to find how much should be invested in each account to satisfy the conditions.

5. *Investment* A man has $400,000 invested in three rental properties. One property earns 7.5% per year on the investment, the second earns 8%, and the third earns 9%. The total annual earnings from the three properties is $33,700 and the amount invested at 9% equals the sum of the first two investments. Let x equal the investment at 7.5%, y equal the investment at 8%, and z represent the investment at 9%.
   a. Write an equation that represents the sum of the three investments.
   b. Write an equation that states the sum of the returns from all three investments is $33,700.
   c. Write an equation that states that the amount invested at 9% equals the sum of the other two investments.
   d. Solve the system of equations to find how much is invested in each property.

6. *Loans* A bank loans $285,000 to a development company to purchase three business properties. If one of the properties costs $45,000 more than the other and the third costs twice the sum of these two properties:

   a. Write an equation that represents the total loaned as the sum of the cost of the three properties, if the costs are x, y, and z, respectively.
   b. Write an equation that states that one cost is $45,000 more than another.
   c. Write an equation that states that the third cost is equal to twice the sum of the other two used.
   d. Use matrices to solve the system of equations formed in parts (a) to (c) to find the cost of each property.

7. *Investment* A brokerage house offers three stock portfolios for its clients. Portfolio I consists of 10 blocks of common stocks, 2 municipal bonds, and 3 blocks of preferred stock. Portfolio II consists of 12 blocks of common stock, 8 municipal bonds, and 5 blocks of preferred stock. Portfolio III consists of 10 blocks of common stocks, 4 municipal bonds, and 8 blocks of preferred stock. If a client wants to combine these portfolios so that she has 290 blocks of common stock, 138 municipal bonds, and 161 blocks of preferred stock, how many units of each portfolio does she need? To answer this question, let x equal the number of units of Portfolio I, y equal the number of units of Portfolio II, and z equal the number of units of Portfolio III.
   a. Write an equation that describes the total number of blocks of common stock.
   b. Write an equation that describes the total number of municipal bonds.
   c. Write an equation that describes the total blocks of preferred stock.
   d. Solve the systems of equations to find how many of each portfolio she needs.

8. *Testing* A professor wants to create a test that has true-false questions, multiple-choice questions, and essay questions. She wants the test to have 35 questions, with twice as many multiple-choice questions as essay questions, and with twice as many true-false questions as multiple-choice questions.
   a. Write a system of equations to represent this problem.
   b. How many of each type question should the professor put on the test? Note that only positive integer answers are useful.

9. *Testing* A professor wants to create a test that has 15 true-false questions, 10 multiple-choice questions, and 5 essay questions. She wants the test to be worth 100 points, with each multiple-choice question worth twice as many points as a true-false question, and with each essay question equal to three times the number of points of a true-false question.

a. Write a system of equations to represent this problem, with x, y, and z equal to the number of points of a true-false, multiple-choice, and essay questions, respectively.

b. How many points should be assigned to each type of problem? Note that only positive integer answers are useful.

10. *Nutrition* The following table gives the calories, fat, and carbohydrates per ounce for three brands of cereal, and the total amount of calories, fat, and carbohydrates required for a special diet.

Cereal	Calories	Fat	Carbohydrates
All Bran	50	0	22.0
Sugar Frosted Flakes	108	0.1	25.5
Natural Mixed Grain	127	5.5	18.0
Total required	393	5.7	91.0

To find the number of ounces of each brand that should be combined to give the required totals of calories, fat, and carbohydrates, let x represent the number of ounces of All Bran, y represent the number of ounces of Sugar Frosted Flakes, and z represent the number of ounces of Natural Mixed Grain.

a. Use the information in the table to write a system with three equations.

b. Solve the system to find the amount of each brand of cereal that will give the necessary nutrition.

11. *Transportation* Ace Trucking Company has an order for three products, A, B, and C, for delivery. The table below gives the volume in cubic feet, the weight in pounds, and the value for insurance in dollars for a unit of each of the products. If the carrier can carry 8000 cubic feet and 12,400 lb, and is insured for $52,600, how many units of each product can be carried?

	Product A	Product B	Product C
Unit Volume (cubic feet)	24	20	40
Weight (pounds)	40	30	60
Value (dollars)	150	180	200

12. *Manufacturing* To expand its manufacturing capacity, Krug Industries borrowed $440,000, part at 6%, part at 8%, and part at 10%. The sum of the money borrowed at 6% and 8% was three times that borrowed at 10%. The loan was repaid in full at the end of 5 years and the annual interest paid was $34,400. How much was borrowed at each rate?

13. *Nutrition* A psychologist studying the effects of good nutrition on the behavior of rabbits feeds one group a combination of three foods, I, II, and III. Each of these foods contains three additives, A, B, and C, that are used in the study. The table below gives the percent of each additive that is present in each food. If the diet being used requires 3.74 g per day of A, 2.04 g of B, and 1.35 g of C, find the number of grams of each food that should be used each day.

	Food I	Food II	Food III
Additive A	12%	15%	28%
Additive B	8%	6%	16%
Additive C	15%	2%	6%

14. *Manufacturing* A manufacturer of swing sets has three models, Deluxe, Premium, and Ultimate, which must be painted, assembled, and packaged for shipping. The table below gives the number of hours required for each of these operations for each type of swing set. If the manufacturer has 55 hours available per day for painting, 75 hours for assembly, and 40 hours for packaging, how many of each type swing set can be produced each day?

	Deluxe	Premium	Ultimate
Painting	.8	1	1.4
Assembly	1	1.5	2
Packaging	.6	.75	1

15. *Investment* A company offers three mutual fund plans for its employees. Plan I consists of 4 blocks of common stocks and 2 municipal bonds. Plan II consists of 8 blocks of common stock, 4 municipal bonds, and 6 blocks of preferred stock. Plan III consists of 14 blocks of common stocks, 6 municipal bonds, and 6 blocks of preferred stock. If an employee wants to combine these plans so that she has 42 blocks of common stock, 20 municipal bonds, and 18 blocks of preferred stock, how many units of each plan does she need?

5.3 Systems of Linear Equations with Nonunique Solutions

Ace Trucking Company has an order for three products, A, B, and C, for delivery. The table below gives the volume in cubic feet, the weight in pounds, and the value for insurance in dollars for a unit of each of the products. If the carrier can carry 30,000 cubic feet and 62,000 lb, and is insured for $276,000, how many units of each product can be carried?

	Product A	Product B	Product C
Unit Volume (cubic feet)	25	22	30
Weight (lb)	25	38	70
Value (dollars)	150	180	300

If the number of units of products *A, B,* and *C* are *x, y,* and *z,* respectively, the system of linear equations that satisfies these conditions is

$$\begin{cases} 25x + 22y + 30z = 30{,}000 & \text{Volume} \\ 25x + 38y + 70z = 62{,}000 & \text{Weight} \\ 150x + 180y + 300z = 276{,}000 & \text{Value} \end{cases}$$

As we will see in Example 3, there is not a unique answer to this question. We saw in Section 5.1 that not all systems of two linear equations in two variables have unique solutions, and this is also true for three linear equations in three variables. In this section, we use matrices to solve systems with infinitely many solutions and see how to determine when the system has no solution.

Nonunique Solution

When a linear equation with three variables is graphed in a three-dimensional coordinate system, its graph is a plane, so a system of three equations in three variables is represented by three planes. Three different planes may intersect in a point, a line, or they may not have a common intersection. If we represent a system of three equations in three variables by a matrix and use matrix row operations to find the reduced row-echelon form, the coefficient matrix will reduce to a 3 × 3 identity matrix if the system has a unique solution. If the coefficient matrix does not reduce to an identity matrix, then there is no solution or there is an infinite number of solutions. The three possibilities follow.

A. If the coefficient matrix associated with a system reduces to the identity matrix, the system has a unique solution. The coefficient matrix reduced to the identity matrix in every problem in Section 5.2, and every system of equations had a unique solution.

When a system of three linear equations in three variables has a unique solution, their graphs will intersect in a point (as when two walls meet a ceiling), as shown in Figure 5.16.

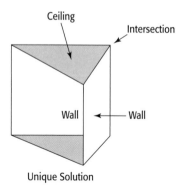

Unique Solution

FIGURE 5.16

B. If a row of the reduced row-echelon coefficient matrix associated with a system contains all 0's and the augment of that row contains a nonzero number, the system has no solution. Such a system is called an **inconsistent system**.

We will see in Example 5 that the matrix representing the system

$$\begin{cases} 5x + 10y + 12z = 6140 \\ 10x + 18y + 30z = 13{,}400 \\ 300x + 480y + 1080z = 214{,}800 \end{cases}$$

has reduced row-echelon form

$$\begin{bmatrix} 1 & 0 & 8.4 & 0 \\ 0 & 1 & -3 & 0 \\ 0 & 0 & 0 & 1 \end{bmatrix}$$

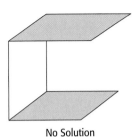

No Solution

FIGURE 5.17(a)

and that the system has no solution because the last row of this matrix represents $0x + 0y + 0z = 1$ or $0 = 1$, which is impossible.

When a system of three linear equations in three variables has no solution, their graphs will not have a common intersection, as shown in Figure 5.17(a).

C. If a row of the reduced 3×3 row-echelon coefficient matrix associated with a system contains all 0's and the augment of that row also contains 0, then there are infinitely many solutions. If any column of the reduced matrix does not contain a leading 1, that variable can assume any value, and the variables corresponding to columns containing leading 1's will have values dependent on that variable. (There may be more than one variable that can assume any value, with other variables depending on them.) Such a system is called a **dependent system**, and it will have infinitely many solutions.

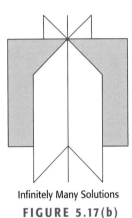

Infinitely Many Solutions

FIGURE 5.17(b)

When a system of three linear equations in three variables has infinitely many solutions, their graphs will intersect in a line, as in the paddle wheel shown in Figure 5.17(b).

Of course, any system with two equations in three variables cannot have a unique solution.

> A system with fewer equations than variables has either infinitely many solutions or no solutions.

Dependent Systems

The following example illustrates the solution of a dependent system.

EXAMPLE 1   A Dependent System

- -

Solve the system

$$\begin{cases} 5x + 10y + 12z = 6140 \\ 10x + 18y + 30z = 13{,}400 \\ 300x + 480y + 1080z = 435{,}600 \end{cases}$$

Solution

The augmented matrix that represents the system is

$$\begin{bmatrix} 5 & 10 & 12 & 6140 \\ 10 & 18 & 30 & 13{,}400 \\ 300 & 480 & 1080 & 435{,}600 \end{bmatrix}.$$

The procedure used in reducing this matrix is the same as that used in Section 5.2.

$$
\begin{bmatrix}
5 & 10 & 12 & 6140 \\
10 & 18 & 30 & 13{,}400 \\
300 & 480 & 1080 & 435{,}600
\end{bmatrix}
\xrightarrow{(1/5)R_1 \to R_1}
\begin{bmatrix}
1 & 2 & 2.4 & 1228 \\
10 & 18 & 30 & 13{,}400 \\
300 & 480 & 1080 & 435{,}600
\end{bmatrix}
$$

$$
\xrightarrow[\substack{-300R_1 + R_3 \to R_3}]{-10R_1 + R_2 \to R_2}
\begin{bmatrix}
1 & 2 & 2.4 & 1228 \\
0 & -2 & 6 & 1120 \\
0 & -120 & 360 & 67{,}200
\end{bmatrix}
$$

$$
\xrightarrow{(-1/2)R_2 \to R_2}
\begin{bmatrix}
1 & 2 & 2.4 & 1228 \\
0 & 1 & -3 & -560 \\
0 & -120 & 360 & 67{,}200
\end{bmatrix}
$$

$$
\xrightarrow[\substack{-2R_2 + R_1 \to R_1}]{120R_2 + R_3 \to R_3}
\begin{bmatrix}
1 & 0 & 8.4 & 2348 \\
0 & 1 & -3 & -560 \\
0 & 0 & 0 & 0
\end{bmatrix}
$$

This reduced matrix corresponds to the reduced system that is equivalent to the original system:

$$
\begin{cases}
x + 0y + 8.4z = 2348 \\
0x + y - 3z = -560 \\
0x + 0y + 0z = 0
\end{cases}
$$

Note that there is no leading 1 in the "z" column of the reduced matrix, and that z occurs in the first two equations. Thus we can solve for x in terms of z in the first equation and for y in terms of z in the second equation. These solutions are

$$
\begin{aligned}
x &= -8.4z + 2348 \\
y &= 3z - 560 \\
z &= z
\end{aligned}
\quad .
$$

This system is a dependent system; the values of x and y are dependent on z in the solution. This system has infinitely many solutions, written here in terms of z. Different values of z give different solutions to the system. Two sample solutions are $z = 0$, $y = -560$, $x = 2348$ and $z = 10$, $y = -530$, $x = 2264$. If $z = a$, the solution is $(-8.4a + 2348, 3a - 560, a)$. ∎

Notice that the row reduction technique is the same for systems that have nonunique solutions as they are for systems that have unique solutions. To solve the system, we attempt to reduce the coefficient matrix to the identity matrix. If we succeed, the system has a unique solution. If we are unable to reduce the coefficient matrix to the identity matrix, either there is no solution or the system is dependent and we can solve it in terms of one or more of the variables.

As with systems of linear equations having unique solutions, we can use technology to assist in the solution of systems having nonunique solutions. Figure 5.18(a) shows the original augmented matrix for the system of Example 1 and Figure 5.18(b) shows the reduced row-echelon form that is equivalent to the original matrix. This is the same as the reduced matrix found in Example 1 and yields the same solutions.

(a) (b)

FIGURE 5.18

| EXAMPLE 2 | Purchasing |

A young man wins $200,000 and (foolishly) decides to buy a new car for each of the seven days of the week. He wants to choose from cars that are priced at $40,000, $30,000, and $25,000. How many cars of each price can he buy with the $200,000?

Solution

If he buys x cars costing $40,000, y cars costing $30,000, and z cars costing $25,000, then

$$x + y + z = 7 \quad \text{and} \quad 40{,}000x + 30{,}000y + 25{,}000z = 200{,}000.$$

Writing these two equations as a system of equations gives

$$\begin{cases} x + y + z = 7 \\ 40{,}000x + 30{,}000y + 25{,}000z = 200{,}000 \end{cases}$$

and creating a corresponding augmented matrix gives

$$\begin{bmatrix} 1 & 1 & 1 & 7 \\ 40{,}000 & 30{,}000 & 25{,}000 & 200{,}000 \end{bmatrix}$$

Because there are three variables and only two equations, the solution cannot be unique. However, we solve the system by reducing the matrix in the same manner as above.

$$\begin{bmatrix} 1 & 1 & 1 & 7 \\ 40{,}000 & 30{,}000 & 25{,}000 & 200{,}000 \end{bmatrix}$$

$$\xrightarrow{-40{,}000R_1 + R_2 \to R_2} \begin{bmatrix} 1 & 1 & 1 & 7 \\ 0 & -10{,}000 & -15{,}000 & -80{,}000 \end{bmatrix}$$

$$\xrightarrow{(-1/10{,}000)R_2 \to R_2} \begin{bmatrix} 1 & 1 & 1 & 7 \\ 0 & 1 & 1.5 & 8 \end{bmatrix} \xrightarrow{-R_2 + R_1 \to R_1} \begin{bmatrix} 1 & 0 & -.5 & -1 \\ 0 & 1 & 1.5 & 8 \end{bmatrix}$$

In this reduced matrix, we see that z can be any value and that x and y depend on z.

$$x = -1 + 0.5z$$
$$y = 8 - 1.5z$$

In the context of this application, the solutions must be positive integers (none of x, y, or z can be negative), so z must be at least 2 (or x will be negative) and z cannot be more than 5 (or y will be negative). In addition, z must be an even number (to get integer solutions). Thus the only possible selections are

$$z = 2, x = 0, y = 5 \quad \text{or} \quad z = 4, x = 1, y = 2.$$

Thus the man can buy two $25,000 cars, zero $40,000 cars, and five $30,000 cars, or he can buy four $25,000 cars, one $40,000 car, and two $30,000 cars. So even though there are an infinite number of solutions to this system of equations, there are only two solutions in the context of this application.

As with other systems of linear equations, technology can be used to reduce the augmented matrix representing the system in this example. (See Figure 5.19).

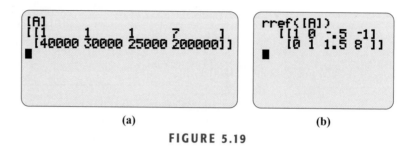

(a)   (b)

FIGURE 5.19

EXAMPLE 3   Transportation

Ace Trucking Company has an order for three products, A, B, and C, for delivery. The table below gives the volume in cubic feet, the weight in pounds, and the value for insurance in dollars for a unit of each of the products. If the company can carry 30,000 cubic feet and 62,000 lb, and is insured for $276,000, how many units of each product can be carried?

	Product A	Product B	Product C
Unit Volume (cubic feet)	25	22	30
Weight (lb)	25	38	70
Value (dollars)	150	180	300

Solution

If we represent the number of units of product A by x, the number of units of product B by y, and the number of units of product C by z, then the data in the table can be written as a system of equations.

$$\begin{cases} 25x + 22y + 30z = 30{,}000 & \text{Volume} \\ 25x + 38y + 70z = 62{,}000 & \text{Weight} \\ 150x + 180y + 300z = 276{,}000 & \text{Value} \end{cases}$$

The Gauss-Jordan elimination method gives

$$\begin{bmatrix} 25 & 22 & 30 & 30{,}000 \\ 25 & 38 & 70 & 62{,}000 \\ 150 & 180 & 300 & 276{,}000 \end{bmatrix} \xrightarrow{(1/25)R_1 \to R_1} \begin{bmatrix} 1 & 22/25 & 6/5 & 1200 \\ 25 & 38 & 70 & 62{,}000 \\ 150 & 180 & 300 & 276{,}000 \end{bmatrix}$$

$$\xrightarrow[\substack{(-150)R_1 + R_3 \to R_3}]{(-25)R_1 + R_2 \to R_2} \begin{bmatrix} 1 & 22/25 & 6/5 & 1200 \\ 0 & 16 & 40 & 32{,}000 \\ 0 & 48 & 120 & 96{,}000 \end{bmatrix}$$

$$(1/16)R_2 \rightarrow R_2 \begin{bmatrix} 1 & 22/25 & 6/5 & 1200 \\ 0 & 1 & 5/2 & 2000 \\ 0 & 48 & 120 & 96{,}000 \end{bmatrix}$$

$$\xrightarrow[\substack{(-48)R_2 + R_3 \rightarrow R_3}]{(-22/25)R_2 + R_1 \rightarrow R_1} \begin{bmatrix} 1 & 0 & -1 & -560 \\ 0 & 1 & 5/2 & 2000 \\ 0 & 0 & 0 & 0 \end{bmatrix}.$$

Reducing this augmented matrix by analytical methods (above) is quite time-consuming. Much time and effort can be saved if we use technology to obtain the reduced row-echelon form. (See Figure 5.20.)

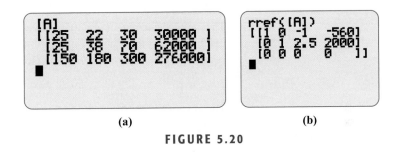

(a)   (b)

FIGURE 5.20

The solution to this system is $x = -560 + z$, $y = 2000 - 2.5z$, with the values of z limited so that all values are positive integers. Because x must be nonnegative, z must be at least 560, and because y must be a nonnegative integer, z must be an even number that is no more 800. This means that the size of the shipments of Products A and B is limited by the size of the shipments of Product C, as the limits on z put limits on x and y. We can write the solution as follows:

Product C: $560 \le z \le 800$, z even

Product B: $y = 2000 - 2.5z$,

Product A: $x = -560 + z$ ■

EXAMPLE 4   Traffic Flow

In an analysis of traffic, a certain city estimates the traffic flow as illustrated in Figure 5.21, where the arrows indicate the flow of the traffic. If x_1 represents the number of cars traveling from intersection A to intersection B, x_2 represents the number of cars traveling from intersection B to intersection C, and so on, we can formulate equations based on the principle that the number of vehicles entering the intersection equals the number leaving it.

a. Formulate an equation for the traffic at each intersection.

b. Solve the system of these four equations, to find how traffic between the other intersections is related to the traffic from intersection D to intersection A.

c. Discuss how the solution can be used to save money while measuring traffic flow during any time period.

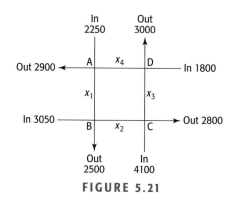

FIGURE 5.21

Solution

a. The arrows indicate that $3050 + x_1$ cars are entering intersection B, and that $2500 + x_2$ cars are leaving intersection B. Thus the equation that describes the number of cars entering and leaving intersection B is

$$3050 + x_1 = 2500 + x_2.$$

The equations that describe the number of cars entering and leaving all of the intersections follow.

$$
\begin{array}{ll}
\text{B:} & 3050 + x_1 = 2500 + x_2 \\
\text{C:} & 4100 + x_2 = x_3 + 2800 \\
\text{D:} & x_3 + 1800 = 3000 + x_4 \\
\text{A:} & x_4 + 2250 = x_1 + 2900
\end{array}
$$

b. Rewriting the equations with the variables on one side gives the following system of equations.

$$
\begin{cases}
x_1 - x_2 = -550 \\
x_2 - x_3 = -1300 \\
x_3 - x_4 = 1200 \\
-x_1 + x_4 = 650
\end{cases}
$$

Writing the augmented matrix and solving gives

$$
\left[\begin{array}{cccc|c}
1 & -1 & 0 & 0 & -550 \\
0 & 1 & -1 & 0 & -1300 \\
0 & 0 & 1 & -1 & 1200 \\
-1 & 0 & 0 & 1 & 650
\end{array}\right]
\xrightarrow{R_1 + R_4 \to R_4}
\left[\begin{array}{cccc|c}
1 & -1 & 0 & 0 & -550 \\
0 & 1 & -1 & 0 & -1300 \\
0 & 0 & 1 & -1 & 1200 \\
0 & -1 & 0 & 1 & 100
\end{array}\right]
$$

$$
\xrightarrow[\begin{array}{c} R_2 + R_1 \to R_1 \\ R_2 + R_4 \to R_4 \end{array}]{}
\left[\begin{array}{cccc|c}
1 & 0 & -1 & 0 & -1850 \\
0 & 1 & -1 & 0 & -1300 \\
0 & 0 & 1 & -1 & 1200 \\
0 & 0 & -1 & 1 & -1200
\end{array}\right]
$$

$$
\xrightarrow[\begin{array}{c} R_3 + R_1 \to R_1 \\ R_3 + R_2 \to R_2 \\ R_3 + R_4 \to R_4 \end{array}]{}
\left[\begin{array}{cccc|c}
1 & 0 & 0 & -1 & -650 \\
0 & 1 & 0 & -1 & -100 \\
0 & 0 & 1 & -1 & 1200 \\
0 & 0 & 0 & 0 & 0
\end{array}\right]
$$

Reducing this matrix using technology gives the same reduced row-echelon matrix. (See Figure 5.22.)

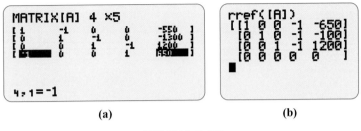

(a)　　　　　　　　　　　　　　　　(b)

FIGURE 5.22

The solution to the system shows the number of vehicles between the intersections as a function of the number traveling from D to A, x_4.

$$x_1 = x_4 - 650$$

$$x_2 = x_4 - 100$$

$$x_3 = x_4 + 1200$$

The variable x_4 must be limited so that all numbers are nonnegative. Thus $x_4 \geq 650$.

c. Using this solution permits the city to determine the number of vehicles traveling between each of the intersections by measuring only the number traveling from intersection D to A (x_4). ■

Inconsistent Systems

The following example shows how we can use matrices to determine that a system has no solution.

EXAMPLE 5　An Inconsistent System

Solve the following system of equations if a solution exists.

$$\begin{cases} 5x + 10y + 12 = 6140 \\ 10x + 18y + 30z = 13{,}400 \\ 300x + 480y + 1080z = 214{,}800 \end{cases}$$

Solution

The augmented matrix for this system is $\begin{bmatrix} 5 & 10 & 12 & 6140 \\ 10 & 18 & 30 & 13{,}400 \\ 300 & 480 & 1080 & 214{,}800 \end{bmatrix}$.

We reduce this matrix as follows.

$$\begin{bmatrix} 5 & 10 & 12 & 6140 \\ 10 & 18 & 30 & 13{,}400 \\ 300 & 480 & 1080 & 214{,}800 \end{bmatrix} \xrightarrow{(1/5)R_1 \rightarrow R_1} \begin{bmatrix} 1 & 2 & 2.4 & 1228 \\ 10 & 18 & 30 & 13{,}400 \\ 300 & 480 & 1080 & 214{,}800 \end{bmatrix}$$

$$
\begin{array}{c}
-10R_1 + R_2 \rightarrow R_2 \\
-300R_1 + R_3 \rightarrow R_3 \\
\xrightarrow{\hspace{2cm}}
\end{array}
\left[
\begin{array}{ccc|c}
1 & 2 & 2.4 & 1128 \\
0 & -2 & 6 & 1120 \\
0 & -120 & 360 & -153{,}600
\end{array}
\right]
$$

$$
\begin{array}{c}
(-1/2)R_2 \rightarrow R_2 \\
\xrightarrow{\hspace{2cm}}
\end{array}
\left[
\begin{array}{ccc|c}
1 & 2 & 2.4 & 1228 \\
0 & 1 & -3 & -560 \\
0 & -120 & 360 & -153{,}600
\end{array}
\right]
$$

$$
\begin{array}{c}
120R_2 + R_3 \rightarrow R_3 \\
-2R_2 + R_1 \rightarrow R_1 \\
\xrightarrow{\hspace{2cm}}
\end{array}
\left[
\begin{array}{ccc|c}
1 & 0 & 8.4 & 2348 \\
0 & 1 & -3 & -560 \\
0 & 0 & 0 & -220{,}800
\end{array}
\right]
$$

This matrix represents the reduced system that is equivalent to the original system.

$$
\begin{cases}
x + 0y + 8.4z = 2348 \\
0x + y - 3z = -560 \\
0x + 0y + 0z = -220{,}800
\end{cases}
$$

The third equation is $0 = -220{,}800$, which is impossible. Thus the system has no solution. ∎

If we use technology to attempt to find the solution of a system that has no solution, the resulting reduced row-echelon matrix will have a row of 0's in the coefficient matrix with a nonzero entry in the augment of that row. This is illustrated in Figure 5.23, which shows the augmented matrix and resulting reduced row-echelon form of this matrix from Example 5 above.

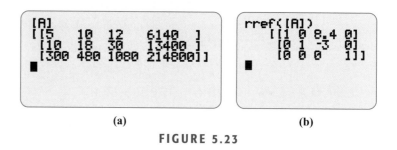

(a)　　　　　　　　　　　　(b)

FIGURE 5.23

Finally, we consider systems with more equations than variables. Such systems of equations may have zero, one, or many solutions, and they are solved in the same manner as other systems.

1. If the reduced augmented matrix contains a row of 0's in the coefficient matrix with a nonzero number in the augment, the system has no solution.

2. If the coefficient matrix in the reduced augmented matrix contains an identity matrix and all remaining rows of the reduced augmented matrix contain all 0's, there is a unique solution to the system.

3. Otherwise, the system has many solutions.

5.3 SKILLS CHECK

1. The reduced matrix associated with a system of linear equations is $\begin{bmatrix} 1 & 0 & 4 & | & 3 \\ 0 & 1 & 2 & | & 2 \\ 0 & 0 & 0 & | & 1 \end{bmatrix}$.

   Does the system have a unique solution, no solution, or many solutions?

2. The reduced matrix associated with a system of linear equations is $\begin{bmatrix} 1 & 0 & 2 & | & 1 \\ 0 & 1 & 3 & | & 5 \\ 0 & 0 & 0 & | & 0 \end{bmatrix}$.

   Does the system have a unique solution, no solution, or many solutions?

3. A row-echelon form of a matrix associated with a system of linear equations is $\begin{bmatrix} 1 & 0 & 2 & | & 1 \\ 0 & 1 & 3 & | & 5 \\ 0 & 0 & 1 & | & 0 \end{bmatrix}$.

   Does the system have a unique solution, no solution, or many solutions?

4. The reduced matrix associated with a system of linear equations is $\begin{bmatrix} 1 & 0 & 3 & | & 2 \\ 0 & 1 & -5 & | & 5 \\ 0 & 0 & 0 & | & 0 \end{bmatrix}$.

   Find the solution of the system, if the columns represent the coefficients of x, y, and z, respectively.

5. The reduced matrix associated with a system of linear equations is $\begin{bmatrix} 1 & 0 & -1 & | & 3 \\ 0 & 1 & 2 & | & -2 \\ 0 & 0 & 0 & | & 0 \end{bmatrix}$.

Find the solution of the system, if the columns represent the coefficients of x, y, and z, respectively.

In Exercises 6 and 7, a system of linear equations and a reduced matrix for the system are given. Use the reduced matrix to find the general solution, if one exists.

6. $\begin{cases} x + 2y + 3z = 3 \\ 2x + 3y + z = -1 \\ 3x + 5y + 4z = 2 \end{cases}$ $\quad \begin{bmatrix} 1 & 0 & -7 & | & -11 \\ 0 & 1 & 5 & | & 7 \\ 0 & 0 & 0 & | & 0 \end{bmatrix}$

7. $\begin{cases} x + y + 7z = 11 \\ 4x - 2y + 10z = 14 \\ 3x - 3y + 3z = 3 \end{cases}$ $\quad \begin{bmatrix} 1 & 0 & 4 & | & 6 \\ 0 & 1 & 3 & | & 5 \\ 0 & 0 & 0 & | & 0 \end{bmatrix}$

In Exercises 8–16, find the solutions, if any exist, to the systems.

8. $\begin{cases} 5x + 7y + 10z = 6 \\ 2x + 5y + 6z = 1 \\ 3x + 2y + 4z = 6 \end{cases}$ 9. $\begin{cases} -2x + 3y + 2z = 13 \\ -2x - 2y + 3z = 0 \\ 4x + y + 4z = 11 \end{cases}$

10. $\begin{cases} 2x + 3y + 4z = 5 \\ x + y + z = 1 \\ 6x + 7y + 8z = 9 \end{cases}$ 11. $\begin{cases} 2x + 5y + 6z = 6 \\ 3x - 2y + 2z = 4 \\ 5x + 3y + 8z = 10 \end{cases}$

12. $\begin{cases} -x + 5y - 3z = 10 \\ 3x + 7y + 2z = 5 \\ 4x + 12y - z = 15 \end{cases}$

13. $\begin{cases} -x - 5y + 3z = -2 \\ 3x + 7y + 2z = 5 \\ 4x + 12y - z = 7 \end{cases}$ 14. $\begin{cases} x - 3y + 2z = 12 \\ 2x - 6y + z = 7 \end{cases}$

15. $\begin{cases} 2x - 3y + 2z = 5 \\ 4x + y - 3z = 6 \end{cases}$ 16. $\begin{cases} 3x + 2y - z = 4 \\ 2x - 3y + z = 3 \end{cases}$

5.3 EXERCISES

1. *Investment* A trust account manager has $500,000 to be invested in three different accounts. The accounts pay 8%, 10%, and 14%, and the goal is to earn $49,000. To solve this problem, assume that x dollars are invested at 8%, y dollars are invested at 10%, and z dollars are invested at 14%. Then x, y, and z must satisfy the equations $x + y + z = 500,000$ and $0.08x + 0.10y + 0.14z = 49,000$.

a. Explain the meaning of each of these equations.
b. Use matrices to solve the systems containing these two equations. Find x and y in terms of z.
c. What limits must there be on z so that all investment values are nonnegative?

2. *Loans* A bank gives three loans totaling $200,000 to a development company for the purchase of three business properties. The largest loan is $45,000 more

than the sum of the other two. Represent the amount of money in each loan as x, y, and z, respectively.

a. Write a system of two equations in three variables to represent the problem.

b. Does the system have a unique solution?

c. Use matrices to solve the system. Find the amount of the largest loan.

d. What is the relationship between the remaining two loans?

3. *Investment* A trust account manager has $400,000 to be invested in three different accounts. The accounts pay 8%, 10%, and 12%, and the goal is to earn $42,400 with a minimum risk. To solve this problem, assume that x dollars are invested at 8%, y dollars are invested at 10%, and z dollars are invested at 12%.

a. Write a system of two equations in three variables to represent the problem.

b. How much can be invested in each account with the largest possible amount invested at 8%?

4. *Investment* A man has $235,000 invested in three rental properties. One property earns 7.5% per year on the investment, a second earns 10%, and the third earns 8%. The annual earnings from the properties total $18,000.

a. Write a system of two equations to represent the problem with x, y, and z representing 7.5%, 10%, and 8%, respectively.

b. Solve this system.

c. If $60,000 is invested at 8%, how much is invested in each of the other properties?

5. *Investment* A brokerage house offers three stock portfolios for its clients. Portfolio I consists of 10 blocks of common stock, 2 municipal bonds, and 3 blocks of preferred stock. Portfolio II consists of 12 blocks of common stock, 8 municipal bonds, and 5 blocks of preferred stock. Portfolio III consists of 10 blocks of common stocks, 6 municipal bonds, and 4 blocks of preferred stock. A client wants to combine these portfolios so that she has 180 blocks of common stock, 140 municipal bonds, and 110 blocks of preferred stock. Can she do this? To answer this question, let x equal the number of units of Portfolio I, y equal the number of units of Portfolio II, and z equal the number of units of Portfolio III, so that the equation $10x + 12y + 10z = 180$ represents the total number of blocks of common stock.

a. Write the remaining two equations to create a system of three equations.

b. Solve the system of equations, if possible.

6. *Investment* A company offers three mutual fund plans for its employees. Plan I consists of 14 blocks of common stock, 4 municipal bonds, and 6 blocks of preferred stock. Plan II consists of 4 blocks of common stock and 2 municipal bonds. Plan III consists of 18 blocks of common stock, 6 municipal bonds, and 6 blocks of preferred stock. If an employee wants to combine these plans so that she has 58 blocks of common stock, 20 municipal bonds, and 18 blocks of preferred stock, how many units of each plan does she need?

7. *Purchasing* A young man wins $100,000 and decides to buy four new cars. He wants to choose from cars that are priced at $40,000, $30,000, and $20,000 and spend all of the money.

a. Write a system of two equations in three variables to represent the problem.

b. Can this system have a unique solution?

c. Solve the system.

d. Use the context of the problem to find how many cars of each price he can buy with the $100,000.

8. *Manufacturing* A company manufactures three types of air conditioners, the Acclaim, the Bestfrig, and the Cool King. The hours needed for the assembly, testing, and packing for each product are given in the table below. The daily number of hours that are available are 300 for assembly, 120 for testing, and 210 for packing. Can all of the available hours be used to produce these three types of air conditioners, and if so, how many of each type can be produced?

	Assembly	Testing	Packing
Acclaim	5	2	1.4
Bestfrig	4	1.4	1.2
Cool King	4.5	1.7	1.3

9. *Transportation* The Hohman Trucking Company has an order for three products, A, B, and C, for delivery. The following table gives the volume in cubic feet, the weight in pounds, and the value for insurance in dollars for a unit of each of the products. If one of the company's trucks can carry 4000 cubic feet and 6000 lb, and is insured to carry $24,450, how many units of each product can be carried on the truck?

	Product A	Product B	Product C
Unit volume (cubic feet)	24	20	50
Weight (pounds)	36	30	75
Value (dollars)	150	180	120

10. *Nutrition* A psychologist studying the effects of good nutrition on the behavior of rabbits feeds one group a combination of three foods, I, II, and III. Each of these foods contains three additives, A, B, and C, that are used in the study. The table below gives the percent of each additive that is present in each food. If the diet being used requires 6.88 g per day of A, 6.72 g of B, and 6.8 g of C, find the number of grams of each food that should be used each day.

	Food I	Food II	Food III
Additive A	12%	14%	8%
Additive B	8%	6%	16%
Additive C	10%	10%	12%

11. *Traffic Flow* In the analysis of traffic, a retirement community estimates the traffic flow on their "town square" at 6 p.m. to be as illustrated in the figure. If x_1 illustrates the number of cars moving from intersection A to intersection B, x_2 represents the number of cars traveling from intersection B to intersection C, and so on, we can formulate equations based on the principle that the number of vehicles entering the intersection equals the number leaving it. For example, the equation that represents the traffic through A is $x_4 + 470 = x_1 + 340$.

a. Formulate an equation for the traffic at each of the four intersections.

b. Solve the system of these four equations, to find how traffic between the other intersections is related to the traffic from intersection D to intersection A.

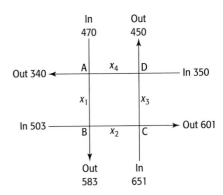

12. *Irrigation* An irrigation system allows water to flow in the pattern shown in the figure below. Water flows into the system at A and exits at B, C, and D with amounts shown. If x_1 represents the number of gallons of water moving from A to B, x_2 represents the number of gallons moving from A to C, and so on, we can formulate equations using the fact that at each point the amount of water entering the system equals the amount exiting. For example, the equation that represents the water flow through C is $x_2 = x_3 + 200,000$.

a. Formulate an equation for the water flow at each of the other three points.

b. Solve the system of these four equations.

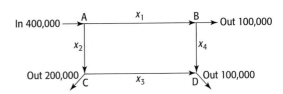

5.4   Matrices; Basic Operations

Table 5.3 gives the years of life expected at birth for male and female blacks and whites born in the United States in the years 1920, 1940, 1960, 1980, and 1998. To determine what this data tells us about the relationship among race, sex, and life expectancy, we can make a matrix W containing the information for whites and a matrix B for blacks, and use these matrices to find additional information. For example, we can find a matrix D that shows how many more years whites in each category are expected to live than blacks.

TABLE 5.3

YEARS	WHITES		BLACKS	
	Males	**Females**	**Males**	**Females**
1920	54.4	55.6	45.5	54.9
1940	62.1	66.6	51.1	45.2
1960	67.4	74.1	61.1	67.4
1980	70.7	78.1	63.8	72.5
1998	74.6	79.9	67.8	75.0

Source: National Center for Health Statistics

We used matrices to solve systems of equations in the last two sections. In this section, we consider the operations of addition, subtraction, and multiplication of matrices, and use these operations in applications like the one above.

Addition and Subtraction of Matrices

If two matrices have the same numbers of rows and columns (the same dimensions) we can add them by adding the corresponding entries of the two matrices.

MATRIX ADDITION

The sum of two matrices with the same dimensions is the matrix that is formed by adding the corresponding entries of the two matrices. Addition is not defined if the matrices do not have the same number of rows and the same number of columns.

EXAMPLE 1   Matrix Addition

Find the following sums of matrices.

a. $\begin{bmatrix} a & b \\ c & d \end{bmatrix} + \begin{bmatrix} w & y \\ x & z \end{bmatrix}$

b. $\begin{bmatrix} 1 & 3 & -8 & 0 \\ 3 & 6 & 1 & -3 \\ -4 & 5 & 3 & 2 \end{bmatrix} + \begin{bmatrix} -1 & 4 & 2 & 4 \\ 5 & -2 & 4 & 2 \\ -5 & 1 & 4 & -1 \end{bmatrix}$

c. $A + B$ if $A = \begin{bmatrix} 2 & -3 \\ -1 & 4 \end{bmatrix}$ and $B = \begin{bmatrix} -2 & 3 \\ 1 & -4 \end{bmatrix}$

Solution

a. $\begin{bmatrix} a & b \\ c & d \end{bmatrix} + \begin{bmatrix} w & y \\ x & z \end{bmatrix} = \begin{bmatrix} a+w & b+y \\ c+x & d+z \end{bmatrix}$

b. $\begin{bmatrix} 1 & 3 & -8 & 0 \\ 3 & 6 & 1 & -3 \\ -4 & 5 & 3 & 2 \end{bmatrix} + \begin{bmatrix} -1 & 4 & 2 & 4 \\ 5 & -2 & 4 & 2 \\ -5 & 1 & 4 & -1 \end{bmatrix}$

$= \begin{bmatrix} 1+(-1) & 3+4 & -8+2 & 0+4 \\ 3+5 & 6+(-2) & 1+4 & -3+2 \\ -4+(-5) & 5+1 & 3+4 & 2+(-1) \end{bmatrix} = \begin{bmatrix} 0 & 7 & -6 & 4 \\ 8 & 4 & 5 & -1 \\ -9 & 6 & 7 & 1 \end{bmatrix}$

c. $A + B = \begin{bmatrix} 2 & -3 \\ -1 & 4 \end{bmatrix} + \begin{bmatrix} -2 & 3 \\ 1 & -4 \end{bmatrix} = \begin{bmatrix} 0 & 0 \\ 0 & 0 \end{bmatrix}$ ∎

The matrix that is the sum in Example 1, part (c) is called a **zero matrix** because each of its elements is 0. Matrix B in Example 1, part (c) is called the **negative of matrix A**, denoted $-A$, because the sum of matrices A and B is a zero matrix. Similarly, matrix A is the negative of matrix B, and can be denoted by $-B$.

MATRIX SUBTRACTION

If matrices M and N are the same size, the difference $M - N$ is found by subtracting the elements of N from the corresponding elements of M. This difference can also be defined as

$$M - N = M + (-N).$$

EXAMPLE 2 Matrix Subtraction

Complete the following matrix operations.

a. $\begin{bmatrix} 2 & -1 & 3 \\ 5 & -4 & 2 \\ 1 & 4 & -2 \end{bmatrix} - \begin{bmatrix} 1 & 2 & 4 \\ 5 & -2 & 3 \\ 6 & 2 & -3 \end{bmatrix}$

b. $\begin{bmatrix} 1 & 4 \\ 3 & 6 \\ -2 & 1 \end{bmatrix} - \begin{bmatrix} -5 & 2 \\ 3 & 5 \\ 12 & 3 \end{bmatrix} + \begin{bmatrix} 2 & 3 \\ -4 & -8 \\ 2 & 0 \end{bmatrix}$

Solution

a. $\begin{bmatrix} 2 & -1 & 3 \\ 5 & -4 & 2 \\ 1 & 4 & -2 \end{bmatrix} - \begin{bmatrix} 1 & 2 & 4 \\ 5 & -2 & 3 \\ 6 & 2 & -3 \end{bmatrix}$

$= \begin{bmatrix} 2-1 & -1-2 & 3-4 \\ 5-5 & -4-(-2) & 2-3 \\ 1-6 & 4-2 & -2-(-3) \end{bmatrix} = \begin{bmatrix} 1 & -3 & -1 \\ 0 & -2 & -1 \\ -5 & 2 & 1 \end{bmatrix}$

b. $\begin{bmatrix} 1 & 4 \\ 3 & 6 \\ -2 & 1 \end{bmatrix} - \begin{bmatrix} -5 & 2 \\ 3 & 5 \\ 12 & 3 \end{bmatrix} + \begin{bmatrix} 2 & 3 \\ -4 & -8 \\ 2 & 0 \end{bmatrix}$

$= \begin{bmatrix} 6 & 2 \\ 0 & 1 \\ -14 & -2 \end{bmatrix} + \begin{bmatrix} 2 & 3 \\ -4 & -8 \\ 2 & 0 \end{bmatrix} = \begin{bmatrix} 8 & 5 \\ -4 & -7 \\ -12 & -2 \end{bmatrix}$ ■

We can use technology to perform the operations of addition and subtraction of matrices, such as those in Examples 1 and 2. Figure 5.24 shows the computations for Example 2, part (b).

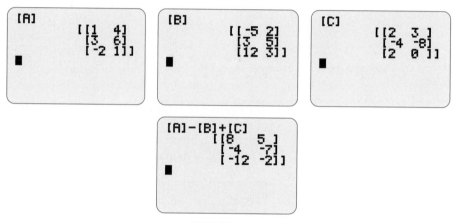

FIGURE 5.24

EXAMPLE 3 Life Expectancy

Table 5.3 gives the years of life expected at birth for male and female blacks and whites born in the United States in the years 1920, 1940, 1960, 1980, and 1998.

TABLE 5.3

YEARS	WHITES		BLACKS	
	Males	Females	Males	Females
1920	54.4	55.6	45.5	54.9
1940	62.1	66.6	51.1	45.2
1960	67.4	74.1	61.1	67.4
1980	70.7	78.1	63.8	72.5
1998	74.6	79.9	67.8	75.0

Source: National Center for Health Statistics

a. Make a matrix W containing the expected life data for whites and a matrix B for blacks.

b. Use these matrices to find matrix $D = W - B$ that represents the difference between the white and black life expectancy.

c. What does this tell us about race and life expectancy?

Solution

a. $W = \begin{bmatrix} 54.4 & 55.6 \\ 62.1 & 66.6 \\ 67.4 & 74.1 \\ 70.7 & 78.1 \\ 74.6 & 79.9 \end{bmatrix}$ $\qquad B = \begin{bmatrix} 45.5 & 54.9 \\ 51.1 & 45.2 \\ 61.1 & 67.4 \\ 63.8 & 72.5 \\ 67.8 & 75.0 \end{bmatrix}$

b. $D = W - B = \begin{bmatrix} 8.9 & 0.7 \\ 11.0 & 21.4 \\ 6.3 & 6.7 \\ 6.9 & 5.6 \\ 6.8 & 4.9 \end{bmatrix}$

c. Because each element in the difference matrix D is positive, we conclude that the life expectancy for whites is longer than that for blacks for all birth years and for both sexes. ∎

We can use electronic spreadsheets as well as calculators to perform operations with matrices. Figure 5.25 shows spreadsheets of matrices W, B, and D from Example 3.

	A	B	C
1	W	54.4	55.6
2		62.1	66.6
3		67.4	74.1
4		70.7	78.1
5		74.6	79.9
6			
7	B	45.5	54.9
8		51.1	45.2
9		61.1	67.4
10		63.8	72.5
11		67.8	75.0
12			
13	D = W − B	8.9	0.7
14		11	21.4
15		6.3	6.7
16		6.9	5.6
17		6.8	4.9

FIGURE 5.25

Multiplication of a Matrix by a Number

As in operations with real numbers, we can multiply a matrix by a positive integer to find the sum of repeated additions. For example, if

$$A = \begin{bmatrix} 1 & -2 & 4 \\ 6 & 3 & -3 \end{bmatrix}$$

then we can find $2A$ in two ways:

$$2A = A + A = \begin{bmatrix} 1 & -2 & 4 \\ 6 & 3 & -3 \end{bmatrix} + \begin{bmatrix} 1 & -2 & 4 \\ 6 & 3 & -3 \end{bmatrix} = \begin{bmatrix} 2 & -4 & 8 \\ 12 & 6 & -6 \end{bmatrix}$$

and

$$2A = \begin{bmatrix} 2 \cdot 1 & 2(-2) & 2 \cdot 4 \\ 2 \cdot 6 & 2 \cdot 3 & 2(-3) \end{bmatrix} = \begin{bmatrix} 2 & -4 & 8 \\ 12 & 6 & -6 \end{bmatrix}$$

In general, we can define multiplication of a matrix by a real number as follows.

PRODUCT OF A NUMBER AND A MATRIX

Multiplying a matrix A by a real number c results in a matrix in which each entry of matrix A is multiplied by the number c.

For example, if $A = \begin{bmatrix} a & b & c & d \\ e & f & g & h \end{bmatrix}$, then $nA = \begin{bmatrix} na & nb & nc & nd \\ ne & nf & ng & nh \end{bmatrix}$.

Note that $-A = (-1)A = \begin{bmatrix} -a & -b & -c & -d \\ -e & -f & -g & -h \end{bmatrix}$.

EXAMPLE 4 Price Increases

Table 5.4 contains the purchase prices and delivery costs (per unit) for plywood, siding, and 2×4 lumber. If the supplier announces a 5% increase in all of these prices and delivery costs, find the matrix that gives the new prices and costs.

TABLE 5.4

	Plywood	Siding	2×4's
Purchase Price (dollars)	32	23	2.60
Delivery Costs (dollars)	3	1	.60

Solution

The matrix that represents the original prices and costs is

$$\begin{bmatrix} 32 & 23 & 2.60 \\ 3 & 1 & .60 \end{bmatrix}$$

To find the prices and costs after a 5% increase, we multiply the matrix by 1.05 (100% of the old prices plus the 5% increase).

$$1.05 \begin{bmatrix} 32 & 23 & 2.60 \\ 3 & 1 & .60 \end{bmatrix} = \begin{bmatrix} 33.60 & 24.15 & 2.73 \\ 3.15 & 1.05 & .63 \end{bmatrix}.$$ ∎

Matrix Multiplication

Suppose that the company Circuitown has made a special purchase for one of its stores, consisting of 22 televisions, 15 washers, and 12 dryers. If the value of each television is $550, each washer is $435, and each dryer is $325, then the value of this purchase is

$$550 \cdot 22 + 435 \cdot 15 + 325 \cdot 12 = 22,525 \text{ dollars.}$$

If we write the value of each item in a 1×3 *row matrix*

$$A = [550 \quad 435 \quad 325]$$

and the number of each of the items in the special purchase in a 3×1 *column matrix*

$$B = \begin{bmatrix} 22 \\ 15 \\ 12 \end{bmatrix}$$

then the value of the special purchase can be represented by the **matrix product**

$$AB = [550 \quad 435 \quad 325] \begin{bmatrix} 22 \\ 15 \\ 12 \end{bmatrix}$$

$$= [550 \cdot 22 + 435 \cdot 15 + 325 \cdot 12] = [22,525]$$

In general, we have the following.

PRODUCT OF A ROW MATRIX AND A COLUMN MATRIX

The product of a $1 \times n$ row matrix and an $n \times 1$ column matrix is a 1×1 matrix given by

$$[a_1 \quad a_2 \ldots a_n] \begin{bmatrix} b_1 \\ b_2 \\ \vdots \\ b_n \end{bmatrix} = [a_1 b_1 + a_2 b_2 + \ldots + a_n b_n]$$

We can expand the multiplication to larger matrices. For example, suppose that Circuitown has a second store and purchases 28 televisions, 21 washers, and 26 dryers for it. Rather than writing two matrices to represent the two stores, we can use a two-column matrix C to represent the purchases for the two stores.

$$C = \begin{matrix} & \text{Store I} & \text{Store II} & \\ & \begin{bmatrix} 22 & 28 \\ 15 & 21 \\ 12 & 26 \end{bmatrix} & \begin{matrix} \text{TVs} \\ \text{Washers} \\ \text{Dryers} \end{matrix} \end{matrix}$$

If these products have the same values as given above, we can find the value of the purchases for each store by multiplying the row matrix times each of the column matrices. The value of the Store I purchases is found by multiplying the row matrix A times the first column of matrix C, and the value of the Store II purchases is found by multiplying matrix A times the second column of matrix C. The result is

$$AC = \begin{bmatrix} 550 & 435 & 325 \end{bmatrix} \begin{bmatrix} 22 & 28 \\ 15 & 21 \\ 12 & 26 \end{bmatrix}$$

$$= \begin{bmatrix} 550 \cdot 22 + 435 \cdot 15 + 325 \cdot 12 & 550 \cdot 28 + 435 \cdot 21 + 325 \cdot 26 \end{bmatrix}$$

$$= \begin{bmatrix} 22{,}525 & 32{,}985 \end{bmatrix}$$

This indicates that the value of the Store I purchase is \$22,525 (the value found before) and the value of the Store II purchase is \$32,985.

The matrix AC above is the **product** of the 1×3 matrix A and the 3×2 matrix C, and the product is a 1×2 matrix. In general, the product of an $m \times n$ matrix and an $n \times k$ matrix is an $m \times k$ matrix, and the product is undefined if the number of columns in the first matrix does not equal the number of rows in the second matrix.

In general, we can define the product of two matrices by defining how each element of the product is formed.

PRODUCT OF TWO MATRICES

The product of an $m \times n$ matrix A and an $n \times k$ matrix B is an $m \times k$ matrix which is the matrix $C = AB$. The element in the ith row and jth column of matrix C has the form

$$c_{ij} = \begin{bmatrix} a_{i1} & a_{i2} & \ldots & a_{in} \end{bmatrix} \begin{bmatrix} b_{1j} \\ b_{2j} \\ \vdots \\ b_{nj} \end{bmatrix} = \begin{bmatrix} a_{i1}b_{1j} + a_{i2}b_{2j} + \ldots + a_{in}b_{nj} \end{bmatrix}$$

We illustrate the product AB in Figure 5.26, with each of the c_{ij} elements found as shown in the box above.

$$C = AB = \begin{bmatrix} a_{11} & a_{12} & \ldots & a_{1n} \\ a_{21} & a_{22} & \ldots & a_{2n} \\ \vdots & \vdots & & \vdots \\ a_{i1} & a_{i2} & \ldots & a_{in} \\ a_{m1} & a_{m2} & \ldots & a_{mn} \end{bmatrix} \begin{bmatrix} b_{11} & b_{12} & \ldots & b_{1j} & \ldots & b_{1k} \\ b_{21} & b_{22} & \ldots & b_{2j} & \ldots & b_{2k} \\ \vdots & \vdots & & \vdots & & \vdots \\ b_{n1} & b_{n2} & \ldots & b_{nj} & \ldots & b_{nk} \end{bmatrix} = \begin{bmatrix} c_{11} & c_{12} & \ldots & c_{1j} & & c_{1k} \\ c_{21} & c_{22} & & c_{2j} & & c_{2k} \\ \vdots & \vdots & & \vdots & & \vdots \\ c_{i1} & c_{i2} & \ldots & c_{ij} & \ldots & c_{ik} \\ \vdots & \vdots & & \vdots & & \vdots \\ c_{m1} & c_{m2} & \ldots & c_{mj} & \ldots & c_{mk} \end{bmatrix}$$

$$m \times n \qquad\qquad n \times k \qquad\qquad m \times k$$

FIGURE 5.26

EXAMPLE 5   Matrix Product

Compute the products AB and BA for the matrices $A = \begin{bmatrix} 1 & 2 \\ 3 & 4 \end{bmatrix}$ and $B = \begin{bmatrix} a & b \\ c & d \end{bmatrix}$.

Solution

$$AB = \begin{bmatrix} 1 & 2 \\ 3 & 4 \end{bmatrix}\begin{bmatrix} a & b \\ c & d \end{bmatrix} = \begin{bmatrix} 1a + 2c & 1b + 2d \\ 3a + 4c & 3b + 4d \end{bmatrix}$$

$$BA = \begin{bmatrix} a & b \\ c & d \end{bmatrix}\begin{bmatrix} 1 & 2 \\ 3 & 4 \end{bmatrix} = \begin{bmatrix} 1a + 3b & 2a + 4b \\ 1c + 3d & 2c + 4d \end{bmatrix}$$ ∎

Note that in Example 5 the product AB is quite different from the product BA. That is, $AB \neq BA$. For some matrices, but not all, $AB \neq BA$. We indicate this by saying that *matrix multiplication is not commutative.*

EXAMPLE 6 Advertising

A business plans to use three methods of advertising: newspapers, radio, and cable TV in each of its two markets, I and II. The cost per ad type in each market (in thousands of dollars) is given by matrix A.

$$A = \begin{matrix} & \text{Mkt I} & \text{Mkt II} & \\ & \begin{bmatrix} 12 & 10 \\ 10 & 8 \\ 5 & 9 \end{bmatrix} & \begin{matrix} \text{Paper} \\ \text{Radio} \\ \text{TV} \end{matrix} \end{matrix} \qquad B = \begin{matrix} & \text{Paper} & \text{Radio} & \text{TV} & \\ & \begin{bmatrix} 3 & 12 & 15 \\ 5 & 16 & 10 \\ 10 & 11 & 6 \end{bmatrix} & \begin{matrix} \text{Teens} \\ \text{Single females} \\ \text{Men 35–50} \end{matrix} \end{matrix}$$

The business has three target groups, teenagers, single women, and men aged 35 to 50. Matrix B gives the number of ads per week directed at each of these groups.

a. Does AB or BA give the matrix that represents the cost of ads for each target group of people in each market?

b. Find this matrix.

c. For what group of people is the most money spent on advertising?

Solution

a. Multiplying matrix B times matrix A gives the total cost of ads for each target group in each market. Note that the product AB is undefined, because multiplying a 3×2 matrix times a 3×3 matrix is not possible.

b. $$BA = \begin{bmatrix} 3 & 12 & 15 \\ 5 & 16 & 10 \\ 10 & 11 & 6 \end{bmatrix}\begin{bmatrix} 12 & 10 \\ 10 & 8 \\ 5 & 9 \end{bmatrix}$$

$$= \begin{bmatrix} 36 + 120 + 75 & 30 + 96 + 135 \\ 60 + 160 + 50 & 50 + 128 + 90 \\ 120 + 110 + 30 & 100 + 88 + 54 \end{bmatrix} = \begin{bmatrix} 231 & 261 \\ 270 & 268 \\ 260 & 242 \end{bmatrix}$$

The columns of this matrix represent the markets and the rows represent the target groups.

$$\begin{matrix} & \text{Mkt I} & \text{Mkt II} & \\ & \begin{bmatrix} 231 & 261 \\ 270 & 268 \\ 260 & 242 \end{bmatrix} & \begin{matrix} \text{Teens} \\ \text{Single females} \\ \text{Men 35–50} \end{matrix} \end{matrix}$$

c. The largest amount is spent on single females, $270,000 in Market I and $268,000 in Market II. ∎

Multiplication with Technology

We can multiply two matrices by using technology. We can find the product of matrix B times matrix A on a graphing utility by entering $[B]*[A]$ (or $[B][A]$) and pressing ENTER . Figure 5.27 shows the product BA from Example 6.

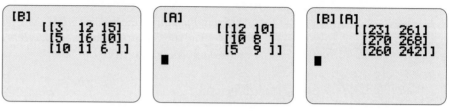

FIGURE 5.27

EXAMPLE 7 Multiplying with Technology

Use technology to compute BA and AB if $A = \begin{bmatrix} 1 & 2 \\ 0 & -1 \\ 3 & -2 \end{bmatrix}$ and $B = \begin{bmatrix} 2 & 4 & -1 \\ 3 & -2 & 1 \\ 2 & 0 & 2 \\ 1 & -3 & 0 \end{bmatrix}$.

Solution

Figure 5.28(a) shows displays with matrix A, matrix B, and the product BA.

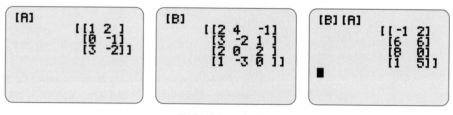

FIGURE 5.28(a)

Figure 5.28(b) shows that the matrix product AB does not exist because the dimensions do not match (that is, the number of columns of A does not equal the number of rows of B). ∎

FIGURE 5.28(b)

Like graphing calculators and software programs, computer spreadsheets can be used to perform operations with matrices. Consider the following example.

EXAMPLE 8 Manufacturing

A furniture company manufactures two products, *A* and *B*, which are constructed using steel, plastic, and fabric. The number of units of each raw material that is required for each product is given by the table below.

	Steel	Plastic	Fabric
Product A	2	3	8
Product B	3	1	10

Because of transportation costs to the company's two plants, the unit costs for some of the raw materials differ. The table below gives the unit costs for each of the raw materials at the two plants. Create two matrices from the information in the two tables and use matrix multiplication to find the cost of manufacturing each product at each plant.

	Plant I	Plant II
Steel	$15	$16
Plastic	$11	$10
Fabric	$ 6	$ 7

Solution

We represent the number of units of raw materials for each product by the matrix *P*, and the costs of the raw materials at each plant by matrix *C*. The cost of manufacturing each product at each plant is given by the matrix product *PC*. The spreadsheet shown in Figure 5.29 gives matrix *P*, matrix *C*, and the matrix product *PC*.

	A	B	C	D
1	Matrix P	2	3	8
2		3	1	10
3				
4	Matrix C	15	16	
5		11	10	
6		6	7	
7				
8	Product PC	111	118	
9		116	128	

FIGURE 5.29

The rows of the matrix PC represent Product A and Product B, respectively, and the columns represent Plant I and Plant II. The entries give the costs of each product at each plant.

$$\begin{array}{cc} \text{Plant I} & \text{Plant II} \end{array}$$
$$\begin{bmatrix} 111 & 118 \\ 116 & 128 \end{bmatrix} \begin{array}{l} \text{Product A} \\ \text{Product B} \end{array}$$

■

Recall that a square $(n \times n)$ matrix with 1's on the diagonal and 0's elsewhere is an identity matrix. For any $n \times k$ matrix A and the $n \times n$ matrix I, $IA = A$, and for any $m \times n$ matrix B and the $n \times n$ matrix I, $BI = B$. If matrix C is an $n \times n$ matrix, then for the $n \times n$ matrix I, $IC = C$ and $CI = C$, so the product of an identity matrix and another square matrix is *commutative*. (Recall that multiplication of matrices, in general, is not commutative.)

EXAMPLE 9   The Identity Matrix

a. Write the 2×2 identity matrix I.

b. For the matrix $A = \begin{bmatrix} 3 & -2 \\ -1 & 5 \end{bmatrix}$, find IA and AI.

Solution

a. $\begin{bmatrix} 1 & 0 \\ 0 & 1 \end{bmatrix}$

b. $\begin{bmatrix} 1 & 0 \\ 0 & 1 \end{bmatrix}\begin{bmatrix} 3 & -2 \\ -1 & 5 \end{bmatrix} = \begin{bmatrix} 3 & -2 \\ -1 & 5 \end{bmatrix} \qquad \begin{bmatrix} 3 & -2 \\ -1 & 5 \end{bmatrix}\begin{bmatrix} 1 & 0 \\ 0 & 1 \end{bmatrix} = \begin{bmatrix} 3 & -2 \\ -1 & 5 \end{bmatrix}$

Thus $IA = A$ and $AI = A$.

■

5.4 SKILLS CHECK

Use the following matrices for Exercises 1–8.

$$A = \begin{bmatrix} 1 & 3 & -2 \\ 3 & 1 & 4 \\ -5 & 3 & 6 \end{bmatrix} \qquad B = \begin{bmatrix} 2 & 1 & -1 \\ 3 & 2 & 4 \end{bmatrix}$$

$$C = \begin{bmatrix} 1 & 3 \\ 2 & 1 \\ 3 & -1 \end{bmatrix} \qquad D = \begin{bmatrix} 2 & 3 & 1 \\ 3 & 4 & -1 \\ 2 & 5 & 1 \end{bmatrix}$$

1. Which pairs of the matrices can be added?

2. Use letters to represent the matrix products that are defined.

3. Find the sum of A and D if it is defined.

4. Find $3A$.

5. Find $2D - 4A$.

6. Find BC and CB if these products exist. Do these products have the same dimensions?

7. Find AD and DA if these products exist. Are these products equal? Do these products have the same dimensions?

8. What is the product of the 3×3 identity matrix and D?

9. Compute the sum of $A = \begin{bmatrix} 1 & 5 \\ 3 & 2 \end{bmatrix}$ and

$B = \begin{bmatrix} 2a & 3b \\ -c & -2d \end{bmatrix}$.

10. Compute the difference $A - B$ if $A = \begin{bmatrix} a & b \\ c & d \\ f & g \end{bmatrix}$ and $B = \begin{bmatrix} 1 & 4 \\ 3 & 2 \\ 5 & 6 \end{bmatrix}$.

11. Compute $3A - 2B$ if $A = \begin{bmatrix} a & b \\ c & d \end{bmatrix}$ and $B = \begin{bmatrix} 1 & 2 \\ 3 & 4 \end{bmatrix}$.

12. Compute $2A - 3B$ if $A = \begin{bmatrix} 1 & -3 & 2 \\ 2 & 2 & -1 \\ 3 & 4 & 2 \end{bmatrix}$ and $B = \begin{bmatrix} 2 & 2 & 2 \\ 3 & -2 & -1 \\ 1 & 1 & 2 \end{bmatrix}$.

13. If an $m \times n$ matrix is multiplied times an $n \times k$ matrix, what is the dimension of the matrix that is the product?

14. If A and B are any two matrices, does $AB = BA$ always, sometimes, or never?

15. If A is a 2×3 matrix and B is a 4×2 matrix:
 a. Which product is defined, AB or BA?
 b. What is the dimension of the product that is defined?

16. Find AB and BA if $A = \begin{bmatrix} a & b & c \\ d & e & f \end{bmatrix}$ and $B = \begin{bmatrix} 1 & 2 \\ 3 & 4 \\ 5 & 6 \end{bmatrix}$.

17. If $A = \begin{bmatrix} 1 & 5 \\ 3 & 2 \end{bmatrix}$ and $B = \begin{bmatrix} 2 & 3 \\ -1 & -2 \end{bmatrix}$, compute AB and BA, if possible.

18. If $A = \begin{bmatrix} 1 & 4 \\ 3 & -1 \\ -2 & 2 \end{bmatrix}$ and $B = \begin{bmatrix} 4 & 2 & 2 \\ -1 & 3 & 1 \end{bmatrix}$, compute AB and BA if possible.

19. If $A = \begin{bmatrix} 1 & -1 & 2 \\ 3 & 4 & 4 \end{bmatrix}$ and $B = \begin{bmatrix} 3 & 1 \\ 1 & 3 \\ -2 & 1 \end{bmatrix}$, compute AB and BA, if possible.

20. Suppose $A = \begin{bmatrix} 1 & -1/2 & -1/4 \\ -1/2 & 0 & 1/2 \\ 0 & 1/2 & -1/4 \end{bmatrix}$ and $B = \begin{bmatrix} 2 & 2 & 2 \\ 1 & 2 & 3 \\ 2 & 4 & 2 \end{bmatrix}$.
 a. Compute AB and BA, if possible.
 b. Are the products equal?

5.4 EXERCISES

1. *Endangered Species* The tables below give the number of some species of threatened and endangered wildlife in the United States and in foreign countries in 1996.
 a. Form the matrix A that contains the number of each of these species in the United States in 1996 and matrix B that contains the number of each of these species outside the United States in 1996.
 b. Write a matrix containing the total number of each of these species, assuming that the U.S. and foreign species are different.

United States	Mammals	Birds	Reptiles	Amphibians	Fishes
Endangered	252	74	14	7	65
Threatened	19	16	19	5	40

Foreign	Mammals	Birds	Reptiles	Amphibians	Fishes
Endangered	252	178	65	8	11
Threatened	19	6	14	1	0

(Source: *Statistical Abstract of the United States*)

2. *Endangered Species*
   a. Use the information and matrices in Exercise 1 to find the matrix $C = B - A$ and tell what the elements of this matrix signify.
   b. What do the negative elements of matrix C mean?

3. *Trade Balances* The table below gives the U.S. exports and imports in three categories for the years 1996 and 1999.
   a. Form the matrix A that contains the number of millions of dollars of U.S. exports in 1996 and 1999.
   b. Form a matrix B that contains the number of millions of dollars of U.S. imports in 1996 and in 1999.
   c. Write the matrix $C = A - B$, which gives the *balance of trade* for these U.S. product categories.
   d. In what categories and years does the U.S. have a positive trade balance?
   e. In which category and year is the trade deficit the greatest? Why?

U.S. Exports (Millions of dollars)	1996	1999
Agricultural	59,385	47,091
Manufactured goods	486,171	565,490
Mineral fuels	12,181	9,880

U.S. Imports (Millions of dollars)	1996	1999
Agricultural	32,575	36,681
Manufactured goods	658,782	882,013
Mineral fuels	78,086	75,803

(Source: U.S. Dept. of Commerce)

4. *Exports*
   a. Use the table in Exercise 3 to create four 3×1 matrices, one for each of the columns in the table.
   b. Use two of these matrices to find a matrix E that gives the balance of trade for 1996.
   c. Use two of these matrices to find a matrix F that gives the balance of trade for 1999.
   d. Find the matrix $E - F$ that compares the balance of trade for the two years.
   e. Did the balance of trade improve in any categories from 1996–1999?

5. *Income* The table below gives the median annual income for different sexes and races in 1999. Create a matrix containing this data and use a matrix operation to create a matrix that contains the median income if the 1999 median incomes are increased by 12% in all categories.

Median Annual Income, 1999		
	Male	Female
White	$39,331	$28,023
Black	30,297	25,142
Hispanic	23,342	20,052

(Source: U.S. Census Bureau)

6. *Exchange Rates* The table below gives the national currency units per U.S. dollar for the United Kingdom (pound) and for the Netherlands (guilder) for the years 1990 and 1999. If the fee for exchanging dollars to these units reduces the national currency returned per U.S. dollar by 5%, place this data in a matrix and use a matrix operation to find the currency units actually returned per dollar.

National Currency Units Per U.S. Dollar		
	United Kingdom	Netherlands
1990	61.2963	6.2957
1999	1.6182	7.7992

(Source: International Monetary Fund)

7. *Advertising* A political candidate plans to use three methods of advertising: newspapers, radio, and cable TV. The cost per ad (in thousands of dollars) for each type of media is given by matrix A. Matrix B shows the number of ads per month in these three media that are targeted to single people, to married males aged 35 to 55, and to married females over 65 years of age. Find the matrix that gives the cost of ads for each target group.

$$A = \begin{bmatrix} 12 \\ 15 \\ 5 \end{bmatrix} \begin{matrix} \text{TV} \\ \text{Radio} \\ \text{Papers} \end{matrix}$$
$$\text{Cost}$$

$$\begin{matrix} \text{TV} & \text{Radio} & \text{Papers} \end{matrix}$$
$$B = \begin{bmatrix} 30 & 45 & 35 \\ 25 & 32 & 40 \\ 22 & 12 & 30 \end{bmatrix} \begin{matrix} \text{Singles} \\ \text{Males 35–55} \\ \text{Females 65+} \end{matrix}$$

8. *Cost* Men and women in a church choir wear choir robes in the sizes shown in matrix *A*. Matrix *B* contains the prices (in dollars) of new robes and hoods according to size.

a. Find the product *BA* and label the rows and columns to show what each row represents.

b. What is the cost of the robes for all the men? For all the women?

$$A = \begin{bmatrix} 10 & 24 \\ 22 & 10 \\ 33 & 3 \end{bmatrix} \begin{matrix} \text{Small} \\ \text{Medium} \\ \text{Large} \end{matrix}$$

$$\begin{matrix} \text{Men} & \text{Women} \end{matrix}$$

$$B = \begin{bmatrix} 45 & 50 & 55 \\ 20 & 20 & 20 \end{bmatrix} \begin{matrix} \text{Robes} \\ \text{Hoods} \end{matrix}$$

$$\begin{matrix} \text{S} & \text{M} & \text{L} \end{matrix}$$

9. *Manufacturing* Two departments of a firm, A and B, need differing amounts of the same products. The following tables give the amounts of the products needed by the departments.

	Steel	Wood	Plastic
Department A	60	40	20
Department B	40	20	40

These three products are supplied by two suppliers, DeTuris and Marriott, with the unit prices given in the following table.

	DeTuris	Marriott
Steel	600	560
Wood	300	200
Plastic	300	400

a. Use matrix multiplication to determine how much these orders will cost each department at each of the two suppliers.

b. From which supplier should each department make its purchase?

10. *Manufacturing* A furniture manufacturer produces three styles of chairs, with the number of units of each type of raw material needed for each style given in the table below.

	Wood	Nylon	Velvet	Springs
Style A	5	20	0	10
Style B	10	9	0	0
Style C	5	10	10	10

The cost in dollars per unit for each of the raw materials is given in the table below.

Wood	15
Nylon	12
Velvet	14
Springs	30

Create two matrices to represent these data and use matrix multiplication to find the price of manufacturing each style of chair.

11. *Politics* In a midwestern state, it is determined that 90% of all Republicans vote for Republican candidates and the remainder for Democratic candidates, while 80% of all Democrats vote for Democratic candidates and the remainder vote for Republican candidates. The percent of each party predicted to win the next election is given by

$$\begin{bmatrix} R \\ D \end{bmatrix} = \begin{bmatrix} .90 & .20 \\ .10 & .80 \end{bmatrix} \begin{bmatrix} a \\ b \end{bmatrix}$$

where *a* is the percent of Republicans and *b* is the percent of Democrats that won the last election. If 50% of those winning the election last time were Republicans and 50% were Democrats, what are the percents of each party that are predicted to win the next election?

12. *Competition* Two phone companies compete for customers in the southeastern region of a state. Company X retains 3/5 of its customers and loses 2/5 of its customers to Company Y, Company Y retains 2/3 of its customers and loses 1/3 to Company X. If we represent the fraction of the market held last year by

$$\begin{bmatrix} a \\ b \end{bmatrix}$$

where *a* is the fraction that Company X had and *b* is the fraction that Company Y had, then the fraction that each company will have this year can be found by

$$\begin{bmatrix} x \\ y \end{bmatrix} = \begin{bmatrix} 3/5 & 1/3 \\ 2/5 & 2/3 \end{bmatrix} \begin{bmatrix} a \\ b \end{bmatrix}$$

If Company X had 120,000 customers and Company Y had 90,000 customers last year, how many customers did each have this year?

13. *Wages* The table below gives the median weekly earnings for nonunion men and women for different age groups in 1999.
    a. Make a matrix containing the data in this table.
    b. Suppose the median weekly earnings of men union workers are found to be 10% more than the earnings given in the table, and the median weekly earnings of women union workers were found to be 25% more than the earnings given in the table. Use matrix multiplication by a 2 × 2 matrix to find the new median weekly earnings by age for the male and female union members.

Age	16–24	25–34	35–44	45–54	55–64
Men	348	648	560	691	751
Women	321	477	486	502	467

(Source: U.S. Dept. of Labor, Bureau of Labor Statistics)

14. *Libraries* The number of libraries and the operating income of public libraries in selected states for 1994 and for 1997 are shown in the table below.
    a. Form the matrix *A* that contains the number of libraries and operating income for 1994.
    b. Form the matrix *B* that contains the number of libraries and operating income for 1997.

c. Use these two matrices to find the increase in the number of libraries and operating expenses from 1994 through 1997.
d. For which state is the increase in the number of libraries largest?
e. For which state is the increase in operating income the largest?

1994	Number of Libraries	Operating Income ($ million)
Illinois	606	358
Iowa	518	49
Kansas	324	50
Alabama	207	48
California	170	566

1997	Number of Libraries	Operating Income ($ million)
Illinois	799	407
Iowa	556	58
Kansas	374	60
Alabama	274	56
California	1039	655

5.5 Inverse Matrices; Matrix Equations

Key Concepts

- Inverse matrices
- Inverses and technology
- Decoding messages
- Solving matrix equations
- Matrix equations and technology
- Encoding messages

Security with credit card numbers on the Internet depends on encryption of the data. Encryption involves providing a way for the sender to encode a message so that the message is not apparent and a way for the receiver to decode the message so it can be read. Throughout history, different military units have used encoding and decoding systems of varying sophistication. In this section we see how to use the **inverse** of the matrix *A* to decode a message that has been encoded with matrix *A*. We also learn how to find the inverse of a matrix, how to use it to decode messages, and how to use it to solve matrix equations that have unique solutions.

Inverse Matrices

If the product of matrices *A* and *B* is an identity matrix, *I*, we say that *B* is the inverse of *A* (and *A* is the inverse of *B*). *B* is called the **inverse matrix** of *A*, and is denoted A^{-1}.

> ### INVERSE MATRICES
>
> Two square matrices, A and B, are called **inverses** of each other if
>
> $$AB = I \text{ and } BA = I, \text{ where } I \text{ is the identity matrix.}$$
>
> We denote this $B = A^{-1}$ and $A = B^{-1}$.

EXAMPLE 1   Inverse Matrices

Show that A and B are inverse matrices if $A = \begin{bmatrix} 1 & -0.6 & -0.2 \\ 0 & 0.4 & -0.2 \\ -1 & 0.4 & 0.8 \end{bmatrix}$ and $B = \begin{bmatrix} 2 & 2 & 1 \\ 1 & 3 & 1 \\ 2 & 1 & 2 \end{bmatrix}$.

Solution

$$AB = \begin{bmatrix} 1 & -0.6 & -0.2 \\ 0 & 0.4 & -0.2 \\ -1 & 0.4 & 0.8 \end{bmatrix}\begin{bmatrix} 2 & 2 & 1 \\ 1 & 3 & 1 \\ 2 & 1 & 2 \end{bmatrix}$$

$$= \begin{bmatrix} 2 - 0.6 - 0.4 & 2 - 1.8 - 0.2 & 1 - 0.6 - 0.4 \\ 0 + 0.4 - 0.4 & 0 + 1.2 - 0.2 & 0 + 0.4 - 0.4 \\ -2 + 0.4 + 1.6 & -2 + 1.2 + 0.8 & -1 + 0.4 + 1.6 \end{bmatrix}$$

$$= \begin{bmatrix} 1 & 0 & 0 \\ 0 & 1 & 0 \\ 0 & 0 & 1 \end{bmatrix}$$

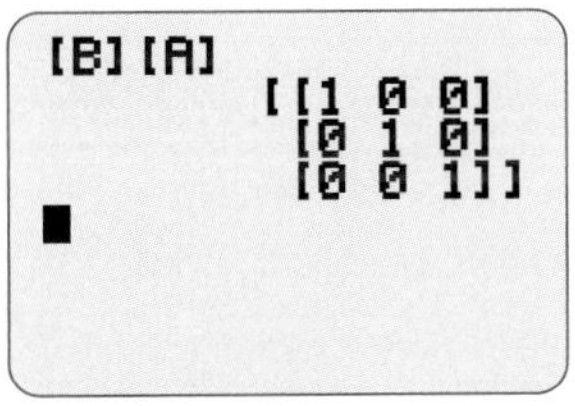

The product AB is the 3×3 identity matrix; we use technology (see Figure 5.30) to see that the product BA is also the identity matrix. Thus A and B are inverse matrices. ∎

```
[B] [A]
      [[1 0 0]
       [0 1 0]
       [0 0 1]]
■
```

FIGURE 5.30

We have used elementary row operations on augmented matrices to solve systems of equations. We can also find the inverse of a matrix A, if it exists, by using elementary row operations. Note that if a matrix is not square, then it does not have an inverse, and that not all square matrices have inverses.

> ### FINDING THE INVERSE OF A SQUARE MATRIX
>
> 1. Write the matrix with the same size identity matrix in its augment, getting a matrix of the form $[A|I]$.
>
> 2. Use elementary row operations on $[A|I]$ until A is transformed into an identity matrix, giving a new matrix of the form $[I|B]$. The matrix B is the inverse of A.
>
> 3. If A does not have an inverse, the reduction process will yield a row of zeros in the left half (representing the original matrix) of the augmented matrix.

EXAMPLE 2 Finding an Inverse Matrix

Find the inverse of $A = \begin{bmatrix} 2 & 2 \\ 2 & 1 \end{bmatrix}$.

Solution

Creating the matrix $[A\,|\,I]$ and performing the operations to convert A to I gives

$$\begin{bmatrix} 2 & 2 & | & 1 & 0 \\ 2 & 1 & | & 0 & 1 \end{bmatrix} \xrightarrow{\left(\frac{1}{2}\right)R_1 \to R_1} \begin{bmatrix} 1 & 1 & | & 1/2 & 0 \\ 2 & 1 & | & 0 & 1 \end{bmatrix}$$

$$\xrightarrow{-2R_1 + R_2 \to R_2} \begin{bmatrix} 1 & 1 & | & 1/2 & 0 \\ 0 & -1 & | & -1 & 1 \end{bmatrix}$$

$$\xrightarrow{-R_2 \to R_2} \begin{bmatrix} 1 & 1 & | & 1/2 & 0 \\ 0 & 1 & | & 1 & -1 \end{bmatrix} \xrightarrow{-R_2 + R_1 \to R_1} \begin{bmatrix} 1 & 0 & | & -1/2 & 1 \\ 0 & 1 & | & 1 & -1 \end{bmatrix}$$

Thus the inverse matrix of A is $A^{-1} = \begin{bmatrix} -1/2 & 1 \\ 1 & -1 \end{bmatrix}$.

We can verify this by observing that $\begin{bmatrix} 2 & 2 \\ 2 & 1 \end{bmatrix}\begin{bmatrix} -1/2 & 1 \\ 1 & -1 \end{bmatrix} = \begin{bmatrix} 1 & 0 \\ 0 & 1 \end{bmatrix}$. ∎

EXAMPLE 3 Inverse of a 3 × 3 Matrix

Find the inverse of $A = \begin{bmatrix} -2 & 1 & 2 \\ 1 & 0 & -1 \\ 4 & -2 & -3 \end{bmatrix}$.

Solution

Creating the matrix $[A\,|\,I]$ and performing the operations to convert A to I gives

$$\begin{bmatrix} -2 & 1 & 2 & | & 1 & 0 & 0 \\ 1 & 0 & -1 & | & 0 & 1 & 0 \\ 4 & -2 & -3 & | & 0 & 0 & 1 \end{bmatrix} \xrightarrow{R_1 \leftrightarrow R_2} \begin{bmatrix} 1 & 0 & -1 & | & 0 & 1 & 0 \\ -2 & 1 & 2 & | & 1 & 0 & 0 \\ 4 & -2 & -3 & | & 0 & 0 & 1 \end{bmatrix}$$

$$\xrightarrow[-4R_1 + R_3 \to R_3]{2R_1 + R_2 \to R_2} \begin{bmatrix} 1 & 0 & -1 & | & 0 & 1 & 0 \\ 0 & 1 & 0 & | & 1 & 2 & 0 \\ 0 & -2 & 1 & | & 0 & -4 & 1 \end{bmatrix}$$

$$\xrightarrow{2R_2 + R_3 \to R_3} \begin{bmatrix} 1 & 0 & -1 & | & 0 & 1 & 0 \\ 0 & 1 & 0 & | & 1 & 2 & 0 \\ 0 & 0 & 1 & | & 2 & 0 & 1 \end{bmatrix}$$

$$\xrightarrow{R_3 + R_1 \to R_1} \begin{bmatrix} 1 & 0 & 0 & | & 2 & 1 & 1 \\ 0 & 1 & 0 & | & 1 & 2 & 0 \\ 0 & 0 & 1 & | & 2 & 0 & 1 \end{bmatrix}$$

Thus we have the inverse of A. $\qquad A^{-1} = \begin{bmatrix} 2 & 1 & 1 \\ 1 & 2 & 0 \\ 2 & 0 & 1 \end{bmatrix}$.

$\blacksquare$

Inverses and Technology

Computer software, spreadsheets, and calculators can be used to find the inverse of a matrix. We will see that if the inverse of a square matrix exists, it can be found easily with technology. Figure 5.31(a) shows the matrix A from Example 3 and Figure 5.31(b) shows the inverse of A found with a graphing calculator.

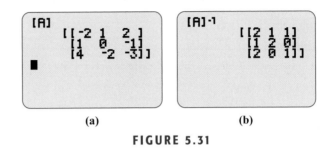

(a) $\qquad\qquad\qquad\qquad\qquad$ (b)

FIGURE 5.31

Figure 5.32 shows a spreadsheet that gives the inverse of the matrix A.

	A	B	C	D
1	Matrix A	−2	1	2
2		1	0	−1
3		4	−2	−3
4				
5	A inverse	2	1	1
6		1	2	0
7		2	0	1

FIGURE 5.32

We can confirm that these matrices are inverses by computing AA^{-1}. (See Figure 5.33.)

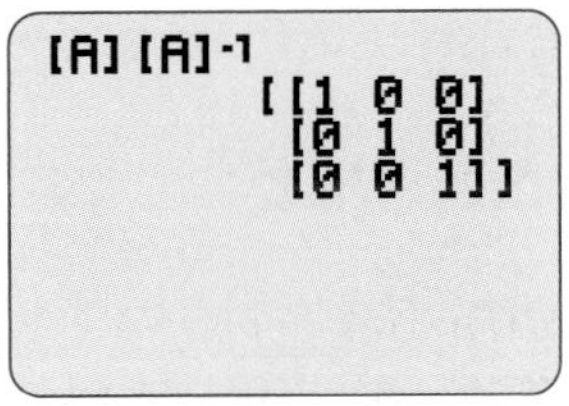

[A] [A]⁻¹
[[1 0 0]
[0 1 0]
[0 0 1]]

FIGURE 5.33

EXAMPLE 4 Does the Inverse Exist?

Find the inverse of $A = \begin{bmatrix} 1 & 2 & -2 \\ 2 & 0 & 2 \\ 6 & 4 & 0 \end{bmatrix}$, if it exists.

Solution

Attempting to find the inverse of matrix A with technology results in an error statement, indicating that the inverse does not exist. (See Figure 5.34.)

Recall that if A does not have an inverse, the reduction process using elementary row operations will yield a row of zeros in the left half of the augmented matrix.*

$$\begin{bmatrix} 1 & 2 & -2 & | & 1 & 0 & 0 \\ 2 & 0 & 2 & | & 0 & 1 & 0 \\ 6 & 4 & 0 & | & 0 & 0 & 1 \end{bmatrix} \xrightarrow[-6R_1 + R_3 \rightarrow R_3]{-2R_1 + R_2 \rightarrow R_2} \begin{bmatrix} 1 & 2 & -2 & | & 1 & 0 & 0 \\ 0 & -4 & 6 & | & -2 & 1 & 0 \\ 0 & -8 & 12 & | & -6 & 0 & 1 \end{bmatrix}$$

$$\xrightarrow{(-1/4)R_2 \rightarrow R_2} \begin{bmatrix} 1 & 2 & -2 & | & 1 & 0 & 0 \\ 0 & 1 & -3/2 & | & 1/2 & -1/4 & 0 \\ 0 & -8 & 12 & | & -6 & 0 & 1 \end{bmatrix}$$

$$\xrightarrow{8R_2 + R_3 \rightarrow R_3} \begin{bmatrix} 1 & 2 & -2 & | & 1 & 0 & 0 \\ 0 & 1 & -3/2 & | & 1/2 & -1/4 & 0 \\ 0 & 0 & 0 & | & -2 & -2 & 1 \end{bmatrix}$$

We see that the left half of the bottom row of the reduced matrix contains all 0's, so it is not possible to reduce the original matrix to the identity matrix, and thus the matrix A does not have an inverse. ■

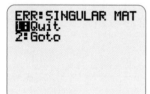

FIGURE 5.34

Encoding and Decoding Messages

In sending messages during military maneuvers, in business transactions, and in sending secure data on the Internet, encoding (or encryption) of messages is important. Suppose that we want to encode the message "Cheer up." The following simple code could be used to change letters of the alphabet to the numbers 1 to 26, respectively, with the number 27 representing a blank space.

a b c d e f g h i j k l m n o p q r s t u v w x y z
1 2 3 4 5 6 7 8 9 10 11 12 13 14 15 16 17 18 19 20 21 22 23 24 25 26 27

Then "Cheer up" can be represented by the numbers 3, 8, 5, 5, 18, 27, 21, 16.

To further encode the message, we put these numbers in pairs and then create a 2×1 matrix for each pair of numbers. Next we choose an *encoding matrix*, like

$$A = \begin{bmatrix} 3 & -2 \\ -1 & 1 \end{bmatrix},$$

---

*The methods used by technology will sometimes yield approximations for numbers that should be zeros.

and multiply each pair of numbers by the encoding matrix, as follows.

$$\begin{bmatrix} 3 & -2 \\ -1 & 1 \end{bmatrix}\begin{bmatrix} 3 \\ 8 \end{bmatrix} = \begin{bmatrix} -7 \\ 5 \end{bmatrix}, \quad \begin{bmatrix} 3 & -2 \\ -1 & 1 \end{bmatrix}\begin{bmatrix} 5 \\ 5 \end{bmatrix} = \begin{bmatrix} 5 \\ 0 \end{bmatrix},$$

$$\begin{bmatrix} 3 & -2 \\ -1 & 1 \end{bmatrix}\begin{bmatrix} 18 \\ 27 \end{bmatrix} = \begin{bmatrix} 0 \\ 9 \end{bmatrix}, \quad \begin{bmatrix} 3 & -2 \\ -1 & 1 \end{bmatrix}\begin{bmatrix} 21 \\ 16 \end{bmatrix} = \begin{bmatrix} 31 \\ -5 \end{bmatrix}$$

Because multiplying a 2×2 matrix times a 2×1 matrix gives a 2×1 matrix, the products are pairs of numbers. Combining the pairs of numbers gives the encoded numerical message

$$-7, 5, 5, 0, 0, 9, 31, -5.$$

Note that another encoding matrix could be used rather than the one used above. We can also encode a message by putting triples of numbers in 3×1 matrices and multiplying each 3×1 matrix by a 3×3 encoding matrix.

EXAMPLE 5   Encoding Messages

Use the encoding matrix $A = \begin{bmatrix} 1 & -3 & 2 \\ 2 & -2 & 2 \\ 3 & -1 & 1 \end{bmatrix}$ to encode the message "Meet me for lunch."

Solution

Converting the letters of the message to triples of numbers gives:

13, 5, 5,   20, 27, 13,   5, 27, 6,   15, 18, 27,   12, 21, 14,   3, 8, 27

with 27 used to complete the last triple.

Placing these triples of numbers in 3×1 matrices and multiplying by matrix A gives

$$\begin{bmatrix} 1 & -3 & 2 \\ 2 & -2 & 2 \\ 3 & -1 & 1 \end{bmatrix}\begin{bmatrix} 13 \\ 5 \\ 5 \end{bmatrix} = \begin{bmatrix} 8 \\ 26 \\ 39 \end{bmatrix}, \quad \begin{bmatrix} 1 & -3 & 2 \\ 2 & -2 & 2 \\ 3 & -1 & 1 \end{bmatrix}\begin{bmatrix} 20 \\ 27 \\ 13 \end{bmatrix} = \begin{bmatrix} -35 \\ 12 \\ 46 \end{bmatrix},$$

$$\begin{bmatrix} 1 & -3 & 2 \\ 2 & -2 & 2 \\ 3 & -1 & 1 \end{bmatrix}\begin{bmatrix} 5 \\ 27 \\ 6 \end{bmatrix} = \begin{bmatrix} -64 \\ -32 \\ -6 \end{bmatrix}, \quad \begin{bmatrix} 1 & -3 & 2 \\ 2 & -2 & 2 \\ 3 & -1 & 1 \end{bmatrix}\begin{bmatrix} 15 \\ 18 \\ 27 \end{bmatrix} = \begin{bmatrix} 15 \\ 48 \\ 54 \end{bmatrix},$$

$$\begin{bmatrix} 1 & -3 & 2 \\ 2 & -2 & 2 \\ 3 & -1 & 1 \end{bmatrix}\begin{bmatrix} 12 \\ 21 \\ 14 \end{bmatrix} = \begin{bmatrix} -23 \\ 10 \\ 29 \end{bmatrix}, \quad \begin{bmatrix} 1 & -3 & 2 \\ 2 & -2 & 2 \\ 3 & -1 & 1 \end{bmatrix}\begin{bmatrix} 3 \\ 8 \\ 27 \end{bmatrix} = \begin{bmatrix} 33 \\ 44 \\ 28 \end{bmatrix}.$$

The products are triples of numbers. Combining the triples of numbers gives the encoded message

$$8, 26, 39, -35, 12, 46, -64, -32, -6, 15, 48, 54, -23, 10, 29, 33, 44, 28. \quad \blacksquare$$

We can also find the encoded triples of numbers for the message of Example 5 by writing the original triples of numbers as columns in one matrix and then multiplying this matrix by the encoding matrix. The columns of the product will be the encoded triples of numbers.

$$\begin{bmatrix} 1 & -3 & 2 \\ 2 & -2 & 2 \\ 3 & -1 & 1 \end{bmatrix} \begin{bmatrix} 13 & 20 & 5 & 15 & 12 & 3 \\ 5 & 27 & 27 & 18 & 21 & 8 \\ 5 & 13 & 6 & 27 & 14 & 27 \end{bmatrix} = \begin{bmatrix} 8 & -35 & -64 & 15 & -23 & 33 \\ 26 & 12 & -32 & 48 & 10 & 44 \\ 39 & 46 & -6 & 54 & 29 & 28 \end{bmatrix}$$

Observe that the columns of this product have the same triples, respectively, as the individual products in Example 5. Figure 5.35 shows a calculator display (in two screens with some columns repeated) of the product.

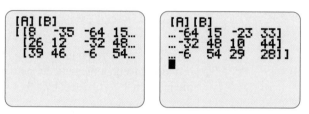

FIGURE 5.35

Sending an encoded message is of little value if the recipient is not able to decode it. If the message is encoded with a matrix, then the message can be decoded with the inverse of the matrix.

Recall that the message "Cheer up" was represented by the numbers 3, 8, 5, 5, 18, 27, 21, 16 and encoded to $-7, 5, 5, 0, 0, 9, 31, -5$ with the encoding matrix

$$A = \begin{bmatrix} 3 & -2 \\ -1 & 1 \end{bmatrix}.$$

To decode the encoded message, we multiply each pair of numbers in the encoded message by the inverse of matrix A.

$$A^{-1} = \begin{bmatrix} 1 & 2 \\ 1 & 3 \end{bmatrix}$$

Multiplying the pairs of the coded message by A^{-1} gives us back the original message.

$$\begin{bmatrix} 1 & 2 \\ 1 & 3 \end{bmatrix} \begin{bmatrix} -7 \\ 5 \end{bmatrix} = \begin{bmatrix} 3 \\ 8 \end{bmatrix}, \quad \begin{bmatrix} 1 & 2 \\ 1 & 3 \end{bmatrix} \begin{bmatrix} 5 \\ 0 \end{bmatrix} = \begin{bmatrix} 5 \\ 5 \end{bmatrix},$$

$$\begin{bmatrix} 1 & 2 \\ 1 & 3 \end{bmatrix} \begin{bmatrix} 0 \\ 9 \end{bmatrix} = \begin{bmatrix} 18 \\ 27 \end{bmatrix}, \quad \begin{bmatrix} 1 & 2 \\ 1 & 3 \end{bmatrix} \begin{bmatrix} 31 \\ -5 \end{bmatrix} = \begin{bmatrix} 21 \\ 16 \end{bmatrix}$$

Thus the decoded numbers are 3, 8, 5, 5, 18, 27, 21, 16, which correspond to the letters in the message "Cheer up."

EXAMPLE 6   Decoding Messages

Suppose we encoded messages using a simple code that changed letters of the alphabet to numbers, with the number 27 representing a blank space, and the encoding matrix

$$A = \begin{bmatrix} 1 & 2 & -1 \\ 2 & 1 & 2 \\ 3 & 2 & -3 \end{bmatrix}.$$

Decode the following coded message:

$$18, 41, 34, 61, 93, 75, 39, 77, 9, 34, 62, 46, 3, 62, -23$$

Solution

To find the numbers representing the message, we place triples of numbers from the coded message into columns of a 3×5 matrix, and multiply that matrix by A^{-1}. The inverse of A is

$$A^{-1} = \begin{bmatrix} \dfrac{-7}{16} & \dfrac{1}{4} & \dfrac{5}{16} \\[2mm] \dfrac{3}{4} & 0 & \dfrac{-1}{4} \\[2mm] \dfrac{1}{16} & \dfrac{1}{4} & \dfrac{-3}{16} \end{bmatrix}$$

Multiplying by A^{-1} on the left gives

$$A^{-1}\begin{bmatrix} 18 & 61 & 39 & 34 & 3 \\ 41 & 93 & 77 & 62 & 62 \\ 34 & 75 & 9 & 46 & -23 \end{bmatrix} = \begin{bmatrix} \dfrac{-7}{16} & \dfrac{1}{4} & \dfrac{5}{16} \\[2mm] \dfrac{3}{4} & 0 & \dfrac{-1}{4} \\[2mm] \dfrac{1}{16} & \dfrac{1}{4} & \dfrac{-3}{16} \end{bmatrix}\begin{bmatrix} 18 & 61 & 39 & 34 & 3 \\ 41 & 93 & 77 & 62 & 62 \\ 34 & 75 & 9 & 46 & -23 \end{bmatrix}$$

$$= \begin{bmatrix} 13 & 20 & 5 & 15 & 7 \\ 5 & 27 & 27 & 14 & 8 \\ 5 & 13 & 20 & 9 & 20 \end{bmatrix}$$

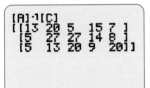

FIGURE 5.36

If we use technology to perform this multiplication, we do not need to display the elements of A^{-1} in the computation. (See Figure 5.36.)

Reading down the columns of the product gives the numbers representing the message:

$$13, 5, 5, 20, 27, 13, 5, 27, 20, 15, 14, 9, 7, 8, 20$$

The message is "Meet me tonight." ∎

Matrix Equations

The system of equations

$$\begin{cases} -2x + y + 2z = 5 \\ x \quad\quad - z = 2 \\ 4x - 2y - 3z = 4 \end{cases}$$

can be written as the matrix equation

$$\begin{bmatrix} -2x + y + 2z \\ x \quad\quad - z \\ 4x - 2y - 3z \end{bmatrix} = \begin{bmatrix} 5 \\ 2 \\ 4 \end{bmatrix}.$$

Because

$$\begin{bmatrix} -2 & 1 & 2 \\ 1 & 0 & -1 \\ 4 & -2 & -3 \end{bmatrix} \begin{bmatrix} x \\ y \\ z \end{bmatrix} = \begin{bmatrix} -2x + y + 2z \\ x - z \\ 4x - 2y - 3z \end{bmatrix},$$

we can write the system of equations as the matrix equation in the form

$$\begin{bmatrix} -2 & 1 & 2 \\ 1 & 0 & -1 \\ 4 & -2 & -3 \end{bmatrix} \begin{bmatrix} x \\ y \\ z \end{bmatrix} = \begin{bmatrix} 5 \\ 2 \\ 4 \end{bmatrix}.$$

Note that the 3×3 matrix on the left side of the matrix equation is the coefficient matrix for the original system. We will call the coefficient matrix A, the matrix containing the variables X, and the matrix containing the constants C, giving the form

$$AX = C.$$

If we multiply both sides of this equation on the left by the inverse of this coefficient matrix, the product is as follows.*

$$A^{-1}AX = A^{-1}C$$

Because $A^{-1}A = I$, and because multiplication of a matrix X by an identity matrix gives the matrix X, we have

$$IX = A^{-1}C, \quad \text{or} \quad X = A^{-1}C.$$

In general, multiplying both sides of the matrix equation $AX = C$ on the left by A^{-1} gives the solution to the system that the matrix equation represents.

EXAMPLE 7 Solution of Matrix Equations with Inverses

Solve the system

$$\begin{cases} -2x + y + 2z = 5 \\ x - z = 2 \\ 4x - 2y - 3z = 4 \end{cases}$$

by writing a matrix equation and using an inverse matrix.

Solution

We can write this system of equations as $AX = C$ where $A = \begin{bmatrix} -2 & 1 & 2 \\ 1 & 0 & -1 \\ 4 & -2 & -3 \end{bmatrix}$,

$X = \begin{bmatrix} x \\ y \\ z \end{bmatrix}$ and $C = \begin{bmatrix} 5 \\ 2 \\ 4 \end{bmatrix}$.

That is,

$$\begin{bmatrix} -2 & 1 & 2 \\ 1 & 0 & -1 \\ 4 & -2 & -3 \end{bmatrix} \begin{bmatrix} x \\ y \\ z \end{bmatrix} = \begin{bmatrix} 5 \\ 2 \\ 4 \end{bmatrix}.$$

---

*Recall that multiplication on the right may give a different product than multiplication on the left.

As we saw in Example 3, the inverse of the matrix A is

$$A^{-1} = \begin{bmatrix} 2 & 1 & 1 \\ 1 & 2 & 0 \\ 2 & 0 & 1 \end{bmatrix}.$$

Thus we can solve the system by multiplying both sides of the matrix equation by A^{-1} as follows.

$$A^{-1}A \begin{bmatrix} x \\ y \\ z \end{bmatrix} = A^{-1} \begin{bmatrix} 5 \\ 2 \\ 4 \end{bmatrix}$$

$$I \begin{bmatrix} x \\ y \\ z \end{bmatrix} = \begin{bmatrix} 2 & 1 & 1 \\ 1 & 2 & 0 \\ 2 & 0 & 1 \end{bmatrix} \begin{bmatrix} 5 \\ 2 \\ 4 \end{bmatrix}$$

$$\begin{bmatrix} x \\ y \\ z \end{bmatrix} = \begin{bmatrix} 16 \\ 9 \\ 14 \end{bmatrix}$$

Thus we have the solution to the system, $x = 16, y = 9,$ and $z = 14,$ or $(16, 9, 14).$ ■

Matrix Equations and Technology

We can use technology to multiply both sides of the matrix equation $AX = C$ by the inverse of A and get the solution to the system that the matrix equation represents. If we use technology to solve a system of linear equations, it is not necessary to display the inverse of the coefficient matrix.

EXAMPLE 8   Manufacturing

The Sharper Technology Company manufactures three types of calculators, a business calculator, a scientific calculator, and a graphing calculator. The production requirements are given in Table 5.5. If during each month, the company has 134,000 circuit components, 56,000 hours for assembly, and 14,000 cases, how many of each type of calculator can it produce each month?

TABLE 5.5

	Business Calculator	Scientific Calculator	Graphing Calculator
Circuit Components	5	7	12
Assembly Time (hours)	2	3	5
Cases	1	1	1

Solution

If the company produces x business calculators, y scientific calculators, and z graphing calculators, then the problem can be represented using the following system of equations.

$$\begin{cases} 5x + 7y + 12z = 134{,}000 \\ 2x + 3y + 5z = 56{,}000 \\ x + y + z = 14{,}000 \end{cases}$$

This system can be represented by the matrix equation $AX = C$, where A is the coefficient matrix and C is the constant matrix.

$$\begin{bmatrix} 5 & 7 & 12 \\ 2 & 3 & 5 \\ 1 & 1 & 1 \end{bmatrix} \begin{bmatrix} x \\ y \\ z \end{bmatrix} = \begin{bmatrix} 134{,}000 \\ 56{,}000 \\ 14{,}000 \end{bmatrix}$$

Multiplying both sides of this equation by the inverse of the coefficient matrix gives the solution. (See Figure 5.37.)

$$\begin{bmatrix} x \\ y \\ z \end{bmatrix} = A^{-1} \begin{bmatrix} 134{,}000 \\ 56{,}000 \\ 14{,}000 \end{bmatrix} = \begin{bmatrix} 2000 \\ 4000 \\ 8000 \end{bmatrix}$$

This shows that the company can produce 2000 business calculators, 4000 scientific calculators, and 8000 graphing calculators.

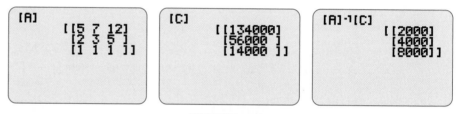

FIGURE 5.37

Recall that we could also solve this system of linear equations by using elementary row operations with an augmented matrix, discussed in Section 5.2. ■

We can also use inverse matrices with Excel to solve a system of linear equations, if a unique solution exists. We can solve such a system by finding the inverse of the coefficient matrix and multiplying this inverse times the matrix containing the constants. The spreadsheet (see Table 5.6) shows the solution of Example 8 found with Excel.

TABLE 5.6

	A	B	C	D
1	Matrix A	5	7	12
2		2	3	5
3		1	1	1
4				
5	Inverse of A	2	−5	1
6		−3	7	1
7		1	−2	−1
8				
9	Matrix C	134,000		
10		56,000		
11		14,000		

TABLE 5.6 *(continued)*

	A	B	C	D
12				
13	Solution X	2000		
14		4000		
15		8000		

We can use the inverse of the coefficient matrix to solve a system of linear equations only if the system has a unique solution. Otherwise, the inverse of the coefficient matrix will not exist, and an attempt to use technology to solve the system with an inverse will give an error message.

5.5 SKILLS CHECK

1. Suppose $A = \begin{bmatrix} 3 & 1 \\ 4 & 2 \end{bmatrix}$ and $B = \begin{bmatrix} 1 & -.5 \\ -2 & 1.5 \end{bmatrix}$.

   a. Compute AB and BA, if possible.
   b. What is the relationship between A and B?

2. Suppose $A = \begin{bmatrix} 1 & -1/2 & -1/4 \\ -1/2 & 0 & 1/2 \\ 0 & 1/2 & -1/4 \end{bmatrix}$ and

   $B = \begin{bmatrix} 2 & 2 & 2 \\ 1 & 2 & 3 \\ 2 & 4 & 2 \end{bmatrix}$.

   a. Compute AB and BA, if possible.
   b. What is the relationship between A and B?

3. Show that A and B are inverse matrices if

   $A = \begin{bmatrix} 1 & -1 & 1 \\ 2 & -1 & 0 \\ -2 & 2 & -1 \end{bmatrix}$ and $B = \begin{bmatrix} 1 & 1 & 1 \\ 2 & 1 & 2 \\ 2 & 0 & 1 \end{bmatrix}$.

4. Find the inverse matrix of $A = \begin{bmatrix} 1 & 3 \\ 2 & 7 \end{bmatrix}$.

5. Find the inverse matrix of $A = \begin{bmatrix} 2 & 4 \\ 2 & 5 \end{bmatrix}$.

6. Find the inverse matrix of $A = \begin{bmatrix} 1/2 & -1 & 1 \\ 1 & -2 & 0 \\ 1 & 2 & -1 \end{bmatrix}$.

7. Find the inverse matrix of $A = \begin{bmatrix} 2 & -2 & 2 \\ 2 & 1 & 2 \\ 2 & 0 & 1 \end{bmatrix}$.

8. Find the inverse of $B = \begin{bmatrix} 2 & 2 & 0 & 1 \\ 2 & 1 & 0 & 2 \\ 1 & 1 & 1 & 0 \\ 0 & 1 & 0 & 1 \end{bmatrix}$.

9. If $A^{-1} = \begin{bmatrix} 1 & 2 \\ 4 & 3 \end{bmatrix}$, solve $AX = \begin{bmatrix} 2 \\ 4 \end{bmatrix}$ for X.

10. If $A^{-1} = \begin{bmatrix} 1 & 2 & -1 \\ 0 & 2 & 1 \\ 2 & 0 & -2 \end{bmatrix}$, solve $AX = \begin{bmatrix} 1 \\ 2 \\ -2 \end{bmatrix}$ for X.

11. Solve the matrix equation
    $$\begin{bmatrix} -1 & 1 & 0 \\ -2 & 3 & -2 \\ 2 & -2 & 1 \end{bmatrix}\begin{bmatrix} x \\ y \\ z \end{bmatrix} = \begin{bmatrix} 3 \\ 5 \\ 8 \end{bmatrix}.$$

12. Solve the matrix equation
    $$\begin{bmatrix} -2 & 1 & 0 \\ -1 & 3 & -2 \\ 2 & -2 & 1 \end{bmatrix}\begin{bmatrix} x \\ y \\ z \end{bmatrix} = \begin{bmatrix} 2 \\ 4 \\ 8 \end{bmatrix}.$$

13. Use an inverse matrix to find the solution to the system
    $$\begin{cases} 4x - 3y + z = 2 \\ -6x + 5y - 2z = -3. \\ x - y + z = 1 \end{cases}$$

14. Use an inverse matrix to solve the system
$$\begin{cases} 5x + 3y + z = 12 \\ 4x + 3y + 2z = 9. \\ x + y + 2z = 2 \end{cases}$$

15. Use an inverse matrix to solve the system
$$\begin{cases} 2x + y + z = 4 \\ x + 4y + 2z = 4. \\ 2x + y + 2z = 3 \end{cases}$$

5.5 EXERCISES

----

1. *Competition* Two phone companies compete for customers in the southeastern region of a state. Each year Company X retains 3/5 of its customers and loses 2/5 of its customers to Company Y, while Company Y retains 2/3 of its customers and loses 1/3 to Company X. If we represent the fraction of the market held last year by

$$\begin{bmatrix} a \\ b \end{bmatrix},$$

where a is the number that Company X had last year and b is the number that Company Y had last year, then the number that each company will have this year can be found by

$$\begin{bmatrix} A \\ B \end{bmatrix} = \begin{bmatrix} 3/5 & 1/3 \\ 2/5 & 2/3 \end{bmatrix} \begin{bmatrix} a \\ b \end{bmatrix}.$$

If Company X has $A = 150{,}000$ customers and Company Y has $B = 120{,}000$ customers this year, how many customers did each have last year?

2. *Competition* A satellite company and a cable company compete for customers on an island. The satellite company retains 2/3 of its customers and loses 1/3 of its customers to the cable company, while the cable company retains 2/5 of its customers and loses 3/5 to the satellite company. If we represent the fraction of the market held last year by

$$\begin{bmatrix} a \\ b \end{bmatrix},$$

where a is the number of customers that the satellite company had last year and b is the number that the cable company had last year, then the number that each company will have this year can be found by

$$\begin{bmatrix} A \\ B \end{bmatrix} = \begin{bmatrix} 2/3 & 2/5 \\ 1/3 & 3/5 \end{bmatrix} \begin{bmatrix} a \\ b \end{bmatrix}.$$

If the satellite company has $A = 90{,}000$ customers and the cable company has $B = 85{,}000$ customers this year, how many customers did each have last year?

3. *Politics* In a midwestern state, it is observed that 90% of all Republicans vote for the Republican candidate for governor and the remainder for the Democratic candidate, while 80% of all Democrats vote for the Democratic candidate and the remainder vote for the Republican candidate. The percent voting for each party in the next election is given by

$$\begin{bmatrix} R \\ D \end{bmatrix} = \begin{bmatrix} .90 & .20 \\ .10 & .80 \end{bmatrix} \begin{bmatrix} r \\ d \end{bmatrix}$$

where r and d are the respective percents voting for each party in the last election. If 55% of the votes in this election were for a Republican and 45% were for a Democrat, what are the percents for each party candidate in the last election?

4. *Politics* Suppose that in a certain city the Democratic, Republican, and Consumer parties always nominate candidates for mayor. The percent of people voting for each party candidate in the next election depends on the percent voting for that party in the last election. The percent voting for each party in the next election is given by

$$\begin{bmatrix} D \\ R \\ C \end{bmatrix} = \begin{bmatrix} 50 & 40 & 30 \\ 40 & 50 & 30 \\ 10 & 10 & 40 \end{bmatrix} \begin{bmatrix} d \\ r \\ c \end{bmatrix}$$

where d, r, and c are the respective percents (in decimals) voting for each party in the last election. If the percent voting for a Democrat in this election is 42%, the percent voting for a Republican in this election is 42%, and the percent voting for the Consumer party candidate in this election is 16%, what were the respective percents of people voting for these parties in the last election?

5. *Loans* A bank gives three loans totaling $400,000 to a development company for the purchase of three business properties. The largest loan is $100,000 more than the sum of the other two, and the smallest loan is one-half of the next larger loan. Represent the amount of money in each loan as *x, y,* and *z,* respectively.
   a. Write a system of three equations in three variables to represent the problem.
   b. Use a matrix equation to solve the system. Find the amount of each loan.

6. *Transportation* Ross Freightline has an order for two products, A and B, to be delivered to a store. The table below gives the volume and weight for each unit of the two products.

	Product A	Product B
Unit Volume (cubic feet)	40	65
Unit Weight (pounds)	320	360

If the truck that can deliver the products can carry 5200 cubic feet and 31,360 lb., how many units of each product can the truck carry?

7. *Investment* An investor has $400,000 in three accounts, paying 6%, 8%, and 10%, respectively. If she has twice as much invested at 8% as she has at 6%, how much does she have invested in each account if she earns $36,000 in interest?

8. *Pricing* A theater has 1000 seats divided into orchestra, main, and balcony. The orchestra seats cost $80, the main seats cost $50, and the balcony seats cost $40. If all the seats are sold, the revenue is $50,000. If all the orchestra and balcony seats are sold and 3/4 of the main seats are sold, the revenue is $42,500. How many of each type of seat does the theater have?

9. *Venture Capital* Suppose a bank draws its venture capital funds annually from three sources of income: business loans, auto loans, and home mortgages. One spreadsheet shows the income from each of these sources for each of the 3 years 2001, 2002, and 2003, and the second spreadsheet shows the venture capital for these years. If the bank uses a fixed percent of its income from each of the business loans, find the percent of income from each of these loans.

	A	B	C	D
1	Income	From Loans	(Millions)	
2	Years	Business	Auto	Home
3	2001	532	58	682
4	2002	562	62	695
5	2003	578	69	722

	A	B
1	Venture	Capital (Millions)
2	2001	483.94
3	2002	503.28
4	2003	521.33

10. *Bookcases* A company produces three types of bookcases, 4-shelf metal, 6-shelf metal, and 4-shelf wooden. The 4-shelf metal bookcase requires 2 hours for fabrication, 1 hour to paint, and one-half hour to package for shipping. The 6-shelf bookcase requires 3 hours for fabrication, 1.5 hours to paint, and one-half hour to package, and the wooden bookcase requires 3 hours to fabricate, 2 hours to paint, and one-half hour to package. The daily amount of time available is 124 hours for fabrication, 68 hours for painting, and 24 hours for packaging. How many of each type of bookcase can be produced each day?

Encoding Messages We have encoded messages by assigning the numbers 1 to 26 to the letters a to z of the alphabet, respectively, and assigning 27 to a blank space.

a	b	c	d	e	f	g	h	i	j	k	l	m
1	2	3	4	5	6	7	8	9	10	11	12	13

n	o	p	q	r	s	t	u	v	w	x	y	z
14	15	16	17	18	19	20	21	22	23	24	25	26

To further encode the messages, we can use an encoding matrix A to convert these numbers into new pairs or triples of numbers. In Exercises 11–14, use the given matrix A to encode the given message.

11. a. Convert "Just do it" from letters to numbers.
    b. Multiply $A = \begin{bmatrix} 4 & 4 \\ 1 & 2 \end{bmatrix}$ times 2×1 matrices created with columns containing pairs of numbers

from part (a) to encode the message "Just do it" into pairs of coded numbers.

12. a. Convert "Call home" from letters to numbers.

b. Multiply $A = \begin{bmatrix} 2 & 3 \\ 2 & 2 \end{bmatrix}$ times 2×1 matrices created with columns containing pairs of numbers from part (a) to encode the message "Call home" into pairs of coded numbers.

13. Use the matrix $\begin{bmatrix} 4 & 4 & 4 \\ 1 & 2 & 3 \\ 2 & 4 & 2 \end{bmatrix}$ to encode the message "Neatness counts" into triples of numbers.

14. Use the matrix $\begin{bmatrix} 4 & 4 & 4 \\ 1 & 2 & 3 \\ 2 & 4 & 2 \end{bmatrix}$ to encode the message "Meet for lunch" into triples of numbers.

Decoding Messages We have encoded messages by assigning the numbers 1 to 26 to the letters a to z of the alphabet, respectively, and assigning 27 to a blank space. We can decode messages of this type by finding the inverse of the encoding matrix and multiplying it times the coded message. Use A^{-1} and the conversion table below to find the messages in Exercises 15–20.

a	b	c	d	e	f	g	h	i	j	k	l	m
1	2	3	4	5	6	7	8	9	10	11	12	13

n	o	p	q	r	s	t	u	v	w	x	y	z
14	15	16	17	18	19	20	21	22	23	24	25	26

15. The encoding matrix is $A = \begin{bmatrix} 3 & -1 \\ -2 & 1 \end{bmatrix}$ and the encoded message is 51, -29, 55, -35, 76, -49, -15 16, 11, 1.

16. The encoding matrix is $A = \begin{bmatrix} 2 & 3 \\ 2 & 4 \end{bmatrix}$ and the encoded message is 59, 74, 72, 78, 84, 102, 81, 90, 121, 148.

17. The encoding matrix is $A = \begin{bmatrix} -1 & 0 & 1 \\ -1 & 1 & 0 \\ 3 & -1 & -1 \end{bmatrix}$ and the encoded message is 1, -4, 16, 21, 23, -40, 3, 6, 6, -26, -14, 67, -9, 0, 23, 9, 1, 8.

18. The encoding matrix is $A = \begin{bmatrix} 1 & 1 & 0 \\ 1 & 3 & 0 \\ 5 & 3 & 3 \end{bmatrix}$ and the encoded message is 16, 34, 128, 32, 86, 145, 32, 86, 109, 29, 33, 195, 6, 8, 61.

19. Use the encoding matrix $A = \begin{bmatrix} 1 & 2 & -1 \\ 2 & 1 & -3 \\ -1 & 4 & 2 \end{bmatrix}$ to decode the message 29, -1, 75, -19, -66, 50, 46, 41, 47, 3, -38, 65.

20. Use the encoding matrix $A = \begin{bmatrix} 2 & 3 & -2 \\ 3 & 2 & -4 \\ -2 & 5 & 3 \end{bmatrix}$ to decode the message 20, -33, 142, 74, 51, 107, 67, 87, -26, 22, -22, 99.

21. *Coded Message*

a. Choose a partner and create a message to send to the partner. Change the message from letters to numbers with the code

a	b	c	d	e	f	g	h	i	j	k	l	m
1	2	3	4	5	6	7	8	9	10	11	12	13

n	o	p	q	r	s	t	u	v	w	x	y	z
14	15	16	17	18	19	20	21	22	23	24	25	26

b. Encode your message with encoding matrix $A = \begin{bmatrix} 1 & 2 & 1 \\ 2 & 1 & -3 \\ -1 & 4 & 2 \end{bmatrix}$.

c. Exchange coded messages.

d. Decode the message that you receive from your partner, and report your original message and your partner's message as you have decoded it.

CHAPTER 5 *Summary*

In this chapter we studied systems of equations and their applications. We first solved systems of linear equations in two variables by using graphical methods, substitution, or elimination. Then we solved systems of three linear equations in three variables by using analytical methods. Matrices were introduced to help solve systems of equations, and other applications of matrices were investigated.

Key Concepts and Formulas

5.1 Systems of Linear Equations in Two Variables

System of equations	A system of linear equations is a set of equations in two or more variables. A solution of the system must satisfy every equation in the system.

Solving a system of linear equations in two variables

• **Graphing**	We graph the equations and find their point of intersection.
• **Substitution**	We solve one of the equations for one variable, and substitute that expression into the other equation, thus giving an equation in one variable.
• **Elimination**	We rewrite one or both equations in a form that allows us to eliminate one of the variables by adding or subtracting the equations.

Possible solutions to a system of linear equations in two variables

• **Unique solution**	Graphs are intersecting lines.
• **No solution**	Graphs are parallel lines; system is **inconsistent**.
• **Many solutions**	Graphs are the same line; system is **dependent**.
• **Modeling systems of equations**	Solution of real problems sometimes requires us to create two or more equations whose simultaneous solution is the solution to the problem.

5.2 Systems of Linear Equations in Three Variables; Matrix Solution

Left-to-right elimination method	An extension of the elimination method used to solve two equations in two variables.

Matrices

• **Dimension**	The number of rows and columns of a matrix.
• **Square matrix**	Equal number of rows and columns.
• **Coefficient matrix**	Coefficients of a system placed in a matrix.
• **Augmented matrix**	A constant column is added to a coefficient matrix.
• **Row-echelon form**	The coefficient part has all 1's or 0's on its diagonal and all 0's below its diagonal.
• **Reduced row-echelon form**	The coefficient part has 1's or 0's on its diagonal and 0's elsewhere.

Matrix row operations	**1.** Interchanging two rows of the matrix.
	2. Multiplying a row by a nonzero constant.
	3. Adding a multiple of one row to another row.
Gauss-Jordan elimination method	Row operations are used to reduce the matrix to reduced row-echelon form. If the coefficient part reduces to the identity matrix, the system has a unique solution.

| Using Technology to Find Reduced Row-Echelon Form | • Graphing utilities and spreadsheets can be used to find the reduced row-echelon Form. |

5.3 Systems of Linear Equations with Nonunique Solutions

Systems of equations can have a unique solution, no solution, or infinitely many solutions.

• Unique solution	The coefficient matrix reduces to the identity matrix.
• No solution	A row of the reduced coefficient matrix contains all 0's and the augment of that row contains a nonzero number.
• Infinite number of solutions	A row of the reduced $n \times n$ coefficient matrix contains all 0's and the augment of that row contains 0 also.

5.4 Matrices; Basic Operations

Addition of matrices	Matrices of the same size are added by adding the corresponding entries of the two matrices.
Zero matrix	Each element is zero.
Negative of a matrix	The sum of a matrix and its negative is the zero matrix.
Subtraction of matrices	If the matrices M and N are the same size, the difference $M - N$ is found by subtracting the elements of N from the corresponding elements of M.
Identity matrix	An $n \times n$ matrix with 1's on the diagonal and 0's elsewhere.
Multiplication of matrices by a number	Multiplying a matrix A by a real number c results in a matrix in which each entry in matrix A is multiplied by the number c.
Matrix multiplication	The product of an $m \times n$ matrix A and an $n \times k$ matrix B is an $m \times k$ matrix $C = AB$ with the element in the ith row and jth column of matrix C given by the product of the ith row of A and the jth column of B.

5.5 Inverse Matrices; Matrix Equations

Inverse matrices	Two square matrices, A and B, are called **inverses** of each other if $AB = I$ and $BA = I$.
Finding the inverse of a square matrix	Placing the same size identity matrix in its augment and reducing the matrix to its reduced row-echelon form will give the inverse matrix in the augment if the inverse exists.
Inverses and technology	Computer software, spreadsheets, and calculators can be used to find the inverse of a matrix, if it exists.
Matrix equations	Multiplying both sides of the matrix equation $AX = C$ on the left by A^{-1}, if it exists, gives the solution to the system that the matrix equation represents.
Encoding messages and decoding messages	Matrix multiplication can be used to encode messages. The inverse of an encoding matrix can be used to decode messages.

Chapter 5 Skills Check

Solve the systems of linear equations in Exercises 1–9.

1. $\begin{cases} 3x + 2y = 0 \\ 2x - y = 7 \end{cases}$

2. $\begin{cases} 3x + 2y = -3 \\ 2x - 3y = 3 \end{cases}$

3. $\begin{cases} -4x + 2y = -14 \\ 2x - y = 7 \end{cases}$

4. $\begin{cases} -6x + 4y = 10 \\ 3x - 2y = 5 \end{cases}$

5. $\begin{cases} x + 2y - 2z = 1 \\ 2x - y + 5z = 15 \\ 3x - 4y + z = 7 \end{cases}$

6. $\begin{cases} -6x + 4y - 2z = 4 \\ 3x - 2y + 5z = -6 \\ x - 4y + z = -8 \end{cases}$

7. $\begin{cases} 2x + 5y + 8z = 30 \\ 18x + 42y + 18z = 60 \end{cases}$

8. $\begin{cases} 9x + 21y + 15z = 60 \\ 2x + 5y + 8z = 30 \\ x + 2y - 3z = -10 \end{cases}$

9. $\begin{cases} x + 3y + 2z = 5 \\ 9x + 12y + 15z = 6 \\ 2x + y + 3z = -10 \end{cases}$

Perform matrix operations, if possible, in Exercises 10–18, with

$$A = \begin{bmatrix} 1 & 3 & -3 \\ 2 & 4 & 1 \\ -1 & 3 & 2 \end{bmatrix}, \quad B = \begin{bmatrix} 1 & 2 & 1 \\ 2 & -1 & 3 \end{bmatrix},$$

$$C = \begin{bmatrix} 2 & 3 \\ -1 & 2 \\ 3 & -2 \end{bmatrix}, \quad \text{and } D = \begin{bmatrix} -2 & 3 & 1 \\ -3 & 2 & 2 \end{bmatrix}.$$

10. $B + D$

11. $D - B$

12. $5C$

13. AB

14. BA

15. CD

16. DC

17. $A^2 = A \cdot A$

18. A^{-1}

19. Solve the system of equations by using the inverse of the coefficient matrix.

$$\begin{cases} x + y - 3z = 8 \\ 2x + 4y + z = 15 \\ -x + 3y + 2z = 5 \end{cases}$$

Chapter 5 Review

1. *Break-Even* A computer manufacturer has a new product with daily total revenue given by $R = 565x$ and daily total cost given by $C = 6000 + 325x$. How many units per day must be produced and sold to give break-even for the product?

2. *Medication* Medication A is given 6 times per day and medication B is given twice per day. For a certain patient, the total intake of the two medications is limited to 25.2 mg per day. If the ratio of dosage of medication A to the dosage of medication B is 2 to 3, how many milligrams are in each dosage?

3. *Market Equilibrium* The demand for a certain brand of women's shoes is given by $3q + p = 340$ and the supply for these shoes is given by $p - 4q = -220$, where p is the price and q is the quantity demanded at price p. Solve the system containing these two equations to find the equilibrium price and the equilibrium quantity.

4. *Market Analysis* Suppose that for a certain product, the supply and demand functions are $p = \dfrac{q}{10} + 8$ and $10p + q = 1500$, respectively. Find the equilibrium price and quantity.

5. *Pricing* A concert promoter needs to make $120,000 from the sale of 2600 tickets. The promoter charges $40 for some tickets and $60 for the others.
   a. If there are x of the $40 tickets and y of the $60 tickets, write an equation that states that the total number of the tickets sold is 2600.
   b. How much money is made from the sale of x tickets for 40 dollars each?
   c. How much money is made from the sale of y tickets for 60 dollars each?
   d. Write an equation that states that the total amount made from the sale is $120,000.

e. Solve the equations simultaneously to find how many tickets of each type must be sold to yield the $120,000.

6. *Rental Income* A woman has $500,000 invested in two rental properties. One yields an annual return of 12% of her investment and the other returns 15% per year on her investment. Her total annual return from the two investments is $64,500. If x represents the 12% investment and y represents the 15% investment:

   a. Write an equation that states that the sum of the investments is $500,000.

   b. What is the annual return on the 12% investment?

   c. What is the annual return on the 15% investment?

   d. Write an equation that states that the sum of the annual returns is $64,500.

   e. Solve these two equations simultaneously to find how much is invested in each property.

7. *Investment* A woman invests $120,000 in two different mutual funds, one that averages 12% per year and another that averages 16% per year. If her average annual return on the two mutual funds is $16,000, how much did she invest in each fund?

8. *Property* A realty partnership purchases three business properties with a total cost of $375,000. One property costs $50,000 more than the second, and the third property costs half the sum of these two properties.

   a. Write an equation that represents the total loaned as the sum of the cost of the three properties, if the costs are x, y, and z, respectively.

   b. Write an equation that states that the cost of one property is $50,000 more than the second.

   c. Write an equation that states that the cost of the third property is equal to half the sum of the first two properties.

   d. Use matrices to solve the system of equations formed in parts (a) to (c).

   e. Find the cost of each property.

9. *Nutrition* A nutritionist wants to create a diet that uses a combination of three foods—I, II, and III. Each of these foods contains three additives, A, B, and C, that are used in the study. The table gives the percent of each additive that is present in each food. If the diet being used requires 12.5 g per day of A, 9.1 g of B, and 9.6 g of C, find the number of grams of each food that should be used each day for this diet.

	Food I	Food II	Food III
Additive A	10%	11%	18%
Additive B	12%	9%	10%
Additive C	14%	12%	8%

10. *Transportation* A delivery service has three types of aircraft, each of which carries three types of cargo. The payload of each type is summarized in the table below. Suppose that on a given day the airline must move 2200 next-day delivery letters, 3860 two-day delivery letters, and 920 units of air freight. How many aircraft of each type should be scheduled?

	Aircraft Type		
Units Carried	Passenger	Transport	Jumbo
Next-day letters	200	200	200
Two-day letters	300	40	700
Air freight	40	130	70

11. *Finance* A financial planner promises a long-term return of 12% from a combination of three mutual funds. The cost per share and average annual return per share for each mutual fund are given in the table below. If the annual returns are accurate, how many shares of each mutual fund will return an average of 12% on a total investment of $210,000?

	Tech Fund	Balanced Fund	Utility Fund
Cost per Share	$180	$210	$120
Annual Return per Share	$18	$42	$18

12. *Transportation* The Marshall Trucking Company has an order for three products, A, B, and C, for delivery. The following table gives the volume in cubic feet, the weight in pounds, and the value for insurance in dollars for a unit of each of the products. If one of the company's trucks can carry 9260 cubic feet and 12,000 lb, and is insured to carry $52,600, how many units of each product can be carried on the truck?

	Product A	Product B	Product C
Unit Volume (cubic feet)	25	30	40
Weight (pounds)	30	36	60
Value (dollars)	150	180	200

Canada (Millions of dollars)		
Year	Exports	Imports
1992	90,594	98,630
1994	114,439	128,406
1996	134,210	155,893
1998	156,603	173,256
1999	166,600	198,711

13. *Nutrition* A biologist is growing three types of slugs (A, B, and C) in the same laboratory environment. Each day, the slugs are given different nutrients (I, II, and III). Each type A slug requires 2 units of I, 6 units of II, and 2 units of III per day. Each type B slug requires 2 units of I, 8 units of II, and 4 units of III per day. Each type C slug requires 4 units of I, 20 units of II, and 12 units of III per day. If the daily mixture contains 4000 units of I, 16,000 units of II, and 8000 units of III, find the number of slugs of each type that can be supported.

14. *Trade Balances* The tables below give the values of U.S. exports and imports with Mexico and with Canada for selected years 1992 to 1999.
  a. Form the matrix A that contains the values of exports to Mexico and to Canada for these years.
  b. Form the matrix B that contains the values of imports from Mexico and from Canada for these years.
  c. Use these two matrices to find the trade balances for the U.S. for these years.
  d. Are the trade balances with these countries improving for the U.S.?
  e. With which country is the trade balance worse in 1999?

Mexico (Millions of dollars)		
Year	Exports	Imports
1992	40,592	36,211
1994	50,844	49,494
1996	56,792	74,297
1998	78,773	94,629
1999	86,909	109,721

15. *Traffic Study* In the analysis of traffic, a city engineer estimates the traffic flow at noon on the town square to be as illustrated in the figure below. If x_1 illustrates the number of cars moving from intersection A to intersection B, x_2 represents the number of cars traveling from intersection B to intersection C, and so on, we can formulate equations based on the principle that the number of vehicles entering the intersection equals the number leaving it. For example, the equation that represents the traffic through A is $x_4 + 2250 = x_1 + 2900$.
  a. Formulate equations for the traffic at the other intersections.
  b. Solve the system of these four equations, to find how traffic between the other intersections is related to the traffic from intersection D to intersection A.

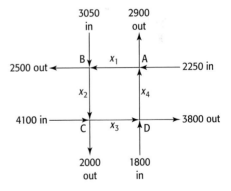

16. *Income and SAT scores* The table on page 484 gives the average verbal and math SAT scores for certain family income levels.
  a. Use linear regression to find a linear equation that gives the verbal SAT score as a function of family income.
  b. Use linear regression to find a linear equation that gives the math SAT score as a function of family income.

c. At what income level do these models predict that the verbal SAT score will reach the math SAT?

Family Income ($ thousands)	Verbal SAT	Math SAT
5	425	445
15	445	460
25	475	480
35	480	490
45	501	505
55	510	515
65	520	525
75	525	530

Family Income ($ thousands)	Verbal SAT	Math SAT
90	535	540
110	560	570

(Source: *Newsweek*, March 5, 2001)

17. *Medication* Suppose that combining x cubic centimeters (cc) of a 20% concentration of a medication and y cc of a 5% concentration of the medication gives 10 cc of a 15.5% concentration.
   a. Write the equation that gives the total amount of solution in terms of x and y.
   b. Write the equation that gives the total amount of medication in the solution.
   c. Use substitution to find the amount of each solution that is needed.

GROUP ACTIVITY/EXTENDED APPLICATION *1*

Salaries

Salaries of College Professors, 1998–99

TEACHING LEVEL	MEN			WOMEN		
	Type of Institution			Type of Institution		
	Public	Private/ Independent	Church-related	Public	Private/ Independent	Church-related
Doctoral level schools						
Professor	$80,379	$99,979	$84,796	$72,885	$90,611	$77,972
Associate	57,653	65,843	60,059	54,322	61,956	56,180
Assistant	48,647	57,296	50,009	45,203	52,521	46,427
Master's level schools						
Professor	64,414	70,643	66,151	61,711	65,593	60,588
Associate	51,812	54,260	52,634	49,615	51,273	48,189
Assistant	42,673	44,511	42,317	41,189	43,002	40,312
General 4-year schools						
Professor	$58,432	68,145	52,945	$57,045	64,089	49,678
Associate	48,643	51,044	43,412	46,808	49,202	41,791
Assistant	40,625	41,551	36,534	39,245	40,634	36,017
2-year schools						
Professor	57,067	45,099	36,422	52,461	40,252	35,496
Associate	48,321	40,515	36,359	44,835	35,513	34,609
Assistant	41,515	35,715	30,342	39,561	34,219	29,774

(Source: American Association of University Professors)

The table above gives the salaries of college professors for men and women who teach at three types of institutions at four different levels of instruction, and at three job designations. To investigate the relationship between salaries of males and females:

1. Create matrix *A* and matrix *B*, which contain the salaries of male and female professors, respectively, in each type of school and at each level of instruction.

2. Subtract the matrix of male professor salaries from the matrix of female professor salaries.

3. What does this show about female salaries versus male salaries in every category? Based on this data, discuss gender bias in educational institutions.

4. Create a scatter plot with male professor salaries as the input and female professor salaries as the output.

5. Find a function that gives a relationship between the salaries of male and female professors from part 4.
6. Graph the scatter plot from part 4 and the function from part 5 on the same axes.
7. The graph of the function $y = x$ would represent the relationship between male and female professors' salaries if they were equal. Graph the function from part 5 and the function $y = x$ on the same axes.

8. Use the two graphs in part 7 to determine how female professors' salaries compare to male salaries as male salaries rise.
9. Solve the two equations whose graphs are discussed in part 8 to determine at what salary level male salaries become greater than female salaries.

GROUP ACTIVITY/EXTENDED APPLICATION 2

Parts-Listing

Wingo Playgrounds, Inc. must maintain numerous parts in its inventory for its Fun-in-the-Sun swing set. The swing set has 4 legs and a top connecting the legs. Each of the legs is made from 1 9-foot pipe connected with 1 brace and 2 bolts, and the top is made of 1 9-foot pipe, 2 clamps, and 6 bolts. The parts-listing for these swing sets can be described by the following matrix.

$$P = \begin{bmatrix} SS & L & T & P & C & Bc & B & \\ 0 & 0 & 0 & 0 & 0 & 0 & 0 & \text{Swing sets} \\ 4 & 0 & 0 & 0 & 0 & 0 & 0 & \text{Legs} \\ 1 & 0 & 0 & 0 & 0 & 0 & 0 & \text{Top} \\ 0 & 1 & 1 & 0 & 0 & 0 & 0 & \text{Pipes} \\ 0 & 0 & 2 & 0 & 0 & 0 & 0 & \text{Clamps} \\ 0 & 1 & 0 & 0 & 0 & 0 & 0 & \text{Braces} \\ 0 & 2 & 6 & 0 & 0 & 0 & 0 & \text{Bolts} \end{bmatrix}$$

Each column indicates how many of each part are required to produce the part shown at the top of the column. For example, column 1 indicates that to produce a swing set requires 4 legs and 1 top, and the second column indicates that each leg (L) is constructed from 1 pipe, 1 brace, and 2 bolts.

Suppose that an order is received for 8 complete swing sets plus the following spare parts: 2 legs, 1 top, 2 pipes, 4 clamps, 4 braces, and 8 bolts. We indicate the number of each part type that must be supplied to fill this order by the matrix

$$X = \begin{bmatrix} x_1 \\ x_2 \\ x_3 \\ x_4 \\ x_5 \\ x_6 \\ x_7 \end{bmatrix}$$

To find the number of each part that is needed, perform the following:

1. Form the 7×1 matrix D that contains the number of items to be delivered to fill the order for the 8 swing sets and the spare parts. Some parts of each type are needed to produce other parts of the swing set. The matrix PX gives the number of parts of each type used to produce other parts, and the matrix

$$X - PX$$

gives the number of parts of each type that are available to be delivered. That is,

$$X - PX = D,$$

where D is the number of parts of each type that is available for delivery.
2. The matrix equation $X - PX = D$ can be rewritten in the form $(I - P)X = D$, where I is the 7×7 identity matrix. Find the matrix $I - P$ that satisfies this equation.
3. Find the inverse of the matrix $I - P$.
4. Multiply both sides of the matrix equation $(I - P)X = D$ by the inverse of $I - P$. This gives the matrix X that contains the number of parts of each type that is needed.
5. The primary assembly parts are the pipes, clamps, braces, and bolts. What is the total number of primary parts of each type needed to fill the order?

6

Special Topics: Systems of Inequalities and Linear Programming; Sequences and Series; Preparing for Calculus

We can solve maximization and minimization problems subject to constraints (inequalities involving the variables) with a technique called **linear programming**, which is an important application of the solution of **systems of linear inequalities** in two or more variables. Linear programming is especially useful to enable a firm to plan its resources in the best possible way.

A function whose domain is a set of positive integers is called a **sequence**. The ordered outputs corresponding to the integer inputs are called the **terms of the sequence**. The sum of the terms of a sequence is called a **series**. Series are useful in finding the future values of annuities if the interest is compounded over discrete time periods. Series are also useful in computing infinite sums, in approximating numbers such as π and e, and in approximating functions.

Courses in **calculus** depend heavily on algebra skills, so we include a Preparing for Calculus section that reviews algebra skills learned earlier in the context of calculus development and applications.

		TOPICS	**APPLICATIONS**
6.1	Systems of Linear Inequalities	Linear inequalities in two variables; systems of inequalities in two variables	Auto purchases, car rental, advertising
6.2	Linear Programming: Graphical Methods	Linear programming; solution with technology	Maximizing car rental profit, cost minimization, profit
6.3	Sequences and Discrete Functions	Sequences; arithmetic sequences; geometric sequences	Football contracts, investments, depreciation, rebounding ball, compound interest
6.4	Series	Finite series; infinite series; arithmetic series; geometric series; infinite geometric series	Football contracts, profit, depreciation, annuities
6.5	Preparing for Calculus	Simplifying expressions; evaluating functions; simplifying fractions; rewriting radicals; equation solution; decomposition of functions; rewriting expressions using negative exponents; rewriting expressions using positive exponents; using logarithmic properties; reducing fractions	Slope of a tangent line, evaluating the difference quotient

6.1 Systems of Linear Inequalities

A rental agency has a maximum of $1,260,000 to invest in the purchase of at most 71 new cars of two different types, compact and midsize. The cost per compact car is $15,000 and the cost per midsize car is $28,000. The number of cars of each type is limited (constrained) by the budget available and the number of cars needed. To get a "picture" of the constraints on this rental agency, we can graph the inequalities determined by the maximum amount of money and the limit on the number of cars purchased. Because these inequalities deal with two types of cars, they will contain two variables. For example, we can denote the number of compact cars by x and the number of midsize cars by y, so the statement that the agency wants to purchase at most 71 cars can be written as the inequality

$$x + y \le 71, \text{ where } x \text{ and } y \text{ are integers.}$$

The collection of all **constraint** inequalities can be expressed by a **system of inequalities** in two variables. Finding the values that satisfy all these constraints at the same time is called solving the system of inequalities.

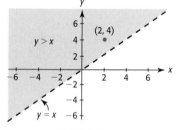

FIGURE 6.1

Linear Inequalities in Two Variables

We solved linear inequalities in one variable in Section 1.8. We now consider linear inequalities in *two* variables. For example, consider the inequality $y > x$. The solutions to this inequality are the ordered pairs (x, y) that satisfy the statement $y > x$. The graph of $y > x$ consists of all points above the line $y = x$ (the shaded area in Figure 6.1). The line $y = x$ is dashed in Figure 6.1 because the inequality does not include $y = x$. This (dashed) line divides the plane into two **half-planes**, $y > x$ and $y < x$. We can determine which half-plane is the solution to the inequality by selecting any point not on the line. If the coordinates of this **test point** satisfy the inequality, then the half-plane containing that point is the graph of the solution. If the test point does not satisfy the inequality, then the other half-plane is the graph of the solution. For example, the point $(2, 4)$ is above the line $y = x$ and $x = 2, y = 4$ satisfies the inequality $y > x$, so the region above the line is the graph of the solution of the inequality. Note that any other point on the half-plane above the line also satisfies the inequality.

EXAMPLE 1 **EXAMPLE 1** Graph of an Inequality

Graph the solution of the inequality $6x - 3y \le 15$.

Solution

First we graph the line $6x - 3y = 15$ (which we can also write as $y = 2x - 5$) as a solid line, because the inequality symbol " $\le$ " indicates that points on the line are part of the solution. (See Figure 6.2(a).) Next we pick a test point that is not on the line. If we use $(0, 0)$, we get $6(0) - 3(0) \le 15$, or $0 \le 15$, which is true, so the inequality is satisfied. Thus the half-plane that contains the point $(0, 0)$ and the line $6x - 3y = 15$ is the graph of the solution, which we call the **solution region** of the inequality.

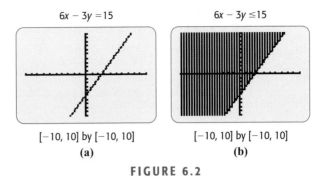

FIGURE 6.2

TECHNOLOGY NOTE

Graphing utilities can also be used to shade the solution region of an inequality. Figure 6.2(b) shows a graphing calculator window with the solution to $6x - 3y \leq 15$ shaded.

EXAMPLE 2 Car Rental Agency

One of the constraint inequalities for the car rental company application given in the introduction is that the sum of x and y (the number of compact and midsize cars, respectively) is at most 71. This constraint can be described by the inequality

$$x + y \leq 71,$$

where x and y are integers. Find the graphical solution of this inequality.

Solution

The inequality $x + y \leq 71$, where x and y are integers, has many integer solutions. The inequality is satisfied if x is any integer from 0 to 71 and y is any integer less than or equal to $71 - x$. For example, some of the solutions are (0, 71), (1, 70), (60, 5), and (43, 20). As we stated in Chapter 1, problems are frequently easier to solve if we treat the variables as continuous rather than discrete. Assuming that this inequality is true for all real numbers permits us to graph the real solutions to the inequality. With this in mind, we graph the inequality as a region on a coordinate plane rather than attempting to write all of the possible integer solutions.

We create the graph of the inequality by first graphing the equation

$$x + y = 71, \text{ or equivalently, } y = 71 - x.$$

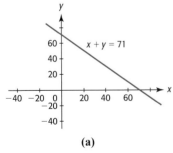

(a)

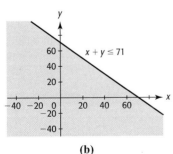

(b)

FIGURE 6.3

(See Figure 6.3(a).) The graph of this line divides the coordinate plane into two half-planes, $y > 71 - x$ and $y < 71 - x$. We can find the half-plane that represents the graph of $y < 71 - x$ by testing the coordinates of a point and determining if it satisfies $y < 71 - x$. For example, the point (0, 0) is in the half-plane below the line, and $x = 0, y = 0$ satisfies the inequality $y < 71 - x$ because $0 < 71 - 0$. Thus the half-plane below the line satisfies the inequality. If we shade the half-plane that satisfies $y < 71 - x$, the line and the shaded region constitute the graph of $x + y \leq 71$. (See Figure 6.3(b).) This is the solution region for the inequality.

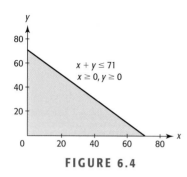

FIGURE 6.4

Note that the number of cars satisfying this constraint has other limitations. In the context of this application, the number of cars cannot be negative, so the values of x and y must be nonnegative. We can show the solution to the inequality for nonnegative x and y by using a window with $x \geq 0$ and $y \geq 0$. Figure 6.4 shows the solution of $x + y \leq 71$ for $x \geq 0$ and $y \geq 0$.

All points in this region with integer coordinates satisfy the stated conditions of this example. ∎

We now investigate the second inequality in the car rental problem.

EXAMPLE 3 Rental Cars

The car rental agency mentioned above has a maximum of $1,260,000 to invest, with x compact cars costing $15,000 each and y midsized cars costing $28,000 each.

a. Write the inequality representing this information.

b. Graph this inequality on a coordinate plane.

Solution

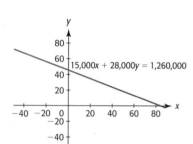

FIGURE 6.5

a. If x represents the number of compact cars, the cost of all the compact cars purchased is $15,000x$ dollars. Similarly, the cost of all the midsized cars purchased is $28,000y$ dollars. Because the total amount available to spend is 1,260,000 dollars, x and y satisfy

$$15,000x + 28,000y \leq 1,260,000.$$

b. To graph the solutions to this inequality, we first graph the equation

$$15,000x + 28,000y = 1,260,000, \text{ or equivalently,}$$

$$y = \frac{1,260,000 - 15,000x}{28,000} = \frac{1260 - 15x}{28}.$$

The graph is shown in Figure 6.5.

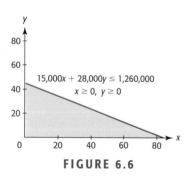

FIGURE 6.6

To determine which of the two half-planes determined by the line satisfies the inequality, we test points in each half-plane. The point $(0, 0)$ is in the half-plane below the line, and it satisfies the inequality because $15,000(0) + 28,000(0) < 1,260,000$. Thus the half-plane containing $(0, 0)$, along with the line, forms the solution region. Because x and y must be nonnegative in this application, the graph of the inequality is shown only in the first quadrant. (See Figure 6.6.) All points in this region with integer coordinates satisfy the stated conditions of this example. ∎

Systems of Inequalities in Two Variables

We now have pictures of the limitations (constraints) on the car purchases of the car rental agency, but how are they related? We know that the number of cars the rental agency can buy is limited by *both* inequalities, so we seek the solution to both inequalities *simultaneously.* That is, we seek the solution to a **system of inequalities**. In general, if we have two or more inequalities in two variables and seek the values of the variables that satisfy both inequalities, we are solving a system of inequalities. The solution to the system can be found by finding the intersection of the solution sets of the inequalities.

EXAMPLE 4   System of Inequalities

Solve the system of inequalities $\begin{cases} 2x - 4y \geq 12 \\ x + 3y > -4 \end{cases}$ by graphing.

Solution

To find the solution graphically, we first solve the inequalities for y.

$$y \leq \frac{1}{2}x - 3 \quad \text{(Inequality 1)}$$

$$y > -\frac{1}{3}x - \frac{4}{3} \quad \text{(Inequality 2)}$$

The "borders" of the inequality region are graphed as the solid line with equation $y = \frac{1}{2}x - 3$ and the dashed line with equation $y = -\frac{1}{3}x - \frac{4}{3}$. One of these "borders" is a solid line because $y = \frac{1}{2}x - 3$ is part of inequality (1) and one is a dashed line because $y = -\frac{1}{3}x - \frac{4}{3}$ is not part of inequality (2). (See Figure 6.7(a).) We find a "corner" of the solution region by finding the intersection of the two lines. We can find this point of intersection by analytically or graphically solving the equations simultaneously. Using substitution and solving gives the point of intersection:

$$\frac{1}{2}x - 3 = -\frac{1}{3}x - \frac{4}{3}$$
$$3x - 18 = -2x - 8$$
$$5x = 10$$
$$x = 2$$
$$y = -2$$

The two "border" lines divide the plane into four regions, one of which is the solution region. We can determine what region satisfies both inequalities by choosing a test point in each of the four regions. For example:

$(0, 0)$ and $(1, -2)$ do not satisfy inequality (1).

$(2, -4)$ does not satisfy inequality (2).

$(6, -1)$ satisfies both inequalities, so the region that contains this point is the solution region.

Figure 6.7(b) shows the solution region for the system of inequalities, with its corner at $(2, -2)$. ∎

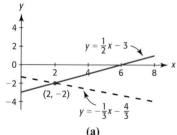

(a)

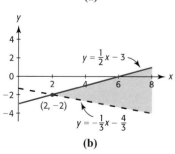

(b)

FIGURE 6.7

We now turn our attention to finding the values of x and y that satisfy both the constraints on the car rental agency.

EXAMPLE 5 Car Rental Agency

The inequalities that satisfy the conditions given in the car rental example form the system

$$\begin{cases} y \le 71 - x \\ y \le \dfrac{1260 - 15x}{28} \\ x \ge 0 \\ y \ge 0 \end{cases}$$

a. Graph the solution to the system.

b. Find the coordinates of the points where the borders intersect and the points where the region intersects the x and y axes.

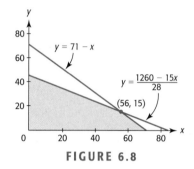

FIGURE 6.8

Solution

a. We graphed the solutions to the two inequalities discussed for the car rental agency application in Examples 2 and 3, respectively. If we graph them both on the same coordinate plane, the region where the two solution regions "overlap" will show the region where both inequalities are true simultaneously. (See Figure 6.8.) Any point in the solution region with integer coordinates is a solution to the application. For example, $x = 50, y = 8$ is a solution, as is $x = 6, y = 40$.

b. The region is bordered on one side by the *line*

$$y = 71 - x$$

and on the other side by the *line*

$$y = \frac{1260 - 15x}{28}.$$

A **corner** of the graph of the solution set occurs at the *point* where these two lines intersect. We find this point of intersection graphically as shown in Figure 6.8 or by solving the two equations simultaneously.

$$\begin{cases} y = 71 - x \\ y = \dfrac{1260 - 15x}{28} \end{cases}$$

$$71 - x = \frac{1260 - 15x}{28}$$

$$1988 - 28x = 1260 - 15x$$

$$728 = 13x$$

$$x = 56 \text{ and } y = 71 - 56 = 15$$

Thus this point of intersection of the borders of the solution region, or the *corner*, occurs at $x = 56, y = 15$.

This is one of many possible solutions to the rental application, and corresponds to purchasing 56 compact cars and 15 midsized cars.

Other corners occur where the region intersects the axes:

- At the origin ($x = 0$, $y = 0$, representing no cars purchased).
- Where $y = 71 - x$ intersects the x-axis ($x = 71$, $y = 0$, representing 71 compact and no midsize cars).
- Where $y = \dfrac{1260 - 15x}{28}$ intersects the y-axis ($x = 0$, $y = 45$, representing 0 compact and 45 midsize cars). ■

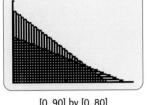

[0, 90] by [0, 80]

FIGURE 6.9

Computer and calculator programs can be used to save considerable time and energy in determining regions that satisfy systems of inequalities. We can also use technology to graph the two lines shown in Figure 6.8 and to find their point of intersection, and we can use ⎡SHADE⎤ to show the region that is the intersection of the inequalities. (See Figure 6.9.)

EXAMPLE 6   Advistising

A candidate for mayor of a city wishes to use a combination of radio and television advertisements in her campaign. Research has shown that each 1-minute spot on television reaches 0.09 million people and each 1-minute spot on radio reaches 0.006 million. The candidate feels that she must reach at least 2.16 million people, and she can buy a total of no more than 80 minutes of advertisement time.

a. Write the inequalities that describe her needs.

b. Graph the region determined by these constraint inequalities.

c. Interpret the solution region in the context of this problem.

Solution

a. If we represent by x the number of minutes of television time, the total number of people reached by television is $0.09x$ million, and if we represent by y the number of minutes of radio time, the total number reached by radio is $0.006y$ million. Thus one condition that must be satisfied is given by the inequality

$$0.09x + 0.006y \geq 2.16.$$

In addition, the statement that the total number of minutes of advertising can be no more than 80 minutes can be written as the inequality

$$x + y \leq 80.$$

Neither x nor y can be negative in this application, so we also have

$$x \geq 0$$
$$y \geq 0$$

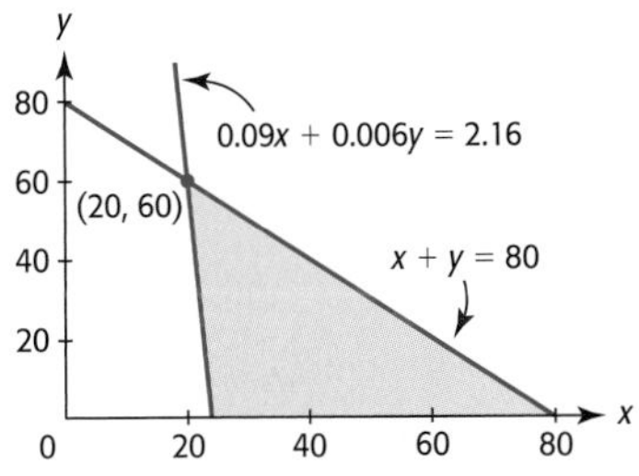

b. Graphing the equations corresponding to these inequalities and shading the regions satisfying the inequalities give the graph of the solution. (See Figure 6.10.)

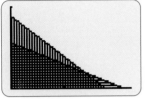

FIGURE 6.10

One corner of the region occurs where the boundary lines intersect. We can write the equations of the boundary lines as

$$0.09x + 0.006y = 2.16 \text{ and } y = 80 - x$$

Their y-values are equal where the lines intersect, so we substitute for y in the second equation and solve.

$$0.09x + 0.006(80 - x) = 2.16$$
$$0.09x + 0.48 - 0.006x = 2.16$$
$$0.084x = 1.68$$
$$x = 20$$
$$y = 60$$

The other corners occur where the region intersects the axes:

- Where $0.09x + 0.006y = 2.16$ intersects the x-axis, at $x = 24, y = 0$.
- Where $x + y = 80$ intersects the x-axis, at $x = 80, y = 0$.

c. Each point in this region represents a combination of television and radio time that the mayor can use in her campaign. No point outside this region can represent a number of minutes used in television and radio advertising for this candidate, because of the constraints given in the problem. ∎

6.1 SKILLS CHECK

In Exercises 1–4, graph each inequality.

1. $y \leq 5x - 4$

2. $y > 3x + 2$

3. $6x - 3y \geq 12$

4. $\dfrac{x}{2} + \dfrac{y}{3} \leq 6$

In Exercises 5–7, the graph of the boundary equations for each system of inequalities is shown with the system. Locate the solution region, and identify it by finding the corners.

5. $\begin{cases} x + y \leq 5 \\ 2x + y \leq 8 \\ x \geq 0, y \geq 0 \end{cases}$

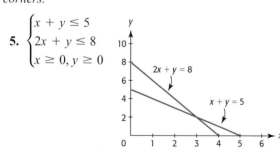

6. $\begin{cases} 2x + y \leq 12 \\ x + y \leq 8 \\ 2x + y \leq 14 \\ x \geq 0, y \geq 0 \end{cases}$

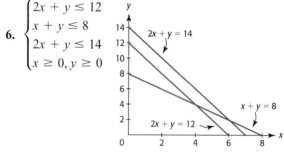

7. $\begin{cases} 2x + 6y \geq 12 \\ 3x + y \geq 5 \\ x + 2y \geq 5 \\ x \geq 0, y \geq 0 \end{cases}$

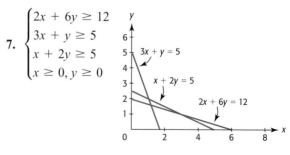

For each system of inequalities in Exercises 8–12, graph the solution region and identify the corners of the region.

8. $\begin{cases} y < 2x \\ y > x + 2 \\ x \geq 0, y \geq 0 \end{cases}$

9. $\begin{cases} y \leq 8 - 3x \\ y \leq 2x + 3 \\ y > 3 \end{cases}$

10. $\begin{cases} 2x + y < 5 \\ 2x - y > -1 \\ x \geq 0, y \geq 0 \end{cases}$

11. $\begin{cases} x + 2y \geq 4 \\ x + y \leq 5 \\ 2x + y \leq 8 \\ x \geq 0, y \geq 0 \end{cases}$

12. $\begin{cases} 2x + y \geq 10 \\ 3x + 2y \geq 17 \\ x + 2y \geq 7 \end{cases}$

6.1 EXERCISES

----

1. *Manufacturing* A company manufactures two types of leaf blowers, an electric "Turbo" model and a gas-powered "Tornado" model. The company's production plan calls for the production of at least 780 blowers per month.
   a. Write the inequality that describes the production plan, if x represents the number of Turbo blowers and y represents the number of Tornado blowers.
   b. Graph the region determined by this inequality in the context of the application.

2. *Manufacturing* A company manufactures two types of leaf blowers, an electric "Turbo" model and a gas-powered "Tornado" model. It costs $78 to produce each of the x Turbo models and $117 to produce each of the y Tornado models, and the company has at most $76,050 per month to use for production.
   a. Write the inequality that describes this constraint on production.
   b. Graph the region determined by this inequality in the context of the application.

3. *Sales* Trix Auto Sales sells used cars. To advertise its cars to target groups, they advertise with x 1-minute spots on cable television, at a cost of $240 per minute, and y 1-minute spots on radio, at a cost of $150 per minute. Suppose the company has at most $36,000 to spend on advertising.
   a. Write the inequality that describes this constraint on advertising.
   b. Graph the region determined by this inequality in the context of the application.

4. *Sales* Trix Auto Sales sells used cars. To advertise its cars to target groups, they advertise with x 1-minute spots on cable television and y 1-minute spots on radio. Research shows that they sell one vehicle for every 4

minutes of cable television advertising and they sell one vehicle for each 10 minutes of radio advertising.
   a. Write the inequality that describes the requirement that at least 33 cars be sold from this advertising.
   b. Graph the region determined by this inequality in the context of the application.

5. *Politics* A political candidate wishes to use a combination of x newspaper and y television advertisements in his campaign. Each 1-minute ad on television reaches 0.12 million eligible voters and each 1-minute ad on radio reaches 0.009 million eligible voters. The candidate feels he must reach at least 7.56 million eligible voters and that he must buy at least 100 minutes of advertisements.
   a. Write the inequalities that describes these advertising requirements.
   b. Graph the region determined by these inequalities in the context of the application.

6. *Housing* A contractor builds two models of homes, the Van Buren and the Jefferson. The Van Buren requires 200 worker-days of labor and $240,000 in capital, while the Jefferson requires 500 worker-days of labor and $300,000 in capital. The contractor has a total of 5000 worker-days and $3,600,000 in capital available per month. Let x represent the number of Van Buren models and y represent the number of Jefferson models and graph the region that satisfies these inequalities.

7. *Manufacturing* A company manufactures two types of leaf blowers, an electric "Turbo" model and a gas-powered "Tornado" model. The company's production plan calls for the production of at least 780 blowers per month. It costs $78 to produce each Turbo model and $117 to manufacture each Tornado model, and the company has at most $76,050 per month to use for

production. Let *x* represent the number of Turbo models and *y* represent the number of Tornado models, and graph the region that satisfies these constraints.

8. *Sales* Trix Auto Sales sells used cars. To advertise its cars to target groups, they advertise with *x* 1-minute spots on cable television at a cost of $240 per minute, and *y* 1-minute spots on radio, at a cost of $150 per minute. Research shows that they sell one vehicle for every 4 minutes of cable television advertising (that is, 1/4 car per minute) and they sell one vehicle for each 10 minutes of radio advertising. Suppose the company has at most $36,000 to spend on advertising and that they sell at least 33 cars per month from this advertising. Let *x* represent the number of cable TV minutes and *y* represent the number of radio minutes, and graph the region that satisfies these constraints. That is, graph the solution to this system of inequalities.

9. *Production* A firm produces three different-size television sets on two assembly lines. The following table summarizes the production capacity of each assembly line and the minimum number of each size TV needed to fill orders.
   a. Write the inequalities that are the constraints.
   b. Graph the constraint region, and identify the corners.

	Assembly Line 1	Assembly Line 2	Number Ordered
19-inch TV	80 per day	40 per day	3200
25-inch TV	20 per day	20 per day	1000
35-inch TV	100 per day	40 per day	3400

10. *Advertising* Tire Town is developing an advertising campaign. The table below indicates the cost per ad package in newspapers and the cost per ad package on radio, and the number of ads in each type of ad package.

	Newspaper	Radio
Cost per ad package	$1000	$3000
Ad per package	18	36

The owner of the company can spend no more than $18,000 per month, and he wants to have at least 252 ads per month.
   a. Write the inequalities that are the constraints.
   b. Graph the constraint region, and identify the corners.

11. *Nutrition* A privately owned lake contains two types of fish, bass and trout. The owner provides two types of food, A and B, for these fish. Trout require 4 units of food A and 6 units of food B, and bass require 10 units of food A and 7 units of food B. If the owner has 1600 units of each food, graph the region that satisfies the constraints.

12. *Manufacturing* The Evergreen Company produces two types of printers, the Inkjet and the Laserjet. It takes 2 hours to make the Inkjet and 6 hours to make the Laserjet. The company can make at most 120 printers per day and has 400 labor-hours available per day.
   a. Write the inequalities that describe this application.
   b. Graph the region that satisfies the inequalities, and identify the corners of the region.

13. *Manufacturing* Easyboy manufactures two types of chairs, Standard and Deluxe. Each Standard chair requires 4 hours to construct and finish, and each Deluxe chair requires 6 hours to construct and finish. Upholstering takes 2 hours for a Standard chair and 6 hours for a Deluxe chair. There are 480 hours available each day for construction and finishing and there are 300 hours available per day for upholstering.
   a. Write the inequalities that describe the application.
   b. Graph the solution of the system of inequalities, and identify the corners of the region.

14. *Manufacturing* The Safeco Company produces two types of chainsaws, the Safecut and the Safecut Deluxe. The Safecut model requires 2 hours to assemble and 1 hour to paint and the Deluxe model requires 3 hours to assemble and 1/2 hour to paint. The daily maximum number of hours available for assembly is 36, and the daily maximum number of hours available for packing is 12.
   a. Write the inequalities that describe the application.
   b. Graph the solution of the system of inequalities, and identify the corners of the region.

15. *Manufacturing* A company manufactures commercial and domestic heating systems at two plant sites. It can produce no more than 1400 units per month, and it needs to fill orders of at least 500 commercial units and 750 domestic units.
   a. Write the system of inequalities that describe the constraints on production for these orders.
   b. Graph the solution set of this system of inequalities.

6.2 Linear Programming: Graphical Methods

Key Concepts

- Linear programming
- Constraints
- Objective function
- Feasible solution
- Feasible region
- Optimum value
  Maximum
  Minimum

A rental agency has a maximum of $1,260,000 to invest in the purchase of at most 71 new cars of two different types, compact and midsize. The cost per compact car is $15,000 and the cost per midsize car is $28,000. The number of cars of each type is limited (constrained) by the budget available and the number of cars needed. We saw in the previous section that purchases of the rental agency were constrained by the number of cars needed and the amount of money available for purchasing the cars. The constraints are given by a system of inequalities, and we found the solution region determined by the constraints.

If the average yearly profit is $6000 for each compact car and $9000 for each midsize car, buying a different number of each type will affect the profit. How do we find the correct number of each type so that the profit is maximized under these conditions? To answer questions of this type without knowledge of special techniques, people must use the "guess and check" method repeatedly to arrive at a solution. And when the number of conditions that enter into a problem of this type increases, guessing is a very unsatisfactory solution method. We can answer these questions with a mathematical technique called **linear programming**. Linear programming is a valuable tool in management; in fact, 85% of all Fortune 500 firms use linear programming. In this section we will see how to use graphical methods to find the **optimal solutions** to linear programming problems such as the one above.

Linear Programming

In any linear programming problem, we seek the point or points in the region determined by the constraint inequalities that give an optimum (*maximum* or *minimum*) value for the **objective function** (the function we seek to optimize).* Any point in the region determined by the constraints is called a **feasible solution** to the linear programming problem, and the region itself is called the **feasible region**.

Recall that the constraints in our car rental application can be written as the system of inequalities

$$\begin{cases} x + y \le 71 \\ 15x + 28y \le 1260 \\ x \ge 0 \\ y \ge 0 \end{cases}$$

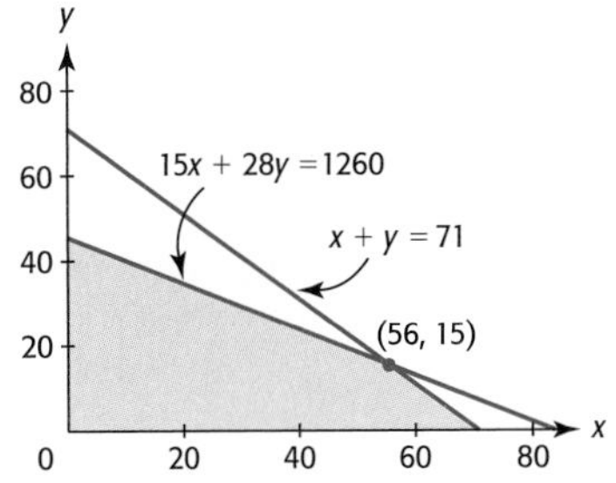

FIGURE 6.11

The solution set of the inequalities for the problem, the feasible region, is shown as the shaded region in Figure 6.11. Any point inside the shaded region or on its boundary satisfies the constraints, so it is a feasible solution. In the context of this application, only feasible solutions with integer coordinates can be solutions to the problem.

Because the profit P is 6 thousand dollars for each of the x compact cars and 9 thousand dollars for each of the y midsize cars, we can model the profit function as

$$P = 6x + 9y, \text{ where } P \text{ is in thousands of dollars.}$$

*The region determined by the constraints must be convex for the optimal solution to exist. A convex region is one such that for any two points in the region, the segment joining those points lies entirely within the region. We restrict our discussion to convex regions.

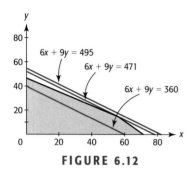

FIGURE 6.12

In this application, we seek to find the maximum profit subject to the constraints listed above. Because we seek to maximize the profit subject to the constraints, we evaluate

$$P = 6x + 9y$$

for points (x, y) in the feasible region. For example, at $(20, 19)$, $P = 291$, at $(50, 10)$, $P = 390$, and at $(50, 15)$, $P = 435$. But there are a large number of points to test, so we will look at a different way to find the maximum value of P. We can graph $P = 6x + 9y$ for different values of P by first writing the equation in the form

$$y = -\frac{2}{3}x + \frac{P}{9}.$$

Then, for each possible value of P, the graph is a line with slope $-2/3$ and y-intercept $P/9$. Figure 6.12 shows the graphs of the feasible region and objective functions for $P = 360, 471$, and 495. Note that the graphs are parallel (with slope $-2/3$), that larger values of P coincide with larger y-intercepts ($P/9$). Observe that if $P = 471$, the graph of $y = -\frac{2}{3}x + \frac{P}{9}$ intersects the feasible region and that values of P less than 471 will give lines that pass through the feasible region, but each of these values of P represent smaller profit. Note also that the value $P = 495$ (or any P-value greater than 471) results in a line that "misses" the feasible region, so that this value of P is not a solution to the problem. Thus the maximum value for P, subject to the constraints, is 471 (thousand dollars). We can verify that the graph of $471 = 6x + 9y$ intersects the region at $(56, 15)$ and that $P = 6x + 9y$ will not intersect the feasible region if $P > 471$. Thus the profit is maximized at \$471,000, when $x = 56$ and $y = 15$; that is, when 56 compact and 15 midsize cars are purchased.

Note that the values of x and y that correspond to the maximum value of P occur at $(56, 15)$, which is a "corner" of the feasible region. In fact, the maximum value of an objective function will always occur at a corner of the feasible region if the feasible region is closed and bounded. (A feasible region is closed and bounded if it is enclosed by, and includes, the lines associated with the constraints.) This gives us the basis for finding the solutions to linear programming problems using the graphical method.

SOLUTIONS TO LINEAR PROGRAMMING PROBLEMS

1. If a linear programming problem has an optimal solution, then the optimum value (maximum or minimum) of an objective function occurs at a corner of the feasible region determined by the constraints.

2. When a feasible region for a linear programming problem is closed and bounded, the objective function has a maximum and a minimum value.

3. If the objective function has the same optimum value at two corners, then it also has that optimum value at any point on the boundary line connecting the two corners.

4. When the feasible region is not closed and bounded, the objective function may have a maximum only, a minimum only, or neither.

Thus, for a closed and bounded region, we can find the maximum and minimum values of an objective function by evaluating the function at each of the corners of the feasible region formed by the graphical solution of the constraint inequalities.

EXAMPLE 1 Finding the Optimum Values of a Function Subject to Constraints

Find (a) the maximum value and (b) the minimum value of $C = 5x + 3y$ subject to the constraints

$$\begin{cases} x + y \leq 5 \\ 2x + y \leq 8 \\ x \geq 0, y \geq 0 \end{cases}.$$

Solution

The feasible region is in the first quadrant because it is bounded by $x \geq 0$ and $y \geq 0$. We can graph the remaining boundary lines with technology if we solve the first two inequalities for y.

$$\begin{cases} y \leq 5 - x \\ y \leq -2x + 8 \\ x \geq 0, y \geq 0 \end{cases}$$

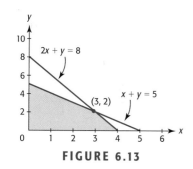

FIGURE 6.13

The solution set for the system of inequalities (that is, the feasible region) is bounded by the lines $y = 5 - x$, $y = -2x + 8$, and the x- and y-axes. The corners of the feasible region occur at the origin $(0, 0)$, one of the x-intercepts of the two boundary lines, one of the y-intercepts of the two boundary lines, and at the point of intersection of the lines $y = 5 - x$ and $y = -2x + 8$. (See Figure 6.13.)

There are two y-intercept points, $(0, 5)$ and $(0, 8)$, and the one that satisfies both inequalities is $(0, 5)$. The x-intercept points are $(4, 0)$ and $(5, 0)$; the one that satisfies both inequalities is $(4, 0)$. We can find the point of intersection of the two boundary lines by solving the system of equations

$$\begin{cases} y = 5 - x \\ y = -2x + 8. \end{cases}$$

Setting the y-values equal gives

$$5 - x = -2x + 8.$$

Solving gives $x = 3$, and substituting gives $y = 2$. Thus the point of intersection of these two boundary lines is $(3, 2)$. This point of intersection can also be found by using a graphing utility.

a. Any point inside the shaded region or on the boundary is a feasible solution to the problem, but the maximum possible value of

$$C = 5x + 3y$$

occurs at a corner of the feasible region. We evaluate the objective function at each of the corner points:

At $(0, 0)$, $C = 5x + 3y = 5(0) + 3(0) = 0$.

At $(0, 5)$, $C = 5x + 3y = 5(0) + 3(5) = 15$.

At $(4, 0)$, $C = 5x + 3y = 5(4) + 3(0) = 20$.

At $(3, 2)$, $C = 5x + 3y = 5(3) + 3(2) = 21$.

Thus the maximum value of the objective function is 21 when $x = 3$, $y = 2$.

b. Observing the values of $C = 5x + 3y$ at each corner point of the feasible region, we see that the minimum value of the objective function is 0 at $(0, 0)$. ∎

EXAMPLE 2   Cost Minimization

The Star Manufacturing Company produces two types of DVD players, which are assembled at two different locations. Plant 1 can assemble 60 units of the Star model and 80 units of the Prostar model per hour, and Plant 2 can assemble 300 units of the Star model and 80 units of the Prostar model per hour. The company needs to produce at least 5400 units of the Star model and 4000 units of the Prostar model to fill an order. If it costs $2000 per hour to run Plant 1 and $3000 per hour to run Plant 2, how many hours should each plant spend on manufacturing DVD players to minimize its cost for this order? What is the minimum cost for this order?

Solution

If we let x equal the number of hours of assembly time at Plant 1 and we let y equal the number of hours of assembly time at Plant 2, then the cost function that we seek to minimize is

$$C = 2000x + 3000y.$$

The constraints on the assembly hours follow:

Star model hours: $60x + 300y \geq 5400$

Prostar model hours: $80x + 80y \geq 4000$

Nonnegative units: $x \geq 0, y \geq 0$

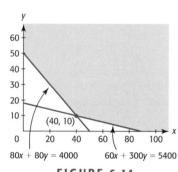

FIGURE 6.14

We can solve the first two inequalities for y, write these constraints as a system of inequalities, and graph the feasible region. (See Figure 6.14.)

$$\begin{cases} y \geq 18 - 0.2x \\ y \geq 50 - x \\ x \geq 0, y \geq 0 \end{cases}$$

The two boundary lines intersect where $18 - 0.2x = 50 - x$, at $x = 40$, which then gives $y = 10$. The boundary lines of the feasible region intersect the axes at $(0, 50)$ and at $(90, 0)$.

 Thus the corners of the feasible region are

$(0, 50)$, the y-intercept that satisfies all the inequalities.

$(90, 0)$, the x-intercept that satisfies all the inequalities.

$(40, 10)$, the intersection of the two boundary lines.

To find the hours of assembly time at each plant that will minimize the cost, we test the corner points:

At $(0, 50)$: $C = 2000(0) + 3000(50) = 150,000.$

At $(90, 0)$: $C = 2000(90) + 3000(0) = 180,000.$

At $(40, 10)$: $C = 2000(40) + 3000(10) = 110,000.$

Thus the cost is minimized when assembly time is 40 hours at Plant 1 and 10 hours at Plant 2. The minimum cost is $110,000.   ■

Note that the feasible region in Example 2 is not closed and bounded, and that there is no maximum value for the objective function even though there is a minimum value. It should also be noted that some applied linear programming problems require discrete solutions; for example, the optimal solution to the car rental problem above required a number of cars, so only positive integer solutions were possible.

EXAMPLE 3   Maximizing Profit

The Smoker Meat Packing Company makes two different types of hot dogs, regular and all-beef. Each pound of all-beef hot dogs requires 0.8 pound of beef and 0.2 pound of spices, and each pound of regular hot dogs requires 0.3 pound of beef and 0.2 pound of spices, with the remainder nonbeef meat products. The company has at most 1020 pounds of beef and has at most 500 pound of spices available for hot dogs. If the profit is $0.90 on each pound of all-beef hot dogs and $1.20 on each pound of regular hot dogs, how many of each type should be produced to maximize the profit?

Solution

Letting x equal the number of pounds of all-beef hot dogs and y equal the number of pounds of regular hot dogs, we seek to maximize the function $P = 0.90x + 1.20y$ subject to the constraints

$$\begin{cases} 0.8x + 0.3y \le 1020 \\ 0.2x + 0.2y \le 500 \\ x \ge 0, y \ge 0 \end{cases}$$

The graph of this system of inequalities is more easily found if the first two inequalities are solved for y.

$$\begin{cases} y \le 3400 - \dfrac{8}{3}x \\ y \le 2500 - x \\ x \ge 0, y \ge 0 \end{cases}$$

Graphing the boundary lines and testing the four regions determines the feasible region. (See Figure 6.15.) The boundary lines intersect where

$$3400 - \frac{8}{3}x = 2500 - x, \text{ or at } x = 540, y = 1960.$$

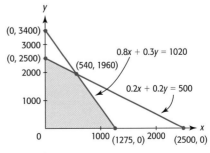

FIGURE 6.15

The x-intercept (1275,0) and the y-intercept (0, 2500) satisfy the inequalities. Thus the corners of this region are (0, 0), (0, 2500), (1275,0), and (540, 1960).

$P = 0$ at $(0, 0)$. Testing the objective function at the other corner points gives:

At $(0, 2500)$: $P = 0.90x + 1.20y = 0.90(0) + 1.20(2500) = 3000.$

At $(1275, 0)$: $P = 0.90x + 1.20y = 0.90(1275) + 1.20(0) = 1147.50.$

At $(540, 1960)$: $P = 0.90x + 1.20y = 0.90(540) + 1.20(1960) = 2838.$

This indicates that profit will be maximized if 2500 pounds of regular hot dogs and no all-beef hot dogs are produced. ∎

EXAMPLE 4   Manufacturing

A company manufactures air conditioning units and heat pumps at its factories in Atlanta, Georgia and Newark, New Jersey. The Atlanta plant can produce no more than 1000 items per day, and the number of air conditioning units cannot exceed 100 more than half the number of heat pumps. The Newark plant can produce no more than 850 units per day. The profit on each air conditioning unit is $400 at the Atlanta plant and $390 at the Newark plant. The profit on each heat pump is $200 at the Atlanta plant and $215 at the Newark plant. If there is an order for 500 air conditioning units and 750 heat pumps:

a. Graph the feasible region and identify the corners.

b. Find the maximum profit that can be made on this order and what production distribution will give the maximum profit.

Solution

a. Because the total number of the air conditioning units needed is 500, we can represent the number produced at Atlanta by x and the number produced at Newark by $(500 - x)$. Similarly, we can represent the number of heat pumps produced at Atlanta by y and the number produced at Newark by $(750 - y)$. Table 6.1 summarizes the constraints.

TABLE 6.1

	Air Conditioning Units	Heat Pumps	Total
Atlanta	x	$+y$	≤ 1000
Newark	$500 - x$	$+(750 - y)$	≤ 850
Other Constraints	x		$\leq 0.5y + 100$
	x		≥ 0
		y	≥ 0
Profit	$400x + 390(500 - x)$	$+200y + 215(750 - y)$	

Using this information gives the following system:

$$\begin{cases} x + y \leq 1000 & \text{(Inequality 1)} \\ (500 - x) + (750 - y) \leq 850 & \text{(Inequality 2)} \\ x \leq 0.5y + 100 & \text{(Inequality 3)} \\ x \geq 0, y \geq 0 \end{cases}$$

We solve each of these inequalities for y and graph the solution set. (See Figure 6.16.)

$$\begin{cases} y \le 1000 - x \\ y \ge 400 - x \\ y \ge 2x - 200 \\ x \ge 0, y \ge 0 \end{cases}$$

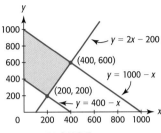

FIGURE 6.16

The intersections of the boundary lines associated with the inequalities follow.

Inequalities (1) and (3): where $1000 - x = 2x - 200$, at $x = 400$, $y = 600$.

Inequalities (2) and (3): where $400 - x = 2x - 200$, at $x = 200$, $y = 200$.

Inequalities (1) and (2): where $1000 - x = 400 - x$, which has no solution, so they do not intersect

The other corners are on the y-axis, at $(0, 400)$ and $(0, 1000)$.

b. The objective function gives the profit for the products.

$P = 400x + 390(500 - x) + 200y + 215(750 - y)$ or $P = 10x - 15y + 356{,}250$.

Testing the profit function at the corners of the feasible region determines where the profit is maximized.

At $(0, 400)$:	$P = 350{,}250$.
At $(200, 200)$:	$P = 355{,}250$.
At $(0,1000)$:	$P = 341{,}250$.
At $(400, 600)$:	$P = 351{,}250$.

Thus the profit is maximized at \$355,250 when 200 air conditioning units and 200 heat pumps are manufactured at Atlanta. The remainder are manufactured at Newark, so 300 air conditioning units and 550 heat pumps are manufactured at Newark. ∎

Solution with Technology

As we have seen in Section 6.2, graphing utilities can be used to graph the feasible region and to find the corners of the region satisfying the constraint inequalities. Note that the graphical method we have been using cannot be used if the problem involves more than two variables, and most real linear programming applications involve more than two variables. In fact, some real applications involve as many as 40 variables. Problems involving more than two variables can be solved with spreadsheet programs such as Lotus 1-2-3, Microsoft Excel, and Quatro Pro. Use of these programs to solve linear programming problems is beyond the scope of this text.

6.2 SKILLS CHECK

In Exercises 1–4, use the given feasible region determined by the constraint inequalities to find the maximum possible value and the minimum possible value of the objective function.

1. $f = 4x + 9y$ subject to the constraints

$$\begin{cases} 2x + 3y \le 134 \\ x + 5y \le 200 \\ x \ge 0, y \ge 0 \end{cases}$$

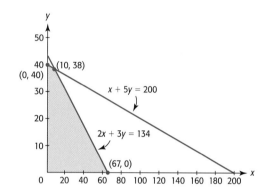

2. $f = 2x + 3y$ subject to the constraints

$$\begin{cases} 6x + 5y \le 170 \\ x + 5y \le 100 \\ x \ge 0, y \ge 0 \end{cases}$$

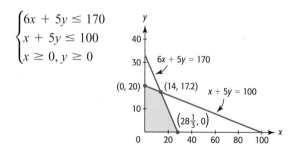

3. $f = 4x + 2y$ subject to the constraints

$$\begin{cases} 3x + y \le 15 \\ x + 2y \le 10 \\ -x + y \le 2 \\ x \ge 0, y \ge 0 \end{cases}$$

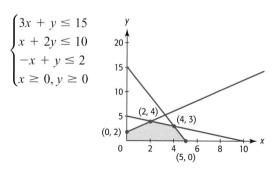

4. $f = 3x + 9y$ subject to the constraints

$$\begin{cases} 2x + y \le 12 \\ x + 3y \le 15 \\ x + y \le 7 \\ x \ge 0, y \ge 0 \end{cases}$$

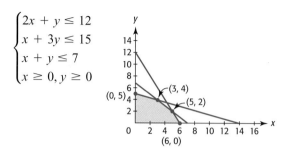

5. Perform the following steps to maximize $f = 3x + 5y$ subject to the constraints $\begin{cases} y \le (54 - 2x)/3 \\ y \le 22 - x. \\ x \ge 0, y \ge 0 \end{cases}$

  a. Graph the region that satisfies the system of inequalities, and identify the corners of the region.

  b. Test the objective function $f = 3x + 5y$ at each of the corners of the constraint region to determine which corner gives the maximum value. Give the maximum possible value and the values of x and y that give the value.

6. Perform the following steps to maximize $f = 3x + 4y$ subject to the constraints $\begin{cases} x + 2y \le 16 \\ x + y \le 10. \\ x \le 0, y \le 0 \end{cases}$

  a. Graph the region that satisfies the system of inequalities, and identify the corners of the region.

  b. Test the objective function $f = 3x + 4y$ at each of the corners of the constraint region to determine which corner gives the maximum value. Give the maximum possible value and the values of x and y that give the value.

7. Perform the following steps to minimize $g = 4x + 2y$ subject to the constraints $\begin{cases} x + 2y \ge 15 \\ x + y \ge 10. \\ x \ge 0, y \ge 0 \end{cases}$

  a. Graph the region that satisfies the system of inequalities, and identify the corners of the region.

  b. Test the objective function $g = 4x + 2y$ at each of the corners of the constraint region to determine which corner gives the minimum value. Give the minimum possible value and the values of x and y that give the value.

8. Find the maximum possible value of $f = 400x + 300y$ and the values of x and y that give the value, subject to the constraints

$$\begin{cases} 3x + 2y \geq 6 \\ 2x + y \leq 7 \\ x \geq 0, y \geq 0 \end{cases}$$

9. Find the maximum possible value of $f = 20x + 30y$ and the values of x and y that give the value, subject to the constraints

$$\begin{cases} y \leq -\dfrac{1}{2}x + 4 \\ y \leq -x + 6 \\ y \leq -\dfrac{1}{3}x + 4 \\ x \geq 0, y \geq 0 \end{cases}$$

10. Find the maximum possible value of $f = 100x + 100y$ and the values of x and y that give the value, subject to the constraints

$$\begin{cases} x + 2y \leq 6 \\ x + 4y \leq 10 \\ 2x + y \leq 9 \\ x \geq 0, y \geq 0 \end{cases}$$

11. Find the maximum possible value of $f = 80x + 160y$ and the values of x and y that give the value, subject to the constraints

$$\begin{cases} 3x + 4y \leq 32 \\ x + 2y \leq 15 \\ 2x + y \leq 18 \\ x \geq 0, y \geq 0 \end{cases}$$

12. Find the minimum possible value of $g = 30x + 40y$ and the values of x and y that give the value, subject to the constraints

$$\begin{cases} x + 2y \geq 16 \\ x + y \geq 10 \\ x \geq 0, y \geq 0 \end{cases}$$

13. Find the minimum possible value of $g = 60x + 10y$ and the values of x and y that give the value, subject to the constraints

$$\begin{cases} 3x + 2y \geq 12 \\ 2x + y \geq 7 \\ x \geq 0, y \geq 0 \end{cases}$$

14. Find the minimum possible value of $g = 30x + 40y$ and the values of x and y that give the value, subject to the constraints

$$\begin{cases} x + 2y \geq 5 \\ x + y \geq 4 \\ 2x + y \geq 6 \\ x \geq 0, y \geq 0 \end{cases}$$

15. Find the minimum possible value of $g = 46x + 23y$ and the values of x and y that give the value, subject to the constraints

$$\begin{cases} 3x + y \geq 6 \\ x + y \geq 4 \\ x + 5y \leq 8 \\ x \geq 0, y \geq 0 \end{cases}$$

6.2 EXERCISES

---

1. *Manufacturing* A company manufactures two types of leaf blowers, an electric "Turbo" model and a gas-powered "Tornado" model. The company's production plan calls for the production of at least 780 blowers per month. It costs $78 to produce each Turbo model and $117 to manufacture each Tornado model, and the company has at most $76,050 per month to use for production. (These constraints were graphed in Exer-

cise 7 of Exercise 6.1.) Find the number of units that should be produced to maximize the profit for the company if the profit on each Turbo model is $32 and the profit on each Tornado model is $45.

2. *Manufacturing* The Evergreen Company produces two types of printers, the Inkjet and the Laserjet. The company can make at most 120 printers per day and has 400 labor-hours available per day. It takes 2 hours to make the Inkjet and 6 hours to make the Laserjet. (These constraints were graphed in Exercise 12 of Exercise 6.1.) If the profit on the Inkjet is $80, and the profit on the Laserjet is $120, find the maximum possible daily profit and the number of each type of printer that gives it.

3. *Manufacturing* The Safeco Company produces two types of chainsaws, the Safecut and the Safecut Deluxe. The Safecut model requires 2 hours to assemble and 1 hour to paint and the Deluxe model requires 3 hours to assemble and 1/2 hour to paint. The daily maximum number of hours available for assembly is 36, and the daily maximum number of hours available for packing is 12. (These constraints were graphed in Exercise 14 of Exercise 6.1.) If the profit is $24 per unit on the Safecut model and $30 per unit on the Deluxe model, how many units of each type will maximize the daily profit?

4. *Production* Two models of riding mowers, the Lawn King and the Lawn Master, are produced on three assembly lines. Producing the Lawn King requires 2 hours on Line I and 1 hour on Line II. Producing the Lawn Master requires 1 hour on Line I and 3 hours on Line II. The number of hours available for production is limited to 60 on Line I and 40 on Line II. If there is $150 profit per mower on the Lawn King and $200 profit per mower on the Lawn Master, producing how many of each model maximizes the profit? What is the maximum possible profit?

5. *Sales* Trix Auto Sales sells used cars. To advertise its cars to target groups, they advertise with x 1-minute spots per month on cable television at a cost of $240 per minute, and y 1-minute spots per month on radio, at a cost of $150 per minute. Research shows that they sell one vehicle for every 4 minutes of cable television advertising (that is, 1/4 car per minute) and they sell one vehicle for each 10 minutes of radio advertising. Suppose the company has at most $36,000 per month to spend on advertising and that they sell at least 33 cars per month from this advertising. Graph the region that satisfies these two inequalities, that is, graph the solution to this system of inequalities.
   a. If the profit on cars advertised on television averages $500 and the profit on cars advertised on radio averages $550, how many minutes per month of advertising should be spent on television advertising and how many should be spent on radio advertising?
   b. What is the maximum possible profit?

6. *Advertising* Tire Town is developing an advertising campaign. The table below indicates the cost per ad package in newspapers and the cost per ad package on radio, and the number of ads in each type of ad package.

	Newspaper	Radio
Cost per Ad Package	$1000	$3000
Ad Exposures per Package	18	36

The owner of the company can spend no more than $18,000 per month, and he wants to have at least 252 ads per month. If each newspaper ad package reaches 6000 people over 20 years of age and each radio ad reaches 8000 of these people, how many newspaper ad packages and radio ad packages should he buy to maximize the number of people over age 20 reached?

7. *Production* A firm produces three different-size television sets on two assembly lines. The following table summarizes the production capacity of each assembly line, the number of each size TV ordered by a retailer, and the daily operating costs for each assembly line. How many days should each assembly line run to fill this order with minimum cost? What is the minimum cost?

	Assembly Line 1	Assembly Line 2	Number Ordered
19-inch TV	80 per day	40 per day	3200
27-inch TV	20 per day	20 per day	1000
35-inch TV	100 per day	40 per day	3400
Daily Cost	$20,000	$40,000	

8. *Nutrition* A privately owned lake contains two types of fish, bass and trout. The owner provides two types of food, A and B, for these fish. Trout require 4 units of food A and 5 units of food B, and bass require 10 units of food A and 4 units of food B. If the owner has 1600 units of food A and 1000 units of food B, find the maximum number of fish that the lake can support.

9. *Housing* A contractor builds two models of homes, the Van Buren and the Jefferson. The Van Buren requires 200 worker-days of labor and $240,000 in capital, while the Jefferson requires 500 worker-days of labor and $300,000 in capital. The contractor has a total of 5000 worker-days and $3,600,000 in capital available

per month. The profit is $60,000 on the Van Buren and $75,000 on the Jefferson. How many of each model will maximize the monthly profit? What is the maximum possible profit?

10. *Manufacturing* Easyboy manufactures two types of chairs, Standard and Deluxe. Each Standard chair requires 4 hours to construct and finish, and each Deluxe chair requires 6 hours to construct and finish. Upholstering takes 2 hours for a Standard chair and 6 hours for a Deluxe chair. There are 480 hours available each day for construction and finishing and there are 300 hours available per day for upholstering. (These constraints were graphed in Exercise 13 of Exercise 6.1.) If the revenue is $178 for each Standard chair and $267 for each Deluxe chair:

a. What is the maximum possible daily revenue?

b. How many of each type should be produced each day to maximize the daily revenue?

11. *Production* The Ace Company produces three models of VCRs, the Ace, the Ace Plus, and the Ace Deluxe models, at two facilities, A and B. The company has orders for at least 4000 of the Ace models, at least 1800 of the Ace Plus, and at least 2400 of the Ace Deluxe models. The weekly production capacity of each model at each facility and the cost per week to operate each facility are given in the table below. For how many weeks should each facility operate to minimize the cost of filling the orders?

	A	B
Ace	400	400
Ace Plus	300	100
Ace Deluxe	200	400
Cost per week	$15,000	$20,000

12. *Politics* A political candidate wishes to use a combination of television and radio advertisements in his campaign. Each 1-minute ad on television reaches 0.12 million eligible voters and each 1-minute ad on radio reaches 0.009 million eligible voters. The candidate feels he must reach at least 7.56 million eligible voters and that he must buy at least 100 minutes of advertisements. If television costs $1000 per minute and radio costs $200 per minute, how many minutes of newspaper and television advertisement does he need to minimize costs?

13. *Nutrition* In a hospital ward, patients are grouped into two general nutritional categories depending on the amount of solid food in their diet, and are provided two different diets with different amounts of the solid foods and with some detrimental substances. The table below gives the patient groups, the weekly diet requirements for each group, and the amount of detrimental substance in each diet. How many servings of each diet will satisfy the nutritional requirements and minimize the detrimental substance? What is the minimum amount of detrimental substance?

	Diet A	Diet B	Minimum Daily Requirements
Group 1	2 oz per serving	1 oz per serving	18 oz
Group 2	4 oz per serving	1 oz per serving	26 oz
Detrimental Substance	.09 oz per serving	.035 oz per serving	

14. *Manufacturing* Kitchen Pride manufactures toasters and can openers, which are manufactured on two different assembly lines. Line 1 can assemble 15 toasters and 20 can openers per hour, and Line 2 can assemble 75 toasters and 20 can openers per hour. The company needs to produce at least 540 toasters and 400 can openers. If it costs $300 per hour to run Line 1 and $600 per hour to run Line 2, how many hours should each line be run to fill all the orders at the minimum cost? What is the minimum cost?

6.3   Sequences and Discrete Functions

Suppose that a football player is offered a chance to sign a contract for 18 games with one of the following salary plans:

Plan A: $10,000 for the first game and a $10,000 raise for each succeeding game.

Plan B: $2 for the first game with his salary doubled for each game.

Which salary plan should he accept if he wants to make the most money for his last game?

   The payments for the games form the **sequences**:

Plan A: 10,000, 20,000, 30,000, ...

Plan B: 2, 4, 8, 16, ...

We will answer this question after we have developed formulas that apply to sequences, and we will use sequences to solve applied problems.

Sequences

In Chapter 3 we found the future values of investments at simple and compound interest. Consider a $5000 investment that pays 1% simple interest for each of 6 months, shown in Table 6.2.

TABLE 6.2

Month	Interest ($I = Prt$) Dollars	Future Value of the Investment (Dollars)
1	(5000)(0.01)(1) = 50	5000 + 50 = 5050
2	(5000)(0.01)(1) = 50	5050 + 50 = 5100
3	(5000)(0.01)(1) = 50	5100 + 50 = 5150
4	(5000)(0.01)(1) = 50	5150 + 50 = 5200
5	(5000)(0.01)(1) = 50	5200 + 50 = 5250
6	(5000)(0.01)(1) = 50	5250 + 50 = 5300

These future values are outputs that result when the inputs are positive integers that correspond to the number of months of the investment. Outputs (such as these future values) that result uniquely from the positive integer inputs define a function whose domain is a set of the positive integers. Such a function is called a **sequence**, and the ordered outputs corresponding to the integer inputs are called the **terms of the sequence**.

   Sequences have the same properties as the other functions we have studied, *except* that the domains of sequences are positive integers, and so sequences are **discrete functions**. Rather than denoting a functional output with y, we denote the output of the sequence f that corresponds to input n with $f(n) = a_n$.

SEQUENCE

The functional values $a_1, a_2, a_3, \ldots$ of a sequence are called the **terms** of the sequence, with a_1 the first term, a_2 the second term, and so on. The general term (or nth term) is denoted by a_n.

If the domain of a sequence is the set of *all* positive integers, the outputs form an **infinite sequence**. If the domain is the set of positive integers from 1 to n, the outputs form a **finite sequence**. Sequences are important because they permit us to apply discrete functions to real problems without having to approximate them with continuous functions.

By looking at the future values in Table 6.2, we can see that the future value at the end of each month is given by

$$a_1 = f(1) = 5000 + 50(1) = 5050$$
$$a_2 = f(2) = 5000 + 50(2) = 5100$$
$$a_3 = f(3) = 5000 + 50(3) = 5150$$
$$a_4 = f(4) = 5000 + 50(4) = 5200$$
$$a_5 = f(5) = 5000 + 50(5) = 5250$$
$$a_6 = f(6) = 5000 + 50(6) = 5300$$

By observing this pattern, we write the general term as $a_n = 5000 + 50n$, which represents

$$f(n) = 5000 + 50n.$$

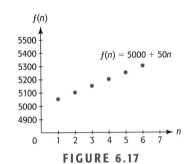

FIGURE 6.17

As with other functions, we can represent sequences in tables and on graphs. Figure 6.17 shows the graph of the function for these values of n.

EXAMPLE 1 Finding Terms of Sequences

Find the first four terms of the sequence defined by:

a. $a_n = \dfrac{8}{n}$ **b.** $b_n = (-1)^n n(n + 1)$

Solution

a. $a_1 = \dfrac{8}{1} = 8, \; a_2 = \dfrac{8}{2} = 4, \; a_3 = \dfrac{8}{3}, \;$ and $a_4 = \dfrac{8}{4} = 2.$

We write these terms in the form $8, 4, \dfrac{8}{3}, 2.$

b. $b_1 = (-1)^1(1)(1 + 1) = -2, \; b_2 = (-1)^2(2)(2 + 1) = 6,$
$b_3 = (-1)^3(3)(3 + 1) = -12, \; b_4 = (-1)^4(4)(4 + 1) = 20.$

Notice that $(-1)^n$ causes the signs of the terms to alternate. The terms are $-2, 6, -12, 20.$ ∎

TECHNOLOGY NOTE

We can find *n* terms of a sequence with a calculator, by putting the calculator in sequence mode. Figure 6.18(a) shows the first four terms defined in Example 2, part (a), in decimal and fractional form. The graph of the sequence and the table of values for the sequence are shown in Figure 6.18(b) and Figure 6.18(c), respectively.

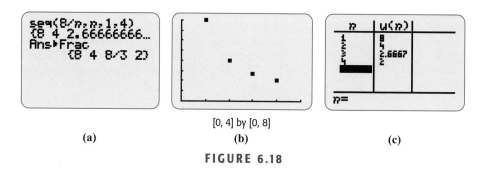

[0, 4] by [0, 8]

(a)   (b)   (c)

FIGURE 6.18

EXAMPLE 2   Depreciation

For tax purposes, a firm depreciates its $900,000 building over 30 years by the straight-line method, which depreciates the value of the building by 900,000/30 = 30,000 dollars each year. Write a sequence that gives the value of the building at the end of each of the first 5 years.

Solution

The value would be reduced each year by 30,000 dollars, so the first five terms of the sequence are

$$870,000, \ 840,000, \ 810,000, \ 780,000, \ 750,000.$$ ■

Arithmetic Sequences

The future values of the investment described in Table 6.2 are the first six terms of the sequence

$$5050, \ 5100, \ 5150, \ 5200, \ 5250, \ 5300, \ \ldots$$

We can define this sequence **recursively**, with each term after the first defined from the previous term, as

$$a_1 = 5050, \ a_n = a_{n-1} + 50 \quad \text{for } n > 1.$$

This sequence is an example of a special sequence called an **arithmetic sequence**. In such a sequence, each term after the first term is found by adding a constant to the preceding term. Thus we have the following definition.

ARITHMETIC SEQUENCES

A sequence is called an **arithmetic sequence** if there exists a number *d*, called the **common difference**, such that

$$a_n = a_{n-1} + d \quad \text{for } n > 1.$$

EXAMPLE 3   Arithmetic Sequences

Write the next three terms of the arithmetic sequences:

a. $1, 4, 7, 10, \ldots$ **b.** $11, 9, 7, \ldots$ **c.** $\dfrac{1}{2}, \dfrac{2}{3}, \dfrac{5}{6}, \ldots$

Solution

a. The common difference that gives each term from the previous one is 3, so the next three terms are 13, 16, 19.

b. The common difference is -2, so the next three terms are 5, 3, 1.

c. The common difference is $\dfrac{1}{6}$, so the next three terms are $1, \dfrac{7}{6}, \dfrac{4}{3}$.   ■

Note that the differences of terms of an arithmetic sequence are constant, so the function defining the sequence is linear, with its rate of change equal to the common difference. That is, an arithmetic sequence is really a linear function whose domain is restricted to the positive integers. Because each term after the first in an arithmetic sequence is obtained by adding d to the preceding term, the second term is $a_1 + d$, the third term is $(a_1 + d) + d = a_1 + 2d$, the fourth term is $a_1 + 3d, \ldots$, and the nth term is $a_1 + (n - 1)d$. Thus we have a formula for the nth term of an arithmetic sequence.

---

nTH TERM OF AN ARITHMETIC SEQUENCE

The nth term of an arithmetic sequence is given by

$$a_n = a_1 + (n - 1)d,$$

where a_1 is the first term of the sequence, n is the number of the term, and d is the common difference between the terms.

Thus the 25th term of the arithmetic sequence $1, 4, 7, 10, \ldots$ (see Example 3) is

$$a_{25} = 1 + (25 - 1)3 = 73.$$

EXAMPLE 4   Depreciation

For tax purposes, a firm depreciates its $900,000 building over 30 years by the straight-line method, which depreciates the value of the building by $900,000/30 = 30,000$ dollars each year. What is the value of the building after 12 years?

Solution

The description indicates that the value of the building is 900,000 dollars at the beginning, that its value is $900,000 - 30,000 = 870,000$ dollars after 1 year, and that it decreases by $30,000 each year for 29 additional years. Because the value of the building is reduced by the same amount each year, after n years the value is given by the *arithmetic* sequence with first term 870,000, common difference $-30,000$, and nth term

$$a_n = 870,000 + (n - 1)(-30,000), \text{ for } n = 1, 2, 3, \ldots, 30.$$

Thus the value of the building at the end of 12 years is $a_{12} = 870,000 + (12 - 1)(-30,000) = 540,000$ dollars. ∎

EXAMPLE 5 Football Contract Plan A

Recall that the first payment plan offered to the football player was

Plan A: 10,000, 20,000, 30,000, ...

a. Do the payments form an arithmetic sequence?

b. What is the 18th payment under this payment plan?

c. If the contract is extended into the postseason, what is the 20th payment under this payment plan?

Solution

a. The payments form a finite sequence with 18 terms. The common difference between terms is 10,000, so the sequence is an arithmetic sequence.

b. Because this is an arithmetic sequence with first term 10,000 and common difference $d = 10,000$, the 18th payment is

$$a_{18} = 10,000 + (18 - 1)10,000 = 180,000 \text{ (dollars)}.$$

c. The 20th payment is

$$a_{20} = 10,000 + (20 - 1)10,000 = 200,000 \text{ (dollars)}. \quad ∎$$

Geometric Sequences

Notice that the payment Plan B for the football player, which pays, $2, $4, $8, . . . , is not an arithmetic sequence, because there is no constant difference between the terms. Each term of this sequence is doubled (that is, multiplied by 2). This is an example of another special sequence, where each term is found by multiplying the previous term by the same number. This sequence is called a **geometric sequence**.

GEOMETRIC SEQUENCE

A sequence is called a **geometric sequence** if there exists a number r, called the **common ratio**, such that

$$a_n = ra_{n-1} \quad \text{for } n > 1.$$

EXAMPLE 6 Geometric Sequences

Write the next three terms of the geometric sequences:

a. 1, 3, 9, . . .   b. 64, 16, 4, . . .   c. 2, −4, 8, . . .

Solution

a. The common ratio that gives each term from the previous one is 3, so the next three terms are 27, 81, 243.

b. The common ratio is $\frac{1}{4}$, so the next three terms are $1, \frac{1}{4}, \frac{1}{16}$.

c. The common ratio is -2, so the next three terms are $-16, 32, -64$. ∎

Note that there is a **constant percent change** in the terms of a geometric sequence, so the function defining the sequence is exponential, and a geometric sequence is really an exponential function with its domain restricted to the positive integers. Because each term after the first in a geometric sequence is obtained by multiplying r times the preceding term, the second term is $a_1 r$, the third term is $a_1 r \cdot r = a_1 r^2$, the fourth term is $a_1 r^3, \ldots$, and the nth term is $a_1 r^{n-1}$. Thus we have a formula for the nth term of a geometric sequence.

---

nTH TERM OF A GEOMETRIC SEQUENCE

The nth term of a geometric sequence is given by

$$a_n = a_1 r^{n-1},$$

where a_1 is the first term of the sequence, n is the number of the term, and r is the common ratio of the terms.

Thus the 25th term of the geometric sequence $2, -4, 8, \ldots$ (see Example 6) is

$$a_{25} = 2(-2)^{25-1} = 33{,}554{,}432.$$

EXAMPLE 7 Rebounding

A ball is dropped from a height of 100 feet and rebounds $\frac{2}{5}$ of the height from which it falls every time it hits the ground. How high will it bounce after it hits the ground the fourth time?

Solution

The first rebound is $\frac{2}{5}$ of $100 = 40$ feet, the second is $\frac{2}{5}$ of $40 = 16$ feet, and the rebounds form a geometric sequence with first term 40 and common ratio $\frac{2}{5}$.

Thus the fourth bounce is

$$40\left(\frac{2}{5}\right)^{4-1} = 2.56 \text{ feet.}$$ ∎

We have found the future value of money invested at compound interest in Chapter 3 by using exponential functions. Because geometric sequences are really exponential functions with domains restricted to the positive integers, the future value of an investment with interest compounded over a number of discrete periods can be found using a geometric sequence.

EXAMPLE 8 Compound Interest

The future value of $1000 invested for 3 years at 6% compounded annually can be found as follows, using the simple interest formula $S = P + Prt$.

Year 1:
$$S = 1000 + 1000(0.06)(1)$$
$$= 1000(1 + 0.06)$$
$$= 1000(1.06)\text{dollars}$$

Year 2:
$$S = 1000(1.06) + [1000(1.06)](0.06)(1)$$
$$= 1000(1.06)(1 + 0.06)$$
$$= 1000(1.06)^2 \text{ dollars}$$

Year 3:
$$S = 1000(1.06)^2 + [1000(1.06)^2](0.06)(1)$$
$$= 1000(1.06)^2[1 + 0.06]$$
$$= 1000(1.06)^3 \text{ dollars}$$

a. Do the future values of this investment form a geometric sequence?

b. What is the value of this investment in 25 years?

Solution

a. The future values of the investment form a sequence with first term $1000(1.06) = 1060$, and common ratio 1.06, so the sequence is a geometric sequence.

b. The future value of the investment is the 25th term of the geometric sequence:

$$1060(1.06)^{25-1} = 4291.87 \text{ dollars.}$$ ∎

EXAMPLE 9 Football Contract Plan B

- -

Recall that the second payment plan offered to the football player was

Plan B: 2, 4, 8, ...

a. Do the payments form a geometric sequence?

b. What is the 18th payment under this payment plan?

Solution

a. The payments form a finite sequence with 18 terms. The common ratio between terms is 2, because each payment is doubled. Thus the sequence is a geometric sequence.

b. Because this is a geometric sequence with first term 2 and common ratio $r = 2$, the 18th payment is

$$a_{18} = 2 \cdot 2^{18-1} = 262{,}144 \text{ (dollars).}$$

Thus the 18th payment ($262,144) from Plan B is significantly larger than the 18th payment ($180,000) from Plan A. ∎

But does the fact that the 18th payment from Plan B is much larger than the payment from Plan A mean that the total payment from Plan B is larger than Plan A? We will answer this question in the next section.

6.3 SKILLS CHECK

1. Find the first 6 terms of the sequence defined by $f(n) = 2n + 3$.

2. Find the first 4 terms of the sequence defined by $f(n) = \dfrac{1}{2n} + n$.

3. Find the first 5 terms of the sequence defined by $a_n = \dfrac{10}{n}$.

4. Find the first 5 terms of the sequence defined by $a_n = (-1)^n 2n$.

5. Find the next 3 terms of the arithmetic sequence $1, 3, 5, 7, \ldots$.

6. Find the next 3 terms of the arithmetic sequence $2, 5, 8, \ldots$.

7. Find the eighth term of the arithmetic sequence with first term -3 and common difference 4.

8. Find the 40th term of the arithmetic sequence with first term 5 and common difference 15.

9. Write 4 additional terms of the geometric sequence $3, 6, 12, \ldots$.

10. Write 4 additional terms of the geometric sequence $8, 20, 50, \ldots$.

11. Find the sixth term of the geometric sequence with first term 10 and common ratio 3.

12. Find the tenth term of the geometric sequence with first term 48 and common ratio $-\dfrac{1}{2}$.

13. Find the first 4 terms of the sequence with first term 5 and nth term $a_n = a_{n-1} - 2$.

14. Find the first 6 terms of the sequence with first term 8 and nth term $a_n = a_{n-1} + 3$.

15. Find the first 4 terms of the sequence with first term 2 and nth term $a_n = 2a_{n-1} + 3$.

16. Find the first 5 terms of the sequence with first term 26 and nth term $a_n = (a_{n-1} + 4)/2$.

6.3 EXERCISES

1. *Salaries* Suppose that you are offered a job with a relatively low starting salary but with a $1500 raise for each of the next 7 years. How much more than your starting salary would you be making in the eighth year?

2. *Depreciation* A new car costing $35,000 is purchased for business and is depreciated with the straight-line depreciation method over a 5-year period, which means that it is depreciated by the same amount each year. Write a sequence that gives the value after depreciation for each of the 5 years.

3. *Landscaping* Grading equipment used for landscaping costs $300 plus $60 per hour or part of an hour thereafter.
   a. Write an expression for the cost of a job lasting n hours.
   b. If the answer to part (a) is the nth term of a sequence, write the first 6 terms of this sequence.

4. *Profit* A new firm loses $2000 in its first month, but its profit increases by $400 in each succeeding month for the rest of the year. What is its profit in the 12th month?

5. *Salaries* Suppose that you are offered two identical jobs, one paying a starting salary of $40,000 with yearly raises of $2000 and a second one paying a starting salary of $36,000 with yearly raises of $2400.
   a. Which job will be paying more in 5 years? By how much?
   b. Which job will be paying more in 10 years? By how much?
   c. What factor is important in deciding which job to take?

6. *Phone Calls* Suppose a long-distance call costs 99 cents for the first minute plus 25 cents for each additional minute.
   a. Write the cost of a call lasting n minutes.
   b. If the answer to part (a) is the nth term of a sequence, write the first 6 terms of this sequence.

7. *Interest* If $1000 is invested at 5% interest, compounded annually, write a sequence that gives the amount in the account at the end of each of the first 4 years.

8. *Interest* If $10,000 is invested at 6% interest, compounded annually, write a sequence that gives the amount in the account at the end of each of the first 3 years.

9. *Depreciation* A new car costing $50,000 depreciates by 20% of its original value each year.
   a. What is the value of the car at the end of the third year?
   b. Write an expression that gives the value at the end of the nth year.
   c. Write the first 5 terms of the sequence of the values after depreciation.

10. *Salaries* If you accept a job paying $32,000 for the first year, with a guaranteed 8% raise each year, how much will you earn in your fifth year?

11. *Bacteria* The size of a certain bacteria culture doubles each hour. If the number of bacteria present initially is 5000, how many will be present at the end of 6 hours?

12. *Bacteria* If a bacteria culture increases by 20% every hour and 2000 are present initially, how many will be present at the end of 10 hours?

13. *Bouncing Ball* A ball is dropped from 128 feet. It rebounds 1/4 of the height from which it falls every time it hits the ground. How high will it bounce after it hits the ground for the fourth time?

14. *Bouncing Ball* A ball is dropped from 64 feet. If it rebounds 3/4 of the height from which it falls every time it hits the ground, how high will it bounce after it hits the ground for the fourth time?

15. *Profit* If changing market conditions cause a company earning a profit of $8,000,000 this year to project decreases of 2% of its profit in each of the next 5 years, what profit does it project 5 years from now?

16. *Pumps* A pump removes 1/3 of the water in a container with every stroke. What amount of water is removed on the fifth stroke if the container originally had 81 cm^3 of water?

17. *Salaries* If you begin a job making $54,000 and are given a $3600 raise each year, use numerical methods to find in how many years your salary will double.

18. *Interest* If $2500 is invested at 8% interest, compounded annually, how long will it take the account to reach $5829.10?

19. *Compound Interest* Find the future value of $10,000 invested for 10 years at 8%, compounded daily.

20. *Compound Interest* Find the future value of $10,000 invested for 10 years at 8% compounded annually.

21. *Rabbit Breeding* The number of pairs of rabbits in the population during the first 6 months can be written in the Fibonacci sequence

$$1, 1, 2, 3, 5, 8, \ldots,$$

in which each term after the second is the sum of the two previous terms. Find the number of pairs of rabbits for the next 4 months.

22. *Bee Ancestry* A female bee hatches from a fertilized egg while a male bee hatches from an unfertilized egg. Thus a female bee has a male parent and a female parent while a male bee has only a female parent. Therefore, the number of ancestors of a male bee follow the Fibonacci sequence

$$1, 2, 3, 5, 8, 13, \ldots$$

   a. Observe the pattern and write three more terms of the sequence.
   b. What do the 1, the 2, and the 3, respectively, represent for a given male bee?

23. *Credit Card Debt* Kirsten has a $10,000 credit card debt. Each month she pays 1% plus a payment of 10% of the monthly balance. If she makes no new purchases, write a sequence that gives the payments for the first 4 months.

6.4 Series

Key Concepts
- Finite series
- Infinite series
- Sigma notation
- Arithmetic series
- Geometric series
- Infinite geometric series

In Section 6.3, we discussed a football player's opportunity to sign a contract for 18 games for either:

Plan A: $10,000 for the first game and a $10,000 raise for each succeeding game.

Plan B: $2 for the first game with his salary doubled for each game.

Which salary plan should he accept if he wants to make the most money?

The payments for the games form the **sequences**:

Plan A: 10,000, 20,000, 30,000, ...

Plan B: 2, 4, 8, 16, ...

We found that Plan B paid more than Plan A for the 18th game, but we did not determine which plan pays the most for all 18 games. To answer this question, we need to find the *sum* of the first 18 terms of these sequences. The sum of the terms of a sequence is called a **series**. Series are useful in computing finite and infinite sums, in approximating numbers such as π and e, and in approximating complex functions.

Finite and Infinite Series

Suppose a firm loses $2000 in its first month of operation, but its profit increases by $500 in each succeeding month for the remainder of its first year. What is its profit for the year? The description indicates that the profit is $-$2000 for the first month and that it increases by $500 each month for 11 months. The monthly profits form the sequence that begins

$$-2000, -1500, -1000, ...$$

The profit for the 12 months is the sum of these terms:

$$(-2000) + (-1500) + (-1000) + (-500) + 0 + 500 + 1000 + 1500 + 2000$$
$$+ 2500 + 3000 + 3500 = 9000$$

The total profit for the year is this sum, $9000.

The sum of the 12 terms of this sequence can be referred to as a **finite series**. In general, a **series** is defined as the sum of the terms of a sequence. If the number of terms is infinite, the series is an **infinite series**.

SERIES

A **finite series** is defined by

$$a_1 + a_2 + a_3 + . . . + a_n$$

where $a_1, a_2, . . . a_n$ are terms of a sequence.

An **infinite series** is defined by

$$a_1 + a_2 + a_3 + . . . + a_n + . . .$$

We can use the Greek letter Σ (sigma) to express the sum of numbers or expressions. For example, we can write

$$\sum_{i=1}^{3} i \text{ to denote } 1 + 2 + 3$$

and we can write the general finite series in the form

$$\sum_{i=1}^{n} a_i = a_1 + a_2 + a_3 + \ldots + a_n.$$

The symbol $\sum_{i=1}^{n} a_i$ may be read as "The sum of a_i as i goes from 1 to n one unit at a time." The general infinite series can be written in the form

$$\sum_{i=1}^{\infty} a_i = a_1 + a_2 + a_3 + \ldots + a_n + \ldots$$

where ∞ indicates that there is no end to the number of terms.

EXAMPLE 1 **Finite Series**

Find the sum of the first six terms of the sequence $6, 3, \dfrac{3}{2}, \ldots$.

Solution

Each succeeding term of this sequence is found by multiplying the preceding term by $\dfrac{1}{2}$, so the sequence is

$$6, 3, \frac{3}{2}, \frac{3}{4}, \frac{3}{8}, \frac{3}{16}, \ldots$$

and the sum of the first six terms is $6 + 3 + \dfrac{3}{2} + \dfrac{3}{4} + \dfrac{3}{8} + \dfrac{3}{16} = \dfrac{189}{16}$. ∎

Arithmetic Series

The sum of the payments to the football player for the first three games under payment Plan A is

$$10{,}000 + 20{,}000 + 30{,}000 = 60{,}000 \text{ dollars,}$$

and the sum of the payments for the first three games under Plan B is

$$2 + 4 + 8 = 14 \text{ dollars,}$$

so the player would be well advised to take Plan A if there were only three games. To find the sum of the payments for 18 games and to answer other questions, it would be more efficient to have formulas to use in finding the sums of terms of sequences. We begin by finding a formula for the sum of n terms of an arithmetic sequence.

Given the first term a_1 in an arithmetic sequence with common difference d, we use s_n to represent the sum of the first n terms and write it as follows:

$$s_n = a_1 + (a_1 + d) + (a_1 + 2d) + \ldots + (a_n - 2d) + (a_n - d) + a_n.$$

Writing this sum in reverse order gives

$$s_n = a_n + (a_n - d) + (a_n - 2d) + \ldots + (a_1 + 2d) + (a_1 + d) + a_1.$$

Adding these two equations term by term gives

$$s_n + s_n = (a_1 + a_n) + (a_1 + a_n) + (a_1 + a_n) + \cdots + (a_1 + a_n)$$
$$+ (a_1 + a_n) + (a_1 + a_n),$$

so $2s_n = n(a_1 + a_n)$ because there are n terms, and we have $s_n = \dfrac{n(a_1 + a_n)}{2}$.

SUM OF n TERMS OF AN ARITHMETIC SEQUENCE

The sum of the first n terms of an arithmetic sequence is given by

$$s_n = \frac{n(a_1 + a_n)}{2}$$

where a_1 is the first term of the sequence and a_n is the nth term.

Recall that Plan A of the football player's contract began with $a_1 = 10,000$ and had 18th payment $a_{18} = 180,000$. Thus the sum of the 18 payments is

$$s_{18} = \frac{18(10,000 + 180,000)}{2} = 1,710,000 \text{ dollars.}$$

EXAMPLE 2 Depreciation

An automobile valued at $24,000 is depreciated over 5 years with the *sum-of-the-years-digits* depreciation method. Under this method, annual depreciation is found by multiplying the value of the property by a fraction whose denominator is the sum of the 5 years' digits and whose numerator is 5 for the first year's depreciation, 4 for the second year's, 3 for the third, and so on.

a. Show that the sum of the depreciations for the 5 years adds to 100% of the value.

b. Find the depreciation for each year and show that the sum of the depreciations is $24,000.

Solution

a. The sum of the 5 years' digits is $1 + 2 + 3 + 4 + 5 = 15$, so the depreciations for the 5 years are the products of $24,000 and each of the numbers

$$\frac{5}{15}, \frac{4}{15}, \frac{3}{15}, \frac{2}{15}, \frac{1}{15}.$$

This is an arithmetic sequence with first term $\dfrac{5}{15}$, common difference $\dfrac{1}{15}$, and 5th term $\dfrac{1}{15}$.

Thus the sum of the terms is $s_5 = \dfrac{5\left(\dfrac{5}{15} + \dfrac{1}{15}\right)}{2} = \dfrac{5}{2} \cdot \dfrac{6}{15} = 1 = 100\%.$

b. The respective annual depreciations for the automobile are

$$\frac{5}{15} \cdot 24{,}000 = 8000, \ \frac{4}{15} \cdot 24{,}000 = 6400, \ \frac{3}{15} \cdot 24{,}000 = 4800,$$

$$\frac{2}{15} \cdot 24{,}000 = 3200, \ \frac{1}{15} \cdot 24{,}000 = 1600.$$

The depreciations form an arithmetic sequence with first term 8000 and fifth term 1600, so the sum of these depreciations is

$$s_n = \frac{5(8000 + 1600)}{2} = 24{,}000 \text{ dollars.}$$

Note that the sum is the original value of the automobile, so it is totally depreciated in 5 years. ∎

Geometric Series

In Section 3.6, we found the future value of the annuity with payments of $1000 at the end of each of 5 years, with interest at 10%, compounded annually, by finding the sum

$$S = 1000 + 1000(1.10)^1 + 1000(1.10)^2 + 1000(1.10)^3 + 1000(1.10)^4$$
$$= 6105.10 \text{ dollars.}$$

We can see that this is a finite geometric series with first term 1000 and common ratio 1.10. We can develop a formula to find the sum of this and any other finite geometric series.

The sum of the first n terms of a geometric sequence with first term a_1 and common ratio r is

$$s_n = a_1 + a_1 r + a_1 r^2 + \ldots + a_1 r^{n-1}$$

To find a formula for this sum, we multiply it by r to get a new expression, and then find the difference of the two expressions.

$$s_n = a_1 + a_1 r + a_1 r^2 + a_1 r^3 + \cdots + a r^{n-1}$$
$$r s_n = a_1 r + a_1 r^2 + a_1 r^3 + \ldots + a_1 r^{n-1} + a_1 r^n$$
$$s_n - r s_n = a_1 - a_1 r^n$$

Rewriting this difference and solving for s_n gives a formula for the sum.

$$s_n(1 - r) = a_1(1 - r^n)$$
$$s_n = \frac{a_1(1 - r^n)}{(1 - r)}$$

SUM OF n TERMS OF A GEOMETRIC SEQUENCE

The sum of the first n terms of a geometric sequence is

$$s_n = \frac{a_1(1 - r^n)}{1 - r}$$

where a_1 is the first term of the sequence and r is the common ratio.

EXAMPLE 3 Sums of Terms of Geometric Sequences

a. Find the sum of the first 10 terms of the geometric sequence with first term 4 and common ratio -3.

b. Find the sum of the first 8 terms of the sequence $2, 1, \dfrac{1}{2}, \ldots$.

Solution

a. The sum of the first 10 terms of the sequence with first term 4 and common ratio -3 is given by

$$s_{10} = \frac{a_1(1 - r^{10})}{1 - r} = \frac{4(1 - (-3)^{10})}{(1 - (-3))} = -59{,}048.$$

b. This sequence is geometric, with first term 2 and common ratio $\dfrac{1}{2}$, so the sum of the first 8 terms is

$$s_8 = \frac{a_1(1 - r^8)}{1 - r} = \frac{2(1 - (1/2)^8)}{1 - (1/2)} = 4\left(1 - \frac{1}{256}\right) = \frac{255}{64}.$$ ∎

EXAMPLE 4 Annuities

Show that if regular payments of $1000 are made at the end of each year for 5 years into an account that pays interest at 10% per year, compounded annually, the future value of this annuity is 6105.10 dollars.

Solution

As we saw above, the future values of the individual payments into this annuity form a geometric sequence with first term 1000 and common ratio 1.10, so the sum of the first five terms is

$$s_5 = \frac{1000(1 - 1.10^5)}{1 - 1.10} = 6105.10.$$ ∎

We found that the total payment to the football player under Plan A is $1,710,000. The payments under Plan B form a geometric sequence with first term 2 and common ratio 2, so the sum of the payments under Plan B is

$$s_n = \frac{2(1 - 2^{18})}{1 - 2} = 524{,}286.$$

Thus Plan A gives total earnings for the 18 games that is more than three times that for Plan B. However, if the season has 20 games, the sum of the payments for Plan A is

$$s_{20} = \frac{20(10{,}000 + 200{,}000)}{2} = 2{,}100{,}000 \text{ dollars,}$$

and the sum of the payments for Plan B is

$$s_{20} = \frac{2(1 - 2^{20})}{1 - 2} = 2{,}097{,}150 \text{ dollars.}$$

In this case, the plans are nearly equal.

Infinite Geometric Series

If we add the terms of an infinite geometric sequence, we have an **infinite geometric series**. It is perhaps surprising that we can add an infinite number of positive numbers and get a finite sum. To see that this is possible, consider a football team that has only 5 yards remaining to score a touchdown. How many penalties on the opposing team will it take for the sum of the penalties to be more than 5 yards (giving them a touchdown), if each penalty is "half the distance to the goal?" The answer is that the team can never reach the goal line because every penalty is half the distance to the goal, which always leaves some distance remaining to travel. Thus the sum of any number of penalties is less than 5 yards. To see this, let's find the sum of the first 10 such penalties. The first few penalties are $\frac{5}{2}, \frac{5}{4}$, and $\frac{5}{8}$, so the sequence of penalties has first term $\frac{5}{2}$ and common ratio $\frac{1}{2}$ and the sum of the first n penalties is

$$s_n = \frac{a_1(1 - r^n)}{1 - r} = \frac{(5/2)(1 - (1/2)^n)}{1 - (1/2)}.$$

The sum of the first 10 penalties is found by letting $n = 10$, getting

$$\frac{(5/2)(1 - (1/2)^{10})}{1 - (1/2)} \approx 4.9951.$$

We can see that this sum is less than 5 yards, but close to five yards. Notice that as n gets larger, $\left(\frac{1}{2}\right)^n$ gets smaller, approaching 0, and the sum approaches $\frac{5/2(1 - 0)}{1 - (1/2)} = 5$. Thus the sum of an *infinite number* of penalties is the finite number 5 yards.

In general, if $|r| < 1$, it can be shown that r^n approaches 0 as n gets large without bound. Thus $1 - r^n$ approaches 1 and $\frac{a_1(1 - r^n)}{1 - r}$ approaches $\frac{a_1}{1 - r}$.

SUM OF AN INFINITE GEOMETRIC SERIES

The sum of an infinite geometric series is

$$S = \frac{a_1}{1 - r}.$$

where a_1 is the first term and r is the common ratio, with $|r| < 1$. If $|r| \geq 1$, the sum does not exist (is not a finite number).

We can use this formula to compute the sum of an infinite number of "half the distance to the goal" penalties from the 5-yard line.

$$S = \frac{5/2}{1 - 1/2} = 5$$

E X A M P L E 5   Infinite Geometric Series

Find the sum of the following series, if possible.

a. $81 + 9 + 1 + \dots$ **b.** $\displaystyle\sum_{i=1}^{\infty} \left(\frac{2}{3}\right)^i$

Solution

a. This is an infinite geometric series with first term $a_1 = 81$ and common ratio $r = 1/9 < 1$, so the sum is

$$S = \frac{81}{1 - 1/9} = \frac{729}{8}.$$

b. This is the sigma notation for an infinite series. Each new term results in one higher power of 2/3 (because i increases by one), so the common ratio is $\frac{2}{3} < 1$. The first term is $\frac{2}{3}$ (the value when $i = 1$) so the sum is

$$S = \frac{2/3}{1 - 2/3} = 2. \qquad \blacksquare$$

6.4 SKILLS CHECK

1. Find the sum of the first 6 terms of the sequence 9, 3, 1, ...

2. Find the sum of the first 7 terms of the sequence 1, 3, 5, ...

3. Find the sum of the first 10 terms of the arithmetic sequence 7, 10, 13, ...

4. Find the sum of the first 20 terms of the arithmetic sequence 5, 12, 19, ...

5. Find the sum of the first 15 terms of the arithmetic sequence with first term -4 and common difference 2.

6. Find the sum of the first 10 terms of the arithmetic sequence with first term 50 and common difference 3.

7. Find the sum of the first 15 terms of the geometric sequence with first term 3 and common ratio 2.

8. Find the sum of the first 12 terms of the geometric sequence with first term 48 and common ratio 1/2.

9. Find the sum of the first 10 terms of the geometric sequence 5, 10, 20, ...

10. Find the sum of the first 15 terms of the geometric sequence 100, 50, 25, ...

11. Find the sum of the following series $81 + 9 + 1 + \ldots$, if possible.

12. Find the sum of the following series $64 + 32 + 16 + \ldots$, if possible.

In Exercises 13–16, evaluate the sums.

13. $\displaystyle\sum_{i=1}^{6} 2^i$

14. $\displaystyle\sum_{i=1}^{5} 4^i$

15. $\displaystyle\sum_{i=1}^{4} \frac{1 + i}{i}$

16. $\displaystyle\sum_{i=1}^{5} \left(\frac{1}{2}\right)^i$

17. Find the sum of the series $\displaystyle\sum_{i=1}^{\infty} \left(\frac{3}{4}\right)^i$ if possible.

18. Find the sum of the series $\displaystyle\sum_{i=1}^{\infty} \left(\frac{5}{6}\right)^i$ if possible.

19. Find the sum of the series $\displaystyle\sum_{i=1}^{\infty} \left(\frac{4}{3}\right)^i$ if possible.

6.4 EXERCISES

1. *Profit* A new firm loses $2000 in its first month, but its profit increases by $400 in each succeeding month for the rest of the year. What is its profit for the first year?

2. *Salaries* Suppose that you are offered a job with a relatively low starting salary but with a $1500 raise for each of the next 7 years. What is the total amount of your raises over the next 7 years?

3. *Bee Ancestry* A female bee hatches from a fertilized egg while a male bee hatches from an unfertilized egg. Thus a female bee has a male parent and a female parent while a male bee has only a female parent. Therefore, the numbers of ancestors of a male bee follow the Fibonacci sequence

$$1, 2, 3, 5, 8, 13, \ldots.$$

Use this sequence to determine the total number of ancestors of a male bee back four generations.

4. *Salaries* Suppose that you are offered two identical jobs, one paying a starting salary of $40,000 with yearly raises of $2000 and one paying a starting salary of $36,000 with yearly raises of $2400.
   a. Which job will pay more over the first 5 years? How much?
   b. Which job will pay more over the first 12 years? How much?
   c. What factor is important in deciding which job to take?

5. *Clocks* A grandfather clock strikes a chime indicating each hour of the day, so that it chimes 3 times at 3:00, 10 times at 10:00, and so on.
   a. How many times will it chime in a 12-hour period?
   b. How many times will it chime in a 24-hour day?

6. *Depreciation* Under the sum-of-the-years depreciation method, annual depreciation for 4 years is found by multiplying the value of the property by a fraction whose denominator is the sum of the 4 years' digits and whose numerator is 4 for the first year's depreciation, 3 for the second year's, 2 for the third, and so on.
   a. Write a sequence that gives the depreciation for each year on a truck with value $36,000.
   b. Show that the sum of the depreciations for the 4 years adds to 100% of the original value.

7. *Profit* Suppose a new business makes a profit of $2000 in its first month, and its profit increases by 10% in each of the next 11 months. How much profit did it earn in its first year?

8. *Profit* If changing market conditions cause a company earning a profit of $8,000,000 this year to project decreases of 2% of its profit in each of the next 5 years, what is the sum of the yearly profits that it projects for the next 5 years?

9. *Pumps* A pump removes 1/3 of the water in a container with every stroke. After 4 strokes, what amount of water is still in a container that originally had 81 cm³ of water?

10. *Salaries* If you accept a job paying $32,000 for the first year, with a guaranteed 8% raise each year, how much will you earn in your first 5 years?

11. *Chain Letters* Suppose you receive a chain letter with 5 names on it, and to keep the chain unbroken, you mail a dollar to the person whose name is at the top of the list, cross out the top name, add your name to the bottom, and mail it to 5 friends.
    a. How many friends will receive your letter?
    b. If your friends cross the second person from the original list off the list and mail out 5 letters each, how many people will receive letters from your friends?
    c. How many "descendants" of your letters will receive letters in the next two levels if no one breaks the chain?
    d. Describe the type of sequence that gives the number of people receiving the letter at each level.

12. *Chain Letters* Mailing chain letters that involve money in the U.S. mail is illegal because many people would receive nothing while a comparative few would profit. Suppose that the chain letter in Exercise 11 were to go through 12 unbroken levels.
    a. How many people would receive a letter asking for money?
    b. What is the significance of this number?

13. *Credit Card Debt* Kirsten has a $10,000 credit card debt. She makes a minimum payment of 10% of the balance at the end of each month. If she makes no new purchases, how much does she owe at the end of 1 year?

14. *Credit Card Debt* Kirsten has a $10,000 credit card debt. She makes a minimum payment of 10% of the balance at the end of each month. If she makes no new purchases, in how many months will she have half of her debt repaid?

15. *Bouncing Ball* A ball is dropped from 128 feet. It rebounds 1/4 of the height from which it falls every time it hits the ground. How far will it have traveled up and down when it hits the ground for the fifth time?

16. *Bouncing Ball* A ball is dropped from 64 feet. If it rebounds 3/4 of the height from which it falls every time it hits the ground, how high will it have traveled up and down when it hits the ground for the fourth time?

17. *Depreciation* A new car costing \$35,000 depreciates by 16% of its current value each year.
  a. Write an expression that gives the sum of the depreciations for the first n years.
  b. What is the value of the car at the end of the nth year?

18. *Loans* An interest-free loan of \$15,000 requires monthly payments of 8% of the unpaid balance. What is the unpaid balance after:
  a. 1 year?
  b. 2 years?

19. *Annuities* Find the 8-year future value of an annuity with a contribution of \$100 at the end of each month into an account that pays 12% per year, compounded monthly.

20. *Annuities* Find the 10-year future value of an ordinary annuity with a contribution of \$300 per quarter

into an account that pays 8% per year, compounded quarterly.

21. *Present Value of Annuities* The present value of the annuity with n payments of \$$R$, at interest rate i per period is

$$A = R(1 + i)^{-1} + R(1 + i)^{-2}$$
$$+ R(1 + i)^{-3} + \ldots + R(1 + i)^{-(n-1)}$$
$$+ R(1 + i)^{-n}$$

Treat this as the sum of a geometric series with n terms, $a_1 = R(1 + i)^{-n}$, and $r = (1 + i)$, and show that the sum is

$$A = R\left[\frac{1 - (1 + i)^{-n}}{i}\right].$$

6.5 Preparing for Calculus

Key Concepts

- Chapter 1 Skills
- Chapter 2 Skills
- Chapter 3 Skills
- Chapter 4 Skills

In this section, we will show how algebra is used in the context of a calculus course; that is, we will show how algebra skills that we have already acquired are applied in the solution of calculus problems. Three fundamental concepts in calculus are limit, derivative, and integral. Although we will not define nor use these concepts here, we will illustrate how algebra skills are necessary for finding, simplifying, and applying these concepts.

Chapter 1 Skills

The following examples demonstrate skills discussed in Chapter 1 that are useful in calculus.

EXAMPLE 1   Simplifying Expressions

The following expressions represent the derivatives of functions. Simplify these expressions by multiplying to remove parentheses and combining like terms.

a. $(x^3 - 3x)(2x) + (x^2 - 1)(3x^2 - 3)$

b. $\dfrac{x^3(20x^4) - (4x^5 - 2)(3x^2)}{(x^3)^2}$

Solution

a. $(x^3 - 3x)(2x) + (x^2 - 1)(3x^2 - 3) = 2x^4 - 6x^2 + 3x^4 - 3x^2 - 3x^2 + 3$

$$= 5x^4 - 12x^2 + 3$$

b. $\dfrac{x^3(20x^4) - (4x^5 - 2)(3x^2)}{(x^3)^2} = \dfrac{20x^7 - (12x^7 - 6x^2)}{x^6}$

$$= \dfrac{8x^7 + 6x^2}{x^6} = \dfrac{8x^5 + 6}{x^4} \qquad \blacksquare$$

EXAMPLE 2   Evaluating Functions

In calculus, the derivative of $f(x)$ is denoted by $f'(x)$, and the value of the derivative at $x = a$ is denoted $f'(a)$. If $f'(x) = 3x^2 - 4x + 2$, find $f'(-1)$ and $f'(3)$.

Solution

$$f'(-1) = 3(-1)^2 - 4(-1) + 2 = 9$$
$$f'(3) = 3(3)^2 - 4(3) + 2 = 17 \qquad \blacksquare$$

EXAMPLE 3   Slope of a Tangent Line

Calculus can be used to find the slope of the tangent to a curve. The slope of the tangent to the curve $y = f(x)$ at $x = a$ is $f'(a)$. If $f'(x) = 4x^3 - 3x^2 + 1$, find the slope of the tangent at $x = 2$.

Solution

The slope of the tangent to the curve at $x = 2$ is $f'(2) = 4(2)^3 - 3(2)^2 + 1 = 21$. $\blacksquare$

EXAMPLE 4   Equation of a Tangent Line

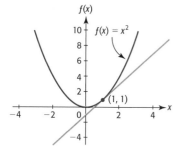

FIGURE 6.19

Calculus can be used to find the slope of the tangent to a curve, and then we can write the equation of the tangent line. If the slope of the line tangent to the graph of $f(x) = x^2$ at $(1, 1)$ is 2, write the equation of this tangent line. (See Figure 6.19.)

Solution

The point-slope form of the equation of a line is

$$y - y_1 = m(x - x_1).$$

Using the given point and the slope of the tangent line gives the equation

$$y - 1 = 2(x - 1), \quad \text{or} \quad y = 2x - 1. \qquad \blacksquare$$

EXAMPLE 5   Simplifying Fractions

a. An important step in finding the derivative of a function involves reducing fractions like

$$\dfrac{(2(x + h) + 1) - (2x + 1)}{h},$$

which is called the difference quotient.

Reduce this fraction, if $h \neq 0$.

b. An important calculation in calculus is the evaluation of the expression $\dfrac{f(x + h) - f(x)}{h}$, where $h \neq 0$. Find $\dfrac{f(x + h) - f(x)}{h}$ if $f(x) = 3x + 2$.

Solution

a. $\dfrac{(2(x + h) + 1) - (2x + 1)}{h} = \dfrac{2x + 2h + 1 - 2x - 1}{h} = \dfrac{2h}{h} = 2$

b. $\dfrac{f(x + h) - f(x)}{h} = \dfrac{(3(x + h) + 2) - (3x + 2)}{h}$

$= \dfrac{3x + 3h + 2 - 3x - 2}{h} = \dfrac{3h}{h} = 3$ ■

EXAMPLE 6   **Solving Equations**

Finding where the derivative is zero is an important skill in calculus.

a. Find where $f'(x) = 0$ if $f'(x) = 3x - 6$.

b. Find where $f'(x) = 0$ if $f'(x) = 8x - 4$.

Solution

a. $f'(x) = 0$ where $0 = 3x - 6$, or where $6 = 3x$, at $x = 2$.

b. $f'(x) = 0$ where $0 = 8x - 4$, or where $4 = 8x$, at $x = \dfrac{1}{2}$. ■

Chapter 2 Skills

The following examples demonstrate skills discussed in Chapter 2 that are useful in calculus.

EXAMPLE 7   **Rewriting Radicals**

In order to perform calculus operations on functions, it is frequently desirable or necessary to write a radical function in the form $y = ax^b$.

a. Convert the function $y = 3\sqrt[3]{x^4}$ to this form.

b. Convert the function $y = 2x\sqrt{x}$ to this form.

Solution

a. $3\sqrt[3]{x^4} = 3x^{4/3}$, so the form is $y = 3x^{4/3}$.

b. $2x\sqrt{x} = 2x \cdot x^{1/2} = 2x^{1+1/2} = 2x^{3/2}$, so the form is $y = 2x^{3/2}$. ■

EXAMPLE 8   **Rewriting Fractional Exponents**

Sometimes a derivative contains fractional exponents, and simplification requires that all fractional exponents be converted to radicals. Write the expression $f'(x) = -3x^{1/3} + 2x^{1/2} + 1$ without fractional exponents.

Solution

$$f'(x) = -3x^{1/3} + 2x^{1/2} + 1 = -3\sqrt[3]{x} + 2\sqrt{x} + 1 \qquad\blacksquare$$

EXAMPLE 9   Evaluating the Difference Quotient

An important calculation in calculus is the evaluation of the difference quotient $\dfrac{f(x + h) - f(x)}{h}$, where $h \neq 0$. Find each of the following, if $f(x) = 3x^2 - 2x$.

a. $f(x + h)$

b. $f(x + h) - f(x)$

c. $\dfrac{f(x + h) - f(x)}{h}$

Solution

a. $f(x + h) = 3(x + h)^2 - 2(x + h) = 3(x^2 + 2xh + h^2) - 2(x - h)$

$\qquad = 3x^2 + 6xh + 3h^2 - 2x - 2h$

b. $f(x + h) - f(x) = (3x^2 + 6xh + 3h^2 - 2x - 2h) - (3x^2 - 2x)$

$\qquad = 6xh + 3h^2 - 2h$

c. $\dfrac{f(x + h) - f(x)}{h} = \dfrac{6xh + 3h^2 - 2h}{h} = 6x + 3h - 2 \qquad\blacksquare$

EXAMPLE 10   Equation Solution

Finding where the derivative is zero is an important skill in calculus.

a. Find where $f'(x) = 0$ if $f'(x) = 3x^2 + x - 2$.

b. Find where $f'(x) = 0$ if $f'(x) = 3x^2 + 5x - 2$.

Solution

a. To find where $f'(x) = 0$, we solve $3x^2 + x - 2 = 0$.

$$3x^2 + x - 2 = 0$$
$$(3x - 2)(x + 1) = 0$$
$$3x - 2 = 0 \quad \text{or} \quad x + 1 = 0$$
$$x = \frac{2}{3} \quad \text{or} \quad x = -1$$

b. To find where $f'(x) = 0$, we solve $3x^2 + 5x - 2 = 0$.

$$3x^2 + 5x - 2 = 0$$
$$(3x - 1)(x + 2) = 0$$
$$3x - 1 = 0 \quad \text{or} \quad x + 2 = 0$$
$$x = \frac{1}{3} \quad \text{or} \quad x = -2$$

$\qquad\blacksquare$

EXAMPLE 11 Simplifying Expressions by Multiplying

After a derivative is found, the expression frequently must be simplified. Simplify by factoring.

$$x[2(6x - 1)6] + (6x - 1)^2$$

Solution

$$
\begin{aligned}
x[2(6x - 1)6] + (6x - 1)^2 &= (6x - 1)[x \cdot 2 \cdot 6 + (6x - 1)] \\
&= (6x - 1)(18x - 1) \qquad \blacksquare
\end{aligned}
$$

EXAMPLE 12 Simplifying Expressions by Factoring

Simplify by factoring.

$$x^4(6(3x^2 - 2x)^5(6x - 2)) + (3x^2 - 2x)^6(4x^3).$$

Solution

Each term contains the factor $(3x^2 - 2x)$ to some power, and it would be difficult to raise this factor to the fifth and sixth power, so we factor $(3x^2 - 2x)^5$ from each term to begin the simplification.

$$
\begin{aligned}
&x^4(6(3x^2 - 2x)^5(6x - 2)) + (3x^2 - 2x)^6(4x^3) \\
&= (3x^2 - 2x)^5[x^4(6(6x - 2)) + (3x^2 - 2x)(4x^3)] \\
&= (3x^2 - 2x)^5[36x^5 - 12x^4 + 12x^5 - 8x^4] = (3x^2 - 2x)^5(48x^5 - 20x^4)
\end{aligned}
$$

Because $4x^4$ is a factor of $(48x^5 - 20x^3)$, we can write the expression as $4x^4(12x - 5)(3x^2 - 2x)^5$ $\qquad \blacksquare$

In calculus, it is sometimes useful to find two functions whose composition gives a function. For example, if $y = (3x - 2)^2$, we can write this function as $y = u^2$, where $u = 3x - 2$. Going backward from a function to two functions whose composition gives the function is called finding the *decomposition* of the function. Another example of decomposition is writing $y = \dfrac{1}{x^3 - 1}$ as $y = \dfrac{1}{u}$, where $u = x^3 - 1$. It is possible to write several decompositions for a given function. For example, $y = \dfrac{1}{x^3 - 1}$ could also be written as $y = \dfrac{1}{u - 1}$ where $u = x^3$.

EXAMPLE 13 Composition of Functions

An equation of the form $y = (ax^k + b)^n$ can be thought of as the composition of functions and can be written as $y = u^n$, where $u = ax^k + b$. Use this idea to write two functions whose composition gives $y = (x^4 + 5)^6$.

Solution

We can let $y = u^6$, where $u = x^4 + 5$. Then $y = (x^4 + 5)^6$. $\qquad \blacksquare$

| EXAMPLE 14 | Function Decomposition |

Find two functions, f and g, such that their composition is $(f \circ g)(x) = \sqrt{x^2 + 1}$.

Solution

One decomposition is found by letting $f(u) = \sqrt{u}$ and $g(x) = x^2 + 1$, then $(f \circ g)(x) = f(g(x)) = \sqrt{x^2 + 1}$.   ■

Chapter 3 Skills

The following examples demonstrate skills discussed in Chapter 3 that are useful in calculus.

| EXAMPLE 15 | Rewriting Expressions Using Negative Exponents |

Many functions must be written in a form with no variables in the denominator in order to take the derivative. Use negative exponents where necessary to write the following function with all variables in the numerator of a term. That is, write each term in the form cx^n.

$$f(x) = \frac{4}{x^2} + \frac{3}{x} - 2x$$

Solution

Using the definition $a^{-n} = \dfrac{1}{a^n}$, $a \neq 0$, we get

$$f(x) = 4\frac{1}{x^2} + 3\frac{1}{x} - 2x = 4x^{-2} + 3x^{-1} - 2x.$$ ■

| EXAMPLE 16 | Rewriting Expressions Using Positive Exponents |

Sometimes a derivative contains negative exponents, and simplification requires that all exponents be positive. Write the expression $f'(x) = -3x^{-4} + 2x^{-3} + 1$ without negative exponents.

Solution

$$f'(x) = -3x^{-4} + 2x^{-3} + 1$$

$$f'(x) = -3\frac{1}{x^4} + 2\frac{1}{x^3} + 1 = -\frac{3}{x^4} + \frac{2}{x^3} + 1$$ ■

| EXAMPLE 17 | Using Logarithmic Properties |

Taking derivatives involving logarithms frequently requires simplifying by converting the expressions using logarithmic properties. Rewrite the following expressions as the sum, difference, or product of logarithms without exponents.

a. $\log[x^2(2x + 3)^2]$

b. $\ln\dfrac{\sqrt{2x - 4}}{3x + 1}$

Solution

a. $\log[x^2(2x + 3)^2] = \log x^2 + \log(2x + 3)^2$

$$= 2 \log x + 2 \log(2x + 3)$$

b. $\ln\dfrac{\sqrt{2x - 4}}{3x + 1} = \ln\sqrt{2x - 4} - \ln(3x + 1)$

$$= \ln(2x - 4)^{1/2} - \ln(3x + 1)$$

$$= \frac{1}{2}\ln(2x - 4) - \ln(3x + 1) \qquad \blacksquare$$

Chapter 4 Skills

The following examples demonstrate skills discussed in Chapter 4 that are useful in calculus.

EXAMPLE 18 Reducing Fractions

Evaluating limits is a calculus skill that frequently requires reducing fractions, like the following. Reducing these fractions may require factoring.

a. Simplify $\dfrac{x^2 - 4}{x - 2}$ if $x \neq 2$

b. Simplify $\dfrac{x^2 - 7x + 12}{x - 4}$ if $x \neq 4$

c. Simplify $\dfrac{x^3 - 4x}{2x^2 - x^3}$ if $x \neq 0, x \neq 2$

Solution

a. $\dfrac{x^2 - 4}{x - 2} = \dfrac{(x - 2)(x + 2)}{x - 2} = x + 2$ if $x \neq 2$

b. $\dfrac{x^2 - 7x + 12}{x - 4} = \dfrac{(x - 3)(x - 4)}{x - 4} = x - 3$ if $x \neq 4$

c. $\dfrac{x^3 - 4x}{2x^2 - x^3} = \dfrac{x(x^2 - 4)}{x^2(2 - x)} = \dfrac{x(x - 2)(x + 2)}{x^2(2 - x)} = \dfrac{x(x - 2)(x + 2)}{-x^2(x - 2)} = -\dfrac{x + 2}{x}$

if $x \neq 0, x \neq 2$ $\qquad \blacksquare$

EXAMPLE 19 Simplifying Expressions

After a derivative is taken, the expression frequently must be simplified. Simplify the following expression by multiplying and combining like terms.

$$\frac{(4x - 3)(2x - 3) - (x^2 - 3x)(4)}{(4x - 3)^2}$$

Solution

$$\frac{(4x - 3)(2x - 3) - (x^2 - 3x)(4)}{(4x - 3)^2} = \frac{8x^2 - 12x - 6x + 9 - 4x^2 + 12x}{(4x - 3)^2}$$

$$= \frac{4x^2 - 6x + 9}{(4x - 3)^2} \qquad ■$$

EXAMPLE 20   Solving Equations

Finding where the derivative is zero is an important skill in calculus.

a. Find where $f'(x) = 0$ if $f'(x) = 2x^3 - 2x$.

b. Find where $f'(x) = 0$ if $f'(x) = 4x^3 - 10x^2 - 6x$.

c. Find where $f'(x) = 0$ if $f'(x) = (x^2 - 1)3(x - 5)^2 + (x - 5)^3(2x)$.

d. Find where $f'(x) = 0$ if $f'(x) = \dfrac{x[3(x - 2)^2] - (x - 2)^3}{x^2}$.

Solution

a. Solving $0 = 2x^3 - 2x$ gives

$$0 = 2x^3 - 2x$$
$$0 = 2x(x^2 - 1)$$
$$0 = 2x(x - 1)(x + 1)$$
$$x = 0, x = 1, \text{ and } x = -1 \text{ are the solutions.}$$

Thus $f'(x) = 0$ at $x = 0$, $x = 1$ and $x = -1$.

b. Solving $0 = 4x^3 - 10x^2 - 6x$ gives

$$0 = 4x^3 - 10x^2 - 6x$$
$$0 = 2x(2x^2 - 5x - 3)$$
$$0 = 2x(2x + 1)(x - 3)$$
$$x = 0, x = -\frac{1}{2}, \text{ and } x = 3 \text{ are the solutions.}$$

Thus $f'(x) = 0$ at $x = 0$, $x = -\dfrac{1}{2}$, and $x = 3$.

c. Each term contains the factor $(x - 5)$ to some power. Rather than raise this factor to the second and third power, we use factoring to simplify this expression. Factoring $(x - 5)^2$ from each term gives the following:

$$0 = (x^2 - 1)3(x - 5)^2 + (x - 5)^3(2x)$$
$$0 = (x - 5)^2[1(3)(x^2 - 1) + (x - 5)(2x)]$$
$$0 = (x - 5)^2(3x^2 - 3 + 2x^2 - 10x)$$
$$0 = (x - 5)^2(5x^2 - 10x - 3)$$

One of the solutions is $x = 5$. The remaining solutions can be found with the quadratic formula.

$$x = \frac{10 \pm \sqrt{100 - 4(5)(-3)}}{2(5)} = \frac{10 \pm 4\sqrt{10}}{10} = \frac{5 \pm 2\sqrt{10}}{5}.$$

Thus $f'(x) = 0$ at $x = 5$, $x = \dfrac{5 + 2\sqrt{10}}{5}$, and $x = \dfrac{5 - 2\sqrt{10}}{5}$.

d. To find where the fraction equals 0, we set the numerator equal to 0 and solve.

$$0 = x[3(x - 2)^2] - (x - 2)^3$$

$$0 = (x - 2)^2[3x - (x - 2)]$$

$$0 = (x - 2)^2(2x + 2)$$

$$0 = (x - 2)^2 \quad \text{or} \quad 0 = 2x + 2$$

$$
\begin{array}{c|c}
0 = x - 2 & -2x = 2 \\
x = 2 & x = -1
\end{array}
$$

Checking these values in the rational function $f'(x) = \dfrac{x[3(x - 2)^2] - (x - 2)^3}{x^2}$ gives

$f'(x) = 0$ at $x = 2$ and at $x = -1$. ∎

6.5 SKILLS CHECK

The expressions in Exercises 1–4 represent the derivatives of functions. Simplify the expressions.

1. $f'(x) = (2x^3 + 3x + 1)(2x) + (x^2 + 4)(6x^2 + 3)$

2. $f'(x) = (x^2 - 3x)(3x^2) + (x^3 - 4)(2x - 3)$

3. $f'(x) = \dfrac{x^2(3x^2) - (x^3 - 3)(2x)}{(x^2)^2}$.

4. Simplify the numerator of
$$f'(x) = \dfrac{(x^2 - 4)(3x^2 - 6x) - (x^3 - 3x^2 + 2)(2x)}{(x^2 - 4)^2}.$$

In calculus, the derivative of $f(x)$ is denoted by $f'(x)$, and the value of the derivative at $x = a$ is $f'(a)$. Evaluate the derivatives in Exercises 5 and 6.

5. Find $f'(2)$ if $f'(x) = 4x^3 - 3x^2 + 4x - 2$.

6. Find $f'(-1)$ if
$$f'(x) = (x^2 - 1)(6x) + (3x^2 + 1)(2x).$$

7. Find $f'(3)$ if $f'(x) = \dfrac{(x^2 - 1)(2) + (2x)(2x)}{(x^2 - 1)^2}$.

8. Calculus can be used to find the slope of the line tangent to a curve. The slope of the tangent to the curve $y = f(x)$ at $x = a$ is $f'(a)$. Find the slope of the tangent to $y = f(x)$ at $x = 2$ if $f'(x) = 8x^3 + 6x^2 - 5x - 10$.

9. The derivative of a function can be used to find the slope of the line tangent to its graph. If the slope of the line tangent to the graph of $f(x) = x^3$ at $(2, 8)$ is 12, write the equation of this tangent line.

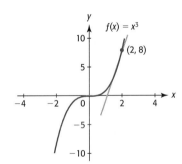

An important calculation in calculus is the simplification of the expression $\dfrac{f(x + h) - f(x)}{h}$, where $h \neq 0$.
In Exercises 10–12, find each of the following.

  a. $f(x + h)$
  b. $f(x + h) - f(x)$
  c. $\dfrac{f(x + h) - f(x)}{h}$

10. $f(x) = 3x - 2$ **11.** $f(x) = 4x + 5$

12. $f(x) = 12 - 2x$

Writing the equation defining the function in the form $y = f(x)$ usually makes it easier to find the derivative. Write each of the equations in Exercises 13–16 in this form by solving for y.

13. $8x^2 + 4y = 12$ **14.** $3x^2 - 2y = 6$

15. $9x^3 + 5y = 18$ **16.** $12x^3 = 6y - 24$

The vertex of a parabola occurs where the derivative of its equation is 0, so an equation associated with the equation of a parabola can be solved to find the x-coordinate of the vertex. For each parabola whose equation and "associated equation" are given in Exercises 17–20:

a. Find the solution of the associated equation.

b. Compare the solution with the *x*-coordinate of the vertex found by the methods of Section 2.1.

c. Find the vertex of the parabola.

Equation of Parabola	**Associated Equation**
17. $y = x^2 - 2x + 5$	$0 = 2x - 2$
18. $y = 5 + x - x^2$	$0 = 1 - 2x$
19. $y = 6x^2 - 24x + 15$	$0 = 12x - 24$
20. $y = -(4 + 3x + x^2)$	$0 = -3 - 2x$

21. Between $x = -3$ and $x = 10$, the largest value of the expression $y = x^3 - 9x^2 - 48x + 15$ occurs where $3x^2 - 18x - 48 = 0$.

a. Find the values of x that satisfy $3x^2 - 18x - 48 = 0$.

b. Find the value of $y = x^3 - 9x^2 - 48x + 15$ at each of the solutions found in part (a).

c. Which of the solutions from part (a) gives the maximum value of $y = x^3 - 9x^2 - 48x + 15$? Test the expression at other values of x to confirm your conclusion.

22. Between $x = -2$ and $x = 4$, the largest value of the expression $y = x^3 - 3x^2 - 9x + 5$ occurs where $3x^2 - 6x - 9 = 0$.

a. Find the values of x that satisfy $3x^2 - 6x - 9 = 0$.

b. Find the value of $y = x^3 - 3x^2 - 9x + 5$ at each of the solutions found in part (a).

c. Which of the solutions from part (a) gives the maximum value of $y = x^3 - 3x^2 - 9x + 5$? Test the expression at other values of x to confirm your conclusion.

In order to perform calculus operations on functions, it is frequently desirable or necessary to write the function in the form $y = cx^b$. Write each of the functions in Exercises 23–26 in this form.

23. $y = 3\sqrt{x}$

24. $y = 6\sqrt[3]{x}$

25. $y = 2\sqrt[3]{x^2}$

26. $y = 4\sqrt{x^3}$

Write each of the functions in Exercises 27–32 without radicals by using fractional exponents.

27. $y = \sqrt[3]{x^2 + 1}$

28. $y = \sqrt[4]{x^3 - 2}$

29. $y = \sqrt[3]{(x^3 - 2)^2}$

30. $y = \sqrt{(5x - 3)^3}$

31. $y = \sqrt{x} + \sqrt[3]{2x}$

32. $y = \sqrt[3]{(2x)^2} + \sqrt{(4x)^3}$

Equations containing y^2 occasionally must be solved for y. Sometimes this can be accomplished with the root method, and sometimes a method such as the quadratic formula is necessary. Solve the equations in Exercises 33–36 for y.

33. $y^2 + 4x - 3 = 0$

34. $5x - y^2 = 12$

35. $y^2 - 6x + y = 0$

36. $2x - 2y^2 + y + 4 = 0$

An equation of the form $y = (ax^k + b)^n$ can be thought as the composition of the functions of $f(u) = u^n$ and $u(x) = ax^k + b$, and can be written as $y = u^n$ where $u = ax^k + b$. This form is very useful in calculus. In Exercises 37–41, use this idea to fill in the blanks.

37. $y = (4x^3 + 5)^2$ can be written as $y = u^-$ where $u(x) = $ _____.

38. $y = $ _____ can be written as $y = u^7$ where $u(x) = 5x^2 - 4x$.

39. $y = \sqrt{\underline{}}$ can be written as $y = u^-$ where $u(x) = x^3 + x$.

40. If $f(u) = u^6$ and $u = x^4 + 5$, then express $6u^5 \cdot 4x^3$ in terms of x alone, and simplify.

41. If $f(u) = u^{3/2}$ and $u = x^2 - 1$, then express $\frac{3}{2}u^{1/2} \cdot 2x$ in terms of x alone, and simplify.

In calculus it is often necessary to write expressions in the form cx^n, where n is a constant. Use negative exponents where necessary to write the expressions in Exercises 42–44 with all variables in the numerator of a term. That is, write each term in the form cx^n if the term contains a variable.

42. $y = 3x + \dfrac{1}{x} - \dfrac{2}{x^2}$

43. $y = \dfrac{3}{x} - \dfrac{4}{x^2} - 6$

44. $y = 5 + \dfrac{1}{x^3} - \dfrac{2}{\sqrt{x}} + \dfrac{3}{\sqrt[3]{x^2}}$

45. Write $y = \dfrac{1}{(x^2 - 3)^3}$ with no denominator.

Sometimes a derivative contains negative exponents, and simplification requires that all exponents be positive. Write each of the expressions in Exercises 46–49 without negative exponents.

46. $f'(x) = 8x^{-4} + 5x^{-2} + x$

47. $f'(x) = -6x^{-4} + 4x^{-2} + x^{-1}$

48. $f'(x) = -3(3x - 2)^{-3}$

49. $f'(x) = (4x^2 - 3)^{-1/2}(8x)$

Rewrite the expressions in Exercises 50–52 as the sum, difference, or product of logarithms.

50. $\ln \dfrac{3x - 2}{x + 1}$

51. $\log x^3 (3x - 4)^5$

52. $\ln \dfrac{\sqrt[4]{4x + 1}}{4x^2}$

Find where the derivative is zero in Exercises 53–56.

53. Find where $f'(x) = 0$ if $f'(x) = 3x^2 - 3x$.

54. Find where $f'(x) = 0$ if $f'(x) = 3x^3 - 14x^2 + 8x$.

55. Find where $f'(x) = 0$ if
$$f'(x) = (x^2 - 4)[3(x - 3)^2] + (x - 3)^3(2x)$$

56. Find where $f'(x) = 0$ if
$$f'(x) = \frac{(x + 1)[3(2x - 3)^2 2] - (2x - 3)^3}{(x + 1)^2}.$$

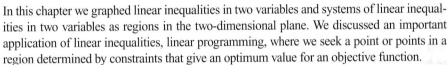

CHAPTER 6 *Summary*

In this chapter we graphed linear inequalities in two variables and systems of linear inequalities in two variables as regions in the two-dimensional plane. We discussed an important application of linear inequalities, linear programming, where we seek a point or points in a region determined by constraints that give an optimum value for an objective function.

We also discussed sequences, which are special discrete functions whose domains are sets of positive integers. Sequences can be used to solve numerous problems involving discrete inputs, including investment problems. We also discussed series, which are the sums of terms of a sequence. Series are useful in finding the future value of annuities if the interest is compounded over discrete time periods.

Finally, we showed how important algebra is in the development and application of calculus.

Key Concepts and Formulas

6.1 Systems of Linear Inequalities

Linear inequalities in two variables	The solution to a linear inequality is the set of ordered pairs that satisfy the inequality.
System of linear inequalities	The solution to a system of linear inequalities in two or more variables consists of the values of the variables that satisfy all the inequalities.
Border	The lines that bound the solution region are called the borders of the system of linear inequalities.
Solution region	The region that includes the half-plane containing the ordered pairs that satisfy an inequality is called the solution region. The border line is included in the solution region if the inequality includes an equal sign.
Corner	A corner of the graph of a solution to a system of inequalities occurs at a point where the boundary lines intersect.

6.2 Linear Programming: Graphical Methods

Linear programming

Linear programming is a technique that can be used to solve problems when the constraints on the variables can be expressed as linear inequalities and a linear objective function is to be maximized or minimized.

Constraints

The inequalities that limit the values of the variables in a linear programming application are called the constraints.

Feasible region

The constraints of a linear programming application form a feasible region in which the solution lies.

Feasible solution

Any point in the region determined by the constraints is called a feasible solution.

Optimum value

A maximum or minimum value of an objective function is called an optimum value.

Solving a linear programming problem

1. If a linear programming problem has a solution, then the optimum value (maximum or minimum) of an objective function occurs at a corner of the feasible region determined by the constraints.
2. When a feasible region for a linear programming problem is closed and bounded, the objective function has a maximum and a minimum value.
3. When the feasible region is not closed and bounded, the objective function may have a maximum only, a minimum only, or neither.

6.3 Sequences and Discrete Functions

Sequence

A sequence is a function whose outputs result uniquely from positive integer inputs. Because the domains of sequences are positive integers, sequences are **discrete functions**.

Terms of the sequence

The ordered outputs corresponding to the integer inputs of a sequence are called the terms of the sequence.

Infinite sequence

If the domain of a sequence is the set of all positive integers, the outputs form an infinite sequence.

Finite sequence

If the domain of a sequence is the set of positive integers from 1 to n, the outputs form a finite sequence.

Arithmetic sequence

A sequence is called an arithmetic sequence if there exists a number d, called the common difference, such that $a_n = a_{n-1} + d$ for $n > 1$.

nth term of an arithmetic sequence

The nth term of an arithmetic sequence is given by $a_n = a_1 + (n - 1)d$, where a_1 is the first term of the sequence, n is the number of the term, and d is the common difference between the terms.

Geometric sequence

A sequence is called a geometric sequence if there exists a number r, the common ratio, such that $a_n = ra_{n-1}$ for $n > 1$.

nth term of a geometric sequence

The nth term of a geometric sequence is given by $a_n = a_1 r^{n-1}$, where a_1 is the first term of the sequence, n is the number of the term, and r is the common ratio of the terms.

6.4 Series

Series

A series is the sum of the terms of a sequence.

Finite series

A finite series is defined by $a_1 + a_2 + a_3 + \ldots + a_n$, where $a_1, a_2, \ldots a_n$ are the terms of a sequence.

Infinite series	An infinite series is defined by $a_1 + a_2 + a_3 + \ldots + a_n + \ldots$.
Sum of n terms of an arithmetic sequence	The sum of the first n terms of an arithmetic sequence is given by the formula $s_n = \dfrac{n(a_1 + a_n)}{2}$ where a_1 is the first term of the sequence and a_n is the nth term.
Sum of n terms of a geometric sequence	The sum of the first n terms of a geometric sequence is $s_n = \dfrac{a_1(1 - r^n)}{1 - r}$, where a_1 is the first term of the sequence and r is the common ratio.
Sum of an infinite geometric series	The sum of an infinite geometric series is $S = \dfrac{a_1}{1 - r}$, where a_1 is the first term and r is the common ratio, with $|r| < 1$. If $|r| \geq 1$, the sum does not exist (is not a finite number).

6.5 Preparing for Calculus

Algebra review   Algebra topics from Chapters 1–4 are presented in calculus contexts.

Chapter 6 Skills Check

Graph the following inequalities.

1. $5x + 2y \leq 10$

2. $5x - 4y > 12$

Graph the systems of inequalities in Exercises 3 and 4. Identify the corners of the solution regions.

3. $\begin{cases} 2x + y \leq 3 \\ x + y \leq 2 \\ x \geq 0, y \geq 0 \end{cases}$

4. $\begin{cases} 3x + 2y \leq 6 \\ 3x + 6y \leq 12 \\ x \geq 0, y \geq 0 \end{cases}$

5. Minimize $g = 3x + 4y$ subject to the constraints
$$\begin{cases} 3x + y \geq 8 \\ 2x + 5y \geq 14. \\ x \geq 0, y \geq 0 \end{cases}$$

6. Maximize the function $f = 7x + 12y$ subject to the constraints $\begin{cases} 7x + 3y \leq 105 \\ 2x + 5y \leq 59 \\ x + 7y \leq 70 \end{cases}$.

Determine whether the sequences in Exercises 7–9 are arithmetic or geometric.

7. $\dfrac{1}{9}, \dfrac{2}{3}, 4, 24, \ldots$

8. $4, 16, 28, \ldots$

9. $16, -12, 9, \dfrac{-27}{4}$

10. Find the fifth term of the geometric sequence with first term 64 and eighth term $1/2$.

11. Find the sixth term of the geometric sequence $\dfrac{1}{9}, \dfrac{1}{3}, 1, \ldots$.

12. Find the sum of the first 10 terms of the geometric sequence with first term 5 and common ratio -2.

13. Find the sum of the first 12 terms of the sequence $3, 6, 9, \ldots$.

14. Find the sum of the series $\displaystyle\sum_{i=1}^{\infty} \left(\dfrac{4}{5}\right)^i$ if possible.

15. Simplify the expression
$(x^3 + 2x + 3)(2x) + (x^2 - 5)(3x^2 + 2)$.

16. Between $x = 0$ and $x = 10$, the smallest value of the expression $y = x^3 - 8x^2 - 9x$ occurs where $x^3 - 8x^2 - 9x = 0$. Find the values of x that satisfy $x^3 - 8x^2 - 9x = 0$.

17. Write $y = x + \dfrac{1}{x^2} - \dfrac{2}{\sqrt{x^3}} + \dfrac{3}{\sqrt[3]{x}}$ with each term in the form cx^n.

18. Write $f'(x) = 8x^{-3} + 5x^{-1} + 4x$ with no negative exponents.

19. Find where $f'(x) = 0$ if $f'(x) = 2x^3 + x^2 - x$.

Chapter 6 Review

1. *Manufacturing* Ace Manufacturing produces two types of DVD players, the Deluxe model and the Superior model, which also plays videotapes. Each Deluxe model requires 3 hours to manufacture and $40 for assembly parts, while each Superior model requires 2 hours to assemble and $60 for assembly parts. The company is limited to 1800 assembly hours and $36,000 for assembly parts each month. If the profit on the Deluxe model is $30 and the profit on the Superior model is $40, how many of each model should be produced to maximize profit?

2. *Manufacturing* A company manufactures two grades of steel, cold steel and stainless steel, at the Pottstown and Ethica factories. The table below gives the daily cost of operation and the daily production capabilities of each of the factories, and the number of units of each type of steel that is required to fill an order.
   a. How many days should each factory operate to fill the order at minimum cost?
   b. What is the minimum cost?

	Units/Day at Pottstown	Units/Day at Ethica	Required for Order
Units of Cold Steel	20	40	At least 1600
Units of Stainless Steel	60	40	At least 2400
Cost per Day for Operation	$20,000	$24,000	

3. *Nutrition* A laboratory wishes to purchase two different feeds, A and B, for its animals. Feed A has 2 units of carbohydrates per pound and 4 units of protein per pound, while Feed B has 8 units of carbohydrates per pound and 2 units of protein per pound. Feed A costs $1.40 per pound and Feed B costs $1.60 per pound. If at least 80 units of carbohydrates and 132 units of protein are required, how many pounds of each feed are required to minimize the cost?

4. *Manufacturing* A company manufactures leaf blowers and weed wackers. One line can produce 260 leaf blowers per day and another line can produce 240 weed wackers per day. The combined number of leaf blowers and weed wackers that shipping can handle is 460 per day. How many of each can be produced daily to maximize company profit if the profit is $5 on the leaf blowers and $10 on the weed wackers?

5. *Profits* A woman has a building with 60 two-bedroom and 40 three-bedroom apartments available to rent to students. She has set the rent at $800 for the two-bedroom units and $1150 per month for the three-bedroom units. She must rent to one student per bedroom and zoning laws limit her to at most 180 students in this building. How many of each type of apartment should she rent to maximize her revenue?

6. *Loans* A finance company has at most $30 million available for auto and home equity loans. The auto loans have an annual return rate of 8% to the company and the home equity loans have an annual rate of return of 7% to the company. If there must be at least twice as many auto loans as home equity loans, how much should be loaned in each type of loan to maximize the profit to the finance company?

7. *Salaries* Suppose you are offered two identical jobs, Job 1 paying a starting salary of $20,000 with yearly raises of $1000 at the end of each year and Job 2 paying a starting salary of $18,000 with yearly increases of $1600 at the end of each year.
   a. Which job will pay more in the fifth year on the job?
   b. Which job will pay more over the first 5 years?

8. *Drug in the Bloodstream* Suppose a 400-mg dose of heart medicine is taken daily. During each 24-hour period, the body eliminates 40% of the drug (so that 60% remains in the body). Thus the amount of drug in the body just after the 21 doses, over 21 days, is given by

$$400 + 400(0.6) + 400(0.6)^2 + 400(0.6)^3 + \ldots + 400(0.6)^{19} + 400(0.6)^{20}$$

Find the level of the drug in the bloodstream at this time.

9. *Chess Legend* Legend has it that when the king of Persia offered to reward the inventor of chess any prize he wanted, the inventor asked for one grain of wheat on the first square of the chessboard, with the number of grains doubled on each square thereafter for the remaining 63 squares.
   a. How many grains of wheat are there for the 64th square?

b. What is the total number of grains of wheat for all 64 squares? (This is enough wheat to cover Alaska more than 3 inches deep in wheat.)

10. *Compound Interest* Find the future value of $20,000 invested for 5 years at 6% compounded annually.

11. *Annuities* Find the 5-year future value of an ordinary annuity with a contribution of $300 per month into an account that pays 12% per year, compounded monthly.

GROUP ACTIVITY/EXTENDED APPLICATION

Salaries

In the 1970s, a local teachers' organization in Pennsylvania asked for a $1000 raise at the end of each year for the next 3 years. When the school board said it could not afford to do this, the teachers' organization then said that the teachers would accept a $300 raise at the end of each 6-month period. The school board initially responded favorably because the members thought it would save them $400 per year per teacher.

If you are an employee, would you rather be given a raise of $1000 at the end of each year (Plan I) or be given a $300 raise at the end of each 6-month period (Plan II)? To answer the question, complete the following table for an employee whose base salary is $40,000 (or $20,000 for each 6-month period), and then answer the other questions.

Year	Period (in months)	Salary Received per Six-Month Period	
		Plan I	Plan II
1	0–6	$20,000	$20,000
	6–12	20,000	20,300
2	12–18	20,500	20,600
	18–24	20,500	20,900
3	24–30		
	30–36		
Total for 3 Years			

1. Find the total additional money earned for the first 3 years from the raises of Plan I.
2. Find the total additional money earned for the first 3 years from the raises of Plan II.
3. Which plan gives more money from the raises, and how much more? Which plan is better for the employee?
4. Find the total additional money earned (per teacher) from raises for each of Plan I and Plan II if they are extended to 4 years. Which plan gives more money in raises, and how much more? Which plan is better for the employee?
5. Did the school board make a mistake in thinking that the $300 raises every 6 months instead of $1000 every year would save money? If so, what was the mistake?
6. If there were 200 teachers in the school district, how much extra would Plan II cost over Plan I for the 4 years?

Answers to Selected Exercises

Chapter 1 Functions, Graphs, and Models; Linear Functions
Toolbox Exercises

1. $\{1, 2, 3, 4, 5, 6, 7, 8\}$; $\{x : x$ is a natural number, $x < 9\}$ **2.** no   **3.** yes   **4.** yes
5. integers, rational numbers   **6.** rational   **7.** irrational   **8.** $x > -3$ **9.** $-3 \le x \le 3$ **10.** $x \le 3$
11. $(-\infty, 7]$ **12.** $(3, 7]$,   **13.** $(-\infty, 4)$ **14.** ———   **15.** ———
16.   **17.**   **18.** 6   **19.** 4

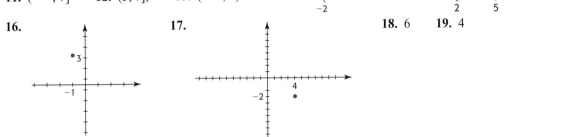

20. $-3x^2, -3; -4x^2, -4$; constant 8   **21.** $5x^4, 5; 7x^3, 7$; constant -3 **22.** $3z^4 + 4z^3 - 27z^2 + 20z - 11$
23. $7y^4 - 2x^3y^4 - 3y^2 - 2x - 9$ **24.** $4p + 4d$ **25.** $-6x + 14y$ **26.** $-ab - 8ac$ **27.** $x - 6y$
28. $x = 2$ **29.** $x = 18$ **30.** $x = 3$ **31.** $x = 3$ **32.** $x = 3/5$

1.1 Skills Check

1. Each input (x-value) results in one output (y-value).   **3.** output   **5.** no, $y = 9$ corresponds to both $x = 12$ and $x = 17$.   **7. a.** $f(2) = -1$ **b.** $f(2) = 10 - 3(2^2) = -2$ **c.** $f(2) = -3$ **9.** yes, by the vertical line test.
11. a. $C(3) = 16 - 2(3)^2 = -2$ **b.** $C(-2) = 16 - 2(-2)^2 = 8$ **c.** $C(-1) = 16 - 2(-1)^2 = 14$ **13.** b
15. no   **17.** $p = 2\pi r$ **19.** the set of inputs of a function   **21.** D is equal to 3 times the square of E, minus 5.

1.1 Exercises

1. a. no   **b.** yes   **3.** yes   **5.** yes   **7.** yes   **9. a.** yes   **b.** the days 1–14 of May
c. $\{171, 172, 173, 174, 175, 176, 177, 178\}$ **d.** May 1, May 3   **e.** May 14   **f.** 3 days   **11. a.** $16,115
b. 23,047; At the end of two years, the couple owes $23,047.   **c.** year 2   **d.** 4   **13. a.** 4.1
b. 4; 4 is the projected ratio of the working-age population (25- to 64-year-olds) to the elderly in 2005
c. $\{1995, 2000, 2005, 2010, 2015, 2020, 2025, 2030\}$ **d.** decreasing   **15. a.** 492,671   **b.** the years 1985 through 1998   **c.** 581,697; 1993   **17.** yes   **b.** 68.6   **c.** 1920   **d.** The percent of U.S. workers in farm occupations was 6.1 in 1960.   **e.** It is decreasing over the years shown.   **19. a.** 103; in 1995, there were 103 pregnancies per 1000 girls aged 15–19.   **b.** 1989 and 1992   **c.** 1987 through 1991   **d.** 1991   **21. a.** yes   **b.** $s \ge -1/4$ **c.** $s \ge 0$
23. a. 8640; 11,664   **b.** $0 < x < 27$ so volume is positive and box exists   **c.** Testing values in the table shows a maximum volume of 11,664 cubic inches when $x = 18$ inches.

The dimensions of the box are 18 in. by 18 in. by 36 in.

x	y
10	6800
15	10,800
20	11,200
21	10,584
19	11,552
18	11,664
17	11,560

1.2 Skills Check

1.

x	−3	−2	−1	0	1	2	3
$y = x^3$	−27	−8	−1	0	1	8	27
(x, y)	(−3,−27)	(−2,−8)	(−1,−1)	(0,0)	(1,1)	(2,8)	(3,27)

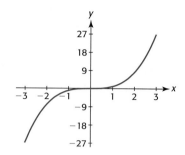

3.

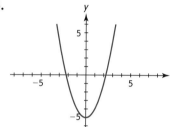

5.

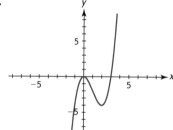

7.

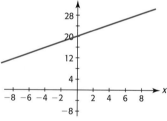

(b) is the better view.

(a)   (b)

9.

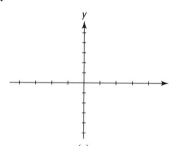

(b) is the better view.

[−20, 20] by [−0.02, 0.02]

(a)   (b)

11. Letting y vary from 0 to 100 gives the view.

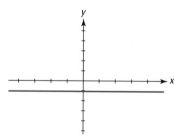

[−3, 3] by [0, 100]

13.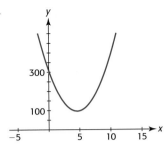

15.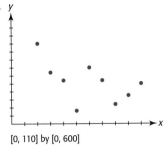

[0, 110] by [0, 600]

17.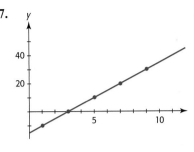

19. 2, 8, 20   **21. a.** $y = 35x^2 + 740x + 1207$ **b.** $x-\text{min} = 0, x-\text{max} = 17$

1.2 Exercises

1. a. 75.57; it means that when W is 80, M is 75.570, or that when the salary for whites is $80,000, minority salary is $75,570 **b.** $94,770.   **3. a.** 4 and 8 **b.** 7032; It means that when x is 8, y is 7032, or that in 1998, the number of welfare cases is 7032.   **c.** 11,721

5. a.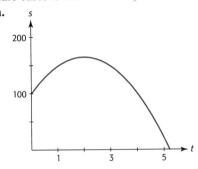

b. 148 ft and 148 ft; ball rising and falling   **c.** 164 ft, in 2 seconds

7. a.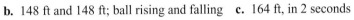

[0, 85] by [0, 65]

b. $51,560.   **9. a.**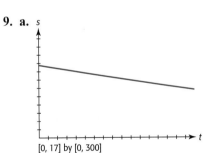

[0, 17] by [0, 300]

b. 152,255

c. 152,255   **11. a.**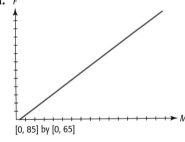

b. increase   **13. a.** $x = 2$ through $x = 16$

b. 0 through 100, because they are percents   **c.** $u(x)$

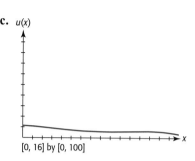

[0, 16] by [0, 100]

d. Graph w/y_{max} at 11 $u(x)$

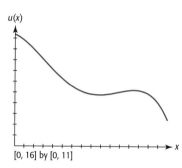

[0, 16] by [0, 11]

e. $u(13) = 6.1278$; 6.13%

15. a.

x	y
0	281
1	242
2	219
3	186
4	200
5	240
6	275
7	292
8	325
9	321
10	421

b. y

17. a. $y = \dfrac{252 + 2x}{5} = 0.4x + 50.4$

b. 88%   **c.** no; the highest possible grade is 90.4 percent   **19.** fails for integer values less than 5°F

1.3 Skills Check

1. b   **3.** undefined   **5.** 0; undefined
7. x: 3; y: −5

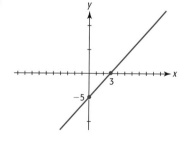

9. x: 3/2; y: 3

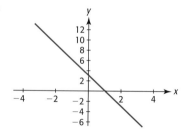

11. 4; 8   **13.** Slope 0; 2/5   **15. a.** 4, 5   **b.** rising

c.

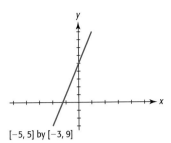

[−5, 5] by [−3, 9]

17. a. −100, 50,000   **b.** falling   **c.**

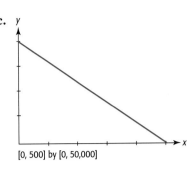

[0, 500] by [0, 50,000]

19. 4   **21. a.** 1   **b.** 1   **23.** 4   **25.** x-intercept

1.3 Exercises

1. linear; rises   **3.** linear; falls   **5. a.** $x = -30/19$ **b.** $p = 1$; 1% of high school seniors were using marijuana daily in 1990.   **c.**

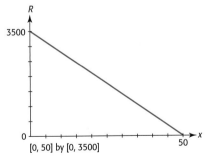

Integer values of $x \geq 0$ on the graph represent years 1990 and after.

7. a. constant function   **b.** $y = 11.81$ **c.** 0   **d.** 0   **9. a.** 26.5   **b.** The percent of Fortune Global 500 firms that recruited on the Internet increased at a rate of 26.5% per year.   **11. a.** positive   **b.** As temperature increases, the number of chirps increases.   **13. a.** yes   **b.** .959   **c.** Minority median annual salaries will increase by $.959 for each $1 increase in median annual salaries for whites.   **15. a.** 19/30   **b.** $19/30 \approx 63.33\%$ per year

17. x: 50; R: 3500

R

3500

0

50

[0, 50] by [0, 3500]

1.4 Skill Check

1. $y = 4x + \dfrac{1}{2}$ **3.** $y = 5x + 9$ **5.** $y = \dfrac{-3}{4}x - 3$ **7.** $y = x + 4$ **9.** $x = 9$ **11.** $y = 3x + 1$

13. 3   **15.** −15   **17.** 1   **19. a.** 30, 30, 30, 30   **b.** Yes; each input is 10 more than the last, the first differences of outputs are constant.   **c.** $y = 3x + 555$ **21.** $y = \dfrac{3}{2}x - 2$

1.4 Exercises

1. $y = 8.95 + 0.0935x$ (dollars)   **3.** $y = -3600t + 36{,}000$ **5. a.** $P = 705x + 198$ **b.** 4428
7. a. $25,000   **b.** $5000   **c.** $s = 26{,}000 - 5000t$ **9. a.** .05   **b.** $f(x) = 0.05x$ **11. a.** $y = 0.504x - 934.46$
b. yes   **c.** They are the same.   **13. a.** 3.42   **b.** 3.42   **c.** no   **d.** no   **15. a.** $61/36 \approx 1.69$ **b.** 1.69

c. The percent is increasing at a rate of approximately 1.69 per year   **d.** $p = 1.69x - 1.94$ **17. a.** 26,612
b. 26,612   **c.** $y = 26,612x - 51,956,196$ **d.** no   **e.** 1990 and 1997   **19. a.** no   **b.** Yes, they appear to lie on a
line.   **c.** -0.085 **d.** $y = -0.085x + 174.75$ **e.** 2056   **21. a.** Increasing at a rate of approximately 2462 thousand
people, or approximately 2.462 million people per year   **b.** $y = 2462x + 152,271$ thousands   **c.** 213,821 thousand, or
213,821,000; no   **d.** The line does not model the data exactly.   **23. a.** yes   **b.** \$0.28   **c.** $y = 0.28x - 3197$
d. Evaluating the function at $x = 30,100$ gives y = 5231 and evaluating it at $x = 30,300$ gives 5287. These values agree
with the income taxes due in the table.   **25.** $s = 900,000 - 30,000t$ **27. a.** \$9.86   **b.** 10.86 cents, or \$.1086
c. 0.1086   **d.** $C = 0.1086K + 9.86$

1.5 Skills Check

1. $x = -37/2$ **3.** $x = 10$ **5.** $x = -13/24$ **7.** $x = -34/5$ **9. a.** -20 **b.** -20 **c.** -20

11. **a.** 4 **b.** 4 **c.** 4   **13. a.** 2 **b.** -34 **c.** 2   **15. a.** 40 **b.** 40   **17. a.** $t = \dfrac{A - P}{P \cdot r}$ **b.** $P = \dfrac{A}{1 + rt}$

19. $x = \dfrac{12a - c}{39}$ **21.** $y = \dfrac{5x - 5}{3}$

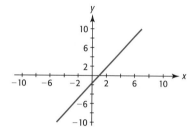

23. $y = \dfrac{6 - x^2}{2}$

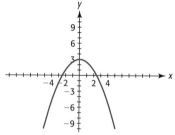

1.5 Exercises

1. 60 months, or 5 years   **3.** $-40°$ **5.** \$6000   **7.** 89   **9.** $x = 24$, so 1994   **11.** 26 years from 1970, or
1996   **13.** \$78.723 billion   **15. a.** $x \approx 46.2$, so in 1996   **b.** no; changing conditions can make models inaccurate
17. 2000   **19.** $x = 4$, so in 1999   **21.** $x = 1$, so in 1991   **23.** $x = 8$, so in 1998   **25.** 1998
27. \$140,000 @ 8%; \$100,000 @ 12%   **29.** 125   **31.** 3 cc more

1.6 Skills Check

1. should not be modeled by a linear function; data points do not lie close to a line
3.   **5.** exactly; first differences of the outputs are equal

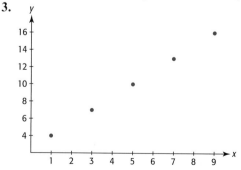

7. $y = 1.5x + 2.5$ **9.**

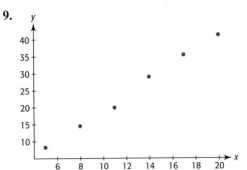

11. $y = 2.419x - 5.571$

13.

15. $y = 1.577x + 1.892$ **17.** $y = -1.5x + 8$

19. because the inputs are not equally spaced

1.6 Exercises

1. a. discrete   **b.** no; the graph of the data has some curvature   **c.** yes   **3. a.** discrete   **b.** $P(x) = 11.81$ where x is the year   **5. a.** 608,688 employees; interpolation   **b.** 720,058 employees; extrapolation   **7. a.** $y = 0.723x + 25.63$
b. sometime during the year 1978   **c.** Because the model is only an approximate fit to the data, the answer to part b is only an approximation.   **9. a.** $y = 0.039x + 1.082$ **b.** Very well; the outputs are very close to the data.   **11. a.** $1.71
b. 12 lb   **c.** 3.9 cents   **d.** 4 cents   **13. a.** $y = -0.762x + 85.284$ **b.** 58.6   **c.** 1997   **d.** The answer is an approximation; the actual data set indicates that the marriage rate fell below 50 in 1996.   **15. a.** $y = 145x + 3230.667$
b. 145 thousand children per year   **17. a.**

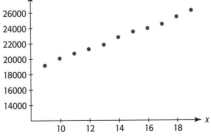

b. $y = 1280.891x + 2096.255$

; the function models the data well

c.

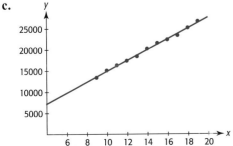

d. i) discrete **ii)** discrete **iii)** continuous   **19. a.** $y = 9.35x + 649.34$ **b.** continuous **c.** increasing until 1990, then decreasing **d.** no, because the world cigarette production appears to be decreasing after 1992.
21. a. $y = 3.303x - 18.874$ **b.** approximately 3.3 million **c.** $x = 11.77$; 2002 **d.** 9/11 terrorist attacks

1.7 Skills Check

1. $41x - 5600$ **3.** 56   **5.** 181

1.7 Exercises

1. a. $R = 1297x$ **b.** \$389,100 **c.** 1297   **3. a.** $C = 32,000 + 432x$ **b.** \$161,600 **c.** \$32,000
5. a. rent, equipment costs, utilities, employee salary **b.** \$1347.50   **7.** \$15,327 million   **9. a.** $P(x) = 66x - 3420$
b. $P(x)$ **c.** \$6480 **d.** 66   **11. a.** $P(x) = 160x - 32,000$ **b.** \$64,000

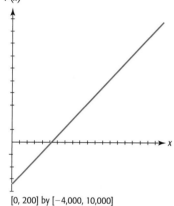

[0, 200] by [−4,000, 10,000]

c. \$160 per unit   **13. a.** 0.56 **b.** \$0.56 per ball   **c.** The cost will increase by \$0.56 for each additional ball produced in a month.   **15. a.** 1.60 **b.** \$1.60 per ball   **c.** The revenue will increase by 1.60 for each additional ball sold in a month.   **17. a.** $P(x) = 19x - 5060$ **b.** $\overline{MR} = 56, \overline{MC} = 37, \overline{MP} = 19$ **19.** 266 units
21. a. demand: 16; supply; 25; surplus **b.** \$50   **23.** 700 units at \$30. When the price is \$30, the amount demanded equals the amount supplied equals 700.

1.8 Skills Check

1.

3.
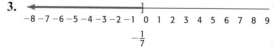

5.
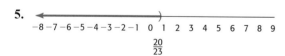

7.
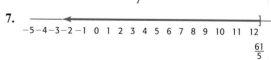

9.
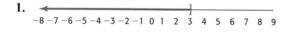

11. $x \leq \dfrac{32}{3}$ **13.** $72 \leq x \leq 146$

1.8 Exercises

1. a. $p \geq 0.1$ **b.** $x \geq 6$ **3.** $C \leq 0$ (degrees)   **5.** more than \$22,000   **7.** $3.83 \leq p \leq 6.37$ **9.** $x \geq 899$
11. $1350 \leq x \leq 1650$ **13. a.** before 1996 **b.** after 2002   **15. a.** $y = 20,000x + 190,000$ **b.** $11 \leq x \leq 14$
17. $x > 2000$ **19.** $x \geq 1500$ **21.** $364x - 345,000 > 0$ **23.** $245 < 0.155x + 244.37 < 248$; between 1974 and 1993   **25. a.** about 59% **b.** $p < -34$ or $p > 106$ **c.** before 1941 or after 2081

Chapter 1 Skills Check

1. Each value of x is assigned exactly one value of y.   **3.** $f(3) = 0$ **5. a.** -2 **b.** 8   **c.** 14

7. [0, 40] by [0, 5000] gives a better graph:

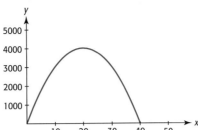

9. $y = 2.8947x - 11.211$

11. No; no, inputs do not have a constant change   **13. a.** $(0, -4)$ and $(6, 0)$ **b.**

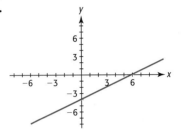

15. $y = \dfrac{1}{3}x + 3$ **17.** $y = x + 4$ **19.** $x = -138$

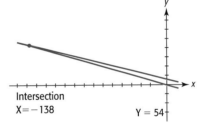

21. $y = \dfrac{4x - 6}{3}$

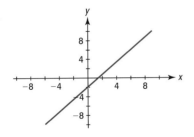

23. $x \le \dfrac{25}{28}$

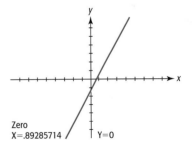

Chapter 1 Review

1. a. yes   **b.** $f(1992) = 82$; in 1992, 82% of African American voters supported Democratic candidates for President.
c. 1964; in 1964, 94% of African American voters supported Democratic candidates for President.

3.

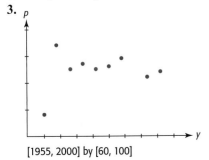

[1955, 2000] by [60, 100]

5. a. As each input value changes by 5000, the output value changes by 89.62.   **b.** $f(25) = 448.11$; the monthly payment to borrow \$25,000 is \$448.11.   **c.** $A = 20,000$ **7. a.** yes   **b.** yes   **9. a.** $f(1960) = 15.9$; in 1960 the average woman was expected to live 15.9 years past the age of 65, or 80.9 years.   **b.** $65 + 19.4 = 84.4$ years   **c.** 1990
11. a. $m = .357$ **b.** Average rate of change $= 0.357$ million users per year   **13. a.** $f(x) = 34.6$ **b.** It is a constant function.   **15.** \$25,440 each   **17.** 1994   **19. a.** $P(x) = 500x - 40,000$ **b.** \$20,000   **c.** 80   **d.** 500
e. marginal revenue minus marginal cost   **21. a.** average rate of change $= \$4.4$ per unit   **b.** slope $= 4.4$
c. $P = 4.4x - 205$ **d.** \$4.4 per unit   **e.** Approximately 47   **23. a.** $y = 0.0655x + 12.3244$
b.

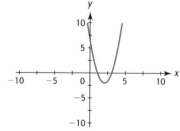

c. $g(130) \approx 20.8$; in the year 2080 (1950 + 130), the average man is expected to

live about 85.8 years   **d.** 2143   **e.** $.0655x + 12.32 \le 16; x \le 51$ **25. a.** A linear equation is a reasonable model.
b. $y = 285.269x + 9875.17$, where x is the number of years past 1980 and y is population in thousands
c. 16,151 thousand   **27. a.** before 1998   **b.** after 2008   **29.** between 17 and 31 months

Chapter 2 Quadratic and Other Nonlinear Functions

Toolbox Exercises

1. a. x^7 **b.** x^5 **c.** $4^4 a^4 y^4 = 256 a^4 y^4$ **d.** $\dfrac{3^4}{z^4} = \dfrac{81}{z^4}$ **e.** $2^5 = 32$ **f.** x^8

2. a. 4   **b.** -4 **c.** not a real number   **d.** 3   **e.** 2   **3. a.** $3\sqrt{2}$ **b.** $6\sqrt{6}$ **c.** $2\sqrt{42}$ **d.** $3a^2 b \sqrt[3]{9ab}$
4. a. $x^{3/2}$ **b.** $x^{3/4}$ **c.** $x^{3/5}$ **d.** $(3^3)^{1/6} y^{9/6} = 3^{1/2} y^{3/2}$ **e.** $27 y^{3/2}$ **5. a.** $\sqrt[4]{a^3}$ **b.** $-15\sqrt[8]{x^5}$ **c.** $\sqrt[8]{(-15x)^5}$
6. a. 2   **b.** $-3, 8$, 1 variable   **7. a.** 4   **b.** $5, -3$, 1 variable   **8. a.** 7   **b.** $-2, -119$, several variables
9. a. 4   **b.** $1, -6$, 1 variable   **10.** $-12 a^2 x^5 y^3$ **11.** $4x^3 y^4 + 8x^2 y^3 z - 6xy^3 z^2$ **12.** $2x^2 - 11x - 21$
13. $k^2 - 6k + 9$ **14.** $16x^2 - 49y^2$ **15.** $3x(x - 4)$ **16.** $12x^3(x^2 - 2)$ **17.** $(3x - 5m)(3x + 5m)$
18. $(x - 3)(x - 5)$ **19.** $(x - 7)(x + 5)$ **20.** $(x - 2)(3x + 1)$ **21.** $(2x - 5)(4x - 1)$

22. $3(2n + 1)(n + 6)$ **23.** $3(y - 2)(y + 2)(y^2 + 3)$ **24.** $(3p + 2)(6p - 1)$ **25.** $y = \dfrac{5x}{4x + 2}$

26. $x = \dfrac{18}{6y^2 - y}$

2.1 Skills Check

1. a. quadratic   **b.** up   **c.** minimum   **3.** not quadratic   **5. a.** quadratic   **b.** down   **c.** maximum
7. a.

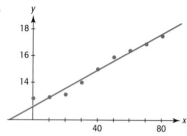

b. yes

9. a. 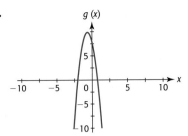   **b.** yes   **11. a.** $(-4, 3)$ **b.**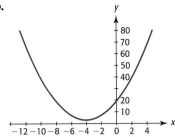

13. a. $(2, 12)$ **b.** 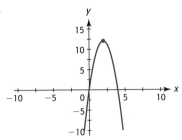   **15. a.** $(-3, -30)$ **b.**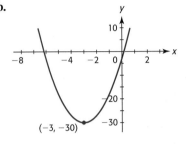

17. 1 and 3   **19.** -2 and 0.8 **21. a.** vertex $(10, -190)$ **b.**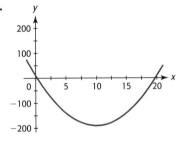

23. a. vertex $(-80, 1282)$ **b.**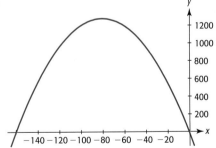

2.1 Exercises

1. a. 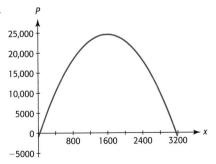   **b.** it increases   **c.** profit decreases

3. The cost function is quadratic, the revenue function is linear   **b.** $P = 1020x - x^2 - 10,000$ **c.** quadratic
5. a. 2000 units   **b.** $37,000   **7. a.** $P = 520x - x^2 - 10,000$ **b.** 260 units   **c.** $57,600
9. a. a parabola opening down   **b.** $t = 3, S = 244$ **c.** The ball reaches its maximum height of 244 feet in 3 seconds.
11. a. yes   **b.** $A = 100x - x^2$; maximum area of 2500 sq ft when x is 50 feet
13. a.   **b.** decreasing   **c.** 25

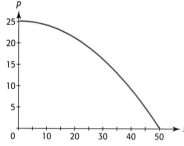

d. When wind speed is 0 mph, amount of pollution is 25 oz per cubic yard.
15. The t-intercepts are 3.5 and -3.75. This means that the ball will strike the ground in 3.5 seconds, -3.75 is meaningless.
17. a.   **b.** 6449.3 million, or 6.4493 billion

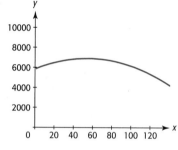

19. a. 444.9   **b.** 25 years from 1980, 2005
21. a.   **b.** 60.8%   **c.** 32 years from 1960, in 1992

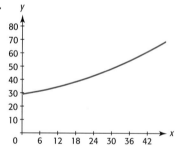

23. a. minimum; positive coefficient of x^2 **b.** $x = 69.8, y = 5.2$ **c.** The percent of the U.S. population that is foreign born was at a minimum of 5.2% in 1970.
d.

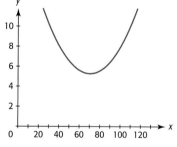

2.2 Skills Check

1. 5, −2   **3.** 8, 3   **5.** −3, 2   **7.** 3/2, 4   **9.** 2/3, 2   **11.** 1/2, −4   **13.** −1/2, 3   **15.** ±3/2
17. 2 ± √13   **19.** 1, 2   **21.** 1 ± √5   **23.** −3, 2   **25.** 2/3 ≈ 0.667, −3/2 = −1.5
27. −1/2 = −1.5, 2/3 ≈ 0.667

2.2 Exercises

1. $t = 2$ sec and 4 sec   **3.** 20 units and 90 units   **5. a.** $P = 520x − 10,000 − x^2$ **b.** − \$ 964; loss of \$964
c. \$5616; profit of \$5616   **d.** 20 units or 500 units   **7. a.** $s = 50, s = −50$ **b.** There is no particulate pollution.
c. $s = 50$, speed is positive   **9. a.** $r = 0$ **b.** $r = 0.05$ **c.** $r = 0.1$; against the wall of the artery   **11.** \$100 is
the price that gives demand = supply = 97 trees
13. a.   **b.** $x = 40, x = 100$; 1940 and 2000

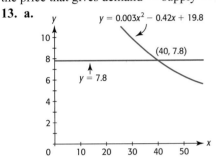

15. a. Set the x-values from 0 to 10, so it includes x-values representing 1989 to 1997 ; use ⎡TRACE⎤ or ⎡TABLE⎤ to find
y-values for these x-values.

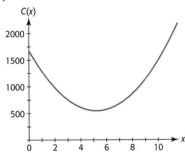

b. $x = 5.2$; testing years shows minimum in 1994.   **c.** $x = 10$; 1999;   **d.** extrapolation, because the model is used for
evaluation beyond the given data
17. $x = 20$; 2007   **19. a.** declined by 262 thousand   **b.** 1994   **c.** 37,851,047; that the model continues to apply.
21. \$3709.3 million increase   **b.** 17.1 years after 1980; in 1998

2.3 Skills Check

1.

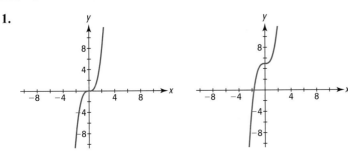

3.

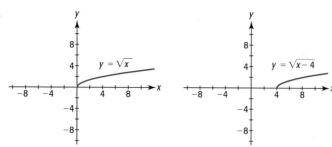

5.

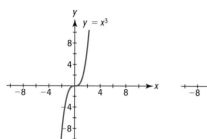

7. a. 4   **b.** 12   **9. a.** −27   **b.** 1   **11. a.** 0   **b.** 2   **13. a.** increasing   **b.** increasing
15. concave down   **17.** $y = (x + 4)^{3/2}$ **19.** Shifted to the right 2 units and up 3 units   **21.** $x = 1$
23. $x = 6$ **25.** $x = -2, x = 6$

2.3 Exercises

1. a. power   **b.** 23,874; 23,874 suicides occurred in 1965   **c.** 28,099
3. a.   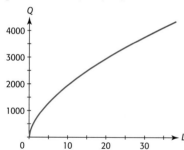   **b.** 3912   **c.** increases; yes   **5. a.** increasing   **b.** concave up   **c.** 2001

7. a.   **b.** 27 inches

9. a. shifted power function   **b.**   **c.** yes; market equilibrium

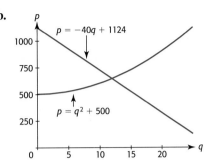

d. (12, 644); 1200 units when the price is $644.

11. a. revenue, cost   **b.**   **c.** at 8000 units, which gives $R(x) = C(x) = \$1420$

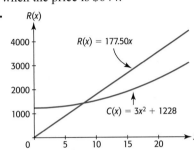

13. $f(x) = 105.095(x + 5)^{1.5307}$

15. a.   **b.** increasing   **c.** may need to expand jail

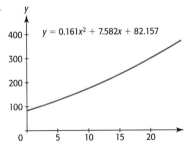

17. a.   **b.** decreasing   **c.** 50.3%   **d.** 49.6%   **e.** no, percent was higher

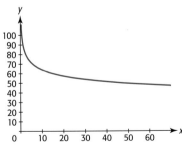

2.4 Skills Check

1. $f(x) = 3x^2 - 2x$ **3.** The x-values are not equally spaced.   **5.** $f(x) = 99.9x^2 + 0.64x - 0.75$
7. a. $y = 3.545x^{1.323}$ **b.** $y = 8.114x - 8.067$ **c.** the power function   **9. a.** $y = 1.292x^{1.178}$
b. $y = 2.065x - 1.565$ **c.** they are both good fits   **11.** $y = 2.98x^{0.614}$

2.4 Exercises

Numerical results are computed with the unrounded models.

1. a. $y = -17.57x^2 + 632.35x + 29909$ **b.** 29,256; 25,479 (from unrounded model)   **c.** No; extrapolating may not give accurate predictions.

3. a.   **b.** yes; $y = 2.142x^2 - 35.118x + 225.618$

c. The y-intercept is $(0, 225)$; this means that in 1980 the total unemployment in South Carolina was approximately 225,000 (from the unrounded model).   **5.** 2023   **7. a.** $y = .0134x^2 - .054x + 1.564$

b. $y = .013x^2 + .080x + 1.630$ **c.** 1988   **d.** The results would be equivalent

9. a.   **b.** yes; $y = 0.144x^2 - 2.48x + 15.47$

c. The y-intercept is $(0, 15.47)$. This means that unemployment in South Carolina was 15.47% in 1980 according to the model.

11. a. no   **b.** power function; $y = \frac{1}{3}x^3$ **13. a.** $y = -0.36x^2 + 38.5x + 5822.9$ **b.** $x \approx 191$, or the year 2181

c. after 2181   **15.** $y = 4700\sqrt{110 - x}$ **17. a.** $y = 154.13x^{-.49}$ **b.** 25%   **19. a.** $y = 52.4x^{-0.29}$ **b.** 16.6

c. decreasing   **d.** eventually the number of students will be 0   **21. a.** $y = 34.7x^{1.58}$ **b.** 7523 million dollars

c. 1990   **d.** increased

23. a.   **b.** yes; $y = .052x^2 - 3.23x + 66.66$ **c.** $y = .011x^{2.04}$

d. the quadratic model   **25. a.** $y = 0.155x^2 - 9.15x + 235.5$; $y = 474.76x^{-0.473}$; both models fit well

b. the discharge rate will be less than the 1996 rate, 102.3   **c.** quadratic model yields 105.6; power model yields 88.4; power model is a better predictor

2.5 Skills Check

1. a. 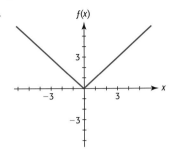 **b.** $2, 5$ **c.** all real numbers **3. a.**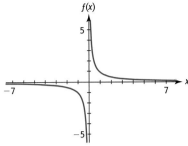

b. $-1/2, 1/5$ **c.** $x \neq 0$

5. a. 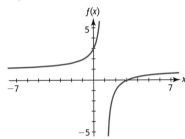 **b.** $5/3, 1/2$ **c.** $x \neq 1$ **7. a.**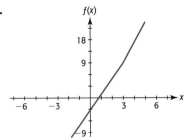

b. $5, 16$ **c.** all real numbers **9. a.** $\sqrt{x} - \dfrac{1}{x}$ **b.** $x > 0$ **11. a.** $\dfrac{4x+1}{x}$ **b.** $x \neq 0$ **13. a.** $\dfrac{4x+1}{\sqrt{x}}$

b. $x > 0$ **15. a.** $4\sqrt{x} + 1$ **b.** $x \geq 0$ **17. a.** $\dfrac{4}{x} + 1$ **b.** $x \neq 0$ **19.** $x = 1, x = 4$ **21.** $x = -5, x = 0$

23. $x = 5/3 \approx 1.667, x = -1$

2.5 Exercises

1. a. $1234; 0.20x$ **b.** the fixed cost

3. a. $\overline{C}(x)$ is $C(x) = 50{,}000 + 105x$ divided by $f(x) = x$. That is $\overline{C}(x) = \dfrac{C(x)}{x}$. **b.** $121.67

5. a. $S(p) + N(p) = 0.5p^2 + 78p + 12{,}900$ **b.** $0 \leq p \leq 100$ **c.** $23{,}970$

7. a. $\overline{C}(x) = \dfrac{3000 + 72x}{x}$ **b.** $102 per printer

9. a. $B(t) + G(t) = 0.014t^2 - 0.32t + 20.825$ **b.** 19.03 million

11. a. $P(x) = \begin{cases} 37 & \text{if } 0 < x \leq 1 \\ 60 & \text{if } 1 < x \leq 2 \\ 83 & \text{if } 2 < x \leq 3 \\ 106 & \text{if } 3 < x \leq 4 \end{cases}$ **b.** 60; it costs 60 cents to mail a 1.2-oz letter by first class

c. $0 < x \leq 4$ **d.** $60, 83$ **e.** 60 cents, 83 cents

13. a. $T(x) = \begin{cases} .15x & \text{if } 0 < x \leq 43{,}850 \\ 6{,}577.50 + .28(x - 43{,}850) & \text{if } 43{,}850 < x \leq 105{,}950 \end{cases}$ **b.** 6300 **c.** $9699.50

d. $6577.50 and $6{,}577.50 + .28(1) = 6577.78$; 28% is paid only on the amount over $43,850, not all income

15. a. reciprocal function, shifted 1 unit to the left   **b.**   **c.** decrease

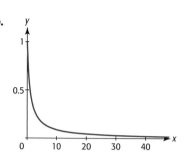

17. a. The graph is shifted to the left 10 units and down 1 unit, is reflected about the x-axis, and is stretched by the factor 1000.   **b.**

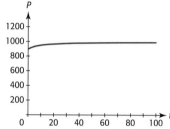

19. a. meat in container   **b.** ground meat   **c.** meat ground and ground again   **d.** ground meat in the container.
e. meat in container, then both ground   **f.** d   **21.** $p(s(x)) = x - 18.5$

23. 50.09 rubles   **25.** $100\dfrac{f(x)}{g(x)}$ **27.** $f(x) + g(x)$ **29.** 40%

2.6 Skills Check

1. Yes;

```
Domain        Range
         f⁻¹
   3 ──────→ 2
   4 ──────→ 5
   1 ──────→ 3
   8 ──────→ 7
```

3. a. x **b.** yes   **5.** yes   **7.**

x	$f(x)$	x	$f^{-1}(x)$
-1	-7	-7	-1
0	-4	-4	0
1	-1	-1	1
2	2	2	2
3	5	5	3

9. (b, a)

11. $f^{-1}(x) = \dfrac{x - 1}{4}$ **13.**

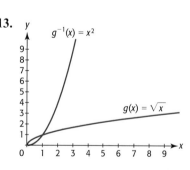

15. yes; yes   **17.** yes   **19.** no

2.6 Exercises

1. a. $d^{-1}(x) = x - 0.5$ **b.** size 8   **3.** $f^{-1}(x) = \dfrac{-t + 75.451}{0.707}$; 1990   **5. a.** $f^{-1}(x) = \dfrac{x - 493.432}{225.304}$; the year

when Ritalin consumption is at a given level x.   **b.** 3 yrs after 1990; 1993

7. a. domain: all real $x \geq -1/4$; range: real $f(x) \geq 0$ **b.** $f^{-1}(x) = \dfrac{x^2 - 16}{64}$

c. domain: all real $x \geq 0$; range: real $f^{-1}(x) \geq -1/4$ **d.** domain: all real $x \geq 0$; range: real $f^{-1}(x) \geq 0$
9. $C^{-1}(x) = x - 3$; THE_REAL_THING   **11.** Yes; each number identifies one exactly person.
13. a. yes   **b.** $f^{-1}(x) = \sqrt[3]{x}$ **c.** domain: $x \geq 0$; range: $f^{-1}(x) \geq 0$ **d.** to find the edge of a cube from its volume

15. a. $f^{-1}(x) = \dfrac{x}{0.66832}$; dividing Canadian dollars by 0.66832 gives U.S. dollars   **b.** $500

17. a. no   **b.** $x \geq 0$; it is a distance   **c.** yes   **d.** $I^{-1}(x) = \sqrt{\dfrac{300{,}000}{x}}$; 2 feet

19. a. $f^{-1}(x) = \dfrac{x}{1.7655}$; converts US dollars to UK pounds   **b.** $1000

21. a. 8 sec   **b.** no   **c.** $0 \leq x \leq 3$ or $3 \leq x \leq 8$ **d.** $f^{-1}(x) = 3 - \dfrac{\sqrt{400 - x}}{4}$; the number of seconds, from 0 sec to

3 sec, to attain height x.

2.7 Skills Check

1. $-4 < x < 0$ **3.** $-3 \leq x \leq 3$ **5.** $4 < x < 5$ **7.** $x \leq 2$ or $x \geq 6$ **9.** $-1 < x < 7$
11. $x \leq \dfrac{7 - \sqrt{33}}{4} \approx 0.314$ or $x \geq \dfrac{7 + \sqrt{33}}{4} \approx 3.186$ **13.** $x \leq \dfrac{1 - \sqrt{31}}{5} \approx -0.914$ or $x \geq \dfrac{1 + \sqrt{31}}{5} \approx 1.314$
15. $0.5 \leq x \leq 4$ **17.** $-1.193 < x < 4.193$

2.7 Exercises

1. $100 < x < 4000$ units   **3.** $100 < x < 8000$ units   **5.** $2 < t < 8.83$ seconds   **7.** $1988 \leq x \leq 1990$
9. $1984 \leq x \leq 2011$ **11.** $1972 \leq x \leq 1994$ **13.** $5 < x < 12$; 1986–1991   **15.** $53 < x < 107$; 1954–2106

Chapter 2 Skills Check

1. $(1, -27)$ **3.** $-2, 4$ **5.** $4, 1$ **7.** $-4/5, 1$ **9.** $3, 1$
11. a.

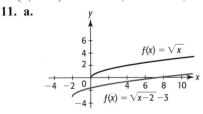

b. The graph of $g(x)$ can be obtained from the graph of $f(x)$ by shifting 3 units down and 2 units to the left.
13. a. increasing   **b.** decreasing

15. $f^{-1}(x) = \dfrac{x + 2}{3}$ **17.**

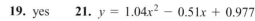

19. yes   **21.** $y = 1.04x^2 - 0.51x + 0.977$

23. c   **25.** $[-2, 9]$

Chapter 2 Review

1. a. $-2, 6$ **b.** no; only the positive value   **c.** at 6 seconds   **3.** 30 and 120   **5. a.** 43,905   **b.** 1990
7. a. 3100   **b.** \$84,100   **9.** 100 and 8000 units   **11.** 1972 through 1994   **13. a.** 1997
b. between 1984 and 2012   **15. a.** the demand function   **b.** 300   **c.** \$80
17. a. 5544 thousand   **b.** 2000; that the model is still valid
19. a. $T(x) = E(x) - I(x) = -191.73x^2 - 21604.55x - 57756.99$

b. \$292,975.49 million deficit   **21. a.** $f^{-1}(x) = \dfrac{x + 2.886}{0.554}$ **b.** the mean sentence length can be found as a function
of the mean time in prison   **23. a.** $y = 11.786x^2 - 142.214x + 493$ **b.**

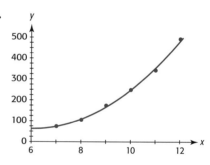

c. yes   **d.** 2002   **25.** Values for table: 10288, 9120, 8712, 8876, 9343; $y = 14.66x^2 - 308.42x + 10,293.89$ where y
is the number of males in thousands and x is the number of years past 1980
27. a. yes, good fit

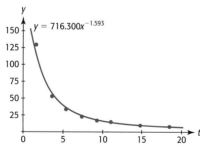

b. Yes, but far in the future, at $x \approx 61.9$ which represents the 2041-42 school year

c. $C(t) = \dfrac{1}{t}$; stretched by a factor of 380, shifted vertically 15 units down, shifted horizontally 0.3 unit to the left

d. the function in part (a); its function values will never be negative   **e.** $y = C(t)$; no
29. a. $y = 0.013x^2 - 0.22x + 7.3$ **b.** $y = -0.26x + 9.34$
c. $f(x) = \begin{cases} 0.013x^2 - 0.22x + 7.3 & \text{if } 0 \le x \le 11 \\ -0.26x + 9.34 & \text{if } 12 \le x \le 16 \end{cases}$
d. i. 6.4; average stay in 1987 was 6.4 days   **ii.** 1992   **iii.** 6.5 days
31. between 200 and 6000 units

Chapter 3 Exponential and Logarithmic Functions

Toolbox Exercises

1. x^{-7} **2.** a **3.** $\dfrac{1}{c^{18}}$ **4.** $x^{1/6}$ **5.** $\dfrac{x^{10}}{4}$ **6.** $\dfrac{6}{ab^2}$ **7. a.** a^{12} **b.** $\dfrac{1}{x^8}$ **8. a.** $3^8 = 6561$ **b.** $\dfrac{1}{8}$

9. a. 10 **b.** 256   **10. a.** x^6 **b.** x^8 **11.** 3.345×10^6 **12.** 1.233×10^{-3} **13.** 437,200

14. 0.00056294   **15.** 3.708×10^6 **16.** 6.461×10^1 **17.** 2.333×10^4 **18.** 6.021×10^{-5}

19. \$600; \$2600   **20.** \$210; \$3710   **21.** \$168; \$5768

3.1 Skills Check

1. c, e   **3. a.** **b.** 2.718; 0.368; 54.598   **c.** x-axis   **d.** 1

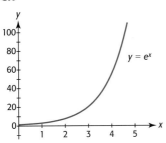

5.    **7.** vertical shift, 2 units up   **9.** reflection across the y-axis

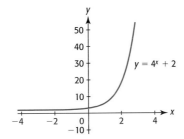   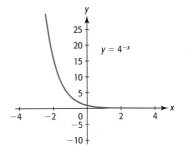

11. vertical stretch by a factor of 3

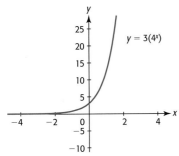

13. The graph of $y = 3 \cdot 4^{(x-2)} - 3$ is the graph of $y = 3(4^x)$ shifted to the right 2 units and shifted down 3 units.

15. a. 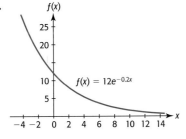   **b.** $f(10) = 1.624; f(-10) = 88.669$ **c.** decay

3.1 Exercises

1. a. $2000 **b.** $1319.51 **c.** no

3. a.

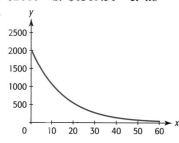

b. $1000 **c.** Sales declined drastically after the end of the ad campaign. (Answers may vary.)

5. a. 14,339.44 **b.** Because of future inflation, they should plan to save money. (Answers may vary.)

7. a. $122,140.28 **b.** in about 14 years

9. a. 376.84g **b.**

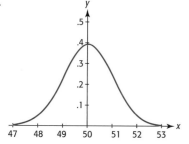

c. about 24.5 years

11. a. about 88.4g **b.** about 18,602 years   **13. a.** increasing **b.** 57,128 **c.** 61,577 **d.** approximately 858 people per year

15. a.

b. 50

3.2 Skills Check

1. $3^y = x$ **3.** $e^y = 2x$ **5.** $\log_4 x = y$ **7.** $\log_2 32 = 5$ **9.** 0.845   **11.** 4.454   **13.** 4.806   **15.** 5
17. 3   **19.** -3 **21.** 1   **23.** 1.6131   **25.** 0.1667
27.

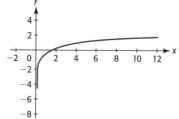

29.

31.

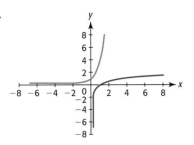

The graphs are symmetric with respect to the line $y = x$.

33. $a^x = a; x = 1$

3.2 Exercises

1. a. $\approx 59; \approx 76$ **b.** Improved health care and better diet. (Answers may vary.)
3. a. 6662.66; 8441.24   **b.** increasing   **c.** Increased cost of living, raise in minimum wage. (Answers may vary.)
5. ≈ 6.9 years   **7. a.** ≈ 35 **b.** 8.75 years   **9.** $t \approx 14.2$; in 15 years   **11.** 7 units
13. a. $\ln x = \dfrac{y - 17.875}{12.734}$ **b.** $x = e^{\frac{y - 17.875}{12.734}}$ **c.** x 88.77; 1989   **d.** 1989; answers agree   **15.** 4.4
17. a. $10^R = \dfrac{I}{I_0}$ **b.** $I = 10^{7.1}I_0 \approx 12{,}589{,}254I_0$ **19.** about 14.1 times more intense   **21.** 43
23. $I = 10{,}000I_0$ **25.** 4.2   **27.** $10^{-14} \le H^+ \le 10^{-1}$

3.3 Skills Check

1. 7.824   **3.** 0.374   **5.** 3.204   **7.** 0.769   **9.** $\ln 2$ **11.** 14   **13.** 2.2146   **15.** $p + q$
17. $2p - 3q$ **19.** $3 \log x + 5 \log(3x - 4)$ **21.** $\log_2 x^3 y$ **23.** $\ln \dfrac{(2a)^4}{b}$
25. $\log_a x = \log_a b^y$
   $\Rightarrow \log_a x = y \log_a b$
   $\Rightarrow \log_a x = (\log_b x)(\log_a b)$
   $\Rightarrow \dfrac{\log_a x}{\log_a b} = \log_b x$

3.3 Exercises

1. 5   **3. a.** $\ln\left(\dfrac{S}{25{,}000}\right) = -.072x$ **b.** 6   **5. a.** 3200   **b.** 9 days   **7. a.** \$1755.358 million   **b.** 1996

9. 13.86 years   **11.** 23.10 years   **13.** $\dfrac{\ln 2}{r} = t$ **15.** $t \approx 17.386$; 17 years, 5 months

17. a. 500 grams   **b.** about 24.5 years   **19.** approximately 7 hours   **21.** about 24,101 years

23. about 2028 years   **25. a.** 16   **b.**   **27.** for the first $7\dfrac{1}{4}$ years

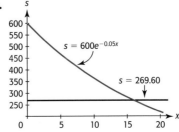

3.4 Skills Check

1. $y = 2(3^x)$ **3.** $f(x) = 4^x$ **5. a.** 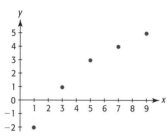 **b.** linear **7.** exponential

9. $y = 0.876(2.494^x)$ **11. a.** 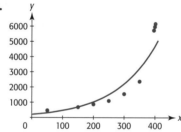 **b.** logarithmic

3.4 Exercises

1. a. $y = 30,000(1.04^t)$ **b.** \$54,028 **3. a.** $y = 20,000(.98^x)$ **b.** \$18,078
5. a. $y = 238.4752(1.0075^x)$ **b.**

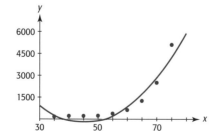

7. a. $y = 4.304(1.096^x)$ **b.** $y = 6.182x^2 - 565.948x + 12810.482$
c. The exponential function appears to be the better fit.

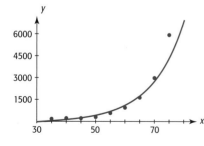

9. a. 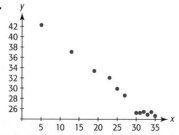 **b.** yes **c.** $y = 47.818(0.981^x)$ **d.** $y = 74.575x^{-.298}$ **e.** exponential

11. a. $y = 17.875 + 12.734 \ln x$ **b.** $y = -0.00249x^2 + 0.568x + 44.472$ **c.** Quadratic function looks like a slightly better fit.   **d.** using the logarithmic function: 77.7; using the quadratic function: 76.8; the logarithmic model is the better predictor because the quadratic model falls after 2014   **13. a.** $y = -681.976 + 251.829 \ln x$ **b.** 31.5%
c. $y = 0.627x^2 - 7.400x - 26.675$ **d.** quadratic

3.5 Skills Check

Approximate answers to 2 decimal places.
1. 49,801.75   **3.** 17,230.47   **5.** 105,966.04   **7.** 58,406.4   **9.** 1,723,331.03
11. 1123.60; 1191.00; 1191.00

3.5 Exercises

1. a. $16,288.19   **b.** $88,551.38

3. a.   **b.** $x \approx 7.27$; 8 years because compounded annually

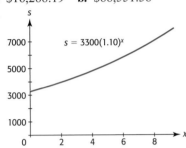

5. $32,620.38   **7. a.** $33,194.62   **b.** more; compounded more frequently   **9.** $49,958.02   **11. a.** $20,544.33
b. $29,446.80

13. a.   **b.** $x \approx 11.45$ years   **15. a.** ≈ 7.27, 8 years   **b.** ≈ 6.93 years

17. a. $2954.91   **b.** $4813.24   **19.** $6152.25

21. a.

Years	0	7	14	21	28
Future Value ($)	1000	2000	4000	8000	16,000

b. $y = 1000(1.104)^x$ **c.** $1640.01, $2826.02

3.6 Skills Check

1. $P = \dfrac{S}{(1 + i)^n}$ **3.** $A = R\left[\dfrac{1 - (1 + i)^{-n}}{i}\right]$ **5.** $R = A\left[\dfrac{i}{1 - (1 + i)^{-n}}\right]$

3.6 Exercises

1. $5583.95   **3.** $4996.09   **5.** $7023.58   **7.** $530,179.96   **9.** $372,845.60
11. a. $702,600.91   **b.** $121,548.38   **c.** The $100,000 plus the annuity gives approximately $452 more per year.

13. a. $198,850.99   **b.** $576,000   **c.** $377,149.01   **15. a.** 2%   **b.** 16   **c.** $736.50
17. a. $1498.88   **b.** $639,596.80   **c.** $289,596.80

3.7 Skills Check

1. 73.83   **3.** 995.51; 999.82   **5. a.** 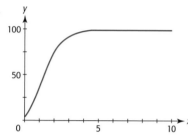   **b.** 25; 99.99   **c.** increasing   **d.** 100

7. a. 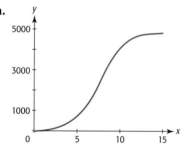   **b.** 5   **c.** 100

3.7 Exercises

1. a. 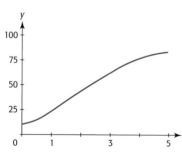   **b.** 5   **c.** 5000

3. a. 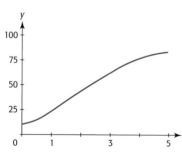   **b.** 29.56%   **c.** 86.93%   **d.** The limiting value is 89.786

5. a. approximately 218 people   **b.** 1970   **c.** the 7th day

7. a. $y = \dfrac{89.786}{1 + 4.6531e^{-0.8256x}}$ **b.** yes   **c.** $y = 14.137x + 17.624$ **d.** the logistic model is the better fit.

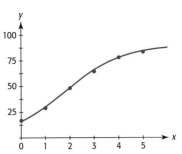

9. a. 4000   **b.** approximately 9985   **c.** upper limit 10,000

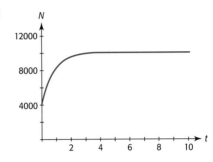

11. a. 21,012 units   **b.**   **c.** 40,000 units

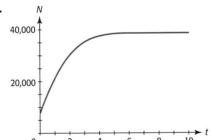

13. a. 10   **b.** 100   **c.** 1000   **d.** 6 years after the company was formed.
15. 10 days   **17.** 11 years

Chapter 3 Skills Check

1. a.

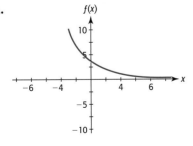

b. ≈ 80.342; ≈ 0.19915 **3.**

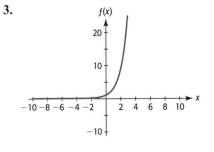

5. The graph of $y = 3^{(x-1)} + 4$ is the graph of $f(x) = 3^x$ shifted to the right 1 unit and up 4 units.
7. a. 500   **b.** $x = 20$ **9.** $3x = \log_7 y$ **11.** $x = 10^y$ **13.** $y = \log_4 x$ **15.** 4.0254   **17.** 4   **19.** -3

21. 1.9358 **23.** 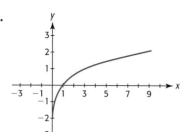 **25.** 0.2012 **27.** 2 **29.** $\log_4 \dfrac{x^6}{y^2}$ **31.** 4926.80

33. a. 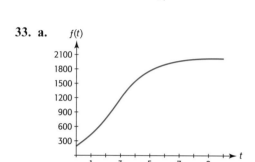 **b.** 222.22; 1973.76 **c.** 2000

Chapter 3 Review

1. 773,760 **3.** $x = 2.802$; 1988 **5.** approximately 1259 times **7. a.** $x = 7 \log_2 \dfrac{S}{1000}$ **b.** 30 years

9. $x = 29$; 1989 **11.** $x \approx 6.3$; 7 weeks **13.** $x \approx 13.9$; 14 years **15.** 15 years

17. a. $y = 57.66(0.978)^x$ **b.** 21.77% **19. a.** $y = 2102.96(1.0267)^x$ **b.** $y = \dfrac{8672.06}{1 + 4.723e^{-0.0623x}}$ **c.** logistic

21. a. $y = 114.198 + 4.175 \ln x$ **b.** approximately 127,466,000 **23.** $20,609.02 **25.** $34,426.47

27. $209,281.18 **29.** $66.43 **31. a.** 67.79%; 76.84% **b.** 96.3641% **33. a.** 1298 **b.** 3997 **c.** 4000

35. a. $C(t) = 5.521(1.349)^t$ **b.** $E(C) = -18.653C^2 + 3226.250C + 3245.657$

c. $E\big(C(t)\big) = -18.653\big(5.521(1.349)^t\big)^2 + 3226.250\big(5.521(1.349)^t\big) + 3245.657$ It is the number of employees E as a function of time t. **d.** 110,476.85 **e.** 110,477 employees

Chapter 4 Higher Degree Polynomial and Rational Functions
Toolbox Exercises

1. a. 4th **b.** 3 **2. a.** 3rd **b.** 5 **3. a.** 5th **b.** -14 **4. a.** 6th **b.** -8 **5.** $4x(x - 7)(x + 5)$

6. $-x^2(2x + 1)(x - 4)$ **7.** $(x - 3)(x + 3)(x - 2)(x + 2)$ **8.** $2(x^2 - 2)^2$ **9.** 1/3 **10.** $\dfrac{x - 2}{x + 4}$

11. $\dfrac{3x + 2}{x - 1}$ **12.** $\dfrac{45xy}{4}$ **13.** $\dfrac{x - 2}{2x}$ **14.** $2x^2 - 7x + 6$ **15.** $\dfrac{1}{2y}$ **16.** $\dfrac{2}{x - 2}$ **17.** $\dfrac{3x^3 + x - 2}{x^3}$

18. $\dfrac{4a - 4}{a^2 - 2a}$ **19.** $\dfrac{x - 2}{x}$ **20.** $\dfrac{2x^2 + 9x - 7}{4x^2 - 2x}$ **21.** $x^4 - x^3 + 2x^2 - 2x + 2 \; R{-}3$ **22.** $a^3 + a^2$

23. $3x^3 - x^2 + 6x - 2 \; R(17x - 5)$ **24.** $x^2 + 1 \; R2$

4.1 Skills Check

1. a.

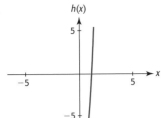

b. window (b) gives a complete graph

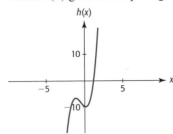

3. a.

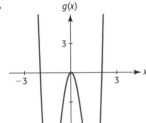

b.

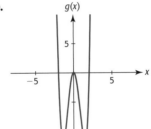

c. window (c) gives a complete graph

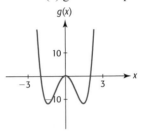

5. a. $-3, -1, 2$ **b.** positive because opening up to right **c.** 3, because has 3 x-intercepts   **7. a.** $-1, 3, 5$
b. negative because falling to right **c.** 3, because of shape and x-intercepts   **9. a.** degree 3, leading coefficient 2
b. opening up to right because of positive leading coefficient, down to left because cubic graphs have end behaviors in opposite directions **c.**

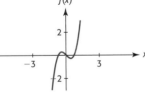

$[-5, 5]$ by $[-3, 3]$

11. a. Expanding shows that this is a 3rd degree function with leading coefficient -2 **b.** opening down to right because of negative leading coefficient, up to left because cubic graphs have end behaviors in opposite directions

c.

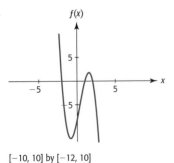

$[-10, 10]$ by $[-12, 10]$

13. a. The graph is shown below.

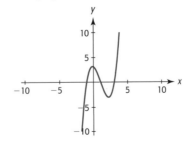

b. Three x-intercepts are shown, along with the y-intercept, so the graph is complete.

15. The graphs are shown below.

a.

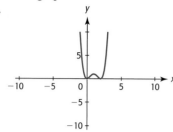

b.

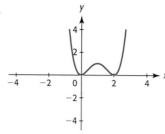

c. The graph shown in part (b).

17. One such graph is the graph of $f(x) = -4x^3 + 4$, shown below.

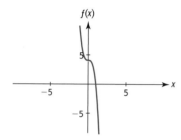

19. One such graph is the graph of $f(x) = 2x^3 - x$, shown below.

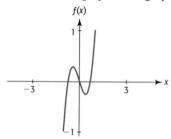

21. a. The graph is shown below.   **b.** The local maximum is shown below.   **c.** The local minimum is shown below.

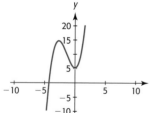

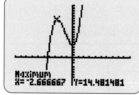

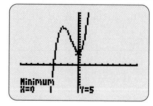

4.1 Exercises

1. a. 2 turning points   **b.** nonnegative; nonnegative   **c.**

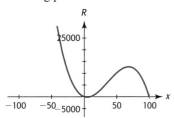

d. $10,000

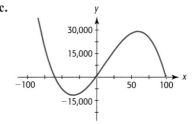

3. a.

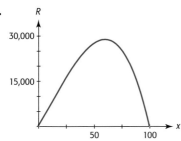

b. 60 units gives $28,800   **c.**

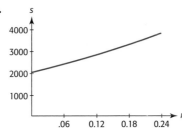

d. part (a)   **e.** $0 < x < 60$

5. a.

r(rate)	S(future value)
0%	2000.00
5%	2315.25
10%	2662.00
15%	3041.75
20%	3456.00

b.

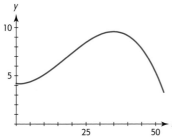

c. $2662.00 at 10% and $3456.00 at 20%; $794   **d.** 10% much more likely

7. a.

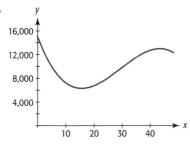

b. 9   **c.** 1985   **d.** increases by .09 per 100,000 people per year

9. a.

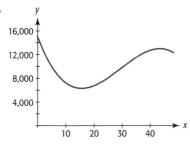

b. 9,690.676 thousand, or 9,690,676   **c.** 1966

11. a.

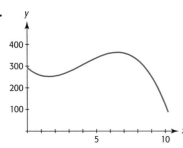

b. The value $x = 8$ represents 1998, so the approximate debt is the corresponding value of y, 324.84 million dollars.

c. The local minimum is (1.55, 251.14) and the local maximum is (6.495, 361.85). The minimum debt is approximately $251 million, and the maximum debt is approximately $362 million.

4.2 Skills Check

1. $y = x^3 - 2x^2$ **3. a.**

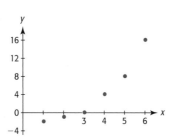

b. cubic

5. a. $y = 35x^3 - 333.667x^2 + 920.762x - 677.714$ **b.** $y = 12.515x^4 - 165.242x^3 + 748x^2 - 1324.814x + 738.286$
7. $y = x^4 - 4x^2 - 3x + 1$ **9.** $y = 0.565x^3 + 2.425x^2 - 4.251x + 0.556$

4.2 Exercises

1. a. $y = 0.197x^3 - 13.706x^2 + 265.064x + 1612.529$ **b.** 2869.492 thousand, or 2,869,492
3. a. $y = -1826x^4 + 16,570x^3 - 47,956x^2 + 49,592x + 51,962$
b. $68,618 thousand, or $68,618,000; it is the same
5. a.

b. $y = -0.0006x^3 + 0.0548x^2 - 0.7588x + 11.854$

c. $y = 1.053x^{0.8878}$ percent, with $x = 0$ in 1950   **d.** The cubic model appears to be a better fit.
7. a. $y = 0.0146x^3 - 0.7992x^2 + 10.7172x + 182.3528$ **b.** 196,701

9. a. $y = 8.629x^3 - 238.217x^2 + 1208.679x + 22{,}281.662$
b. $y = -0.371x^4 + 21.970x^3 - 395.717x^2 + 1882.410x + 21{,}521.213$
c. The quartic model appears to be a slightly better fit for the data than the cubic model, but both models are relatively good fits for the data.

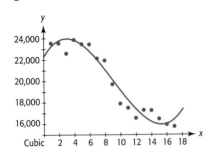

11. a. $y = 3.283x^3 + 0.016x^2 - 0.080x + 4.258$ **b.** $x = 11; 2001$
13. a.   **b.** $f(x) = -0.357x^3 + 9.417x^2 - 51.852x + 361.208$

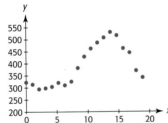

c.

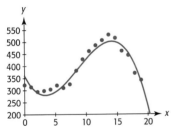

d. $f(19) \approx 327$ and $f(10) \approx 427$, so the ratio of juvenile violent crimes in 1999 to those in 1990 is approximately $327/427 \approx 3/4$.

4.3 Skills Check

1. $0, 4, -4$ **3.** $0, 2$ **5.** $4, 3, -3$ **7.** $0, 1, -1$ **9.** 2 **11.** $2, -2$ **13.** $0, \pm\sqrt{2}$ **15.** $0, 5, -5$
17. $\pm\sqrt{3}$ **19.** $-3, 1, 4$

4.3 Exercises

1. a. $0, 20$ **b.** yes   **3. a.** $0, 1000$ **b.** yes   **5. a.**   **b.** 10% **c.** 20%

7. a. x inches   **b.** $18 - 2x$ inches by $18 - 2x$ inches   **c.** $V = (18 - 2x)(18 - 2x)x = 324x - 72x^2 + 4x^3$
d. $x = 0, x = 9$ **e.** neither   **9.** 2 units or 20 units
11. a.

t	0	.1	.2	.3
s (cm/sec)	810	240	30	0

   b. .3 second: yes

13. a. $y = 779,365$ **b.** 1998   **15. a.** $x = 7.11; 1998$ **b.** yes, in 1999

4.4 Skills Check

1. $x^3 - x^2 - 3x - 6 - \dfrac{8}{x - 3}$ **3.** $2x^3 - x^2 - x - \dfrac{7}{x - 1}$ **5.** no   **7.** no   **9.** 1 (a double solution)
11. $-2, 1, 4$ **13.** $4, -2, -5$ **15.** $-2, 1, 1/3$ **17.** $\pm 1, \pm 2, \pm 1/2, \pm 1/4$ **19.** $-1, 3, 4$ **21.** $1/3, -1, -4/3$

4.4 Exercises

1. a. $-0.2x^2 + 56x + 1200$ **b.** $x = 300, x = -20$, so 300 units gives break-even
3. a.

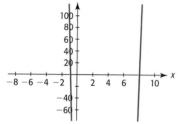

b. 8   **c.** $-0.1x^2 + 49.9x + 50$ **d.** $x = 8, x = 500, x = -1$

e. 8 units or 500 units   **5.** $x = 9, x = 10, x = -100$; 9 units or 10 units
7. a. $y = 0.4566x^3 - 14.3085x^2 + 117.2978x - 136.1544 = 0$
b. $x = 12$

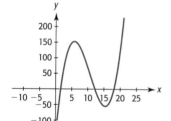

c. $0.4566x^2 - 8.8293x + 11.3462$ **d.** $x = 17.935, x = 1.384$

e. 1982, 1992, 1998   **9.** $x = 15.81$ $x = 10.35$; 1976 and 1971

4.5 Skills Check

1. imaginary   **3.** real   **5.** $a = 4, b = 0$ **7.** $a = 15, b = -3$ **9.** $-1 + 9i$ **11.** 0   **13.** i **15.** 5
17. $-3 - 4i$ **19.** $\dfrac{7}{6} + \dfrac{1}{6}i$ **21.** $6i$ **23.** -3 **25.** $-3\sqrt{10}$ **27.** $-1 + 3i\sqrt{2}$ **29.** $-5 - 12i$
31. $x = \pm 5i$ **33.** $x = 1 \pm 2i$ **35.** $x = -2 \pm 2i$ **37.** $x = \dfrac{4 \pm i\sqrt{2}}{2}$ **39.** $x = \dfrac{5 \pm i\sqrt{23}}{8}$
41. $x = 0, x = \dfrac{-7 \pm \sqrt{89}}{2}$ **43.** $w = 1, w = 2 \pm \sqrt{2}$ **45.** $z = 2, w = -1 \pm i\sqrt{3}$

4.6 Skills Check

1. a. $x = 5$　**b.** $y = 0$　　**3. a.** $x = 5/2$　**b.** $y = -1/2$　　**5. a.** $x = 1, x = -1$　**b.** none

7. (c) no value of x makes the denominator 0

9. a. $x = 2$　**b.** $y = 1$　**c.**

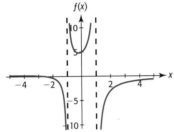

11. a. $x = -1, x = 1$　**b.** $y = 0$　**c.**

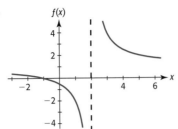

13. a hole in the graph　　　　　**15.** $(0, 0)$ and $(2, 4)$

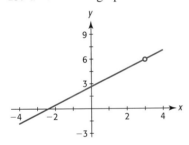

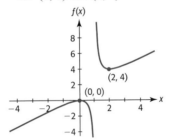

17. a.　　　　　　　　　　　**b.** $0; -8$　**c.** $x = 4, x = 3.5$　**d.** $x = 4, x = 3.5$

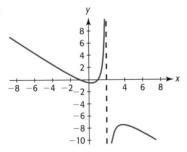

19. a.　　　　　　　　　　　**b.** $-3, -1$　**c.** $x = -1$　**d.** $x = -1$　　**21.** $x = \frac{1}{2}$

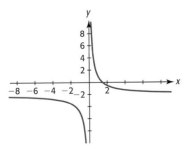

4.6 Exercises

1. a. $55.80 per unit **b.** $57.27 per unit **c.** $55 per unit **d.** no; for example, at 600 units the average cost is $56.67 per unit. **3. a.** $80,000 **b.** -20; no
5. a.

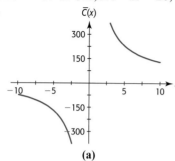

(a)

b.

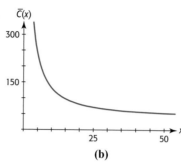

(b)

c. [0, 50] by [0, 300]

d. $50 per unit when 10,000 units produced
7. a.

b. 6; 6 employees when the company starts

c. 17.586; the number of employees after 12 months; approximately 18 employees
9. a. at $p = 100$ **b.** it is impossible to remove 100% of the impurities
11. a. yes, -2

b.

Price per Unit	5	20	50	100	200	500
Weekly Sales Volume	13,061	1322	237	61	16	3

c. nonnegative values of p; no **d.** $y = 0$; weekly sales approach 0 as price increases
13. a.

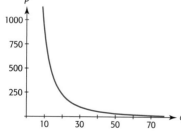

b. $y = 0$ **c.** As the price drops, the quantity demanded increases.

15. a. $S = \dfrac{x^2 + 40x + 160}{4x}$ **b.** 4 hours or 40 hours

17. a.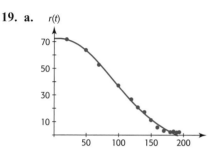

b. $y = 3/2$; as the hours of training increase, the time a production team takes to assemble one unit approaches 1.5 hours   **c.** 17 days

19. a.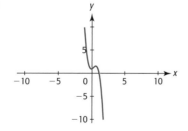

b. 21.2% from the table and approximately 21.0% from the model

4.7 Skills Check

1. $-4 \le x \le 4$ **3.** $x < 0$ or $x > 2$ **5.** $-1 \le x \le 1$ or $x \ge 3$ **7.** $0 < x < 1$ **9.** $x < -3$ or $-1 < x \le 2$
11. $x > 4$ **13.** $x < 5$ **15.** $x = -5, x > 0$ **17. a.** $x < 3$ **b.** $x \ge 3$ **19. a.** $x < -3$ or $0 < x < 2$
b. $-3 \le x \le 0$ or $x \ge 2$

4.7 Exercises

1. $0 < x < 20$; more than 0 and less than 20 units   **3. a.** $0 < x < 18$ or $x > 18$ **b.** more than 0 cm and less than
18 cm   **5.** $0 \le x \le 2$ or $x \ge 10$; between 0 and 2 units, or at least 10 units   **7.** $x \ge 20$; at least \$20,000
9. $.10 \le r \le .20$; between 10% and 20%   **11.** $10 \le x \le 195$; between 10 and 195 units
13. $p > 80$; if the price is above \$80 per unit   **15. a.** $0 \le t < 15$ **b.** the first 14 months

Chapter 4 Skills Check

1. 4   **3.** yes

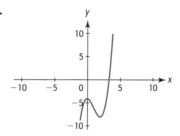

5. a.

b. $(0, -4)$ **c.** $(2, -8)$

7. $x = -1, x = 1, x = 5$ **9.** $x = -2, x = 2, x = -3, x = 3$ **11.** $4x^3 + 5x^2 + 10x + 22\ R36$
13. $2, 1/3, -2$ **15.** local maximum at $(0, 0)$; local minimum at $(8, 16)$ **17.** $5 + i$ **19.** $-12 - i$
21. $z = 2 - i\sqrt{2}, z = 2 + i\sqrt{2}$ **23.** $x = 1, x = -1 - i\sqrt{3}, x = -1 + i\sqrt{3}$ **25.** $x = 0, x \ge 5$
27. $x > 3$ or $x < 0$

Chapter 4 Review

1. a.

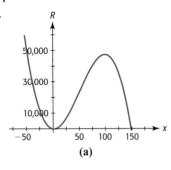

(a)

b.

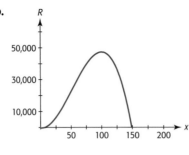

c. $23,750

3. a.

r (rate)	S (future value)
1%	$5307.60
5%	6700.48
10%	8857.81
15%	11,565.30

b.

c. $6072.11

5. a.

b. $400.844 million   **c.** $521.768 million

7. The local maximum is (4, 640). Thus an intensity of 4 allows the maximum amount of photosynthesis.

9. a. $x = 0, x = 9$ **b.** no sides if $x = 0$; no box material left if $x = 9$ **c.** $0 < x < 9$

11. a. $C = 0$ **b.** as t increases without bound, concentration of drug approaches 0.   **c.** 0.15 when $t = 1$

13. a. $\bar{C}(x)$ **b.** 20 units; $1200

15. a. $C = \dfrac{200x + 20x^2 + 180}{x}$ **b.** 5 plates   **c.** 3 plates   **17.** 50 or more hours

19. at least 50 or more units, but no more than 605 units   **21.** $4134 or more

Chapter 5 Systems of Equations and Matrices

Toolbox Exercises

1. 12   **2.** 1/10   **3.** no   **4.** yes   **5.** yes   **6.** yes   **7.** 99   **8.** 99   **9.** $(-4, -8)(0, -3)(4, 2)$ (there are many solutions)   **10.** $(-2, 10), (1, 5), (4, 0)$ (there are many solutions)   **11. a.** no   **b.** parallel lines
12. a. no   **b.** neither   **13.** parallel planes   **14.** same plane   **15.** neither

5.1 Skills Check

1. $(1, 1)$ **3.** many solutions; the two equations have the same graph   **5.** $x = -14, y = -54$
7. $x = -4/7, y = 4/7$ **9.** $x = 2, y = -2$ **11.** $x = 2, y = 1$ **13.** $x = -2/3, y = 7/3$ **15.** $x = 2, y = 4$
17. $x = 2, y = 1$

5.1 Exercises

1. 60 units   **3. a.** 8 supplied, 17 demanded   **b.** shortfall   **c.** $80; 12 units   **5. a.** 1987   **b.** about 521,787
7. a. $x = 37$ **b.** 2027   **c.** yes, the scales are different   **9.** high: $52,69; low: $30.81   **11. a.** $x + y = 2400$
b. $30x$ **c.** $45y$ **d.** $30x + 45y = 84,000$ **e.** 1600 $30 tickets and 800 $45 tickets
13. a. $75,000 at 8% and 25,000 at 12%   **b.** 12% account is more risky; might lose money
15. 3 glasses of milk and 2 servings of meat   **17.** 8 cc of the 5% solution and 12 cc of the 10% solution
19. a. $p = -\dfrac{1}{2}q + 155$ **b.** $p = \dfrac{1}{4}q + 50$ **c.** $85   **21.** no solution

5.2 Skills Check

1. $\begin{bmatrix} 1 & 1 & -1 & 4 \\ 1 & -2 & -1 & -2 \\ 2 & 2 & 1 & 11 \end{bmatrix}$ **3.** $x = -1, y = 4, z = -2$ **5.** $x = 3, y = 2, z = 1$ **7.** $x = 3, y = -2, z = 1$
9. $x = -25/3, y = -25/3, z = 14/3$ **11.** $x = 0, y = -1, z = 2$ **13.** $x = 40, y = 22, z = -19, w = 1$

5.2 Exercises

1. 30 compact, 20 midsize, 10 luxury   **3. a.** $x + y = 2600$ **b.** $40x$ **c.** $60y$ **d.** $40x + 60y = 120,000$
e. 1800@$40, 800@$60   **5. a.** $x + y + z = 400,000$ **b.** $.075x + .08y + .09z = 33,700$ **c.** $z = x + y$
d. $60,000@7.5%, $140,000@8%, $200,000@9%   **7. a.** $10x + 12y + 10z = 290$ **b.** $2x + 8y + 4z = 138$
c. $3x + 5y + 8z = 161$ **d.** 5 units of Portfolio I, 10 units of Portfolio II, 12 units of Portfolio III
9. a. $\begin{cases} 15x + 10y + 5z = 100 \\ y = 2x \\ z = 3x \end{cases}$ **b.** 2 points for T-F, 4 points for MC, 6 points for essay
11. A: 100 units; B: 120 units; C: 80 units   **13.** 5 g of Food I, 6 g Food II, 8 g Food III
15. 3 units of Plan I, 2 units of Plan II, 1 unit of Plan III

5.3 Skills Check

1. no solution   **3.** unique solution   **5.** $x = 3 + z, y = -2 - 2z, z = z$ **7.** $x = 6 - 4z, y = 5 - 3z, z = z$
9. $x = 0, y = 3, z = 2$ **11.** $x = 32/19 - 22z/19, y = 10/19 - 14z/19, z = z$
13. $x = 11/8 - 31z/8, y = 1/8 + 11z/8, z = z$ **15.** $x = 23/14 + z/2, y = -4/7 + z, z = z$

5.3 Exercises

1. a. First equation: sum of investments is $500,000; 2nd equation: sum of interest earned is $49,000
b. $x = 50,000 + 2z, y = 450,000 - 3z$ **c.** $z \le 150,000$

3. a. $\begin{cases} x + y + z = 400{,}000 \\ .08x + .10y + 12z = 42{,}400 \end{cases}$ **b.** \$260,000 at 12%, \$0 at 10%, and \$140,000 at 8%

5. a. $\begin{cases} 2x + 8y + 6z = 140 \\ 3x + 5y + 4z = 110 \end{cases}$ **b.** not possible   **7. a.** $\begin{cases} x + y + z = 4 \\ 40{,}000x + 30{,}0000y + 20{,}000z = 100{,}000 \end{cases}$

b. no   **c.** $x = -2 + z, y = 6 - 2z$ **d.** z is limited by $2 \le z \le 3$, so can have

\$40,000 cars	\$30,000 cars	\$20,000 cars
0	2	2
1	0	3

9. If $x = $ number of units of product A, $y = $ number of units of product B, and $z = $ number of units of product C, then z is limited by $3 \le z \le 35$, and $x = 175 - 5z, y = 3.5z - 10$.

11. a. $\begin{cases} x_4 + 470 = x_1 + 340 \text{ (intersection A)} \\ x_1 + 503 = x_2 + 583 \text{ (intersection B)} \\ x_2 + 651 = x_3 + 601 \text{ (intersection C)} \\ x_3 + 350 = x_4 + 450 \text{ (intersection D)} \end{cases}$ **b.** $x_1 = 130 + x_4, x_2 = 50 + x_4, x_3 = 100 + x_4$

5.4 Skills Check

1. A and D **3.** $\begin{bmatrix} 3 & 6 & -1 \\ 6 & 5 & 3 \\ -3 & 8 & 7 \end{bmatrix}$ **5.** $\begin{bmatrix} 0 & -6 & 10 \\ -6 & 4 & -18 \\ 24 & -2 & -22 \end{bmatrix}$

7. $AD = \begin{bmatrix} 7 & 5 & -4 \\ 17 & 33 & 6 \\ 11 & 27 & -2 \end{bmatrix}, DA = \begin{bmatrix} 6 & 12 & 14 \\ 20 & 10 & 4 \\ 12 & 14 & 22 \end{bmatrix}$; not equal; yes

9. $\begin{bmatrix} 1 + 2a & 5 + 3b \\ 3 - c & 2 - 2d \end{bmatrix}$ **11.** $\begin{bmatrix} 3a - 2 & 3b - 4 \\ 3c - 6 & 3d - 8 \end{bmatrix}$ **13.** $m \times k$ **15. a.** BA **b.** 4×3

17. $AB = \begin{bmatrix} -3 & -7 \\ 4 & 5 \end{bmatrix}, BA = \begin{bmatrix} 11 & 16 \\ -7 & -9 \end{bmatrix}$ **19.** $AB = \begin{bmatrix} -2 & 0 \\ 5 & 19 \end{bmatrix}, BA = \begin{bmatrix} 6 & 1 & 10 \\ 10 & 11 & 14 \\ 1 & 6 & 0 \end{bmatrix}$

5.4 Exercises

1. a. $A = \begin{bmatrix} 252 & 74 & 14 & 7 & 65 \\ 19 & 16 & 19 & 5 & 40 \end{bmatrix}; B = \begin{bmatrix} 252 & 178 & 65 & 8 & 11 \\ 19 & 6 & 14 & 1 & 0 \end{bmatrix}$ **b.** $\begin{bmatrix} 504 & 252 & 79 & 15 & 76 \\ 38 & 22 & 33 & 6 & 40 \end{bmatrix}$

3. a. $A = \begin{bmatrix} 59385 & 47091 \\ 486171 & 565490 \\ 12181 & 9880 \end{bmatrix}$ **b.** $B = \begin{bmatrix} 32575 & 36681 \\ 658782 & 882013 \\ 78086 & 75803 \end{bmatrix}$ **c.** $C = \begin{bmatrix} 26810 & 10410 \\ -172611 & -316523 \\ -65905 & -65923 \end{bmatrix}$

d. 1996: agricultural; 1999: agricultural   **e.** 1999: manufactured goods; this is the largest negative entry in the matrix

5. $\begin{bmatrix} 44{,}050.72 & 31{,}385.76 \\ 33{,}932.64 & 28{,}159.04 \\ 26{,}143.04 & 22{,}458.24 \end{bmatrix}$ **7.** $BA = \begin{bmatrix} 1210 \\ 980 \\ 594 \end{bmatrix}$ **9. a.**

	DeTuris	Marriott
Dept A	54000	49600
Dept B	42000	42400

b. Dept A should purchase from Marriott; Dept B should purchase from DeTuris.   **11.** 55% Republican; 45% Democrat

13. a.

	16–24	25–34	35–44	45–54	55–64
Men	348	648	560	691	751
Women	321	477	486	502	467

b. The 2×2 matrix is $\begin{bmatrix} 1.1 & 0 \\ 0 & 1.25 \end{bmatrix}$

	16–24	25–34	35–44	45–54	55–64
Men	382.80	712.80	616	760.10	826.10
Women	401.25	596.25	607.50	627.50	583.75

5.5 Skills Check

1. a. $AB = BA = \begin{bmatrix} 1 & 0 \\ 0 & 1 \end{bmatrix}$ **b.** They are inverses. **3.** $AB = BA = \begin{bmatrix} 1 & 0 & 0 \\ 0 & 1 & 0 \\ 0 & 0 & 1 \end{bmatrix}$ **5.** $\begin{bmatrix} 2.5 & -2 \\ -1 & 1 \end{bmatrix}$

7. $A^{-1} = \begin{bmatrix} -1/6 & -1/3 & 1 \\ -1/3 & 1/3 & 0 \\ 1/3 & 2/3 & -1 \end{bmatrix}$ **9.** $X = \begin{bmatrix} 10 \\ 20 \end{bmatrix}$ **11.** $x = 24, y = 27, z = 14$ **13.** $x = 1, y = 1, z = 1$

15. $x = 2, y = 1, z = -1$

5.5 Exercises

1. Company X: 225,000; Company Y: 45,000 **3.** 50% Republican, 50% Democrat

5. a. $\begin{cases} x + y + z = 400,000 \\ x + y - z = -100,000 \\ x - \dfrac{1}{2}y = 0 \end{cases}$ **b.** \$50,000; \$100,000; \$250,000

7. \$50,000 at 6%; \$100,000 at 8%; \$250,000 at 10% **9.** 47% from business loans, 27% from auto loans, 32% from home loans

11. a. j u s t _ d o _ i t
    10 21 19 20 27 4 15 27 9 20

b. 124, 52, 156, 59, 124, 35, 168, 69, 116, 49

13. 80, 27, 50, 156, 63, 106, 260, 138, 168, 156, 96, 108, 212, 111, 146 **15.** Vote early **17.** Mind your manners

19. Monday night **21.** Answers will vary.

Chapter 5 Skills Check

1. $x = 2, y = -3$ **3.** many solutions **5.** $x = 3, y = 1, z = 2$ **7.** $x = 41z - 160$
    $y = -18z + 70$
    $z = z$

9. no solution **11.** $\begin{bmatrix} -3 & 1 & 0 \\ -5 & 3 & -1 \end{bmatrix}$ **13.** not possible

15. $\begin{bmatrix} -13 & 12 & 8 \\ -4 & 1 & 3 \\ 0 & 5 & -1 \end{bmatrix}$ **17.** $\begin{bmatrix} 10 & 6 & -6 \\ 9 & 25 & 0 \\ 3 & 15 & 10 \end{bmatrix}$ **19.** $x = 2, y = 3, z = -1$

Chapter 5 Review

1. 25   **3.** $p = 100, q = 80$ **5. a.** $x + y = 2600$ **b.** $40x$ **c.** $60y$ **d.** $40x + 60y = 120,000$
e. 1800 \$40 tickets and 800 \$60 tickets   **7.** \$80,000 at 12% and \$40,000 at 16%   **9.** 20 g of Food I, 30 g of Food II, 40 g of Food III
11. If $x = $ Tech, $y = $ Balanced, $z = $ Utility, $x = (2800 - z)/3$ $y = 200 - (2/7)z$, $0 \leq x \leq 700$
13. $x = 2z, y = 2000 - 4z$ for $0 \leq z \leq 50$; $A = $ twice number of C, $B = 2000 - 4$ times number of $C, 0 \leq C \leq 500$
15. a. $x_1 + 3050 = x_2 + 2500$ **b.** $x_1 = x_4 - 650$
$\qquad x_2 + 4100 = x_3 + 2000$ $\qquad x_2 = x_4 - 100$
$\qquad x_3 + 1800 = x_4 + 3800$ $\qquad x_3 = x_4 + 2000$
$\qquad\qquad\qquad\qquad\qquad\qquad\quad x_4 \geq 650$
17. $x + y = 10$ **a.** $0.20x + 0.05y = 0.155(10)$ **b.** 7 cc of 20% and 3 cc of 5%

Chapter 6 Special Topics

6.1 Skills Check

1. $y \leq 5x - 4$ **3.** $6x - 3y \geq 12$ **5.** Corners $(0, 0), (0, 5), (4, 0), (3, 2)$

7. Corners $(0, 5), (1, 2), (3, 1), (6, 0)$ **9.** $\begin{cases} y \leq 8 - 3x \\ y \leq 2x + 3 \\ y > 3 \end{cases}$

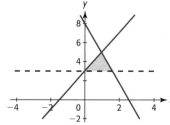

Corners: $(0, 3), (1, 5), (5/3, 3)$

11. $\begin{cases} x + 2y \geq 4 \\ x + y \leq 5 \\ 2x + y \leq 8 \\ x \geq 0, y \geq 0 \end{cases}$

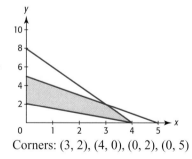

Corners: $(3, 2), (4, 0), (0, 2), (0, 5)$

6.1 Exercises

1. a. $x + y \geq 780$ **b.**

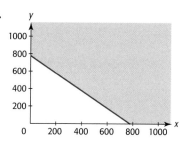

3. a. $240x + 150y \leq 36,000$ **b.**

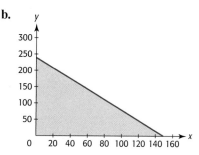

5. a. $\begin{cases} .12x + .009y \geq 7.56 \\ x + y \geq 100 \\ x \geq 0, y \geq 0 \end{cases}$

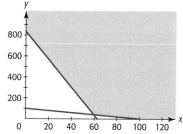

b. Corners: $(0, 840)$, $(60, 40)$, $(100, 0)$

7. $\begin{cases} x + y \geq 780 \\ 78x + 117y \leq 76,050 \\ x \geq 0, y \geq 0 \end{cases}$

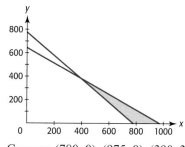

Corners: $(780, 0)$, $(975, 0)$, $(390, 390)$

9. a. $\begin{cases} 80x + 40y \geq 3200 \\ 20x + 20y \geq 1000 \\ 100x + 40y \geq 3400 \end{cases}$

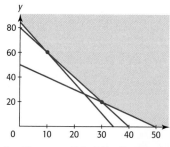

b. Corners: $(30, 20)$, $(0, 85)$, $(50, 0)$, $(10, 60)$

11. $\begin{cases} 4x + 10y \le 1600 \\ 6x + 7y \le 1600 \\ x \ge 0, y \ge 0 \end{cases}$

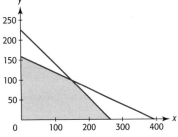

Corners: $(0, 0)$, $(0, 160)$, $(150, 100)$, $(266\ 2/3, 0)$

13. a. $\begin{cases} 4x + 6y \le 480 \\ 2x + 6y \le 300 \\ x \ge 0, y \ge 0 \end{cases}$

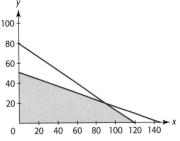

b. Corners: $(90, 20)$, $(0, 50)$, $(120, 0)$

15. a. $\begin{cases} x + y \le 1400 \\ x \ge 500 \\ y \ge 750 \\ x \ge 0, y \ge 0 \end{cases}$

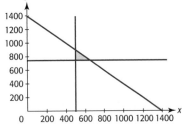

b. Corners: $(500, 750)$, $(500, 900)$ $650, 750)$

6.2 Skills Check

1. max: 382 @ $(10, 38)$ min: 0 @ $(0, 0)$ **3.** max: 22 @ $(4, 3)$ min: 0 @ $(0, 0)$

5. a.

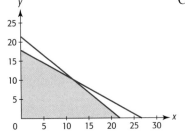

Corners: $(0, 0)$, $(0, 18)$, $(22, 0)$, $(12, 10)$ **b.** max is 90 at $x = 0, y = 18$

7. a. 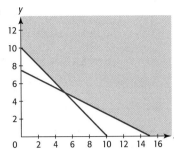   Corners: $(0, 10), (5, 5), (15, 0)$ **b.** minimum is 20 at $x = 0, y = 10$

9. 140 @ (4, 2)   **11.** 1200 at all points on the line segment from (2, 6.5) to (0, 7.5)   **13.** 70 @ (0, 7)
15. 161 @ (3, 1)

6.2 Exercises

1. 975 Turbo and 0 Tornado models gives max profit of $31,200.   **3.** Max daily profit is $396 with 9 Safecut and
6 Safecut Deluxe.   **5.** 100 minutes on TV and 80 minutes on radio gives maximum profit of $94,000.   **7.** Using
Assembly Line 1 for 50 days and Assembly Line 2 for 0 days gives a minimum cost of $1,000,000.   **9. b.** Profit is
maximized at $900,000 with 15 Van Buren and 0 Jefferson models, with 5 Van Buren and 8 Jefferson models, or with
10 Van Buren and 4 Jefferson models.   **11.** Operating Facility A for 8 weeks and Facility B for 2 weeks gives minimum
cost, $160,000.   **13.** 4 servings of diet A and 10 servings of diet B; min is .71 oz.

6.3 Skills Check

1. 5, 7, 9, 11, 13, 15   **3.** $10, 5, \dfrac{10}{3}, \dfrac{5}{2}, 2$ **5.** 9, 11, 13   **7.** 25   **9.** 24, 48, 96, 192   **11.** 2430

13. 5, 3, 1, −1   **15.** 2, 7, 17, 37

6.3 Exercises

1. $10,500   **3. a.** $300 + 60n$ **b.** 360, 420, 480, 540, 600, 660   **5. a.** first by $2000   **b.** They are the same.
c. Years on job; if less than 10 years, choose first.   **7.** 1050, 1102.50, 1157.63, 1215.51
9. a. $20,000   **b.** $50,000 − 10,000n$ **c.** 40,000, 30,000, 20,000, 10,000, 0   **11.** 320,000
13. $\frac{1}{2}$ foot   **15.** $7,231,366.37   **17.** 15 years   **19.** $22,253.46   **21.** 13, 21, 34, 55
23. 1100, 990, 891, 801.90

6.4 Skills Check

1. $\dfrac{364}{27}$ **3.** 205   **5.** 150   **7.** 98301   **9.** 5115   **11.** $\dfrac{729}{8}$ **13.** $2^1 + 2^2 + 2^3 + 2^4 + 2^5 + 2^6 = 126$

15. $2 + \dfrac{3}{2} + \dfrac{4}{3} + \dfrac{5}{4} = \dfrac{73}{12}$ **17.** 3   **19.** not possible, infinite

6.4 Exercises

1. $2400   **3.** 11   **5. a.** 78   **b.** 156   **7.** $42,768.57   **9.** $16 \, \text{cm}^3$ **11. a.** 5   **b.** 25   **c.** 125; 625
d. geometric, with $r = 5$ **13.** $2824.29   **15.** 213 ft   **17. a.** $s_n = 35,000(1 - .84^n)$ **b.** $35,000(0.84)^n$
19. $15,992.73

21. n^{th} term: $R(1 + i)^{-n}(1 + i)^{n-1}$

$$S_n = \frac{R(1 + i)^{-n}(1 - (1 + i)^n)}{1 - (1 + i)}$$

$$= R\left[\frac{(1 + i)^{-n} - (1 + i)^0}{-i}\right]$$

$$= R\left[\frac{(1 + i)^{-n} - 1}{-i}\right]$$

$$= R\left[\frac{1 - (1 + i)^{-n}}{i}\right]$$

6.5 Skills Check

1. $10x^4 + 33x^2 + 2x + 12$ **3.** $\dfrac{x^3 + 6}{x^3}$ **5.** 26   **7.** $0.8125 = 13/16$ **9.** $y = 12x - 16$

11. a. $4(x + h) + 5$ **b.** $4h$ **c.** 4   **13.** $y = 3 - 2x^2$ **15.** $y = \dfrac{18 - 9x^3}{5}$ **17. a.** $x = 1$ **b.** same

c. $(1, 4)$ **19. a.** $x = 2$ **b.** same   **c.** $(2, -9)$ **21. a.** $x = -2, x = 8$ **b.** $y = 67$ at $x = -2$; $y = -433$ at $x = 8$ **c.** $x = -2$ **23.** $y = 3x^{1/2}$ **25.** $y = 2x^{2/3}$ **27.** $y = (x^2 + 1)^{1/3}$ **29.** $y = (x^3 - 2)^{2/3}$

31. $y = x^{1/2} + (2x)^{1/3}$ **33.** $y = \pm\sqrt{3 - 4x}$ **35.** $y = \dfrac{-1 \pm \sqrt{1 + 24x}}{2}$ **37.** $2; 4x^3 + 5$

39. $x^3 + x; \frac{1}{2}$ **41.** $3x\sqrt{x^2 - 1}$ **43.** $y = 3x^{-1} - 4x^{-2} - 6$ **45.** $y = (x^2 - 3)^{-3}$

47. $f'(x) = -\dfrac{6}{x^4} + \dfrac{4}{x^2} + \dfrac{1}{x}$ **49.** $f'(x) = \dfrac{8x}{(4x^2 - 3)^{1/2}}$ **51.** $3 \log x + 5 \log(3x - 4)$ **53.** $x = 0, x = 1$

55. $x = 3, x = \dfrac{3 + \sqrt{69}}{5}, x = \dfrac{3 - \sqrt{69}}{5}$

Chapter 6 Skills Check

1.

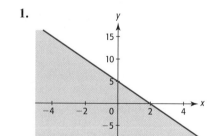

3. Corners: $(0, 0), (0, 2), (1, 1), (1.5, 0)$

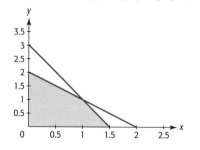

5. min is 14 at $(2, 2)$ **7.** geometric; $r = 6$ **9.** geometric; $r = -3/4$ **11.** 27   **13.** 234

15. $5x^4 - 9x^2 + 6x - 10$ **17.** $y = x + x^{-2} - 2x^{-3/2} + 3x^{-1/3}$ **19.** $x = 0, x = -1, x = \frac{1}{2}$

Chapter 6 Review

1. \$25,200 profit at 360 units each   **3.** minimum cost is \$48 with 32 units of Feed A and 2 units of Feed B
5. maximum profit of \$71,000 with 60 two-bedroom and 20 three-bedroom apartments   **7. a.** Job 2   **b.** Job 1
9. a. 9.22337×10^{18} grains   **b.** 1.84467×10^{19} grains   **11.** \$24,500.90

Index of Applications

Index